W9-CTN-111

ELECTRICAL E
FOR PROFESSI
ENGINEERS' EX

Electrical Engineering for Professional Engineers' Examinations

JOHN D. CONSTANCE

Engineering Registration Consultant; Registered Professional Engineer, New York and New Jersey; Certificate of Qualification (Committee on National Engineering Certification)

THIRD EDITION

McGraw-Hill Book Company

New York St. Louis San Francisco Auckland Düsseldorf Johannesburg Kuala Lumpur London Mexico Montreal New Delhi Panama Paris São Paulo Singapore Sydney Tokyo Toronto

Library of Congress Cataloging in Publication Data

Constance, John Dennis, date.
Electrical engineering for professional engineers' examinations.

Bibliography: p.
Includes indexes.
1. Electric engineering—Problems, exercises, etc.
2. Electric engineering—Examinations, questions, etc.
I. Title.
TK168.C65 1975 621.3'076 75-14339
ISBN 0-07-012448-5

Copyright © 1975, 1970, 1959 by McGraw-Hill, Inc. All rights reserved. Printed in the United States of America. No part of this publication may be reproduced, stored in a retrieval system, or transmitted, in any form or by any means, electronic, mechanical, photocopying, recording, or otherwise, without the prior written permission of the publisher.

567890 MUBP 7843210987

To My Parents

CONTENTS

FOREWORD

One of the implications of the new emerging human resources development concept is Engineers' Registration with its attendant broadening of the scope of the individual engineer. Boards of examiners predict that the number of licensed professional engineers in the United States and its territorial possessions will exceed 500,000 in the 1970s and by the end of this decade may approach the three-quarter-million mark.

Engineers' Registration, with its legal right to practice, will be required in an ever-increasing array of new industrial and community activities brought about by the need for consumer protection and a clean environment. In the ultimate pursuit of the public's protection, its health, its safety, and its well-being, we will find the registered professional engineer actively participating in design safety studies, air, water, and noise pollution control, urban planning and renewal, as well as the traditional engineering fields of design: bridges, dams, office and industrial buildings, computers, nuclear and fossil-fuel power-generating plants, distribution of electricity, and so on.

No engineering discipline, including the traditional electrical, chemical, mechanical, and civil fields, will be exempted, since each state registration law that regulates Engineers' Registration is primarily for the protection of the public weal. Practically every design, every operation, and every process developed by engineers will ultimately involve the public. Every design for a bridge, a new instrument, a steam or gas turbine, or an intricate control system must take the safety and welfare of the public into account. It should be evident that the profession of engineering, because it must be accountable to the public, must be regulated by the states. Without exception, state legislatures have confirmed this reasoning: registration laws have been enacted by every state and territorial possession.

John D. Constance

PREFACE

Surely, there are many well-qualified practicing electrical engineers who have talked at one time or other about seeking licensure but who have not taken the necessary steps to become registered. Perhaps they have thought licensure would somehow "disappear," or "go away," but were left with the distinct feeling they would have to do something about it some day. They argue, "who needs it?" that licensure is solely for civil and mechanical engineers, that electrical engineers simply ran conduit to motor locations or selected switchgear that serviced mechanical equipment or ran elevators.

Suddenly, these same electrical engineers find themselves in a "bind." The promotional ladder is closed off to them; they may miss a lifetime opportunity to become a department head or chief electrical engineer; they may be affected by a reduction in the work force. Finally that impulse to become a registered professional is triggered into action.

The aim of this book is to provide the needed encouragement for these engineers and to implement the impulse and urge to become licensed.

Every state board of professional engineers examines in the major field of electrical engineering. While the first day's examination for Engineer-in-Training (New York's Intern Engineer) is designed to test the candidate's facility in mathematics and engineering theory, the second day's examination requires evidence of proper judgment in the practical and economical approaches to the more advanced field of electrical engineering. This part of the examination is professional in character and is designed for persons with several years of experience. It encompasses the application of scientific principles to the problems the electrical engineer meets in everyday practice in the design engineering office and field. Equally, the book provides the electrical engineering student just completing undergraduate studies an insight into the type and style of problems required at the beginning of the internship in engineering.

In the true practice of the profession, the electrical engineer must be qualified to plan, design, construct, test, maintain, and operate electrical power machinery, including all devices used in the generation, transmission, distribution, measurement, and utilization of this form of energy. In electronics, the engineer must be competent in design principles and use of circuitry, vacuum tubes, transistors, antennas, wave propagation, communications, including radio, television, radiotelephone, and the like; in illumination, he must be qualified in light-source measurement, interior and exterior applications, and photoelectric systems; in acoustics, he must be qualified in sound systems and their applications.

The examination in electrical engineering is not comprehensive in each of these areas. It is intended to be a test of whether or not the applicant can apply the required engineering knowledge to the solution of several well-sampled typical problems in his major field.

While the written examination is not the complete test of an applicant's qualifications, it is one of those devices in universal use by all the professions to assist a state board of examiners in determining whether or not an applicant has the necessary technical knowledge. Boards realize that replies from the applicant's references do not and are never intended to tell the full story about capabilities, and the board is required to determine, as far as possible, an applicant's qualifications to practice. The written examination has been found to be a reliable means of supplying a part of the information needed. It will be used until a better way is found.

Thus, in this third edition the specific purpose and original intent of this book is retained: to help prepare the license candidate to pass the electrical engineering examination for professional engineers. Many candidates who have used the book report that it serves this purpose in a uniquely useful manner.

The third edition has been amplified in many areas. Improvements have been made in certain solutions, to reflect greater accuracy, and in notation. Additions have been made to the chapters on wire sizing, dc machinery, dc motors, condensers, ac motors, power circuits, transformers, lighting, electronic tubes and circuits. The supplemental sections on direct current, transmission lines, short-circuit calculations and faults, and transient analysis have been enlarged. A new supplement has been added covering problems and solutions in aerials. The use and application of Laplace transforms has been expanded in a number of solutions in transient analysis.

As in the first and second editions, the aim of the earlier chapters is to provide direction and assistance for the electrical section in the Engineer-in-Training and Intern Engineer (first day's) examination.

Problems and solutions, mainly from past examinations of recent vintage from a number of states, are explained and illustrated in sufficient detail for a good understanding of the particular material at hand. The book continues to be more than a mere compilation of problems and solutions, which makes it more than an ordinary refresher publication. In addition to representative problems with solutions, a quick review of theory is included to provide the candidate with the sense of mastery of the entire range of test material.

The reference book list has not only been updated in this third edition but enlarged and sectionalized for better-organized examination preparation and exam-room use.

In reviewing the problem material and before making the addition of new material, a number of guidelines were used: (1) What are the most common examination problems that have been solved in electrical engineering? (2) What are the most accurate as well as convenient methods for solving these problems? (3) What other problem material can be expected over the long haul for examinations of the future? (4) What range of tested and realistic problems in electrical engineering could be presented?

When the answers to these and other related questions were obtained, the discussion material, procedures, and worked-out problems were chosen. Thus, this book represents a distillation of theory and basic formulas with maximum emphasis on problems, their analyses, and their solution.

To those who claim that the question and answer approach to examination study makes engineering "too easy," the author wishes to point out that for many years engineering educators have recognized the importance and value of problem solving in the development of engineering judgment and experience. Problems courses have been popular in numerous engineering schools for many years and are still given in many schools.

Although most of the problems have been taken from actual state board examinations, many others have been included in anticipation of examinations to come.

A previously presented problem may be rewritten by the board so that it may not be quickly recognized as one in the book. Obviously the basic theory must be understood. However, sufficient background material is presented to make it possible for the candidate to augment this with standard references and texts.

An important feature of the book is that the solution is worked out directly following the problem, a timesaver whose value has been repeatedly demonstrated in over 30 years of guidance and preparation of applicants for registration. Another important feature is presentation of short-cut methods of solution to speed up performance and help the engineer over the test hurdle the first time, thus avoiding time-consuming and costly reexaminations.

Detailed problem and subject indexes are included so that the scope of the subject matter and problem coverage become immediately apparent for ready and rapid use.

This book can also serve you:

- In testing for the Electrical Engineer Group
- In Civil Service Examinations for engineers
- In Power Engineer examinations
- In Engineer-in-Training or Intern Engineer examinations
- In a reference capacity—at your desk, in the plant, in the field.

Following the precedent of both first and second editions, this new third edition, in summary:

- Gives you problems from recent state examinations
- Forecasts future trends in examinations
- Tacks solutions onto problems to avoid your paging back and forth
- Backs up answers with explanation and theory
- Provides short-cut solution methods to speed up your exam performance
- Spells out how boards credit solutions and related details on the exam itself

Offers you such thorough grounding that it can enable you to approach the exam with total confidence

Technical school graduates and engineering technicians will find the book helpful as they seek guidance in achieving professional status.

The author acknowledges with gratitude his many readers who have directly or indirectly helped him to formulate new ideas for the continued improvement of this work, truly a labor of love.

The author also wishes to thank Robert J. Kelemen, P. E., for his contributions to the first edition. As to the second and third editions, Shih-Rong Wang, P. E., has been most helpful in an advisory capacity. To others who have also assisted and contributed by means of personal contact or written communication, thank you.

John D. Constance

INTRODUCTION—HOW YOU CAN PASS THE FIRST TIME

There is an art to taking the professional engineers' registration examination, and following a rather simple pattern of self-discipline and working out many problems can help you *pass the first time*. Preparing for and passing the examination will also help you fight off the specter of engineering obsolescence.

The importance of and the "hows" of filling out application forms and writing up experience records appear elsewhere.[1] The main purpose here is to show the way by presenting subject matter and material that have appeared in many states' licensing examinations, together with worked-out solutions that are acceptable to boards of examiners. How do you go about preparing for the examinations? How can you pass the first time? What do you take into the examination room? Let's see what the answers to these questions and others are.

Some states conduct closed-book examinations. For these, textbooks are barred, but a preselected list of approved aids will be sent the candidate (note that "candidate" and "examinee" are used synonymously in this work) before examination time. Most states, however, conduct open-book examinations, for which the examinee will be allowed to use reference books, textbooks, personal notebooks, and any other material that he feels will help in meeting the challenge of the test. Follow board instructions.

Boards of examiners are favorably impressed by the performance of many candidates, some of whom score perfect marks. Factors such as lack of time, misunderstood requirements, anxiety, and

[1] J. D. Constance, "How to Become a Professional Engineer," 2d ed., McGraw-Hill Book Company, New York, 1966.

perhaps poor preparation can cause candidates to stumble and prevent many from performing at their best. Most boards give liberal part credit, but this is not in direct proportion to the mass of written material unless the approach to the problem in its solution is the proper one. Boards are aware that pages of calculations on a wrong basis are not worth much in partial credit.

Boards seek to offer considerable choice on all parts of the examination, but problems to be solved should be selected with care. No one candidate is ever able to answer all problems, for some are more directly related to the individuals' experience and educational background than others. Since problems most often bear equal weight, candidates are advised not to spend more than equal time. This is the area where many candidates are neglectful and become so engrossed in the solution that they forget time, even the existence of time.

Boards advise candidates to read the problems carefully, since all information contained in the statement of the problem may not be needed to work out an acceptable solution. An important part of the grade an examinee receives depends upon the soundness of judgment displayed. Mere routine numerical solutions, without adequate explanation of the assumptions made to define a realistic problem situation, are not considered by boards. to demonstrate that the candidate has sufficient judgment and experience to justify a passing grade. Careful definition of the problem and its method of solution are as important as numerical answers. The examinee is also expected to demonstrate habits of thoroughness, neatness, and accuracy.

Slide-rule accuracy is acceptable. Many candidates lose valuable time by "feather edging" an answer and indicating it as, say, "1,425.3, approximately," when a value of 1,400 would have been acceptable, consistent with the accuracy of the initial given data.

At times the use of specific tables or a specific approach technique is required in the problem statement. Unless the candidate is familiar with the technique or has with him the set of tables, he should select a more familiar problem to solve. For example, in a power factor problem a nomograph or chart solution would not be acceptable. However, if the solution involves trial and error, the first trial value may be selected by *any* means at the candidate's disposal. On the other hand, a nomograph or chart may be used to check the results of a calculation.

Many engineering problems may be categorized: A problem in circuitry may require a transient or a steady-state analysis; the solution to a problem may require simplification by use of Laplace transforms. Preparing for the examination should make full use of these criteria.

The candidate is cautioned not to go off the deep end by making assumptions that are meaningless.

A numerical value, alone, without a "handle" is in most cases insufficient. If the final answer requires resistance, the candidate must state the result in, say, 20 ohms or 75 megohms. If an equation is used, the examinee should be familiar with its source and the homogeneity of units.

There is also the "order of magnitude" to be considered. In any calculation, look for factors that will provide a rough check on the indicated results. In electric heating roughly 3 to 5 watts per cubic foot of heated space is needed.

Often, not all data are provided in the statement of the problem and the candidate finds that he does not have access to a crucial bit of information. For example, if he needs the resistivity of a metal and he does not have it, he should clearly state that he realizes this value is needed and should at least indicate the solution to the problem, assuming a certain reasonable value. The part credit he would receive thereby would be much greater than if he ignored the problem altogether as unsoluble.

Some problems can be solved by trial and error only. In such situations, the candidate can achieve satisfactory results after two or three trials. To achieve greater accuracy by further trials would be unnecessary, and he should indicate that successive applications of the method would eventually yield the desired results. The time spent in "feather-edging" could be put to greater advantage by working out the solution to some simpler problem, for usually problems have equal weight in solution.

Do not copy down the problem statement and diagram. This is not necessary and wastes valuable problem-solving time. Merely extract the necessary data and apply them, but remember to show any diagram that would help. For example, if the problem is to determine the value of current flow X microseconds after a switch is thrown, a calculation that gives the steady-state value of the current is not worth much.

Recent trends show the curricula of many engineering schools have placed greater emphasis on the basic sciences and mathematics, and most state examinations reflect this trend. This can have a startling effect on the candidate, for he can be faced with problems that require knowledge of a subject that he may never have studied in his undergraduate days on campus. Such problems often require the setting up of a differential equation and the subsequent solution of it. He may be required to apply the Laplace transform method of circuit analysis in order to solve transient problems by purely algebraic means. There is little substitute for specific study and training in such areas. The best advice here is not to go over your depth.

In the professional part (second day) of the examination, essay-type questions are often offered. These require specialized knowledge, and unless an examinee has specific experience in the given area, the suggestion is to avoid it. A generalized answer copied from a reference manual will be of little value.

Practice solving problems. You may use the following systematic approach which has been found helpful:

1. Read the statement of the problem carefully and decide exactly what is required. The better a person understands a problem, the better he is able to marshal a multitude of approaches and methods when tackling it. Strip away side problems or conditions.

2. Draw a suitable diagram or sketch of the action or process and list the data given. This will provide the examiner with full appreciation of your thinking process. Remember, a picture is worth 10,000 words.

3. Identify the problem and list the physical principles which seem relevant to its solution. These may be expressed concisely as algebraic equations. Thus, in a problem involving a capacitor composed of several parallel circular plates its capacitance may be calculated by $C = 0.0885K(N - 1)S/t$.

4. Determine if the data given in the problem are adequate. If not adequate, you must determine what is missing and decide

how to get it. You may have to consult references or tables, make reasonable assumptions, or draw upon your general knowledge for this information.

5. Break down the parameters or variables of the problem through analysis, keeping in mind the total problem situation and the proper relationship of the parts to the whole. Organize facts; underline the key facts.

6. Decide whether in the particular problem at hand it is more expeditious to substitute numerical values immediately or first carry out an algebraic solution of the equations listed.

7. Substitute the numerical data in the equations obtained from physical principles. Include the units for each quantity, making sure that all are in the same system in any one problem.

8. Calculate the numerical value of the unknown, preferably with the aid of the slide rule. Determine the units in which the answer is expressed.

9. Examine the reasonableness of the answer. Ask yourself if it can be obtained by an alternate method to check the results. Can it possibly be checked by use of nomographs?

10. Reread the problem before going on to the next.

Such orderly procedure helps to avoid errors and saves time. Most important, it enables you to analyze and eventually solve the more complex questions whose solution is not immediately apparent.

During problem-solving practice there are a number of techniques which can increase your power of observation and association. Don't give up if ideas are slow in coming. A sense of leisure is an important factor. Always carry a notebook. Ideas strike at any hour and under the strangest circumstances. They may disappear into the unconscious if not permanently recorded. Do not trust to memory.

Establish proper mood, and this is not done by simply waiting for the mood to occur. Pick up a pencil and begin writing down the different parts of the problem, the different approaches you might take, and the directions you might explore. Detach yourself from the problem after a period of involvement and view it objectively. An effective process requires continuous shifting between involvement and detachment. If you are not making headway, drop the problem completely and do something entirely different. Unremitting pressure sometimes interferes with the unconscious formation of new configurations. Remember that creative insight occurs most often in relaxed or dispersed attention. The closure appears frequently after you have left the problem in despondency or disgust.

In problem-solving practice determine the physical conditions during which you regularly do your best work. If you find that certain physical postures, e.g., pacing the floor, lying down, or relaxing in an easy chair, are conducive to your best work, don't hesitate to use them. In fact, you should deliberately make an effort to determine what sort of physical activity accompanies your most productive efforts, then deliberately assume it when attempting to solve problems in practice.

Avoid distractions and intrusions as much as possible. Choose a time when you can stay with your problem hours on end and without interruptions from others. Develop a retrospective awareness of the periods when you solved your problems most effectively. Note the methods that were most successful and those that failed. Try to retrace and rehearse the routes that were successful.

When writing for the application to file for the license, request a sample set of questions. Most states provide these. Some have a centrally located file that may be used for inspection only, no permission being granted for their removal for personal use. Sets of past examinations are available from outside sources, and your state board secretary should be willing to supply this information.

For an open-book examination, carry only a minimum number of books into the examination room. Many examinees carry suitcases filled with books to their places and then waste valuable time juggling through them from knee to knee, hunting for answers. Not only is time wasted, but such conduct can annoy those about you. The crowded examination room of today accents this condition.

But whether the examination is open-book or closed-book, all boards allow the use of slide rule or battery-operated electronic calculator. Detailed instructions on what is permitted in the examination room will be listed for each candidate. The following suggested list may be useful as a guide:

Time watch
Supply of sharpened pencils and a fountain pen
Supply of ink
Slide rule or battery-operated electronic calculator
Graph paper ($8\frac{1}{2}$ by 11 inches)
Scratch paper
Rule, scale, triangles, and protractor
Reference books—a minimum

When you get into the examination room, seek out a seat in a well-lighted area of the room, if seats are not previously assigned. This will help ward off fatigue later during the examination.

Rules of conduct of the examination set forth by the board must be followed implicitly. Such rules are rigidly adhered to and enforced by the proctors. There are no substitutes.

The time you have prepared for is now at hand. You are ready. You have gone through the material in this book and you have worked out all the problems, checking the author's calculations and approach. You have looked through the reference list and compiled a set of your own reference books with which you are most familiar. You have an intimate knowledge of the types of problems that have been presented in the past. You have worked these out to your satisfaction, paying attention to neatness and clarity of presentation.

When the examination papers are handed you, read the instructions carefully. Before attempting to answer any question, *read all questions over carefully and check those which appear most familiar.* Usually a candidate's name is not allowed to appear on the work sheets, but you will be given a number. This number should be written in ink on *every* piece of work paper, whether or not you expect to use them all, *before* beginning the examination.

After reading all of the questions, divide the remaining time equally among the questions you have elected to answer, and *adhere strictly to these allotments.* The instructions will tell how many problems are to be answered. On the average, 20 to 40 minutes per problem should be your timing. It is prudent to allocate a certain time interval to the solution of each problem. Some problems are more difficult than others and the easy problems will require less time than allocated. Time saved on the easier ones will be available to go back and work some more on the tough ones. But a repeated word of caution is in order: Avoid spending an inordinate time on one or two problems. Experience shows that the perfect solution of three problems on a six-problem examination will result in failure.

In most cases, solutions receive equal weight. Then why spend more than the allotment? If it appears that one answer is taking more than its share of time, drop it and go on to the next selected. Chances are that you will have time to go back and finish. Of course, if you are near the end of your answer, complete it if it takes but a few minutes; but guard against falling into a trap by becoming too involved in solving one at the expense of another, perhaps simpler, problem. Some boards are including hints and suggestions to help the candidate to complete his paper successfully.

First work out answers to those questions that appear simplest, and work out one answer or any part thereof on a *single sheet* of work paper. This will save time and will permit you to skip around and rearrange papers at the end of the examination.

Strive toward solving the required number of problems, as the board will see a reflection of breadth and scope thereby. Most boards give part credit to mere arithmetic but greater part credit to a correct method of solution. If time is running out, avoid the risk of no credit by writing down the method step by step. To save still more time, when the final expression for the answer takes on the form of a complicated exponential equation, merely fill in the terms with the proper numerical values and write down the dimension of the answer.

Should tension mount during the examination, relax by taking a break. Stretching the arms and legs does help relieve nervousness. Working those questions you understand best will always nurture confidence and make you feel better, but careful preparation that builds confidence is probably your best defense against emotional block. A good working knowledge of examination questions is the best insurance against a sense of false security.

Preparation Check List

1. Learn scope of examination.
2. Obtain sets of past examinations.

3. Become familiar with nomenclature, symbols, etc.
4. Practice to work two to three questions per hour.
5. Practice using pen and ink.
6. Work with slide rule.
7. Work with nomographs and charts to check calculations.
8. Include at least one sketch or diagram per solution.
9. Train to use a minimum of textbooks or references.
10. Learn to make assumptions when needed.
11. Work out all questions in this refresher book to your personal satisfaction.

Examination-time Check List

1. Read instructions carefully.
2. Look over examination problems carefully.
3. Check off those you feel you can solve.
4. Work problems you can do readily first.
5. Allot time to each problem and keep close check on time.
6. Include sketch or diagram with each solution.
7. Aim toward clarity and neatness.
8. When required, make assumptions and proceed on that basis.
9. Solve the required number of problems.
10. Remember that the method of solution is given most weight.
11. Strive for reasonable final answer.
12. Check all work.
13. Don't be a "cookbook" engineer.
14. Avoid panic

The author extends to you his best wishes and offers encouragement for success in your quest for licensure and legal recognition in your chosen profession. Remember, others before you have done it. Good luck.

John D. Constance

ELECTRICAL ENGINEERING FOR PROFESSIONAL ENGINEERS' EXAMINATIONS

Part I

DIRECT CURRENT

Chapter **1**

ELECTRICAL UNITS AND OHM'S LAW

1-1. Electrical Units. The following electrical units are those most often used in working out problems in electricity. The "IEEE Standard Dictionary of Electrical and Electronic Terms," Wiley-Interscience (1972), gives the following definitions:

(*a*) *Ampere,* that constant current which, if maintained in two straight parallel conductors of infinite length, of negligible circular cross section, and placed 1 meter apart in vacuum, would produce between these conductors a force equal to 2×10^{-7} newton per meter of length.

(*b*) *Coulomb,* the unit of electric charge in SI units (International System of Units). The coulomb is the quantity of electric charge that passes any cross section of a conductor in 1 second when the current is maintained constant at 1 ampere.

(*c*) *Farad,* the unit of capacitance in SI units. The farad is the capacitance of a capacitor in which a charge of 1 coulomb produces 1 volt potential difference between its terminals.

(*d*) *Henry,* that unit of inductance in SI units for which the induced voltage in volts is numerically equal to the rate of change of current in amperes per second.

(*e*) *Joule,* the unit of work and energy in SI units. The joule is the work done by a force of 1 newton acting through a distance of 1 meter.

(*f*) *Microfarad,* equal to 10^{-6} farad.

(*g*) *Millihenry,* equal to 10^{-3} henry; microhenry, equal to 10^{-6} henry.

(*h*) *Volt,* the unit of voltage or potential difference in SI units. The volt is the voltage between two points of a conducting wire carrying a constant current of 1 ampere, when the power dissipated between these points is 1 watt.

(*i*) *Watt,* the unit of power in SI units. The watt is the power required to do work at the rate of 1 joule per second.

(*j*) *Ohm*, the unit of resistance (and of impedance) in SI units. The ohm is the resistance of a conductor such that a constant current of 1 ampere in it produces a voltage of 1 volt between its ends.

(*k*) *Gram-calorie*, the energy required to raise the temperature of 1 gram of water 1 centigrade degree. One gram-calorie is equal to 4.18 joules (nearly).

(*l*) *Horsepower*, used to rate electrical equipment and equal to 746 watts; also equal to 2,546 Btu per hour.

(*m*) *Kilowatt*, equal to 1,000 watts.

1-2. Ohm's Law. When a current flows in an electric circuit the magnitude of the current flowing is determined by the emf in the circuit and the resistance of the circuit. The resistance to current flow is dependent upon the cross section of the conductor and the length of the conductor.

This relationship between voltage E, current I, and resistance R is expressed by Ohm's law and may appear in one of three familiar forms as follows:

$$I = \frac{E}{R} \tag{1-1}$$

or

$$E = IR \tag{1-2}$$

and

$$R = \frac{E}{I} \tag{1-3}$$

Q1-1. A circuit has a resistance of 5 ohms (Ω). If a voltmeter connected across its terminals reads 10 volts, how much current is flowing through the circuit? See Fig. 1-1 for circuit.

ANSWER. $$I = \frac{E}{R} = \frac{10}{5} = \textbf{2 amp}$$

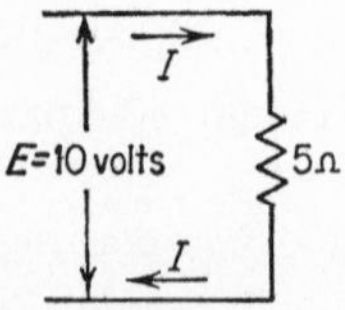

Fig. 1-1

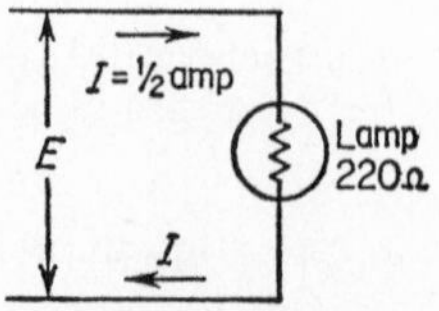

Fig. 1-2

Q1-2. The hot resistance of an incandescent carbon lamp is 220 ohms. It requires ½ amp to cause it to glow. What voltage must be impressed across it? See Fig. 1-2.

ANSWER. Using Eq. (1-2) we see that

$$E = \tfrac{1}{2} \times 220 = \mathbf{110\ volts}$$

Q1-3. An electromagnet has a resistance of 62.5 ohms. It will lift a certain piece of iron when the current through its windings is 3.8 amp, but will not drop the load until the current is reduced to 2.5 amp. What voltage is required (*a*) to lift, and (*b*) to release the load?

ANSWER. Pressure = current × resistance [another form of Eq. (1-2)].

(*a*) To lift: $E = 3.8 \times 62.5 =$ **237.5 volts**

(*b*) To release: $E = 2.5 \times 62.5 =$ **156.3 volts**

Q1-4. The pressure across a pure resistance circuit is 40 volts. What is the value of the resistance to permit the flow of 5 amp? See Fig. 1-3.

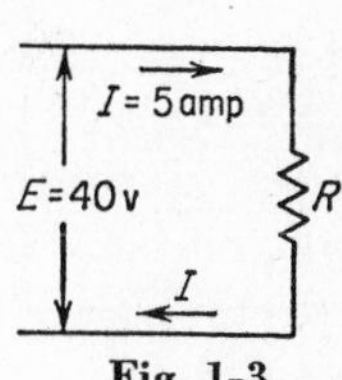

Fig. 1-3

ANSWER. Using Eq. (1-3)

$$R = 40/5 = \mathbf{8\ ohms}$$

1-3. Ohm's Law in Circuits. Ohm's law may be applied to an electric circuit as a whole, or it may be applied to any part of it. The two following statements of the law should be kept in mind.

(*a*) The current in the entire circuit equals the voltage across the entire circuit divided by the resistance of the entire circuit.

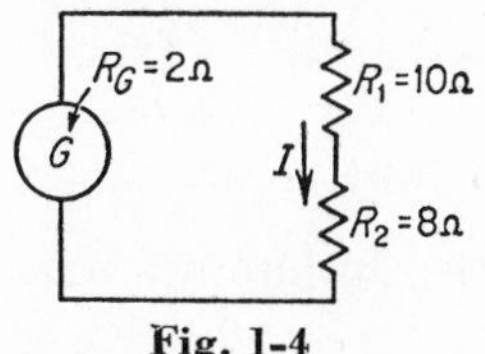

Fig. 1-4

(*b*) The current in a certain part of the circuit equals the voltage across that part divided by the resistance of that part.

Q1-5. The generator *G* (Fig. 1-4) has a resistance of 2 ohms and generates 150 volts pressure. This is not the voltage across its brushes. What current flows through the circuit shown?

ANSWER.

Resistance of entire circuit: $R = 2 + 10 + 8 = 20$ ohms
Voltage across entire circuit: $E = 150$ volts
Current passing through each part: I

Since there is no split, there is but one value of current.

$$I = 150/20 = \mathbf{7.5\ amp}$$

Voltage across an 8-ohm resistance: $E_2 = IR_2 = 7.5 \times 8 = 60$ volts
Voltage across a 10-ohm resistance: $E_1 = IR_1 = 7.5 \times 10 = 75$ volts
Voltage across generator resistance: $E_3 = IR_3 = 7.5 \times 2 = 15$ volts

These voltages are those required to drive the 7.5 amp through the respective parts. The voltage across the brushes is

$$150 - 15 = 135 \text{ volts}$$

This is also known as the brush potential or terminal voltage under the above conditions. This brush potential of a generator must be distinguished carefully from the total voltage generated, or emf, of the generator. This may be compared to the voltage across the terminals of a dry cell connected in a circuit, remembering that the cell or any other battery has an internal resistance.

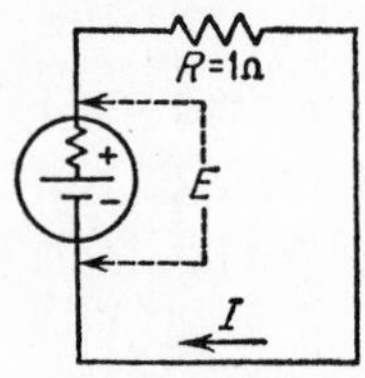

Fig. 1-5

Q1-6. If the current from a short-circuited 1.5-volt dry cell is 25 amp, what is the internal resistance of the cell? Assume resistance of the short circuit is negligible. If this cell is connected through a 1-ohm resistance external circuit, find the current and the reading of a voltmeter connected to its terminals (see Fig. 1-5).

ANSWER. By application of Ohm's law and Eq. (1-3), the internal resistance is

$$R = 1.5/25 = \mathbf{0.06\ ohm}$$

And from Eq. (1-1) the current for the 1-ohm circuit is

$$I = 1.5/(0.06 + 1.00) = \mathbf{1.414\ amp}$$

The voltage across the terminals of the cell will be the internal voltage less the IR drop

$$E = 1.5 - (1.414 \times 0.06) = 1.414 \text{ volts}$$

From the figure of the circuit we can see that the voltmeter also reads the drop across the resistor.

$$E = 1.414 \times 1 = \mathbf{1.414\ volts}$$

1-4. Diagrams of Electric Circuits. Note that the symbol ı| is used to represent a battery (see Figs. 1-5 and 1-6), the longer line representing the + terminal, or the point from which the current flows out of the cell.

In Fig. 1-6 a simple electric circuit is shown. The current flow is out and away from the (+) battery terminal and clockwise through the external circuit. In drawing diagrams we mark the higher potential (level) positive (+) and the lower potential negative (−), thus indicating the direction of current flow (higher to lower potentials). Sometimes arrowheads are placed on the conductors to indicate direction of flow. Except for generators and batteries, the current always flows from the positive point to the negative. In the figure point B is positive with respect to point C but negative with respect to A. A given point is then (+) to all points below its level and (−) to all points above its level.

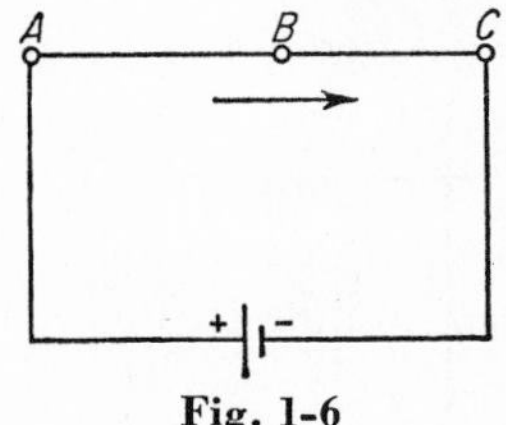

Fig. 1-6

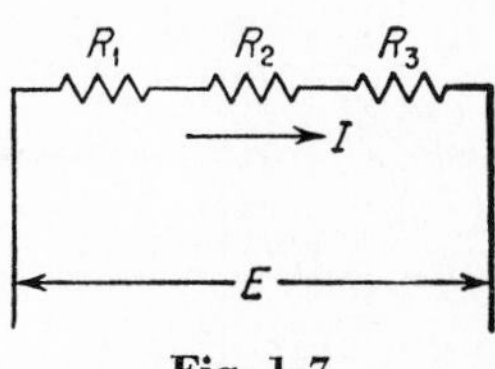

Fig. 1-7

1-5. Series Circuits. In series circuits resistances are connected in a continuous run, and the same current passes through each (refer to Fig. 1-7). The total resistance is equal to the sum of all the resistances. The total potential drop is equal to the sums of the drops due to each individual resistance.

Q1-7. In order to determine the voltage of a d-c source, three resistors of 10, 15, and 30 ohms are connected in series. If the current flowing is 2 amp, what is the potential of the source?

ANSWER.

$$E = I(R_1 + R_2 + R_3) = 2(10 + 15 + 30) = \mathbf{110\ volts}$$

Q1-8. If a relay is designed to operate properly from a 6-volt d-c source, and if the resistance of the winding is 120 ohms, what value of resistance should be connected in series with the winding if the relay is to be used with a 120-volt source?

ANSWER. The resistor must drop voltage, 120 − 6 = 114 volts. Then the current through the relay is

$$I = 6/120 = 0.05 \text{ amp}$$
$$\text{New resistor} = R = 114/0.05 = \mathbf{2{,}280\ ohms}$$

1-6. Parallel Circuits. In a parallel circuit, such as that shown in Fig. 1-8, the voltage across the group is the same, and the current flowing through each resistor varies inversely as the value of it. The sum of all currents, however, is equal to the total current leaving the cell. Thus, $I = E \div R_t = E \times (1/R_1 + 1/R_2 + 1/R_3)$. For resistances in parallel the total is

$$\frac{1}{R_t} = \frac{1}{R_1} + \frac{1}{R_2} + \frac{1}{R_3} + \cdots + \frac{1}{R_n} \tag{1-4}$$

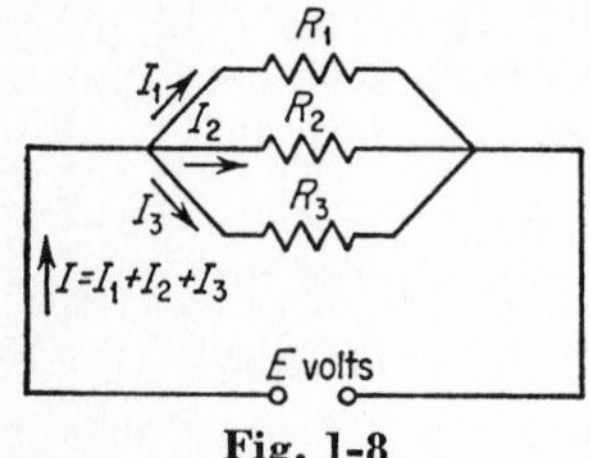

Fig. 1-8

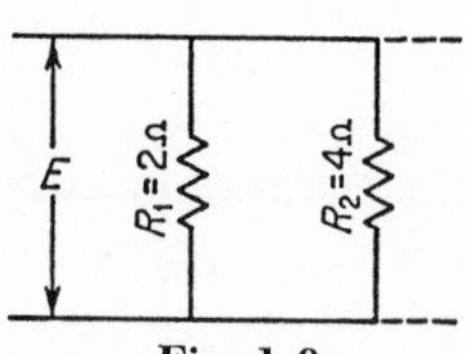

Fig. 1-9

Q1-9. What is the equivalent resistance of the circuit in Fig. 1-9.

ANSWER. From Eq. (1-4) we obtain

$$\frac{1}{R_t} = \frac{1}{2} + \frac{1}{4} = \frac{3}{4} \qquad \text{or} \qquad R_t = \mathbf{1\tfrac{1}{3}\ ohms}$$

Q1-10. Electric resistances of 7 and 11 ohms are connected in parallel. This combination is then placed in series with a single resistance of 15 ohms and the entire combination is placed across 110-volt d-c mains. What current passes through each resistor? See Fig. 1-10.

ANSWER. The total equivalent resistance is

$$R_t = 15 + (7 \times 11)/(7 + 11) = 19.28 \text{ ohms}$$

The total current from Eq. (1-1) is

$$I = 110/19.28 = 5.7 \text{ amp}$$

After passing through the 15-ohm resistance, the current divides inversely as the resistances through each branch of the parallel combination. Or the current through the 7-ohm branch is

$$5.7 \times 11/(7 + 11) = \mathbf{3.48\ amp}$$

The current through the 11-ohm resistance is

$$5.7 \times 7/(7 + 11) = \mathbf{2.22\ amp}$$

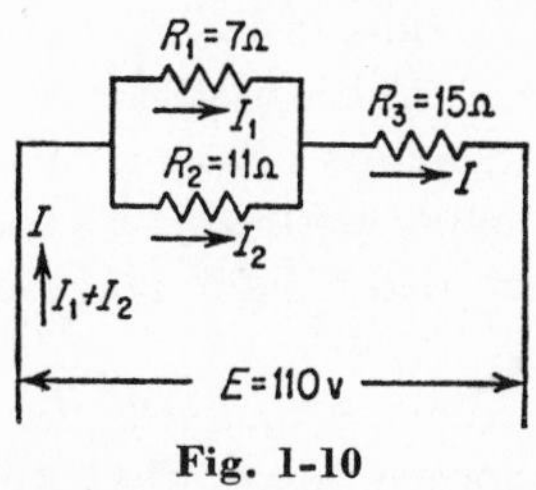

Fig. 1-10

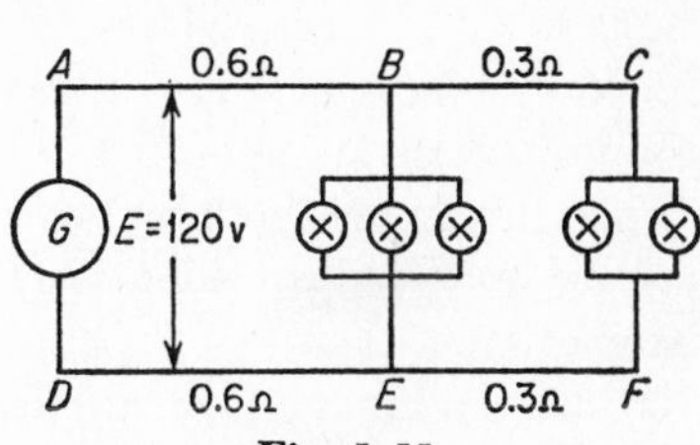

Fig. 1-11

1-7. Simple Parallel Lighting Systems. Incandescent lamps are usually installed in parallel for interior lighting. The resistance of all the lamps of any make with the same rating is not the same; nor is the voltage the same across all the lamps when they are installed. Nevertheless, for convenience in calculating the line drop, the efficiency of transmission, etc., each lamp is assumed to take the same current. The error introduced by this assumption is usually too small to be taken into account.

Q1-11. A lighting circuit is arranged with two groups of lamps, as shown in Fig. 1-11. One group has two lamps, the other three. Each lamp takes 2 amp. The generator has a terminal voltage of 120 volts. The resistance of line wires AB and DE equals 0.60 ohm each and that of line wires BC and EF equals 0.30 ohm each. Find the volts dropped in line wires and the voltage across each group of lamps.

ANSWER. The current distribution throughout the circuit is

$$\begin{aligned} \text{Current through } BC &= 2 \times 2 = 4 \text{ amp} \\ \text{Current through } AB &= (2 \times 3) + (2 \times 2) = 10 \text{ amp} \\ \text{Current through } EF &= 2 \times 2 = 4 \text{ amp} \\ \text{Current through } DE &= (2 \times 3) + (2 \times 2) = 10 \text{ amp} \end{aligned}$$

Now by Ohm's law solve for the drop in the line wires.

$$\text{Drop in } AB = 10 \times 0.6 = \textbf{6 volts}$$
$$\text{Drop in } DE = 10 \times 0.6 = \textbf{6 volts}$$

Therefore, total volts used in line between the generator and the first group of lamps is the $6 + 6 = 12$ volts. Then the generator *minus* the drop in the lead wires equals the voltage at lamps *BE*, or $120 - 12 = 108$ volts across *BE*.

Volts lost in *BC* is current through *BC* times resistance of *BC*, or $4 \times 0.3 = 1.2$ volts. Thus, total voltage lost between the two groups of lamps is $1.2 + 1.2 = 2.4$ volts. Voltage at the second group of lamps is $108 - 2.4 = 105.6$ volts across *CF*.

This simple method by which this simple problem was solved may be followed to solve any problem in d-c light and power distribution.

Q1-12. If the voltage applied to a circuit is doubled and the resistance is increased to three times its former value, what will be the final current value?

ANSWER. Assume the following initial conditions: emf of 1 volt and resistance of 1 ohm. Then the current flowing is 1 amp and

$$I = (1 \times 2)/(1 \times 3) = 2/3, \text{ or } \textbf{two-thirds of initial value}$$

Chapter 2

BATTERY CIRCUITS

2-1. Series Connections of Batteries. By connecting the positive (+) terminal of one cell to the negative (−) of the other, a series connection results (see Fig. 2-1). Thus, a series connection adds to the voltage of each cell and the results are additive.

$$E_t = E_1 + E_2 + E_3 + \cdots + E_n \tag{2-1}$$

As for their internal resistances they are also additive. However, the resulting current capacity is that of a single cell.

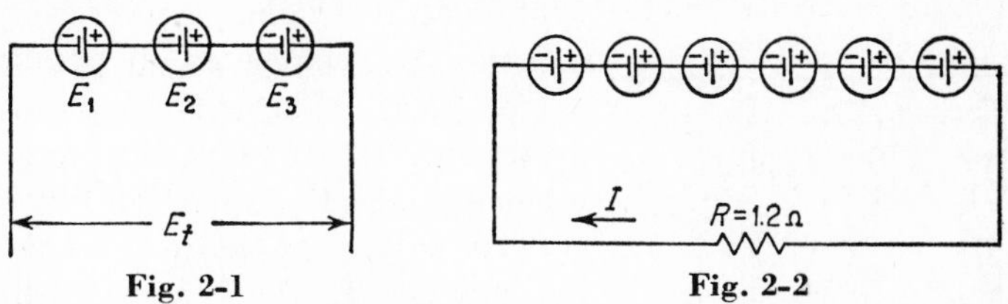

Fig. 2-1

Fig. 2-2

Q2-1. A battery of dry cells consists of six cells connected in series, each having an emf of 1.5 volts and an internal resistance of 0.5 ohm. If the outside terminals are connected by a resistor of 1.2 ohms, what is the current flowing?

ANSWER. Refer to Fig. 2-2. We see that

$$I = (1.5 \times 6)/[1.2 + (0.5 \times 6)] = \mathbf{2.14\ amp}$$

2-2. Parallel Connections of Batteries. A parallel connection consists in connecting terminals, as shown in Fig. 2-3. Then

$$I = \frac{E}{R + r/n} \tag{2-2}$$

where n = number of cells. As for their internal resistances, $r = r_1 = r_2 = r_3 = r_4$, where the individual r's are the internal resistances of the numbered cells. For satisfactory operation of cells in parallel it is necessary for the cells to have the same emf.

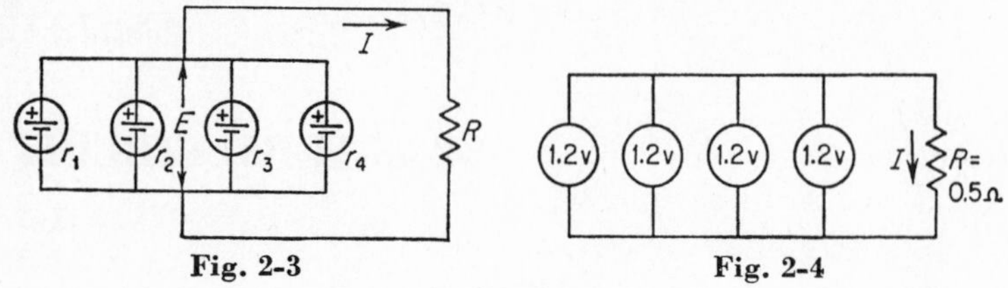

Fig. 2-3 Fig. 2-4

Q2-2. If each cell in Fig. 2-4 with an emf of 1.2 volts and internal resistances of 0.5 ohm is connected to an external resistance of 0.5 ohm, what current will flow?

ANSWER. $I = 1.2/(0.5 + 0.5/4) =$ **1.92 amp**

2-3. Series-parallel Connection of Cells. It is sometimes convenient to arrange the cells in such a manner that the terminals of several similar series groups are themselves connected in parallel (see Fig. 2-5). The effect of such a grouping is to give an emf three times that of one cell and a current twice as large as that of a simple series grouping. If there are n cells with internal resistances r' and emf's E', then for each series grouping

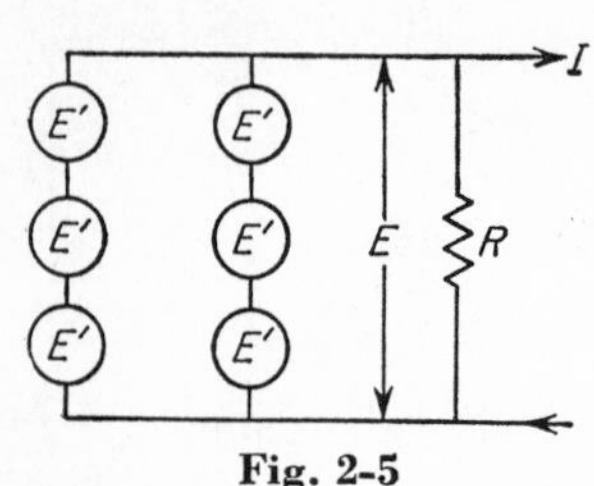

Fig. 2-5

$$E = nE' \tag{2-3}$$

$$r_1 = nr' \tag{2-4}$$

For m parallel-connected groups, the resistance of the combination

$$r = \frac{r_1}{m} = \frac{nr'}{m} \tag{2-5}$$

The current through the external resistance is found by

$$I = \frac{nE'}{(nr'/m) + R} \tag{2-6}$$

It has been found that the maximum current will be obtained when cells are so grouped that their internal resistances equal the external resistance of the circuit. Then

$$\frac{n}{m}r' = R \tag{2-7}$$

Q2-3. A small lighting plant requires a current of 2 amp at a potential of 10 volts. If the cells to be used have an internal resistance of 0.75 ohm and an emf of 1 volt, how should the cells be arranged?

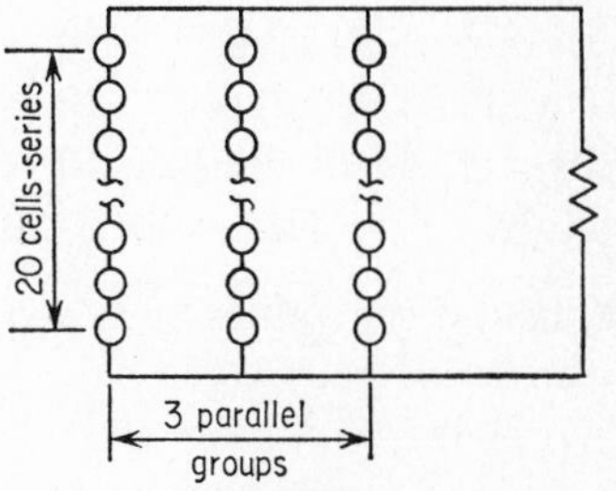

ANSWER. Refer to Fig. 2-6. The external resistance

$$R = \frac{E}{I} = \frac{10}{2} = 5 \text{ ohms}$$

20 cells-series

3 parallel groups

Fig. 2-6

The cells should be arranged so that their internal resistance equals 5 ohms. Then the emf of the cells is

$$nE' = E + IR = 10 + (2 \times 5) = 20 \text{ volts}$$

And $n = {}^{20}\!/_{1} = 20$ cells, since $n = 20/E'$. The internal resistance of the circuit may be written

$$r = \frac{nr'}{m} = \frac{20 \times 0.75}{m} = 5$$

and

$$m = \frac{20 \times 0.75}{5} = 3$$

Thus, the arrangement is **20 cells in series** with **three** such **parallel groups,** or a total of $20 \times 3 =$ **60 cells.**

2-4. Battery Emf's and Kirchhoff's Laws. These laws are of special value when dealing with networks of conductors. The laws follow:

(*a*) At any point in a circuit there is as much current flowing away from the point as there is flowing to it.

(*b*) The summation of the *IR* drops around any one path of an electric circuit equals the summation of the emf's impresssed on that same path.

Use care to get the algebraic signs of the emf's correct. If there is no source of emf in any given path, then the sum of the *IR* drops in one direction equals the sum of the *IR* drops in the other direction.

Q2-4. The circuit illustrated in Fig. 2-7 has resistances as shown. What is the value of the current in the 100-ohm resistance?

ANSWER. This is an unbalanced Wheatstone bridge circuit.

Branch *ABC* gives: $10 = 10I_1 + 10(I_1 + I_3)$ (1)
Branch *ADC* gives: $10 = 10I_2 + 30(I_2 - I_3)$ (2)
Branch *ADBC* gives: $10 = 10I_2 + 100I_3 + 10(I_1 + I_3)$ (3)
Now, $I_1 = (10 - 10I_3)/20$ and $I_2 = (10 + 30I_3)/40$

Then, if the values of I_1 and I_2 are substituted in Eq. (3) above and the terms rearranged

$$2.5 = 112.5I_3$$

Solving for I_3 we obtain

$$I_3 = 2.5/112.5 = \mathbf{0.0222\ amp}$$

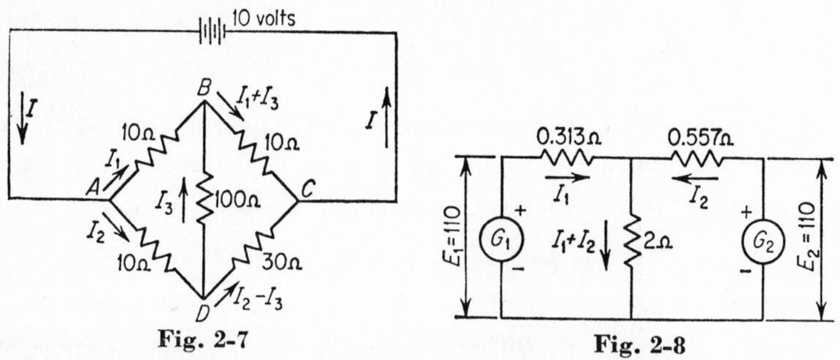

Fig. 2-7

Fig. 2-8

Q2-5. Determine the current supplied by each of the sources of power G_1 and G_2 in Fig. 2-8 when the potential across each source is 110 volts.

ANSWER. Let's look at the left-hand portion of the circuit. According to Kirchhoff's law the equation represented is

$$110 = 0.313I_1 + 2(I_1 + I_2)$$

For the right-hand side of the circuit we, in like manner, obtain

$$110 = 0.557I_2 + 2(I_1 + I_2)$$

From the first equation above we obtain

$$I_2 = \frac{110 - 2.313I_1}{2}$$

which if substituted in the second equation gives

$$110 = \frac{2.557(110 - 2.313I_1)}{2} + 2I_1$$

Thus $\qquad I_1 = \mathbf{32\ amp}$

A substitution in the first equation gives us $I_2 =$ **18 amp,** and the total current is $I_1 + I_2 = 32 + 18 = 50$ amp.

2-5. Parallel Combinations of Unlike Generators or Batteries. When battery cells of either primary or storage type are used in series and parallel combinations, they are ordinarily considered to have the same internal resistance and emf. This is only approximately true, and the process of finding resulting voltage and current is a matter of simple addition. Even if the cells are unlike and are joined in series, the emf's and resistances merely add together. We shall not consider this simple case, however. The following problem will show us what will happen if we have two or more cells of unlike emf and like or unlike resistance joined in parallel and feeding a line. Such combinations represent actual conditions in many telegraph and telephone circuits.

Q2-6. Two batteries in different states of charge are connected in parallel to send current through an external resistance of 1 ohm. The emf's of the batteries are 6.3 and 5.4 volts and their internal resistances are 0.004 and 0.008 ohm, respectively. How much current is flowing in each battery and in the external resistance?

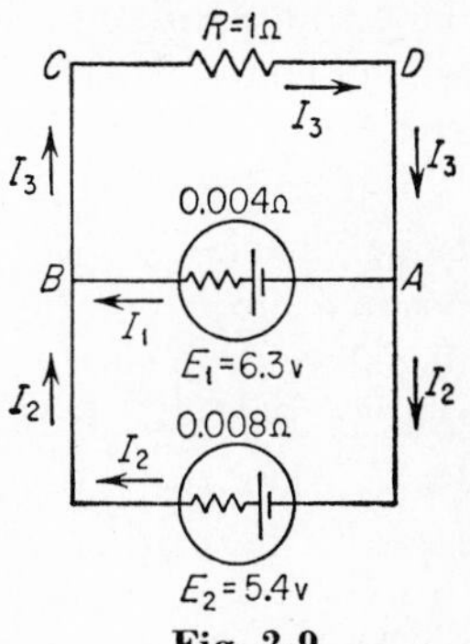

Fig. 2-9

ANSWER. This is an application of Kirchhoff's laws with respect to batteries. There is probably a reversed current through E_2 because of the higher emf of battery E_1. We shall assume that I_2 flows in the direction shown, and if the current value for I_2 comes out a negative quantity, we have merely to reverse the arrowhead. Now let's proceed with the answer. Please refer to Fig. 2-9.

Loop E_1BCDAE_1:

$$6.3 = 0.004I_1 + 1(I_1 + I_2) = 1.004I_1 + I_2 \qquad (1)$$

Loop E_2BCDAE_2:

$$5.4 = 0.008I_2 + 1(I_1 + I_2) = I_1 + 1.008I_2 \qquad (2)$$

In order to solve for I_2 subtract Eq. (2) from Eq. (1) after multiplying Eq. (2) by 1.004 in order to eliminate I_1. Now

$6.3 = 1.004I_1 + I_2$ or corrected to $-5.42 = -1.004I_1 - 1.012I_2$

Then $I_2 = -73$ **amp.** Next, solving for I_1, using Eq. (2) above,

$$5.4 = I_1 + 1.008I_2 = I_1 + [1.008(-73)]$$

And thus

$$I_1 = 5.4 + 73.6 = \mathbf{79\ amp}$$

The current through the external resistance of 1 ohm is found:

$$I_3 = I_1 + I_2 = 79 + (-73) = \mathbf{6\ amp}$$

From the results of our computations we see that the current was backing up through the battery E_2 on account of its lower emf. This shows what is likely to happen in a battery of storage cells when one cell or set of cells, joined in parallel with others, becomes worn out before the others. It is common practice to operate both shunt and compound generators in parallel. The necessity of using machines of approximately equal emf's and resistances becomes apparent from an inspection of the above solution. The distribution of current in line and generators so used is similar to that in the case of batteries. Thus, if in the above problem and its solution, the expression "generators" were substituted for "batteries," the method, computation, and results would be the same.

Chapter 3

UNITS OF AREA AND RESISTANCE

3-1. General. In the design of an electric appliance it is often necessary to determine two things: the resistance of the electric circuit and the current-carrying capacity for a given temperature rise. The first of these is usually computed and the second is obtained from tables.

3-2. Circular Mil. This is the unit of cross section used in the American wire gauge. The term "mil" means one-thousandth of an inch (0.001 in.). It is the area of a circular wire having a diameter of one mil. To find the number of circular mils in a circle of a given diameter, we merely have to square the number of mils in the diameter. Thus, let d equal the diameter of a circle in mils. Then

$$\text{Area of circle} = 0.785(d \times d) \qquad \text{sq mils} \tag{3-1}$$

In square measure

$$\text{Area of cir mils} = 0.785 \times 1 \times 1 = 0.785 \text{ sq mils}$$
$$\text{Cir mils} = 0.785d^2/0.785 = d^2 \tag{3-2}$$

Q3-1. A certain switchboard needs a conversion from a circular conductor to a rectangular conductor, or bus bar, having the same cross-sectional area. If the diameter of the circular wire is 0.846 in., calculate (*a*) width of an equivalent bus bar if the thickness of the bar is to be ¼ in., (*b*) its area in circular mils, (*c*) the current-carrying capacity if 1 sq in. of copper carries 1,000 amp.

ANSWER.

(*a*) Area of wire, sq in.: $A = 0.785 \times 0.846^2 = 0.562$ sq in.

Area of bus bar = 0.562 and width = 0.562/0.25 = **2.248 in.**

(*b*) Area in cir mils = 846^2 = **715,716 cir mils**

(*c*) $0.562 \times 1{,}000$ = **562 amp**

Q3-2. A certain 115-volt 100 hp d-c motor has an efficiency of 90 per cent and takes a starting current of 150 per cent of full-load

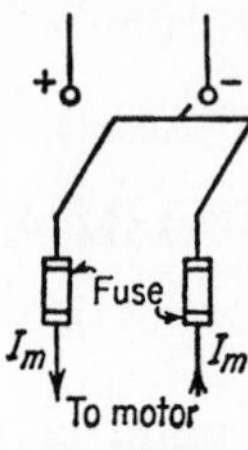

Fig. 3-1

current. Determine (*a*) fuse size, (*b*) copper requirements of switch (1 sq in. carries 1,000 amp). See Fig. 3-1.

ANSWER.

Motor current $= I_m = \text{hp} \times 746/(E \times \text{eff}) = 100 \times 746/(115 \times 0.9)$

This when calculated out equals 721 amp.

(*a*) Fuse amperes $= 721 \times 1.5 = 1{,}081.5$ or **1,100 amp**

(*b*) Copper area $= 721/1{,}000 =$ **0.721 sq in.**

3-3. Circular Mil–foot. The circular mil–foot (mil-foot, for short) is a unit circular conductor one foot in length and one mil in diameter. This is a value which is very important when determining the resistance of a certain number of feet of wire with a definite cross section.

$$R = K \frac{l}{d^2} \qquad \text{ohms} \tag{3-3}$$

where K = resistivity, ohms per mil foot
l = length of conductor, ft
d = diameter of conductor, mils or cir mils

See Table 3-1 for values of K.

Q3-3. What size will an aluminum wire be which has the same resistance as a No. 4 copper wire?

ANSWER. Please refer to Table 3-2, page 20. Number 4 copper wire has a resistance of 0.2485 ohm per 1,000 ft. The resistance of the aluminum wire is found from

$$R = \frac{17l}{d^2} = 0.2485 = \frac{17{,}000}{d^2}$$

Solving for d^2 we find this value to be 68,400 cir mils. This would require an aluminum wire of **No. 1 gauge** (see Tables 3-1 and 3-2). Thus, this wire would be the equivalent of the copper wire.

Q3-4. What copper wire (*B* and *S* gauge) should be used to transmit electric power a total distance of 2 miles? The resistance is not to exceed 2.7 ohms, and the temperature assumed is 20°C.

ANSWER. Ohms per ft = 2.7/(2 × 5,280) × 1,000 = 0.256 ohm per 1,000 ft is required. Now refer to Table 3-2. Number 5 wire offers 0.3133 ohm per 1,000 ft; No. 4 wire, 0.2485. Therefore, use **No. 4 wire** for least resistance. Distance is 1 mile out and 1 mile back.

TABLE 3-1. RESISTIVITY OF METALS AT 32°F

Metal	Resistivity	Metal	Resistivity
Aluminum	17.0	Gold	14.7
Constantan	295.0	Iron (commercial)	66.4–81.4
Annealed copper	10.4	Iron (cast)	590.0
Hard-drawn copper	10.7	Lead	132.5
Pure copper	10.2	Mercury	577.0
German silver	199.0	Nickel	47.0
Monel metal	253.0	Steel (hard)	275.0
Platinum	60.2	Steel (soft)	95.7
Tungsten	33.2	Calorite	720.0
Nichrome	600.0	Zinc	37.4
Silver	9.89–11.2	Tin	69.2
Brass (annealed)	42.0		

Q3-5. If the resistance of a copper wire ⅛ in. in diameter is found to be 0.125 ohm, what is the length of this wire?

ANSWER. Arranging Eq. (3-3)

$$l = \frac{Rd^2}{K} = \frac{0.125 \times 125^2}{10.4} = \textbf{188 ft} \text{ approximately}$$

Q3-6. A 1,000-ft wire having a diameter of 0.2 in. also has a resistance of 0.26 ohm. What is the resistance of another wire of the same material but half this length and with a diameter of 0.04 in.?

ANSWER. First determine the resistance per mil-foot by rearranging Eq. (3-3).

$$K = (0.26 \times 200^2)/1{,}000 = 10.4$$

This is for both wires. And the resistance of the shorter one

$$R = (10.4 \times 500)/40^2 = \textbf{3.25 ohms}$$

TABLE 3-2. WIRE TABLE, STANDARD ANNEALED COPPER
American Wire Gauge (B. & S.) English Units

Gauge number	Diameter in mils at 20°C	Cross section at 20°C		Ohms per 1,000 ft[1]			
		Cir mils	Sq in.	0°C (32°F)	20°C (68°F)	50°C (122°F)	75°C (167°F)
0000	460.0	211,600	0.1662	0.04516	0.04901	0.05479	0.05961
000	409.6	167,800	0.1318	0.05695	0.06180	0.06909	0.07516
00	364.8	133,100	0.1045	0.07181	0.07793	0.08712	0.09478
0	324.9	105,500	0.08289	0.09055	0.09827	0.1099	0.1195
1	289.3	83,690	0.06573	0.1142	0.1239	0.1385	0.1507
2	257.6	66,370	0.05213	0.1440	0.1563	0.1747	0.1900
3	229.4	52,640	0.04134	0.1816	0.1970	0.2203	0.2396
4	204.3	41,740	0.03278	0.2289	0.2485	0.2778	0.3022
5	181.9	33,100	0.02600	0.2887	0.3133	0.3502	0.3810
6	162.0	26,250	0.02062	0.3640	0.3951	0.4416	0.4805
7	144.3	20,820	0.01635	0.4590	0.4982	0.5569	0.6059
8	128.5	16,510	0.01297	0.5788	0.6282	0.7023	0.7640
9	114.4	13,090	0.01028	0.7299	0.7921	0.8855	0.9633
10	101.9	10,380	0.008155	0.9203	0.9989	1.117	1.215
11	90.74	8,234	0.006467	1.161	1.260	1.408	1.532
12	80.81	6,530	0.005129	1.463	1.588	1.775	1.931
13	71.96	5,178	0.004067	1.845	2.003	2.239	2.436
14	64.08	4,107	0.003225	2.327	2.525	2.823	3.071
15	57.07	3,257	0.002558	2.934	3.184	3.560	3.873
16	50.82	2,583	0.002028	3.700	4.016	4.489	4.884
17	45.26	2,048	0.001609	4.666	5.064	5.660	6.158
18	40.30	1,624	0.001276	5.883	6.385	7.138	7.765
19	35.89	1,288	0.001012	7.418	8.051	9.001	9.792
20	31.96	1,022	0.0008023	9.355	10.15	11.35	12.35
21	28.45	810.1	0.0006363	11.80	12.80	14.31	15.57
22	25.35	642.4	0.0005046	14.87	16.14	18.05	19.63
23	22.57	509.5	0.0004002	18.76	20.36	22.76	24.76
24	20.10	404.0	0.0003173	23.65	25.67	28.70	31.22
25	17.90	320.4	0.0002517	29.82	32.37	36.18	39.36
26	15.94	254.1	0.0001996	37.61	40.81	45.63	49.64
27	14.20	201.5	0.0001583	47.42	51.47	57.53	62.59
28	12.64	159.8	0.0001255	59.80	64.90	72.55	78.93
29	11.26	126.7	0.00009953	75.40	81.83	91.48	99.52
30	10.03	100.5	0.00007894	95.08	103.2	115.4	125.5
31	8.928	79.70	0.00006260	119.9	130.1	145.5	158.2
32	7.950	63.21	0.00004964	151.2	164.1	183.4	199.5
33	7.080	50.13	0.00003937	190.6	206.9	231.3	251.6
34	6.305	39.75	0.00003122	240.4	260.9	291.7	317.3
35	5.615	31.52	0.00002476	303.1	329.0	367.8	400.1
36	5.000	25.00	0.00001964	382.2	414.8	463.7	504.5
37	4.453	19.83	0.00001557	482.0	523.1	584.8	636.2
38	3.965	15.72	0.00001235	607.8	659.6	737.4	802.2
39	3.531	12.47	0.000009793	766.4	831.8	929.8	1,012
40	3.145	9.888	0.000007766	966.5	1,049	1,173	1,276

[1] Resistance at the stated temperatures of a wire whose length is 1,000 ft at 20°C.

Chapter 4

TEMPERATURE COEFFICIENT OF RESISTANCE

4-1. Effect of Temperature Changes on Resistance. The resistance of most metals increases with an increase in temperature, and thus they have positive temperature coefficients of resistance. This is expressed as the number of ohms by which the resistance changes in temperature of one centigrade degree.

The relation of resistance changes due to changes in temperature is expressed as follows:

$$R_t = R_0(1 + \alpha t) \qquad (4\text{-}1)$$

where R_t = resistance at temperature t, °C
R_0 = resistance at 0°C
α = temperature coefficient of resistance (see Table 4-1).

TABLE 4-1. TEMPERATURE COEFFICIENT OF RESISTANCE α

Material	α	Material	α
Aluminum	0.00420	German silver	0.00036
Constantan	Nearly zero	Monel metal	0.00208
Annealed copper	0.00426	Platinum	0.0037
Hard-drawn copper	0.00413	Tungsten	0.0049
Pure copper	0.00410	Nichrome	0.00044
Silver	0.00411	Brass (annealed)	0.00208
Gold	0.00365	Iron (commercial)	0.00618
Lead	0.00466	Mercury	0.00088
Nickel	0.006	Steel (hard)	0.0016
Steel (soft)	0.00458	Zinc	0.0040
Tin	0.00458		

Q4-1. If the resistance of a copper conductor is 10 ohms at 0°C, what would be its resistance at 20°C?

ANSWER. $R_t = 10(1 + 0.00427 \times 20) =$ **10.845 ohms**

Since in most practical cases it is necessary to calculate a change from one temperature other than 0°C to some other temperature, we may make use of the following expression:

$$\frac{R_1}{R_2} = \frac{234.5 + t_1}{234.5 + t_2} \qquad (4\text{-}2)$$

Equation (4-2) is particularly useful since it may be applied to problems for both resistance and temperature calculations of conductors and wire.

Q4-2. The resistance of a coil of wire at 20°C was found to be 49.122 ohms. When the temperature was increased to 65°C, the resistance increased to 50 ohms. Calculate the temperature coefficient of the resistance α.

ANSWER.

$$\alpha = \frac{R_2 - R_1}{R_1 t_2 - R_2 t_1} = \frac{50 - 49.122}{(49.122 \times 65) - (50 \times 20)} = \mathbf{0.0004}$$

Q4-3. The resistance of a coil of copper wire at 20°C is 2.21 ohms. After a current of 10 amp has been flowing in the wire for 2 hr, the resistance of the coil is found to be 2.59 ohms. What is the temperature rise?

ANSWER. Rearranging Eq. (4-2) to read as follows,

$$t_2 = \frac{R_2}{R_1}(234.5 + t_1) - 234.5 \qquad (4\text{-}3)$$

Then

$$t_2 = \frac{2.59}{2.21}(234.5 + 20) - 234.5 = 63.8°\text{C}$$

$$\text{Temperature rise of coil } t_2 - t_1 = 63.8 - 20 = \mathbf{43.8°C}$$

Q4-4. The field resistance of a d-c machine is 80 ohms at 25°C. The allowable temperature rise is 50°C. At the end of a test run in which all parts reach an average temperature the field resistance increases by 7 ohms. Does the machine exceed its allowable temperature rise?

ANSWER. This problem is same as previous one. Use Eq. (4-3).

$$t_2 = (87/80)(234.5 + 25) - 234.5 = 47.5°\text{C}$$

Then temperature rise of field resistance is 47.5 − 25 = 22.5. *The machine does not exceed its allowable temperature rise.*

Q4-5. Design an electric unit heater to take 1,200 watts for its coil on a 240-volt d-c circuit. Using a Nichrome wire with K equal to 660 for the resistors and having a wire cross section of 40 cir mils, determine the length of wire needed.

ANSWER.

$$\text{Area} = \frac{0.040^2}{0.001^2} \times \frac{0.785}{0.785} = 1{,}600 \text{ cir mils} = d^2$$

$$\text{Current flowing} = \frac{\text{power}}{\text{volts}} = \frac{1{,}200}{240} = 5 \text{ amp}$$

$$\text{Resistance of wire} = \frac{E}{I} = \frac{240}{5} = 48 \text{ ohms}$$

and from Eq. (3-3)

$$l = \frac{Rd^2}{K} = \frac{48 \times 1{,}600}{660} = \mathbf{116.5\ ft}$$

Q4-6. An electric pressing iron has a resistance composed of commercial iron wire of 80 ohms when cold (20°C). When hot, the resistance rises to 150 ohms. What temperature is reached if a constant temperature coefficient α is assumed?

ANSWER. See Table 4-1. For commercial iron the reciprocal of α is $1/\alpha$, or 161.8. Using Eq. (4-2) as guide

$$t_2 = (161.8 + 20)(^{150}/_{80}) - 161.8 = \mathbf{177.2°C}$$

Chapter 5

POWER, ENERGY, AND THERMAL UNITS

5-1. Power in Electric Circuits. Power is the rate of energy transfer. Its unit is the watt. It may be expressed as

$$P \text{ (power)} = E \text{ (volts)} \times I \text{ (amp)} \tag{5-1}$$

or

$$P = I^2R \tag{5-2}$$

$$P = \frac{E^2}{R} \tag{5-3}$$

In practice, the kilowatt and the horsepower are more convenient.

Q5-1. An electric iron takes 3½ amp. If the heating element has a resistance of 40 ohms, what is its power consumption?

ANSWER. Equation (5-2) provides us with the clue

$$P = 3.5^2 \times 40 = \mathbf{490 \text{ watts}}, \text{ or } \mathbf{0.490 \text{ kw}}$$

5-2. Electric Energy. This is equal to power × time. Thus, if a generator delivers electricity at the rate of 25 kw for 10 hr, it has delivered 25 × 10, or 250 kwhr of electric energy. The practical unit of electric energy is the kilowatt hour.

The joule is another. For example,

$$\text{Watt-sec} = EIt \tag{5-4}$$

where t = time in seconds. The kilowatt hour is 1,000 × 60^2 times longer than the joule or watt-second.

Q5-2. If 10 amp flow through a circuit of 5 ohms resistance per hour, what is the equivalent energy in the circuit?

ANSWER. $I^2Rt = 10^2 \times 5 \times 60^2 = \mathbf{1.8 \times 10^6}$ **joules**

5-3. Thermal Units. The amount of heat required to raise the temperature of one gram of water one centigrade degree is the gram-calorie. It has been found that 1 g-cal equals 4.2 watt-sec, or joules.

$$W = \frac{1}{4.2} \times I^2Rt = 0.24I^2Rt \quad \text{g-cal} \qquad (5\text{-}5)$$

This is the expression for Joule's law. Now the relation between the electrical energy in a circuit and the heat in Btu is

$$H = 0.057I^2Rt \qquad \text{Btu} \qquad (5\text{-}6)$$

Here t is time in minutes as compared to t for time in seconds in Eq. (5-5).

Q5-3. What current flow can be expected to operate a water-heating device containing 0.6 liter of water, when the resistance of the heating coil is 5 ohms, and the water is to be heated 80 centigrade degrees in 10 min? Neglect the losses.

ANSWER. From Eq. (5-5) we can determine gram-calories

$$W = 600 \times 80 = \text{g-cal} = 0.24I^2Rt$$

Rearranging and solving for I

$$I = \sqrt{W/0.24Rt} = \mathbf{8.16\ amp}$$

5-4. Determining Power in an Electric Circuit. To find the power that is being consumed in an electric circuit, we merely need to insert an ammeter to measure the current in that part of the circuit and apply a voltmeter to measure the voltage across that part of the circuit. Then we multiply the ammeter reading in amperes by the voltmeter reading in volts. It is necessary to measure both voltage and current for the circuit at the same time, and their product is equal to the power consumed in that part of the circuit alone.

Q5-4. What power does an electric motor consume when taking 20 amp at 220 volts?

ANSWER. Watts consumed = 20 × 220 = **4,400 watts.** This is equivalent to 4,400/1,000, or **4.4 kw.**

Q5-5. What electric power is required to drive a pump which must raise water 96 ft and supply 2,600 gpm? One gallon water weighs 8.3 lb.

ANSWER. Horsepower required is given by the expression

$$\frac{\text{Height} \times \text{ppm}}{33{,}000} = \frac{96 \text{ ft} \times 2{,}600 \times 8.3}{33{,}000} = 62.5 \text{ hp}$$

This is equivalent to

$$62.5 \times 0.746 = \mathbf{46.6\ kw} \text{ of power}$$

5-5. Efficiency of Electric Equipment. When current flows through electric equipment, some heat is generated. Unless this heat is utilized, it is wasted, and the equipment becomes inefficient. No electric machine utilizes all the energy it receives. The percentage of energy it gives out, (input − losses)/input, is called its efficiency. Thus, any electric apparatus which gives out 9 watts for every 10 watts received has an efficiency of 90 per cent; if only 7 watts output, 70 per cent. In other words, the efficiency of any device is always less than 100 per cent. We must remember to express input, losses, and output in the same units, either kilowatts or horsepower.

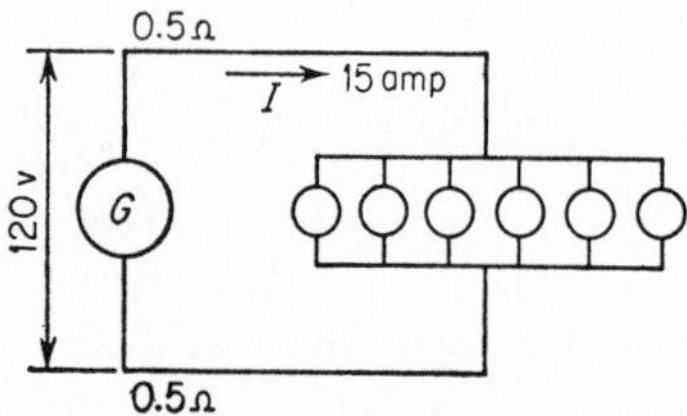

Fig. 5-1

Q5-6. A 25-hp engine drives a d-c generator. If the generator has an efficiency of 84 per cent, how many kilowatts and horsepower does it deliver?

ANSWER. The energy transmitted to the generator shaft is

$$25 \times 0.84 = \mathbf{21\ hp}$$

or

$$21 \times 0.746 = \mathbf{15.67\ kw}$$

5-6. Efficiency of Power Transmission. In like manner, when energy is transmitted through conductors, it suffers a loss in the form of heat because of the resistance in the conductors. The efficiency of transmission is that percentage of the input which finally reaches its destination.

Q5-7. The lamp bank shown in Fig. 5-1 uses 15 amp. What is its efficiency of transmission? Assume total resistance = 0.5 ohm.

ANSWER.

Total power: 120 volts × 15 amp = 1,800 watts
Line loss: 15 × 15 × 0.5 = 112.5 watts
Watts for lamps: 1,800 − 112.5 = 1,687.5 watts
Efficiency of transmission: 1,687.5/1,800 = 0.94, or **94 per cent**

Chapter 6

WIRE SIZING

6-1. Safe Capacity for Copper Wire. Copper wiring is most commonly used when wiring up a building or an industrial plant for lighting or power. In such design it is necessary to take into account another factor besides voltage drop. Because an electric current heats a conductor through which it is flowing, the heat generated must be dissipated as quickly as it is generated or the temperature will rise. Remember, too, that conductor resistance increases with temperature.

It is necessary, therefore, to select a wire size so that its temperature will not rise so high as to cause overheating and resultant deterioration of the insulation. Accordingly, the National Board of Fire Underwriters has published a table of safe carrying capacities of copper wire. Wherever local regulations do not specify otherwise, the currents should not exceed the values listed under Table 6-1.

6-2. Interior Wire Sizing. First determine the current to be carried by each section of a circuit and then select a wire size from the table. Next check the voltage drop in line. This must not exceed 5 per cent for lamp loads or 10 per cent for motor loads. Voltage drops greater than these will cause excessive motor currents with resultant overheating of motors and low output of light from lamps with resultant poor lighting. Where excessive voltage drops are found, larger wires must be used.

Q6-1. A panel board is located 150 ft away from the main switch in an industrial plant. From this board there run three branches each 50 ft long. Each branch is supplied with six outlets for 100-watt lamps. Size the mains, if the lamps are 110-volt units (see Fig. 6-1).

TABLE 6-1. ALLOWABLE CURRENT-CARRYING CAPACITIES OF INSULATED CONDUCTORS IN AMPERES[1]
Not More Than Three Conductors in Raceway or Cable
(Based on Room Temperature of 30°C, 86°F)

Size AWG MCM	Rubber, type R type RW type RU type RUW (14-2) / Thermo-plastic, type T type TW	Rubber, type RH	Paper / Thermo-plastic asbestos, type TA / Var-Cam, type V / Asbestos Var-Cam, type AVB	Asbestos Var-Cam, type AVA type AVL	Impreg-nated asbestos, type AI (14-8) type AIA	Asbestos, type A (14-8) type AA
14	15	15	25	30	30	30
12	20	20	30	35	40	40
10	30	30	40	45	50	55
8	40	45	50	60	65	70
6	55	65	70	80	85	95
4	70	85	90	105	115	120
3	80	100	105	120	130	145
2	95	115	120	135	145	165
1	110	130	140	160	170	190
0	125	150	155	190	200	225
00	145	175	185	215	230	250
000	165	200	210	245	265	285
0000	195	230	235	275	310	340
250	215	255	270	315	335	
300	240	285	300	345	380	
350	260	310	325	390	420	
400	280	335	360	420	450	
500	320	380	405	470	500	
600	355	420	455	525	545	
700	385	460	490	560	600	
750	400	475	500	580	620	
800	410	490	515	600	640	
900	435	520	555			
1,000	455	545	585	680	730	
1,250	495	590	645			
1,500	520	625	700	785		
1,750	545	650	735			
2,000	560	665	775	840		

[1] Also refer to Pender, Del Mar, and McIlwain, "Electrical Engineers' Handbook," John Wiley & Sons, Inc., New York, 1947, pp. 14-192 to 14-214.

TABLE 6-2. CORRECTION FACTOR FOR ROOM TEMPERATURES OVER 30°C, 86°F

°C	°F						
40	104	0.82	0.88	0.90	0.94	0.95	
45	113	0.71	0.82	0.85	0.90	0.92	
50	122	0.58	0.75	0.80	0.87	0.89	
55	131	0.41	0.67	0.74	0.83	0.86	
60	140	....	0.58	0.67	0.79	0.83	0.91
70	158	....	0.35	0.52	0.71	0.76	0.87
75	167	....	....	0.43	0.66	0.72	0.86
80	176	....	....	0.30	0.61	0.69	0.84
90	194	....	....	....	0.50	0.61	0.80
100	212	....	....	....	....	0.51	0.77
120	248	....	....	....	....	....	0.69
140	284	....	....	....	....	....	0.59

ANSWER. Each branch carries 6 × 100, or 600, watts. This means the equivalent of $600/110$, or 5.45 amp, in each branch wire.

Branch wires: Size according to the table with no size smaller than No. 14 wire for strength. This size is apparently ample.

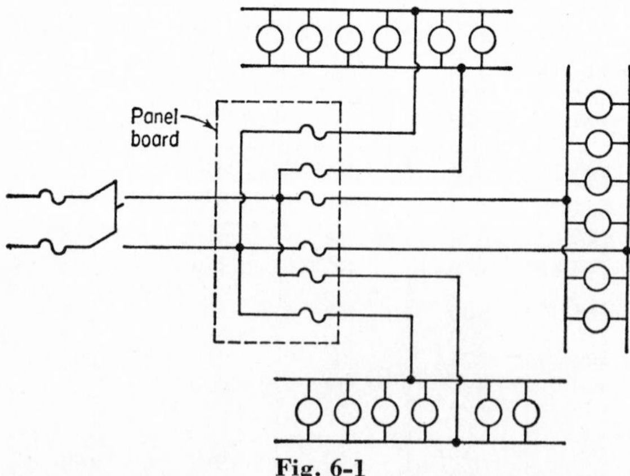

Fig. 6-1

Mains: Size according to the table. Each main must carry the current in all three branches, 3 × 5.45, or 16.35 amp. According to the table, No. 12 wire must be used, as the next size in common use would have only a 15-amp capacity. Thus, No. 14 wire does not have sufficient copper cross section.

A check for voltage drop will show that the line drop is over 5 per cent and that No. 12 wire is still insufficient. By trial and error it will be found that No. 8 wire is adequate. If smaller wires were used, the brightness of the lamps would vary through wide ranges depending on how many were in use at one time.

Q6-2. Find the regulation of a line supplying a 50-kw d-c load at 220 volts. The line is 200 ft long and consists of two 0000 wires.

ANSWER. The regulation for a direct-current circuit may be easily calculated, as the voltage change from full load to no load is the *IR* drop of the circuit. The resistance of the circuit (two wires) may be obtained from standard tables; the current is obtained from the load. Then the ampere flow is 50,000 watts/220 volts = 227.3. The resistance of 1,000 ft 0000 wire = 0.049 ohm. The resistance

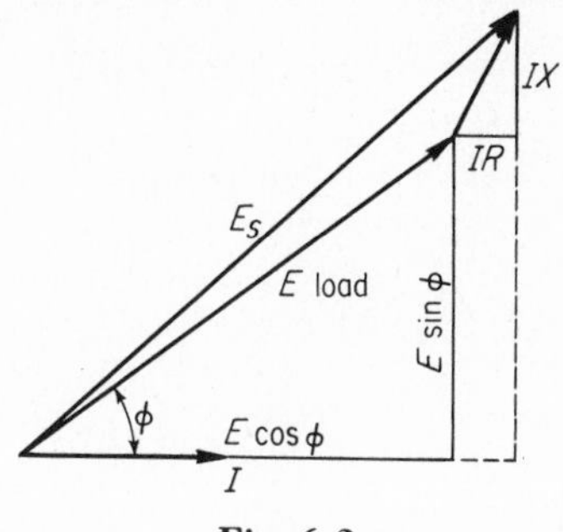

Fig. 6-2

of 0000 wire (400 ft) = 0.0196 ohm. The *IR* drop is 227.3 × 0.0196 = 4.46 volts. Therefore, the regulation is

$$\frac{4.46 \times 100}{220} = \textbf{2 per cent}$$

In an alternating circuit, the power factor of the load and reactance of the circuit must also be taken into account. See Q6-3.

Q6-3. Find the regulation of a three-phase line supplying 440 volts to a 90-kva load of 0.8 power factor. Line is 2,000 ft long and consists of three 0 wires with 24-in. spacing of wires.

ANSWER. From tables the resistance of 0 wire is 0.0983 per 1,000 ft. Also from tables the reactance of 0 wire is 0.121 per 1,000 ft at the given spacing. Then resistance of 2,000 ft = 0.197 and the reactance of 2,000 ft = 0.242. Calculations should be per phase. Assume Y connection. Kilovolt-amperes per phase is 90/3 = 30. Volts per phase, $440/\sqrt{3} = 254$. Amperes per phase, 30,000/254 = 118. Then *IR* drop per phase, $0.196 \times 118 = 23.2$. The *IX* drop per phase is $0.240 \times 118 = 28.3$. Now refer to vector diagram, Fig. 6-2. We note that $E \cos \phi = 203.2$ and $E \sin \phi = 152.4$. We can now calculate sending voltage per phase to be

$$\sqrt{(226.4)^2 + (180.7)^2} = 289.5$$

Sending voltage $= 289.5 \times \sqrt{3} = 501.4$. And the regulation is

$$\frac{(501.4 - 440) \times 100}{440} = \textbf{14 per cent}$$

Chapter 7

MEASUREMENT OF DIRECT-CURRENT ELECTRICITY

7-1. The Galvanometer. This is an instrument for measuring a small electric current. A d-c galvanometer consists essentially of a permanent magnet and a coil of wire through which the current may flow. The d'Arsonval galvanometer, of all galvanometers in use, has been found to be the most practical instrument for measurement of small currents. The principles of design of the d'Arsonval galvanometer are shown in Fig. 7-1.

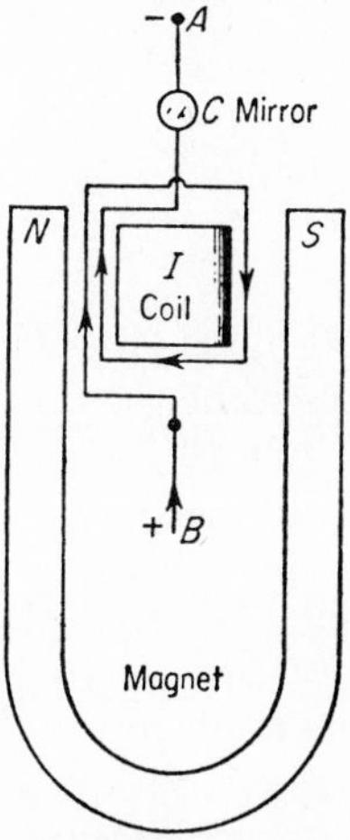

Fig. 7-1

7-2. Galvanometer Shunts. The extreme sensitivity of the galvanometer may be used to measure rough values of current. For this purpose the instrument may be *shunted* by connecting any desired resistance across its terminals so that the current will divide between the galvanometer and its shunt. Figure 7-2 shows this arrangement. If the resistance of this bypass (galvanometer shunt) is $\frac{1}{9}$, $\frac{1}{99}$, or $\frac{1}{999}$ of the galvanometer resistance, the current sensitivity will thereby be reduced to $\frac{1}{10}$, $\frac{1}{100}$, or 1/1,000 of its original value, respectively.

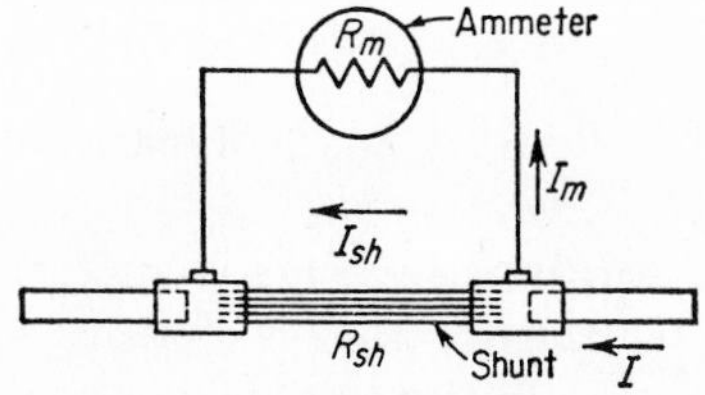

Fig. 7-2

Q7-1. A galvanometer, having a constant of 0.00005 of 1 amp per division, shows a reading of 300 divisions when shunted by a $\frac{1}{999}$-ohm shunt. Find the current flowing.

ANSWER. The current passing through the galvanometer for the deflection of 300 divisions is 300 × 0.00005 amp. The total current is determined by

$$\frac{1}{1{,}000} = \frac{300 \times 0.00005}{I}$$

$$I = 300 \times 0.05 = \mathbf{15\ amp}$$

Q7-2. A galvanometer with a resistance of 500 ohms must be provided with a shunt so that only 0.001 of the total current to be measured will pass through it. Find the resistance of the shunt.

ANSWER. The current flow through the galvanometer will be 1/1,000 of the total current flow. The remainder, or 999/1,000 of the total, will flow through the shunt. From the expression

$$R_s = \frac{R_g I_g}{I_s} \tag{7-1}$$

$$R_s = \frac{500 \times 1/1{,}000}{999/1{,}000} = \mathbf{0.5005\ ohm}$$

Q7-3. A galvanometer with a resistance of 1,000 ohms is to be used to test a circuit. What should be the shunt resistance so that the deflection of the galvanometer may be reduced 100:1?

ANSWER. For a reduction of 100:1 we can use the expression

$$R_s = \tfrac{1}{99} \times 1{,}000 = \mathbf{10.10\ ohms}$$

7-3. The Ammeter. This is another form of galvanometer and shunt arrangement. It is customary to build ammeters with *calibrated* shunts. With the ammeter connected directly in *series* in a circuit which carries larger quantities of current, it is necessary to have an extremely large coil in order to obtain the magnetic field required for measuring purposes, and this becomes heavy and bulky. With the calibrated shunt, however, the ammeter proper passes only a small proportion of the current, the rest passing around the working part of the meter by way of the shunt, resulting in protection against coil burnout and accuracy of reading.

7-4. The Voltmeter. The principal difference between the d-c voltmeter and ammeter is one of resistance. As noted before, the ammeter is *connected in series* in the circuit. By making the coil of an ammeter of many more turns of fine wire and connecting it across the portion of the circuit to be measured, with an additional resistance in series to protect the coil, we have a means of measuring the electric pressure of a circuit (voltage).

7-5. Ammeter Shunts. We saw that the difference between a voltmeter and an ammeter lies in the amount of current sent through each. The ammeter may be used to measure voltage if a high-resistance coil is connected in series with it. When used to measure current, a low resistance (shunt) is placed across it.

The shunt is used to simplify the construction of an ammeter. Thus, all ammeters for use in d-c measurements may be designed to pass similar amount of amperes, although the actual amount of current in the circuit may differ greatly. The shunt consists essentially of a low resistance usually made of manganin strips mounted on heavy copper blocks. The shunt is inserted in the main circuit, as shown in Fig. 7-2.

The shunt law is given by the following equation:

$$\frac{R_m}{R_{sh}} = \frac{I_{sh}}{I_m} \tag{7-2}$$

where R_m = resistance of meter coil
R_{sh} = resistance of shunt
I_{sh} = current passing through shunt
I_m = current passing through meter coil

Q7-4. A certain ammeter has a coil resistance of 2.5 ohms. If the shunt resistance is 0.0005 ohm and the instrument current is 0.02 amp, what is the total current in the circuit?

ANSWER. Rearranging Eq. (7-2) we can find the shunt current.

$$I_{sh} = I_m \frac{R_m}{R_{sh}} = 0.02 \times \frac{2.5}{0.0005} = 100 \text{ amp}$$

and $$I = I_m + I_{sh} = 0.02 + 100 = \mathbf{100.02\ amp}$$

7-6. Measurement of Resistance. There are three general classes into which resistance measurements may be divided. They are:

(*a*) Low resistance (below 5 ohms)
(*b*) Medium resistance (5 to 50,000 ohms)
(*c*) High resistance (above 50,000 ohms)

Measurement of low resistances may be accomplished by the voltage-drop method, using an ammeter and a voltmeter by the comparison or substitution method and by various bridge arrangements.

Measurement of medium resistances may be made by using the Wheatstone bridge ohmmeters.

Measurement of high resistances such as insulation resistances of electrical machinery and materials is made with high-resistance voltmeters (meggers) and other devices.

7-7. Voltmeter-Ammeter Method of Resistance Measurement. Simultaneous readings on each instrument are taken, and from these readings the unknown resistance R_x is calculated using Ohm's law. Meters should be so selected that the readings are made in the middle of the meter scale. The current used should be sufficient to give good deflection, but not great enough to heat the device under test and thereby change its resistance. This is a very important point and may be easily overlooked; the greater the temperature coefficient of the material the more important this factor becomes.

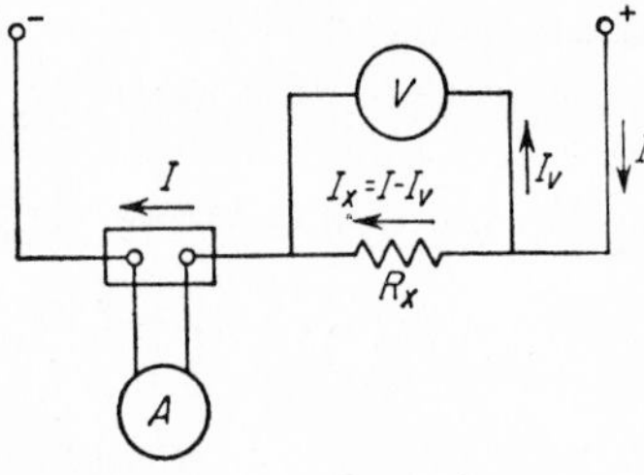

Fig. 7-3

The ratio of the voltmeter resistance to that of the device under test affects the results. If the ratio is large, 2,000 or more, the law of divided circuits can be neglected. Therefore, high-resistance voltmeters should always be used.

Q7-5. It is desired to measure the value of a resistor R_x by means of the arrangement shown in Fig. 7-3. The simultaneous readings of the meters show I equal to 100 amp and E equal to 0.5 volt. If the resistance of the voltmeter R_v equals 100 ohms, what is the value of R_x?

ANSWER. By Ohm's law

$$R_x = \frac{E}{I} = \frac{0.5}{100} = \mathbf{0.005\ ohm}$$

If the shunting current through the voltmeter is taken into account, we obtain R_x equal to 0.00500025 or 0.005 per cent difference—too small to be considered. Thus,

$$I_v = \frac{E}{R_v} = \frac{0.5}{100} = 0.005 \text{ amp}$$

The current through R_v is $I_x = I - I_v$, or 99.995 amp. And

$$R_x = 0.5/99.995 = \mathbf{0.00500025\ ohm}$$

If the resistance of the voltmeter were, on the other hand, 1 ohm, then

$$R_x = 0.005025 \text{ ohm}$$

by the above method of calculation—a difference too great to be neglected.

7-8. Measurement of Resistance by Substitution Method. If a known or standard resistance is connected in series with a resistance to be measured and a current is passed through the series circuit, the voltage drop across each resistor would be proportional to its resistance, and the value of the unknown resistance may be computed. See Fig. 7-4 for a convenient hook-up.

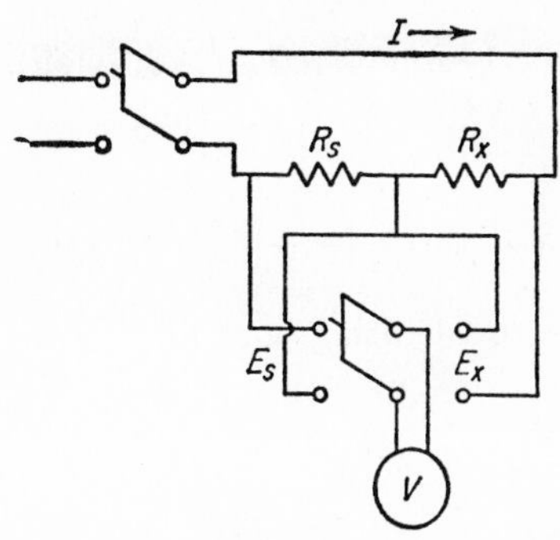

Fig. 7-4

The unknown resistance is again R_x and the known resistance is R_s. The relation between the pertinent voltages and respective resistances is

$$R_x = R_s \frac{E_x}{E_s} \tag{7-3}$$

In this equation R_s, E_x, and E_s are known, and R_x is easily determined.

The resistance is proportional to the voltage drop. Actually the currents in both resistances will differ, unless the resistances are equal. Of course, the same voltmeters and scales must be used. However, (*a*) The current need not be known.

(*b*) The method depends on accurate comparison of voltages, and voltmeter used need not be accurately calibrated.

(*c*) With a constant applied voltage no error is produced by using a low-resistance voltmeter.

7-9. The Wheatstone Bridge. In its simplest form it is shown in Fig. 7-5. A and B are called ratio arms and usually are resistances whose values are 1, 10, 100, etc. R is the rheostat arm. It gives, in up to four or five significant figures, resistance values from 1 to 10,000 ohms.

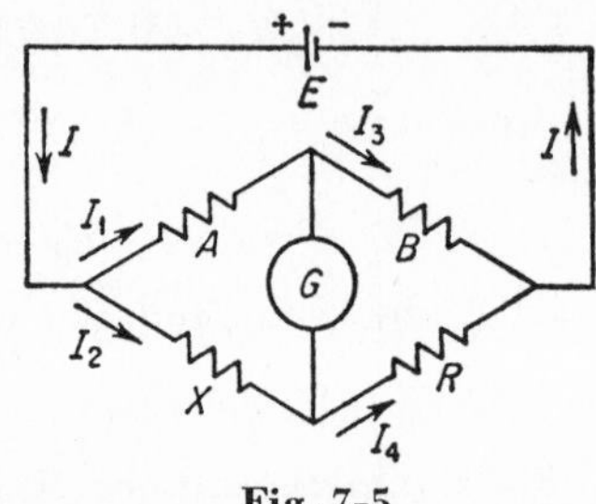

Fig. 7-5

G is the galvanometer whose resistance should be of the order of the magnitude of the resistance between the points to which it is connected.

When the bridge is in balance, the current through the galvanometer is zero. Therefore, $I_1 = I_3$ and $I_2 = I_4$. And thus

$$I_1A = I_2X \tag{7-4}$$
$$I_3B = I_4R \tag{7-5}$$

Dividing Eq. (7-4) by Eq. (7-5) and substituting current values, we obtain

$$\frac{I_1A}{I_1B} = \frac{I_2X}{I_2R} \tag{7-6}$$

or

$$X = R\frac{A}{B} \tag{7-7}$$

This is the general equation of the Wheatstone bridge under stated conditions. It should be noted, however, that although the adjustment to zero current through the galvanometer is the simplest method of obtaining the value of the unknown resistance, this value may also be obtained in the unbalanced circuit by a simple application of Kirchhoff's laws (see Q2-4, p. 14).

Q7-6. In a Wheatstone bridge the ratio arms have a resistance of 20 ohms each, and the other two arms have two resistances of 20 and 30 ohms. If the galvanometer has a resistance of 100 ohms, how much current will flow through the galvanometer if the bridge is connected to a 6-volt d-c battery of negligible resistance? See Fig. 7-6.

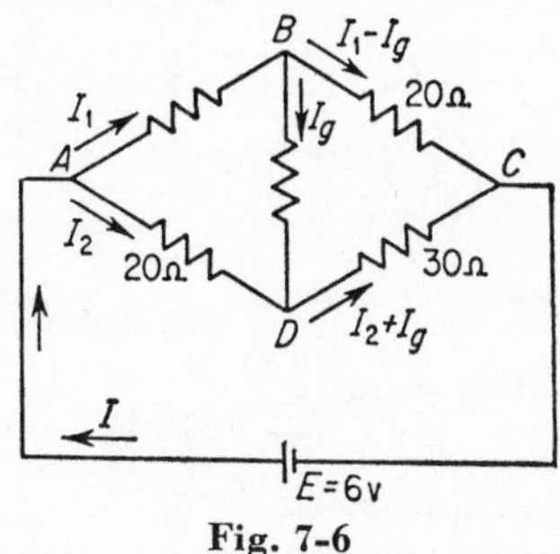

Fig. 7-6

ANSWER. Also refer to Q2-4, p. 14. By Kirchhoff's laws we obtain the following equations:

Circuit *ABC*:

$$6 = 20I_1 + 20(I_1 - I_g) = 40I_1 - 20I_g \tag{1}$$

Circuit *ADC*:

$$6 = 20I_2 + 30(I_2 + I_g) = 50I_2 + 30I_g \tag{2}$$

Circuit *ABDC*:

$$6 = 20I_1 + 100I_g + 30(I_2 + I_g) \tag{3}$$

or $6 = 20I_1 + 100I_g + 30I_2 + 30I_g = 20I_1 + 30I_2 + 130I_g$

From Eq. (1) above

$$I_1 = \frac{6 + 20I_g}{40}$$

Equation (2) above gives

$$I_2 = \frac{6 - 30I_g}{50}$$

Substituting values of I_1 and I_2 in Eq. (3) above and rearranging,

$$I_g = \text{minus } (-)\mathbf{0.0047\ amp}$$

Therefore, reverse the arrow direction in Fig. 7-6.

7-10. High-resistance Measurements. In measuring a high resistance such as that of insulation in electric machinery and cables, the method shown in Fig. 7-7 is commonly used. A d-c voltmeter of high resistance and a potential of constant value of

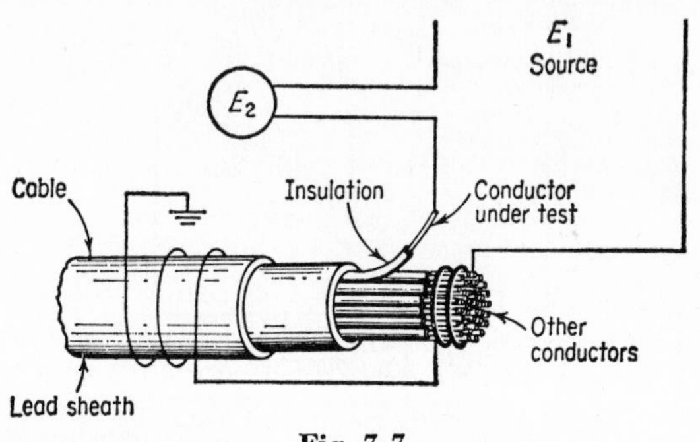

Fig. 7-7

500 volts are used. The same voltmeter is used to read E_1 and E_2. If R_x is the insulation resistance and R_v the meter resistance, then

$$R_x = \frac{R_v(E_1 - E_2)}{E_2} \tag{7-8}$$

Q7-7. A 50,000-ohm voltmeter is used in measuring the insulation resistance of a motor. When connected in series with the insulation across a 250-volt line, the instrument reads 1 volt. What is the value of the insulation resistance of the motor?

ANSWER. Substituting in Eq. (7-8)

$$R_x = 50{,}000[(250 - 1)/1] = 12{,}450{,}000 \text{ ohms, or } \mathbf{12.45\ megohms}$$

The voltmeter method is based on the principle of comparing the currents flowing when a fixed emf is impressed on a known resistance and when impressed on the same resistance in series with the unknown. In its application, the internal resistance of a suitable voltmeter is used as the known resistance and the indication of its pointer as proportional to the current flowing. Two readings are taken: one with the voltmeter only across the power supply, the other with the unknown resistance added between the meter and the power supply.

7-11. Power Measurements. There are two generally employed methods for d-c power measurements, voltmeter and wattmeter. In the former the currents through the voltage across the circuit are recorded directly on the instruments, and their product IE equals watts. In other words

$$\text{Watts (in any part of circuit)} = \text{volts (across that part of circuit)} \times \text{amp (through that part of circuit)}$$

The same precautions must be observed in the use of this equation as in the use of Ohm's law (Sec. 1-3).

Q7-8. A generator G is furnishing a current of 4 amp to the line at a pressure of 120 volts. There is in the circuit a resistance R which requires 5 volts to force the current through it and a motor M which requires 115 volts. How much power does the resistance R consume, and how much does the motor consume?

ANSWER. Refer to Fig. 7-8. The resistance R consumes

$$P_r = I_rE_r = 4 \times 5 = \mathbf{20\ watts}$$

The motor consumes

$$P_m = I_mE_m = 4 \times 115 = \mathbf{460\ watts}$$

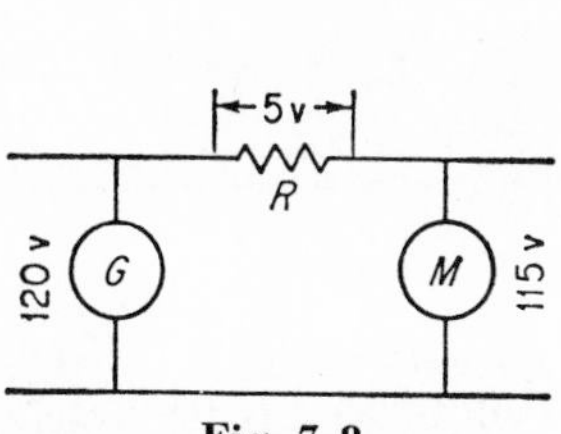

Fig. 7-8

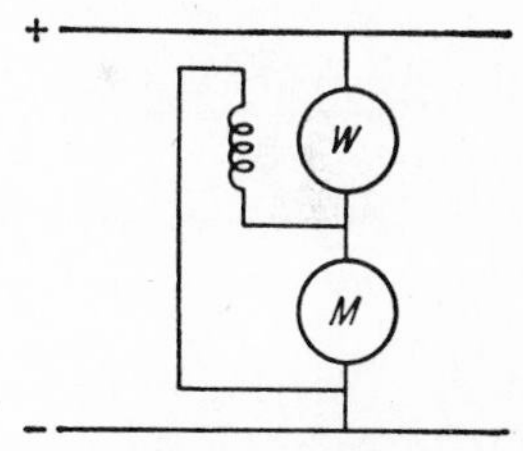

Fig. 7-9

7-12. The Wattmeter. A single instrument, the wattmeter, is used to measure d-c power directly. This instrument is, in effect, a combination voltmeter and ammeter, the former of high resistance, the latter of low resistance. The wattmeter usually has four terminals, two for the voltmeter and two for the ammeter. Figure 7-9 shows the correct method of connecting up a wattmeter in a circuit to measure power taken by the motor M. A current coil is placed in series with the motor, and a voltage coil is connected in shunt across the motor.

7-13. Measurement of Electric Energy. The instrument most commonly used to measure d-c energy is the Thomson watt-hour meter shown in Fig. 7-10.

Q7-9. Show how a three-wire d-c watt-hour meter is connected to measure the energy drawn from an electric load. Indicate how the current and voltage connections are made. Explain why the speed of rotation of this meter is proportional to the power drawn by the load.

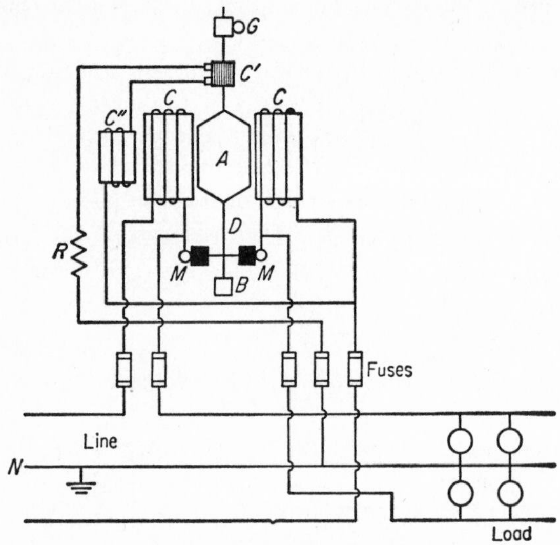

Fig. 7-10

ANSWER. See Fig. 7-10. The Thomson watt-hour meter is essentially a d-c motor. The two coils *CC*, which act as field coils, are connected in series with the outside wires; hence they must be designed to carry the rated current of the meter. The revolving element consists of a spherical armature *A* with a commutator *C′* and an aluminum disk *D*, all mounted on a shaft which revolves in a cup-shaped jeweled bearing *B*. The shaft rpm is registered to the gears *G*, which, in turn, are connected to a series of dials. The aluminum disk *D* revolves between two permanent magnets *MM*. Eddy currents proportional to its angular velocity are induced in the disk; hence the retarding torque is proportional to the angular velocity of the disk.

To compensate for the ever present friction, an auxiliary coil C'' is provided, connected in series with the armature circuit, its function being to supply an additional torque which is independent of the load and equal to the frictional torque. In most types of meters the effect of coil C'' is changed by changing positions with respect to the armature. As there is no iron in the magnetic circuit, the magnetic flux due to coils CC is proportional to the load current, and the armature voltage is proportional to the load voltage. Therefore, the meter torque is proportional to the power taken by the load. As the accuracy of registration depends on the voltages to the neutral being in balance, a slight error will be introduced in the measurement when the neutral wire is carrying current. In most cases the unbalancing is small, and thus the error is small.

7-14. Testing Watt-hour Meters. When testing, the revolutions of the disk are counted over a period of time which is measured with a stop watch. The relation between watt-hours and the revolutions of the disk in most meters is

$$\text{Watts} \times \text{hr} = K \text{ (meter constant)} \times N \text{ (revolutions of disk)} \tag{7-9}$$

When checking the meter, the time t is measured in seconds and

$$\text{Watts} \times t = KN \tag{7-10}$$

A run of about 1 min should give good results. Several runs averaged out would be acceptable, and average watts are

$$W = \frac{KN \times 3{,}600}{t} \tag{7-11}$$

A voltmeter and an ammeter are required to read average line voltage and current.

Q7-10. In the test of a 10-amp watt-hour meter having a constant of 0.4 the disk makes 40 revolutions in 53.6 sec. The average voltage and current during this period are 116 volts and 9.4 amp. What is the per cent accuracy of the meter at this load?

ANSWER.

True watts = 116 volts × 9.4 amp = 1,090 watts
Test watts = (0.4 × 40 × 3,600)/53.6 = 1,074 watts
Per cent accuracy = (1,074/1,090) × 100 = 98.5.

Meter is 1.5 per cent slow

Chapter 8

DIRECT-CURRENT TRANSMISSION

8-1. General. In long line transmission of direct current the generator voltage suffers a drop in the line. The current flowing is given by

$$I_L = \frac{E_g \pm \sqrt{E_g^2 - 4RP_r}}{2R} \quad \text{amp} \tag{8-1}$$

where E_g = generated voltage
R = resistance of both lines
P_r = power at receiving end

As a result there is a loss in power. Every attempt is made to keep these losses at a minimum.

8-2. Efficiency of Transmission. This is the ratio of the power transmitted to the power delivered to the line before transmission, expressed as

$$\eta = \frac{E_r I_L}{E_g I_L} = \frac{E_r}{E_g} \tag{8-2}$$

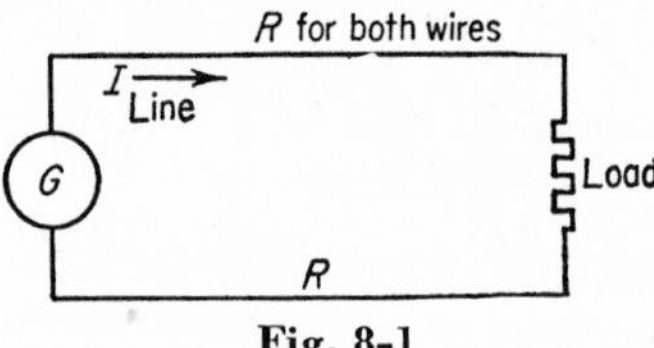

Fig. 8-1

8-3. Transmission-line Problems. When designing transmission lines, there are four important points to bear in mind (see Fig. 8-1):

(*a*) There must be no undue heating of the lines.

(*b*) The voltage drop shall not be so great as to impair the operation of motors and other electrical apparatus.

(*c*) The wires mounted on poles must be strong enough to withstand mechanical stresses. Such are caused by wind, ice, and sleet. Wire stresses within themselves must be considered because of their own weight.

(*d*) Not considering voltage regulation or any of the above, the most economical weight of conductor must be evaluated.

8-4. Cross Section of the Most Economical Conductor. The most economical cross section of a conductor is that for which the annual lost-energy cost is equal to the interest, tax, and depreciation on that portion of the capital outlay. This can be considered to be the weight of copper used in the transmission line.

The most economical cross section is given by

$$A = 593I\sqrt{\frac{c_1 h}{cp}} = \text{cir mils} \tag{8-3}$$

where I = current, amp, for h hr per year
c_1 = cost of electric energy, dollars per kwhr
c = cost of wire, dollars per lb
p = annual per cent interest on capital invested in line wires, including depreciation and taxes

Q8-1. It is required to deliver 22 kw at 220 volts direct current at the end of a transmission line, 1,000 ft long. If the copper wire in the line is 0.5 in. in diameter, calculate (*a*) the current in the line, (*b*) the resistance of the line, (*c*) the voltage at the generator end, (*d*) the power loss in the line, and (*e*) the efficiency of transmission.

ANSWER.

(*a*) $I_L = \dfrac{P_r}{E_r} = \dfrac{22{,}000}{220} = $ **100 amp**

(*b*) $R = \dfrac{Kl}{A} = \dfrac{10.4 \times 2{,}000}{500^2} = $ **0.0832 ohm**

(*c*) $E_g = E_r + I_L R = 220 + (100 \times 0.0832) = $ **228.32 volts**

(*d*) $I_L^2 R = 100^2 \times 0.0832 = $ **832 watts loss**

(*e*) $\eta = 220/228.32 = 0.964$, or **96.4 per cent**

Q8-2. A group of d-c motors is to operate at 220 volts at a place that is 800 ft away from 230-volt mains. (*a*) What diameter of copper wire is to be used in connecting the motors to the main, if the motors take a total of 75 amp? (*b*) What is the power loss in the wires? Assume resistivity of copper may be taken as 10.4 (see Fig. 8-2).

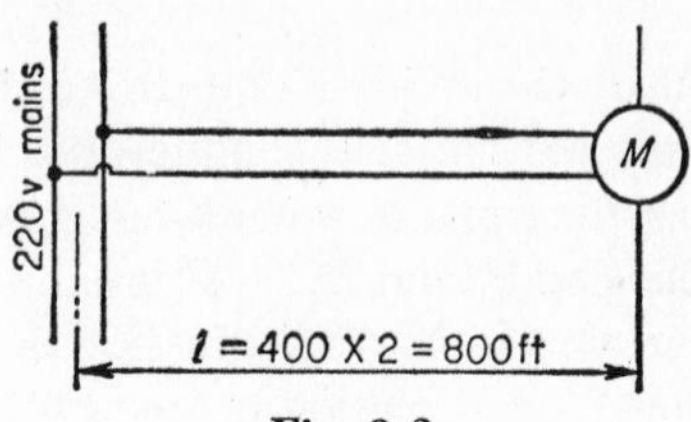

Fig. 8-2

ANSWER.

(*a*) Line resistance $R_L = \frac{E}{I} = \frac{230 - 220}{75} = 0.1335$ ohm

$$d^2 = \frac{Kl}{R} = \frac{10.4 \times 800}{0.1335} = 62{,}300 \text{ cir mils, or } \mathbf{0.2576\ in.}$$

Use a No. 2 wire.

(*b*) Power loss $= EI = (230 - 220) \times 75 =$ **750 watts loss**

8-5. Comparison of Copper Requirements. In two- or three-wire systems, if it is to be assumed that a definite amount of power P is to be transmitted through a line with length L, resistance R_1 (both wires), and with voltage at the load E_1, the current I_1 in the line is then

$$I_1 = \frac{P}{E_1}$$

and the power loss in the line is

$$P_1 = I_1^2R_1$$

Assume further that the voltage is increased to a value E_2 and that the power, distance, and loss remain the same as before. The current is then

$$I_2 = \frac{P}{E_2}$$

and the power loss in the line is

$$P_2 = I_2^2R_2$$

and $$I_2^2R_2 = I_1^2R_1$$

follows. Then it may be shown that

$$\frac{R_1}{R_2} = \frac{E_1^2}{E_2^2} \tag{8-4}$$

Thus, we see that the line resistance varies directly as the square of the voltage. Now since the resistance in the line varies inversely as the cross-sectional area and the volume, it follows that the weight of copper under the above condition is given as follows with Q being weight:

$$\frac{Q_1}{Q_2} = \frac{E_2^2}{E_1^2} \tag{8-5}$$

And the weight of the line wires varies inversely as the square of the voltages, when power, distance, and losses remain the same—or

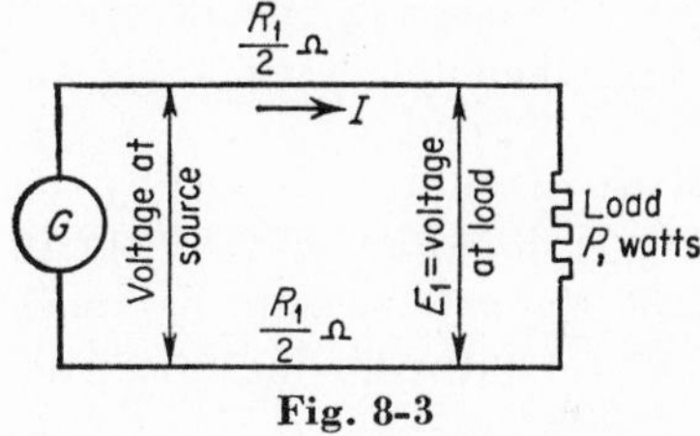

Fig. 8-3

fixed. Further, when the line voltage is doubled, the area used need only be one-quarter, other conditions remaining the same (see Fig. 8-3).

In a three-wire system with the load between the two outer wires equally balanced with no current flow in the neutral, the total amount of copper in the line would be only 25 per cent of that required for transmission of an equal load over an equal distance with the two-wire system, when the voltage is only one-half of that in the three-wire system.

However, this is true under ideal conditions (no current flow in neutral) when no imbalance is present. In practice, it is customary to give the neutral wire the same area as the two outer wires; hence the total copper weight for the three-wire system is 37.5 per cent of that required for an equal two-wire system. The copper saving is, therefore, 100 − 37.5, or 67.5 per cent.

Q8-3. A three-wire 220-volt system with a neutral wire one-half of the size of the line wires is often used. How does the copper economy of such a system compare with that of a two-wire 110-volt system delivering the same amount of power?

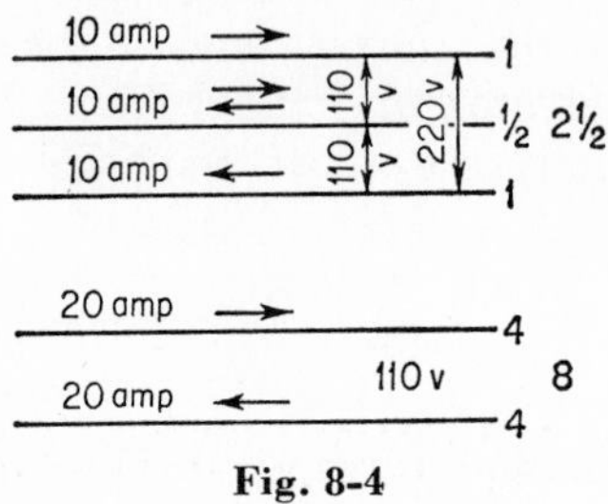

Fig. 8-4

ANSWER. Please refer to Fig. 8-4. Note that in a great many cases with moderate unbalancing the neutral wire can be made somewhat smaller than the outside wires, **2½ to 8.**

8-6. The Three-wire System. By merely doubling the voltage with attendant savings of power and copper in the transmission line, the wide use of 220-volt circuits has been established. To elucidate further, since 110-volt incandescent lamps are most

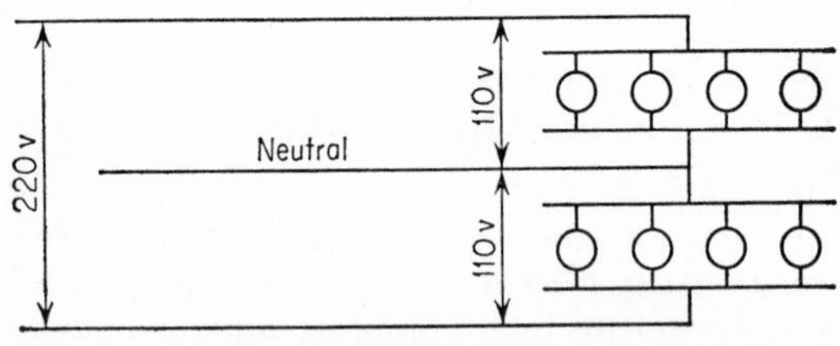

Fig. 8-5

common because of cheapness and great durability, two of these lamps would have to be put in series if connected up to a 220-volt line. This would necessitate burning two lamps at one time. If the consumer needed five lamps, he would have to burn six, three parallel sets of two in series. To permit the use of single lamps as required and still retain the economic advantage of a 220-volt system, a third wire called the neutral was added, as shown in Fig. 8-5.

8-7. Balanced and Unbalanced Three-wire System. When the loading (lamps) are the same on each side of the neutral, as shown in Fig. 8-6, the neutral carries no current. It is only when the system is unbalanced that the neutral is of use and carries current, as shown in Fig. 8-7. In other words, the system is said

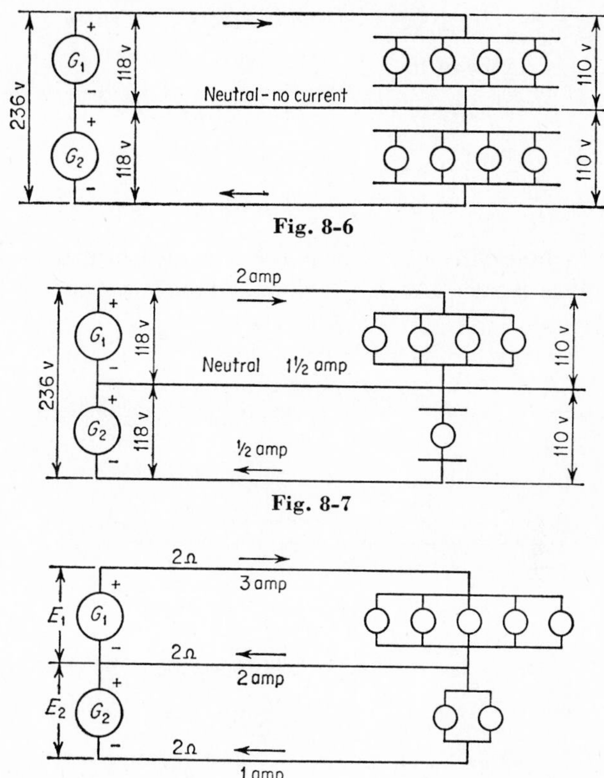

Fig. 8-6

Fig. 8-7

Fig. 8-8

to be unbalanced when the load on one side of the neutral wire is greater (carrying more current) than on the other side. Under these conditions the neutral carries the surplus current. In practice an effort is made to keep the load as nearly balanced as possible.

Q8-4. A three-wire system has an unbalanced load, as shown in Fig. 8-8. The line potential across each load is 110 volts.

Calculate (*a*) the potential across each generator, (*b*) the resistance of each load combined.

ANSWER.

(*a*) The generator potential E_1 is

$$E_1 = 110 + (3 \times 2) + (2 \times 2) = \mathbf{120\ volts}$$
$$E_2 = 110 + (2 \times 1) - (2 \times 2) = \mathbf{108\ volts}$$

(*b*) The resistance of load 1

$$R_1 = {}^{110}\!/_{3} = \mathbf{36.67\ ohms}$$

The resistance of load 2

$$R_2 = {}^{110}\!/_{1} = \mathbf{110\ ohms}$$

Q8-5. A three-wire system supplies a load shown in Fig. 8-9. The resistance of each lamp is 100 ohms and the motor takes 25 amp.

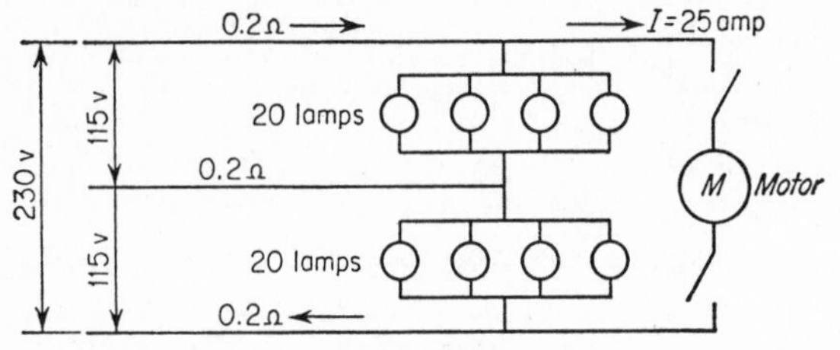

Fig. 8-9

Calculate the voltage across each group of lamps, when the motor is "off," and when the motor is "on."

ANSWER. The resistance of the 20 lamps is found by the following:

$$R = {}^{100}\!/_{20} = 5 \text{ ohms}$$

Because the load is balanced, the current flowing in the neutral wire is zero. With the motor disconnected the current through the lamps is

$$I = 230/10.0 = 23 \text{ amp}$$

Chapter 9

MAGNETISM AND THE MAGNETIC CIRCUIT

9-1. Magnetic Field. If a compass needle is placed above or below a wire carrying an electric current, the needle will turn on its pivot so as to set itself as nearly parallel as possible to the lines of force or at right angles to the wire. The region around the wire through which magnetic forces act is defined as the magnetic field of the wire. This magnetic field is composed of *lines of force* (see Fig. 9-1).

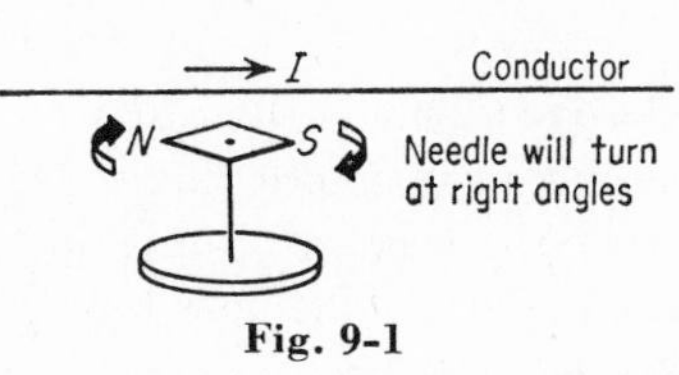

Fig. 9-1

9-2. The Magnet. Everyone is familiar with the action of a magnet, how it will attract needles or steel filings and cause them to adhere to it until removed. While, in an unmagnetized state, iron or steel does not attract other steel or iron, it does attract other magnets. A compass needle will turn toward a piece of iron or steel nearby and be deflected from its natural position of pointing north or south.

Upon studying the action of a bar magnet, we find that its attraction is much stronger at the ends than at the sides, and that two magnets if placed together in one way will repel each other and if placed together the opposite way will attract each other. The poles of a magnet are the ends where the attraction is greatest.

9-3. Magnetic Lines of Force. We find that a magnet has constantly passing through it a magnetic impulse, which we call *lines of force*, or *magnetic flux*, or *field*. The lines of force are believed to move through the magnet, out at one end, through the space immediately surrounding the magnet, and return to the magnet at the other end. These lines come out of the *north pole* of the magnet, going around the magnet in space, and back into the magnet through the *south pole,* and finally going through the magnet itself back to the north pole. Thus we see that the lines of force form a complete circuit (see Fig. 9-2).

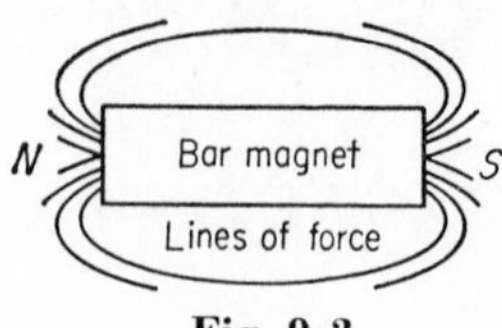

Fig. 9-2

An iron ring may be magnetized so that all the lines of force lie inside the ring, thus producing no poles. Most electric machines consist of magnets, the force lines of which lie partly in air (air gap) and partly in iron and steel. In each case there is a definite circuit much like an electric circuit. This is the *magnetic circuit.*

9-4. Reluctance of a Magnetic Circuit. This is the resistance a substance offers to the force line grouping or magnetic flux. The reluctance of iron and steel is low. Thus, a piece of iron in a magnetic field offers less reluctance than the air. The lines become more dense in the iron making it magnetic.

9-5. Thumb Rule for Straight Wire. It is easily proved that magnetic lines of force exist around a wire through which an electric current is moving. If a wire is connected to a dry cell so that current flows, and a pocket compass is held over the wire, we will see (as previously stated) that the compass needle will turn at right angles to the wire, with the north pole of the needle pointing in the direction of the lines of force around the wire.

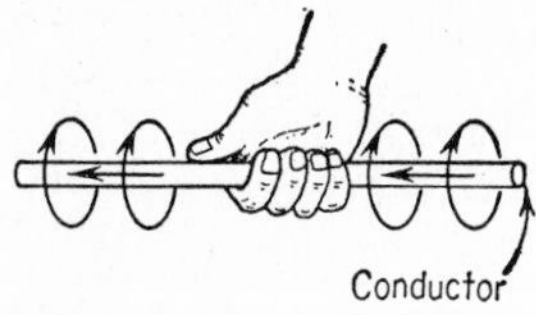

Fig. 9-3

By coiling the *right hand* around the wire with the palm upward, and the fingers pointing in the direction indicated by the north pole of the compass, the negative end of the circuit is indicated by the thumb. This is known as the *right-hand rule* for the direction of current through the conductor (see Fig. 9-3).

9-6. The Solenoid. When we coil a wire, it is called a *helix.* When the helix is charged with electric current, it is called a *solenoid.* A solenoid unites the lines of force around the individual turns of wire composing it, and a complete magnetic flux is produced similar to that around a magnet. When an iron core is inserted into the solenoid, it provides a path for the lines of force passing through the center of the coil, and we have a temporary magnet, or *electromagnet* (see Fig. 9-4).

9-7. Thumb Rule for a Coil. The polarity of the magnet within the solenoid depends upon the polarity of the solenoid. The lines of force pass along the coil and form magnetic lines of force from one loop to another, until at the last loop they enter the space in the center of the coil and pass back through it. Therefore, the polarity of the coil depends upon the *direction of the winding*—the direction of the current being the same.

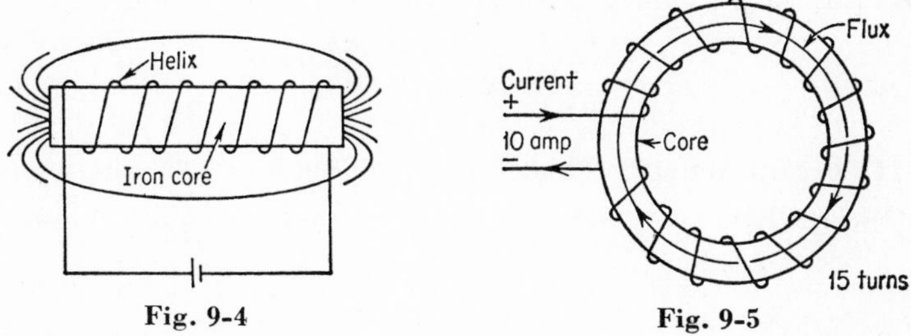

Fig. 9-4

Fig. 9-5

This can be determined by the *right-hand rule for the polarity of a solenoid*, as follows: Grasp the coil with the fingers turned in the direction of the winding, when looking into the coil from the positive end of the source of current. The thumb will then point to the north pole of the solenoid, which will be nearest to the negative end of the circuit if the coil is wound clockwise, or nearest to the positive end of the circuit if the coil is wound counterclockwise.

9-8. Ampere Turns. When a coil passes around a core several times, its magnetizing power is proportional to both the current flowing through the wire and the number of turns made by the wire. If I is current amperes, and N the number of turns, then IN is known as *ampere turns*. In Fig. 9-5 IN is 10×15, or 150 amp turns.

9-9. Magnetomotive Force. Designated as F, it is necessary to drive the flux through the coil, and it is directly proportional to the number of ampere turns on it. Thus,

$$F = 0.4\pi IN = 1.257IN \qquad \text{gilberts} \tag{9-1}$$

We can measure the electric pressure necessary to drive the flux through a magnetic circuit in the same manner we designate volts in an electric circuit.

9-10. Reluctance. This is a property of a magnetic circuit to resist the flow of a magnetic flux. This is also known as magnetic resistance. If we represent reluctance by the script letter $\mathcal{R}$, the length of the magnetic path as l in centimeters, A is the cross section in square centimeters and μ the permeability of the material with air equal to unity

$$\mathcal{R} = \frac{l}{A\mu} \text{ (just plain reluctance)} \tag{9-2}$$

9-11. Permanence. This is the reciprocal of reluctance and is the expression of the circuit's ability to conduct magnetic flux.

$$P = \frac{1}{\mathcal{R}} \tag{9-3}$$

9-12. Permeability. This is a measure of the ease with which magnetic lines of force pass through any substance. It may be defined as the ratio between the number of lines of force per unit area passing through a magnetizable substance and the magnetomotive force producing it. It may be written

$$\mu = \frac{B}{H} \tag{9-4}$$

where B = lines of force per sq cm (gausses)

H = force required per cm of magnetic circuit, gilberts

Thus, in order to determine the ampere turns necessary per inch of magnetic circuit it is only necessary to know the flux density and μ. If for a particular material a curve is plotted to show the relationship between flux density B and ampere turns required per unit length of various magnetic materials H, we will see curves as shown in Fig. 9-6.

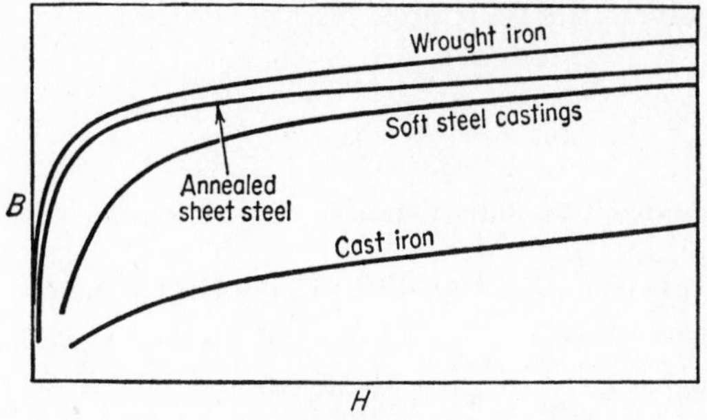

Fig. 9-6

9-13. Flux Density. Designated as B, it is the number of lines per square centimeter, is called *flux density*, and may be obtained

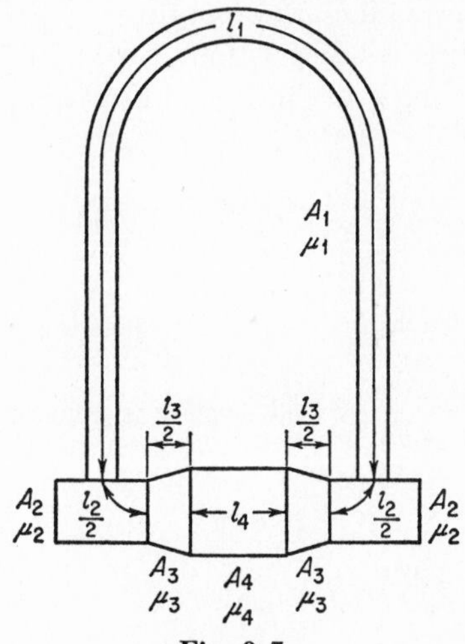

Fig. 9-7

by dividing the total lines of flux ϕ by the cross-sectional area of the magnetic circuit A:

$$B = \frac{\phi}{A} \tag{9-5}$$

9-14. Reluctance of a Series Magnetic Circuit. Please refer to Fig. 9-7. When only one magnetic circuit was involved, the reluctance was merely written $\mathcal{R} = l/A\mu$. But when more than one are involved, the reluctance may be written

$$\mathcal{R} = \mathcal{R}_1 + \mathcal{R}_2 + \cdots + \mathcal{R}_n = \frac{l_1}{A_1\mu_1} + \frac{l_2}{A_2\mu_2} + \cdots + \frac{l_n}{A_n\mu_n} \quad (9\text{-}6)$$

Figure 9-7 shows but four reluctances in series.

9-15. Reluctance of a Parallel Magnetic Circuit.

$$\frac{1}{\mathcal{R}} = \frac{1}{\mathcal{R}_1} + \frac{1}{\mathcal{R}_2} + \cdots + \frac{1}{\mathcal{R}_n} \quad (9\text{-}7)$$

Q9-1. A ring of cast steel has an outside diameter (OD) of 10 in., an inside diameter (ID) of 6 in., and a thickness of 3 in. It is completely encircled by a winding having 600 turns distributed evenly over the circumferential length of the ring. A current of 4.5 amp through this winding is found to produce a magnetic flux of 700,000ϕ in the ring cross section. Find μ at the magnetization shown above.

ANSWER.

$$H = \frac{0.4\pi NI}{2\pi r} = \frac{0.4NI}{2r}$$

and since r is 4×2.54

$$H = \frac{0.4 \times 600 \times 4.5}{2 \times 4 \times 2.54} = 53.2 \text{ gilberts per cm}$$

$$B = \frac{\phi}{A} = \frac{700{,}000}{2.54 \times 3 \times 2 \times 2.54} = 18{,}000 \text{ maxwells per cm}^2$$

$$\mu = \frac{18{,}000}{53.2} = \mathbf{340}$$

9-16. Ohm's Law for the Magnetic Circuit. Like Ohm's law for the electric circuit Ohm's law for magnetic circuits can have the three forms

$$\phi = \frac{F}{R} \quad \text{or} \quad \text{magnetic lines} = \frac{\text{gilberts}}{\text{reluctance}} \quad (9\text{-}8)$$

$$R = \frac{F}{\phi} \quad \text{or} \quad \text{reluctance} = \frac{\text{gilberts}}{\text{magnetic lines}} \quad (9\text{-}9)$$

$$F = \phi R \quad \text{or} \quad \text{gilberts} = \text{magnetic lines} \times \text{reluctance} \quad (9\text{-}10)$$

Q9-2. The reluctance of a wooden ring (Fig. 9-8) is 16 units. What magnetic flux would be set up in this ring if 3 amp were sent through the 800 turns of the coil?

ANSWER. Using Eq. (9-8),

$$\phi = (1.257 \times 3 \times 800)/16 = \mathbf{188\ magnetic\ lines}$$

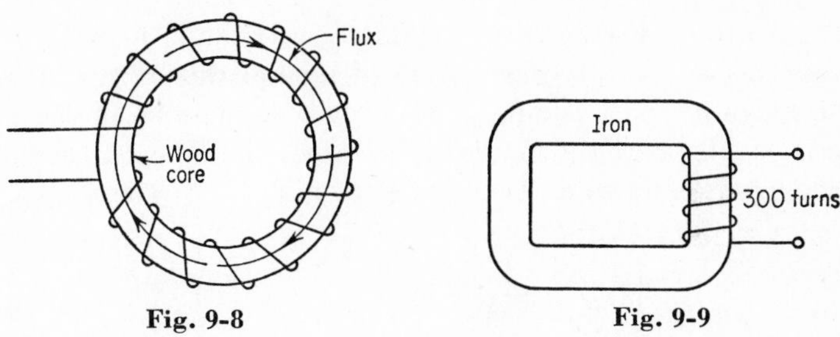

Fig. 9-8 **Fig. 9-9**

Q9-3. A magnetic circuit composed of a 300-turn coil and iron has 4 amp flowing through the coil. Under these conditions a flux of 3,500 lines is set up in the iron circuit. Find the reluctance of the magnetic circuit.

ANSWER. Refer to Fig. 9-9.

$$R = (1.257 \times 4 \times 300)/3{,}500 = \mathbf{0.432\ unit}$$

9-17. Tractive Force of Magnets. In order to calculate the "pull" of a magnet on a piece of steel, we make use of the formula

$$f = \frac{8.94B^2A}{10^8} \tag{9-11}$$

where f = pull of magnet, lb

B = flux density in air gap, gausses

A = area of air gap, sq cm

Q9-4. Calculate the pull of a horseshoe magnet with sufficient current to cause a flux of 500,000 lines. The length of the horseshoe path of metal circuit is 48 cm, the path through the steel bar is 20 cm, and the air gaps, two of them, are each 0.15 cm in length. The number of turns in each coil shown is 2,000 (refer to Fig. 9-10).

ANSWER. On the basis of Eq. (9-11), let us first determine the area A.

$$A = 5 \times 5 \times 0.7854 = 19.6 \text{ sq cm}$$
$$B = 500{,}000/19.6 = 25{,}500 \text{ gausses}$$
$$f = (8.94 \times 25{,}500^2 \times 19.6)/10^8 = 1{,}140 \text{ lb}$$

Total force for the two ends: 2 × 1,140 = **2,280 lb.**

9-18. Laminated Magnets. In many cases it is found advantageous to use a magnet composed of several thin layers of steel pressed together or a bundle of fine iron wires instead of the solid material. These are called laminated magnets. They are stronger than magnets composed of one piece of material. The reason for

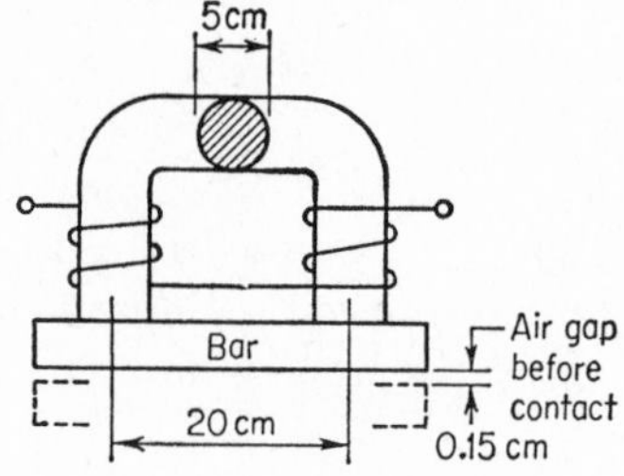

Fig. 9-10

this is believed to be due to the magnetism which is artificially induced into them being stronger nearer the surface of the magnet than through the center. Thus, if we provide more surface, we have more magnetism. It also reduces the tendency of little stray currents, or *eddy currents*, to move in the metal, when the magnet is energized by a solenoid, and it reduces thus the heating of the metal with attendant loss of energy.

Laminations also reduce heat in electromagnets due to *hysteresis*, which is the opposition to a change of the position of the molecules in the metal, when the current in the solenoid is alternating current, constantly reversing the polarity of the solenoid, and consequently of the magnet which is its core.

This lagging effect is the cause of a certain loss in every a-c machine and in the armatures of d-c machines. The student is requested to review the formation of the *hysteresis loop* in any standard text.

Chapter **10**

DIRECT-CURRENT MACHINERY

10-1. D-C Generator. This is a machine used for continuous conversion of mechanical energy into electric energy. The power delivered is called its load. Generators are usually driven at constant speed at all loads, whereas in motors the speed may vary with the load, depending upon the particular type of motor used and the character of the load.

10-2. Electromagnetic Induction. The d-c generator consists fundamentally of a number of loops of insulated wires revolving in a strong magnetic field in such a way that these wires cut across the lines of force set up between the poles of a magnet. This cutting of lines of force sets up an electromotive force (emf) along the wires.

10-3. Direction of This Induced Emf. The simplest form of a d-c generator is represented in Fig. 10-1. Here S and N represent

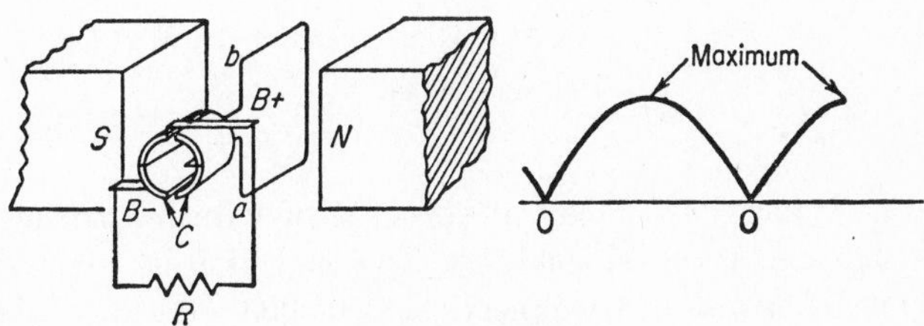

Fig. 10-1

the south and north poles, respectively, ab a revolving loop of wire, $B+$ and $B-$ the brushes, c a two-segmental commutator, and R an external circuit. Now, let us explore this further.

When the coil is perpendicular to the magnetic field, then the maximum flux ϕ links the coil. If it is rotated in the direction (counterclockwise) another quarter turn, no lines link the coil. Continuing the rotation, the coil will thus generate the wave pattern indicated, and the average induced emf during t sec is given by

$$e = \frac{N\phi}{t \times 10^8} \quad \text{volts} \tag{10-1}$$

where N = turns in coil
t = seconds per quarter revolution

Where conductors cut flux by moving mechanically through it as in a generator (rotating machinery), we use

$$e = \frac{Blv}{10^8} \quad \text{volts per conductor (instantaneous)} \tag{10-2}$$

where B = flux density, gausses
l = length of conductor, cm
v = speed of conductor, cm per sec

Q10-1. A metal transport plane has a wing spread of 88 ft. What difference of potential exists between the extremities of the wings when the plane moves in a horizontal plane with a speed of 150 mph? The value of the vertical component of the earth's magnetic field is 0.65 gauss at the plane.

ANSWER. This may be considered as a moving conductor cutting lines of force in the earth's magnetic field. Using Eq. (10-2)

$$e = \frac{0.65 \times 88 \times 12 \times 2.54 \times 150 \times 5{,}280 \times 12 \times 2.54}{10^8 \times 3{,}600}$$

$$= \mathbf{0.117\ volt}$$

Q10-2. A rectangular coil of 20 turns of wire measuring 10 in. on one side and 6 in. on the other is placed with its plane parallel to the flux in a uniform magnetic field of 200 gausses. It is then turned to a position which makes its plane perpendicular to the flux. If this movement occurs in 0.01 sec, what average emf will be induced in the coil?

ANSWER. See Fig. 10-3. For every centimeter rise of the coil side a, the flux cut by one conductor is

$$1 \times 10 \times 2.54 \times 200 = 5{,}080 \text{ lines}$$

Since a rises 3 in., the total flux cut is given by

$$5{,}080 \times 3 \times 2.54 = 38{,}700 \text{ lines}$$

For side b included and the 20 conductors

$$38{,}700 \times 2 \times 20 = 1{,}548{,}000 \text{ lines of force cut}$$

Now since these lines are cut in 0.01 sec, the average voltage generated is determined from

$$1{,}548{,}000/(0.01 \times 10^8) = \mathbf{1.55\ volts}$$

Note that this average is $E_{av} = 0.637 \times E_{max}$. If we use the rule for finding the direction of the induced emf, the *right-hand rule* will

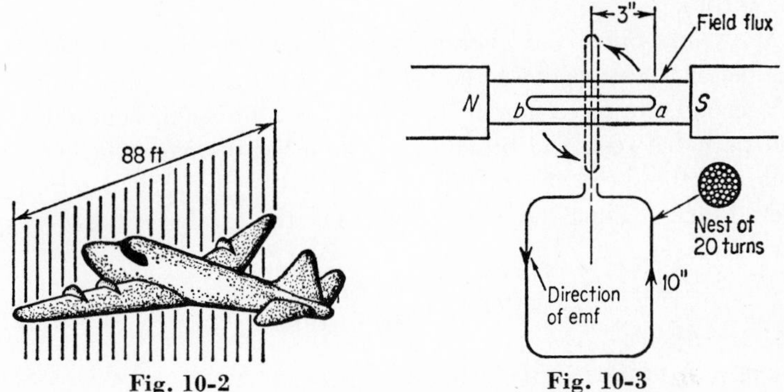

Fig. 10-2 **Fig. 10-3**

show the middle finger or forefinger in the direction of the flux, the thumb in the direction of the motion, and the third finger in the direction of the induced emf. Check for sides a and b.

Q10-3. What is the maximum voltage generated in a drum armature with concentrated windings consisting of 300 series conductors in each path? The armature speed is 1,200 rpm. Each conductor cuts twice through a field of 1.5×10^6 lines of force per revolution.

ANSWER.

$$e = \frac{1.5 \times 10^6 \times 2 \times 1{,}200 \times 300}{60 \times 10^8} = 180 \text{ volts average}$$

$$E_{max} = \frac{180}{0.637} = \mathbf{282\ volts}$$

Q10-4. A conductor passes 50 times per sec across the face of the pole of a field magnet having a flux density of 15,000 lines per sq in. The dimensions of the pole face are 20 by 20 in. What average emf is induced in the conductor?

ANSWER. The total flux is $20 \times 20 \times 15{,}000 = 6{,}000{,}000$ lines. Lines cut per second $6 \times 10^6 \times 50 = 300 \times 10^6 = 3.0 \times 10^8$

$$\text{Voltage induced} = (3.0 \times 10^8)/10^8 = \mathbf{3.0\ volts}$$

Q10-5. A d-c generator has four poles with faces 6 in. square and a flux density in the air gap of 40,000 lines per sq in. The machine has 360 conductors arranged in four parallel paths through the armature. Compute the emf of the generator when driven at 1,200 rpm.

ANSWER. Since the average voltage generated in an armature conductor is always one-hundred-millionth of the magnetic lines cut per second, we have merely to know the number of conductors in one path between the brushes, the number of magnetic lines per pole, and the speed of rotation of the armature in order to determine the voltage. Then the average value of the total emf between the brushes is

$$\text{emf} = \frac{p\phi ZN}{60 \times 10^8 m}$$

where p = number of field poles
ϕ = total useful magnetic flux per pole
Z = total number of armature conductors
N = rpm of armature
m = number of parallel paths through the armature

$$\text{emf} = \frac{4 \times 40{,}000 \times 360 \times 1{,}200 \times 6 \times 6}{60 \times 10^8 \times 4} = \mathbf{103.7\ volts}$$

10-4. Types of D-C Generators. Electrically speaking there is little difference between a d-c motor and a d-c generator. In a motor, electric energy is supplied and the rotation of the shaft is the result. In the generator the shaft is rotated by means of a prime mover, and the electric energy becomes available at the terminals This presupposes that a magnetic field is present through which the coils are rotated.

This matter is simplest when the field coils which produce the magnetic field are energized from an outside source, as in Fig. 10-4. This machine would be called a *separately excited d-c generator.*

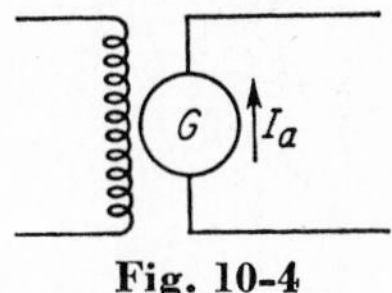

Fig. 10-4

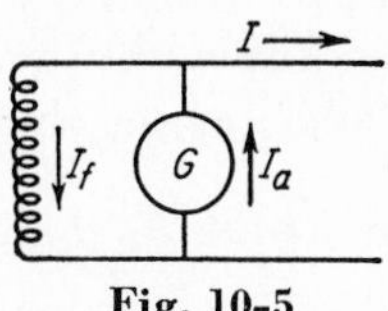

Fig. 10-5

However, it is also possible to obtain this excitation current from the generator itself. It should be emphasized here that the separately excited generator is seldom used in practice.

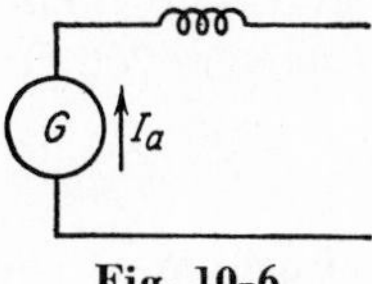

Fig. 10-6

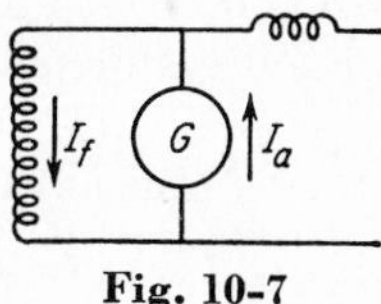

Fig. 10-7

In this way it is possible to build a *shunt generator* (Fig. 10-5), a *series generator* (Fig. 10-6), and a *compound machine* (Figs. 10-7 and 10-8).

This is possible for only one reason, which is that a certain small amount of magnetism remains in the poles even after the electric current is interrupted. When the coils are rotated in this weak magnetic field, a small amount of current is induced in the armature coils and a part of this current is utilized to strengthen the field so that more current can be induced in the armature coils, and so on. In this way the field strength is gradually built up until the desired magnitude has been reached.

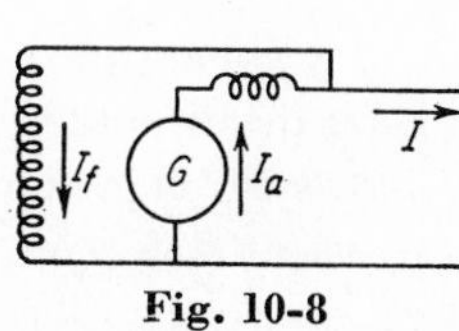

Fig. 10-8

If the poles become demagnetized, an outside source of current is required to magnetize the field in the proper direction. This is known as "priming the fields."

10-5. Shunt Generator. The student should refer to a standard text for detailed discussion of this machine. Now refer back to

Fig. 10-5. Note that the field coils are connected in parallel with the armature. Kirchhoff's law applies, and generally speaking the field current is approximately 2 per cent of the armature current. The terminal voltage drops with the load (see Fig. 10-9). The following relation holds for the shunt generator.

$$E_t = E_g - I_a R_a \qquad (10\text{-}3)$$

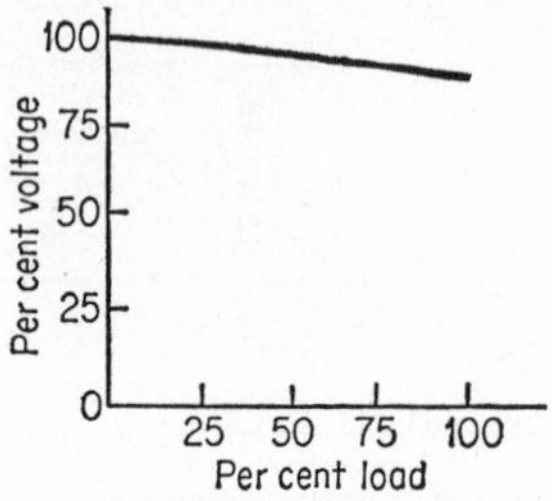

where E_t = terminal voltage
E_g = generated voltage
I_a = armature current
R_a = armature resistance

Fig. 10-9

Q10-6. The generated voltage included in the armature of a shunt generator is 600 volts. The armature resistance is 0.1 ohm. What is the terminal voltage when the armature delivers 200 amp?

ANSWER. $E_t = 600 - (200 \times 0.1) = 600 - 20 =$ **580 volts**

10-6. Voltage Generated by Shunt Generator. The voltage generated is directly proportional to the magnetic field strength and the speed of rotation.

$$E_g = K\phi S \qquad (10\text{-}4)$$

where E_g = generated voltage
K = constant
ϕ = magnetic field intensity

Q10-7. The total effective flux of a six-pole d-c generator is 3.19 megalines per pole. There are 324 conductors on the armature, and the winding is such that there are as many paths through the armature as there are poles. If the generator runs at 750 rpm, what is the average emf generated?

ANSWER. One megaline is equal to 10^6 maxwells. Now let

p = number of pole pieces
ϕ = effective flux per pole
Z = number of armature conductors
N = armature speed, rpm
m = number of parallel paths

$$\text{emf} = \frac{6 \times 3.19 \times 10^6 \times 324 \times 750}{60 \times 10^8 \times 6} = \mathbf{129\ volts}$$

Note: The values of m for different types of windings are:

Type	Lap winding	Wave winding
Simplex.	$m = p$	$m = 2$
Duplex.	$m = 2 \times p$	$m = 4$
Triplex.	$m = 3 \times p$	$m = 6$
Quadruplex.	$m = 4 \times p$	$m = 8$

Since for a given machine all factors except ϕ and N are constant, then let

$$K = \frac{PZ}{60 \times 10^8 m} \tag{10-5}$$

Then $$\text{emf average} = K\phi N$$

This shows that the emf generated is directly proportional to the flux per pole and the speed.

10-7. Shunt Generator–Current Considerations. In the shunt generator the armature current differs from the total current by the current flowing in the field.

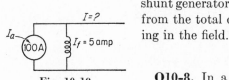

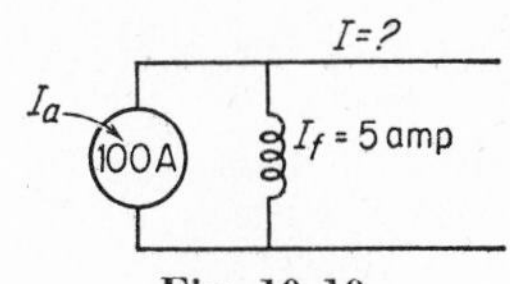

Fig. 10-10

$$I_a = I + I_f \tag{10-6}$$

Q10-8. In a shunt generator the armature current is 100 amp. The field current is 5 amp. What is the load current? See Fig. 10-10.

ANSWER. $I = I_a - I_f = 100 - 5 =$ **95 amp**

10-8. Shunt Generator Losses and Efficiency. These losses include: armature losses, core losses, field losses, and brush losses.

The *armature losses* are given by the following formula:

$$I_a^2 R_a \tag{10-7}$$

Core losses result from heating of the iron core in the armature because of the alternating magnetic field. This is due partly to the rearrangement of the molecular structure (hysteresis loss) but more to induced voltages in the iron core.

Field losses for the shunt field may be determined from

$$I_s^2 R_s \tag{10-8}$$

where I_s = shunt field current
R_s = shunt field resistance

It may also be given by E_s^2/R_s, where E_s is voltage across shunt field.

Brush losses are the result of resistance at the point of brush contact with armature. These brush losses are given by the formula

$$I_a^2 \times \text{brush resistance (ohms)} \tag{10-9}$$

Efficiency is equal to the output divided by the input. It is also equal to the output divided by output plus losses.

Q10-9. A 20-kw 220-volt shunt generator has a brush resistance of 0.005 ohm, an armature resistance of 0.065 ohm, and a shunt field resistance of 200 ohms. What power is developed in the armature when it delivers its rated load? See Fig. 10-11.

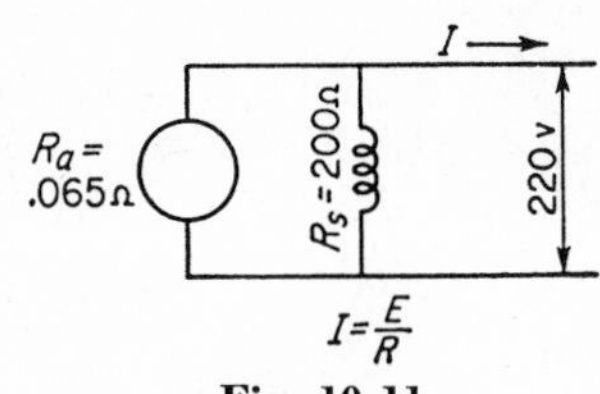

Fig. 10-11

ANSWER.

Rated current: $I = 20{,}000/220 = 90.9$ amp
Field current: $I_f = 220/200 = 1.1$ amp
Armature current: $I_a = 90.9 + 1.1 = 92$ amp
Generated voltage: $E = 220 + (92 \times 0.07) = 226.4$ volts
Power developed in armature:

$$P = EI = 226.4 \times 92 = 20{,}830 \text{ watts, or } \mathbf{20.83\ kw}$$

The same results may be obtained by adding power losses.

Field loss: $$P_f = \frac{E_s^2}{R_f} = \frac{220^2}{200} = 242 \text{ watts}$$

Armature and brush losses:

$$I_a^2(R_a + R_b) = 92^2(0.065 + 0.005) = 592 \text{ watts}$$

Power developed in armature:

$$20{,}000 + 242 + 592 = 20{,}834 \text{ watts, or } 20.834 \text{ kw}$$

If the stray power losses, core losses, windage, and friction, etc., are 700 watts, what is the efficiency?

$$\text{Efficiency} = \frac{\text{output}}{\text{output} + \text{losses}}$$

$$= \frac{20{,}000}{592 + 700 + 20{,}834} = 0.93, \text{ or } \mathbf{93\ per\ cent}$$

Q10-10. Explain whether shunt generators are operated above or below the knee of the saturation curve for the machine. Give your reasons for your answer.

ANSWER. The saturation curve is a line expressing the relation between flux density and magnetomotive force for a given material. In a shunt generator operated at no load and constant speed, the saturation curve *ac* in Fig. 10-12 is obtained by plotting the terminal voltage E, which is proportional to the flux density, against the field current I_f, which is proportional to the magnetomotive force.

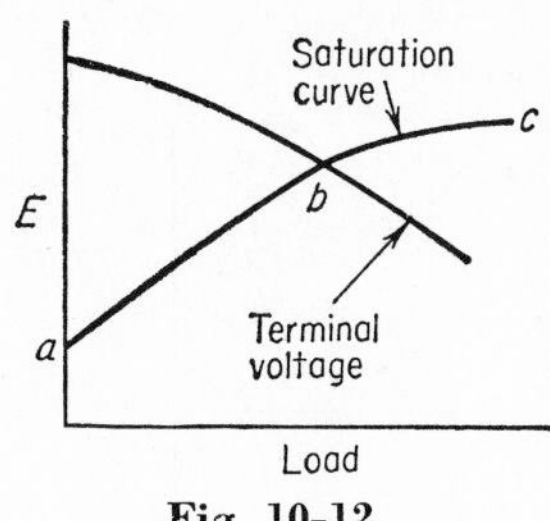

Fig. 10-12

Shunt generators are normally operated just below the knee of the saturation curve (point *b*) in order to improve their regulation. If E_0 and E_f represent, respectively, the no-load and full-load terminal voltage of the generator, its regulation is given by

$$\frac{E_0 - E_f}{E_f} \times 100 \qquad \text{per cent}$$

Because of the armature reaction and the armature resistance drop, the terminal voltage decreases as a load is placed on the generator. A drop in the terminal voltage reduces the current flowing through the shunt field, causing the voltage to fall off still more. The full-load voltage E_f can be increased by cutting out resistance in the shunt field, thus increasing I_f. A larger field current will result in an appreciably higher value of E_f only if operation takes place in the steep part of the saturation curve. Beyond the knee, toward point *c*, no great rise in E_f can be expected because the saturation curve is flat. An increase in E_f naturally improves the regulation.

10-9. Voltage Regulation. A generator may be characterized by its *voltage regulation.* When we say voltage regulation, we mean some change that automatically takes place when the load on the generator is changed. This change is usually assumed to take place anywhere between no load and full load. For example, should the voltage of a generator change from 25 volts at zero load to 20 volts at full load, its voltage regulation would be 5 volts.

Percentagewise, the generator would have $\frac{5}{20}$, or 25 per cent, regulation. The smaller the regulation percentage the better the generator.

Q10-11. The armature of a four-pole 20-hp 220-volt shunt generator consists of 100 conductors arranged as a lap winding. If the

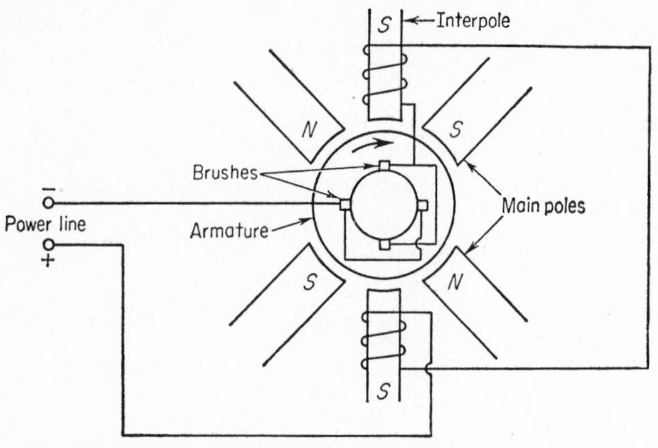

Fig. 10-13

field structure of this machine is to include two interpoles, how many turns should be wound on each? Show by means of a diagram where you would place the interpoles and how the main poles and the interpoles should be magnetized. Assume appropriate values for the efficiency and shunt field resistance of this machine

ANSWER. Let Z = number of lap-wound conductors

p = number of main poles

n = number of turns on each interpole

Please refer to Fig. 10-13. Where there are as many interpoles as there are main poles, a condition more frequently encountered, $n = Z/p$. However, with only one interpole for each pair of main poles, the field strength produced by each interpole must be twice as great or the number of turns twice as large; thus

$$n = \frac{2Z}{p} = \frac{2 \times 100}{4} = \mathbf{50}$$

In practice 40 per cent more turns are provided together with a shunt across the interpole field winding. The correct commutation effect is then found experimentally by varying the resistance of the shunt.

10-10. Parallel Operation of Generators. In order to obtain economical power distribution, two or more generators are frequently operated in parallel. In general parallel operation is favored rather than single-unit operation because of the following reasons:

(*a*) The total load may be larger than the capacity of a single generator.

(*b*) One may be used as a stand-by unit for the other.

(*c*) Generators may be added to correspond with the growth of the power demand.

(*d*) Units may be added (or shut down) according to the diurnal power demand, thus increasing the efficiency of operation.

It is quite necessary that the no-load voltages of both generators be adjusted to the same value and that their external characteristics be identical. Should two shunt generators with different no-load characteristics be connected in parallel, a current will circulate through the two machines, and no current is being delivered to the load.

Should two shunt generators with different characteristics be connected in parallel, each will not take its correct fraction of the load as the load changes. Often the machine with the better regulation will heat up beyond normal operating temperatures. The two generators need not have the same capacity, but each must have identical no-load voltage and identical full-load voltage.

Q10-12. Two shunt generators, each with a no-load rating of 125 volts, are operated in parallel. Both machines have external characteristics which are straight lines over their operating ranges. Machine No. 1 is rated at 250 kw, and its full-load voltage is 118 volts. Machine No. 2 is rated 200 kw at 114 volts. Calculate the operating voltage when the two machines supply a total current of 3,400 amp. How is the load divided between the two?

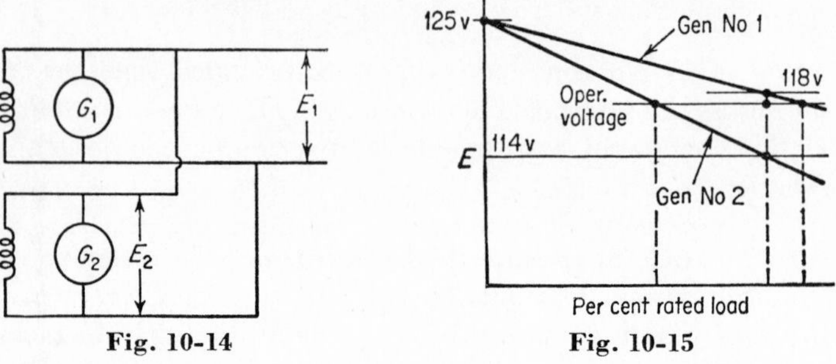

Fig. 10-14 **Fig. 10-15**

ANSWER. Please refer to Figs. 10-14 and 10-15. Now let

E = generator terminal voltage, volts
I = current from each generator, amp
x = load carried by each generator, per cent of rated load
P = load carried by each generator, watts

Also let subscript (1) refer to the 250-kw generator, and subscript (2) to the 200-kw generator. Then, referring to Fig. 10-15, it will be seen that the voltage of each generator can be related to its load, or

$$E_1 = 125 - \left[(125 - 118)\left(\frac{x_1}{100}\right)\right]$$

$$E_2 = 125 - \left[(125 - 114)\left(\frac{x_2}{100}\right)\right]$$

With the generators connected in parallel, their terminal voltages are equal, or $E_1 = E_2$, or equating and clearing

$$125 - \left[7\left(\frac{x_1}{100}\right)\right] = 125 - \left[11\left(\frac{x_2}{100}\right)\right]$$

or

$$x_2 = 7/11 x_1$$

Since in d-c circuits the power delivered, in watts, is given by EI and 1 kw is equal to 1,000 watts, the load on *both* generators is

$$\left[x_1(250) \times \frac{1{,}000}{100}\right] + \left[x_2(200) \times \frac{1{,}000}{100}\right] = E_1 \times 3{,}400$$

Replacing E_1 and x_2 with terms involving x_1, we have as a result

$$\left[x_1(250) \times \frac{1,000}{100}\right] + \left[\frac{7}{11}(x_1) \times 200 \times \frac{1,000}{100}\right] = \left(125 - x_1 \times \frac{7}{100}\right) \times 3,400$$

Solving for the unknown, $x_1 = 106.2$ per cent. Therefore, the operating voltage is

$$E_2 = E_1 = 125 - (7/100 x_1) = 125 - (7/100 \times 106.2) = \mathbf{117.6\ volts}$$

Now let us determine the division of the load.

$$x_1 = P_1 \times \frac{100}{250,000} \qquad \text{and} \qquad x_2 = P_2 \times \frac{100}{200,000}$$

$$\frac{x_1}{x_2} = \frac{11}{7} = \frac{P_1 \times 100 \times 200,000}{P_2 \times 100 \times 250,000}$$

This is all equal to

$$\frac{E_1 I_1 \times 200,000}{E_2 I_2 \times 250,000} = \frac{200,000 I_1}{250,000 I_2}$$

Since the total current delivered is $I_1 + I_2 = 3,400$, or

$$I_2 = 3,400 - I_1$$

also

$$\frac{11}{7} = \frac{200,000 I_1}{250,000(3,400 - I_1)}$$

Or it follows that

$$I_1 = \mathbf{2,250\ amp}$$
$$I_2 = 3,400 - 2,250 = \mathbf{1,150\ amp}$$

10-11. Compound Generators. The drop in voltage with load, which is characteristic of a shunt generator, makes it undesirable where a constant output voltage is necessary. A generator may be made to produce a substantially constant voltage or even a rise in voltage as the load increases, by placing on the field core a few turns which are connected in series with the load or the armature. These turns are connected so as to aid the shunt turns when the generator delivers current.

The manner in which these turns are connected may be as shown in Figs. 10-16 and 10-17. The former (Fig. 10-16) shows a long-shunted generator, the latter a short-shunted generator.

The machine in Fig. 10-16 is connected with the shunt field across the armature and series field. The machine in Fig. 10-17 shows a short-shunt connection. In the long-shunt machine, when no current is being delivered by the machine, the same current flows

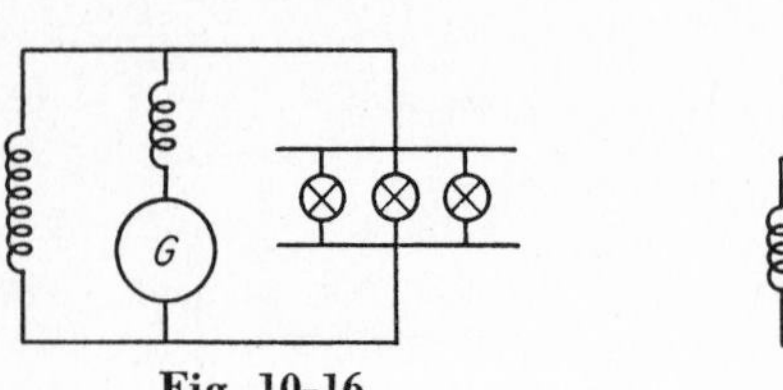

Fig. 10-16

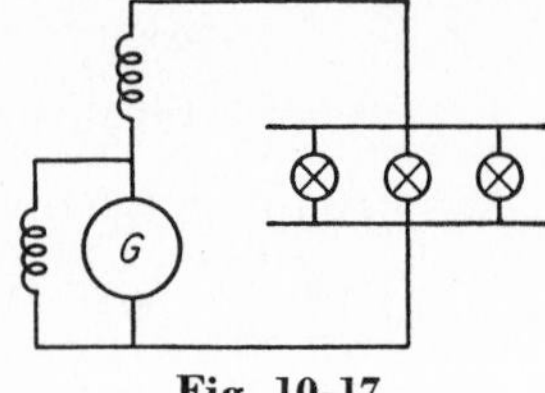

Fig. 10-17

through the series and shunt coils. The shunt field then maintains the voltage at the normal value.

However, when the lamps draw current, this current must also flow through the series field coils. The turns in these series coils are so designed that the terminal voltage is kept at practically a constant value throughout the load range, from zero to full load. The operating characteristics of both long- and short-shunt generators are practically the same.

10-12. Types of Compounding. Actually, in practice, most compound machines have an operating characteristic curve as shown in Fig. 10-18. As you can see, voltages at no load and at full load are the same. However, at intermediate loading points the voltages rise convexly. Such a generator is said to be *flat compounded.* Flat-compounded generators built for commercial operation are often built to deliver rated voltage at one-third load and at full load.

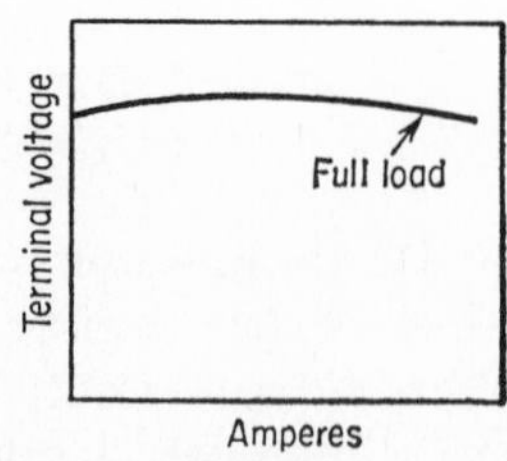

Fig. 10-18

More often generators are built for overcompounding so that the terminal voltage rises a little as the load increases. This may be done for certain reasons. When a line drop is considered, an overcompounded machine does the job nicely, since a line drop of several volts may be compensated for and the voltage at the load can be kept practically constant.

When connected to a driving engine that decreases speed as the load increases, there would be a tendency for the output voltage to drop with the load increase. Unless the generator were overcompounded commensurate with the drop in speed, a corresponding drop in the terminal voltage would result. Thus, machines may be purchased with overcompounding to suit actual operating conditions. The characteristic curve of such a machine would be as shown in Fig. 10-19.

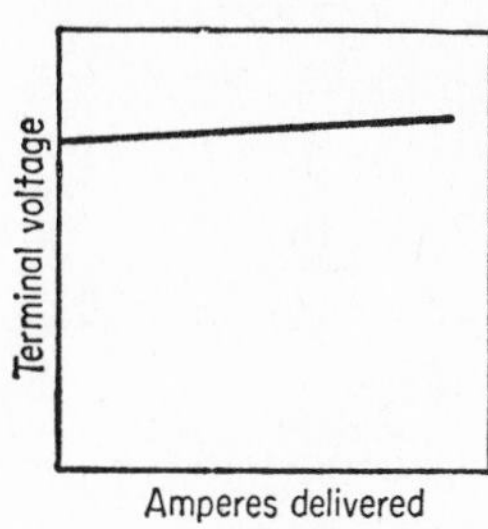

Fig. 10-19

Undercompounding results in a drooping characteristic curve similar to that of a shunt motor. Here the full-load voltage is less than that at the no-load condition.

10-13. Compound Generator–Generated Voltage. In a compound generator the generated voltage (emf) in the armature is given by

$$E = E_t + I_s R_s + I_a R_a \tag{10-10}$$

where E_t = terminal voltage
I_s = series field current
I_a = armature current
R_s = resistance of series field
R_a = resistance of armature

In a long-shunt generator I_s equals I_a. In a short-shunt generator I_s equals $I_a - I_f$ (see Fig. 10-21).

Q10-13. In a compound generator connected short-shunt, the terminal voltage is 230 volts when the generator delivers 150 amp. The shunt field current is 2.5 amp, the armature resistance 0.032 ohm, and the series field resistance 0.010 ohm (see Fig. 10-21).

Series field current: I_s = load current = 150 amp
Voltage drop in series field: $E = IR = 150 \times 0.01 = 1.5$ volts
Armature current: $I_a = 150 + 2.5 = 152.5$ amp

Generated voltage: $E = 230 + 1.5 + (152.5 \times 0.032) = 236.4$ volts
Total power generated: $P_a = 236.4 \times 152.5 = 36{,}050$ watts, or 36.05 kw

Armature loss: $152.5^2 \times 0.032 = 744$ watts
Series field loss: $P_s = 150^2 \times 0.01 = 225$ watts
Shunt field loss: $P_{sh} = (230 + 1.5) \times 2.5 = 579$ watts
Power delivered: $P = 230 \times 150 = 34{,}500$ watts, or 34.5 kw

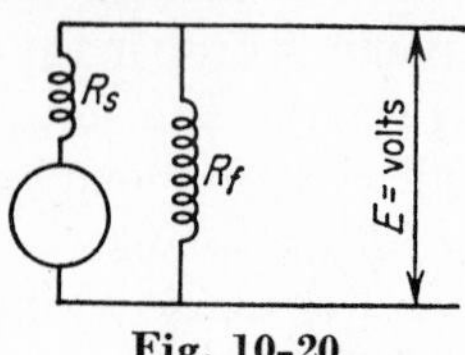

Fig. 10-20

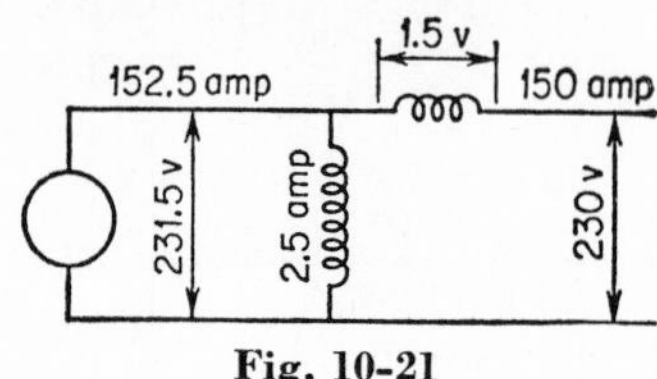

Fig. 10-21

Check by addition of output plus losses:

$$34{,}500 + 744 + 225 + 579 = 36{,}048 \text{ watts}$$

10-14. Series Generator. As we saw in Fig. 10-6, the series generator has its field winding connected in series with the armature and the external load. This machine is used for constant current applications, but the generated emf will increase with the load (see Fig. 10-22). This is so since

$$I_s = I_L = I_a \qquad (10\text{-}11)$$

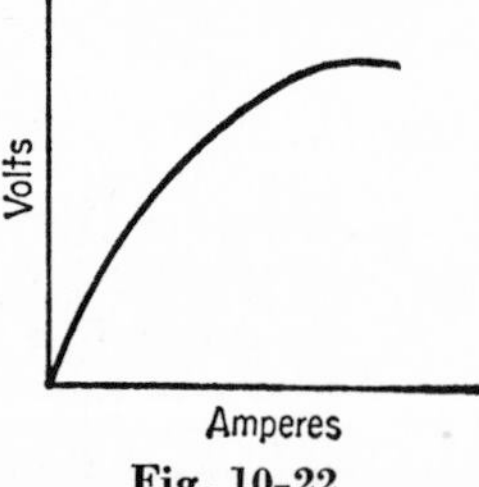

Fig. 10-22

The terminal voltage for any value of the load will be less than the armature voltage induced by $I_aR_a + I_sR_s$.

Because of its inherent instability, this generator has few applications except in certain series lighting systems operated at constant current and variable potential.

$$\text{Generated voltage: } E = E_t + [I_a(R_a + R_s)] \qquad (10\text{-}12)$$

Q10-14. An industrial plant is purchasing two 1,000-kw compound-wound d-c generators, driven from a single synchronous motor and operated essentially at constant speed. The compounding is to be flat to give the same voltage at full load as at no load. If these generators are to be operated in parallel under manual control, and if they are to be assumed to be duplicate machines, how should they be connected together for parallel operation? Explain by circuit diagram and give reasons for your method of connection.

ANSWER. Please refer to Fig. 10-23 for a circuit diagram showing connections of generators to the bus bars. The circuit breakers are inserted for protection in case of overload, while the equalizer bus

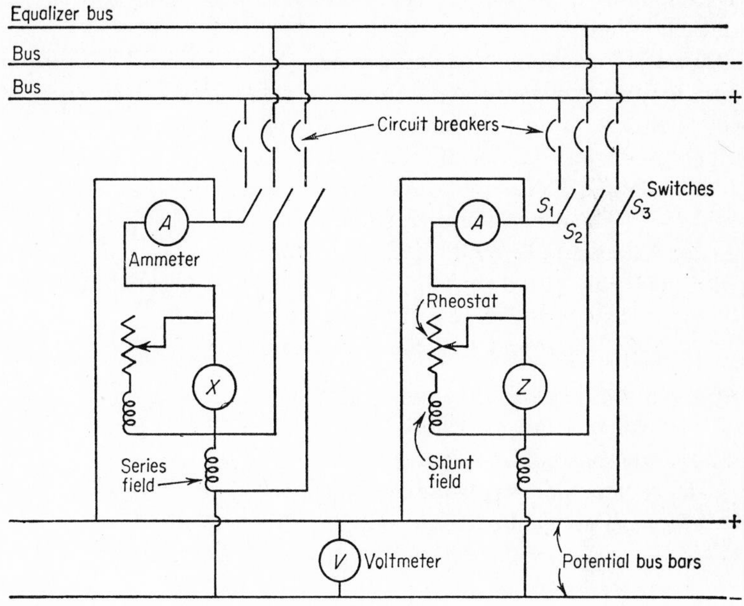

Fig. 10-23

stabilizes the action of the generators when both are on line by maintaining the IR drop in the two series fields equal at all loads; i.e., it prevents the terminal voltage of one generator from changing with a change in load.

The generator X is brought up to synchronous speed, and its field rheostat is adjusted manually to give the voltage desired at the bus bar.

The next step consists in connecting the generator Z to the same bus bars. In order to do this, the generator Z is started up and the voltmeter read to make sure that the generator has the proper polarity. The voltmeter must not read backwards when the polarity is such that the positive terminal of the generator is connected to the positive bus bar.

The generator is now brought up to speed with the circuit breakers "in" and switches S_2 and S_3 closed so that the series field will be excited. The voltage of the incoming generator is next adjusted so as to be equal to that across the bus bars, and S_1 is closed. By varying the shunt field strength the machine is now made to share the load as desired.

Q10-15. A compound-wound generator has an armature resistance of 0.0025 ohm, a shunt field resistance of 7.1 ohms, and a series field resistance of 0.001 ohm. The stray power losses are 5 kw. The generator operating at full load delivers 90 kw at 125 volts and 1,200 rpm. Find the efficiency of the machine (*a*) at rated load, (*b*) at one-half of full load. Assume short shunt.

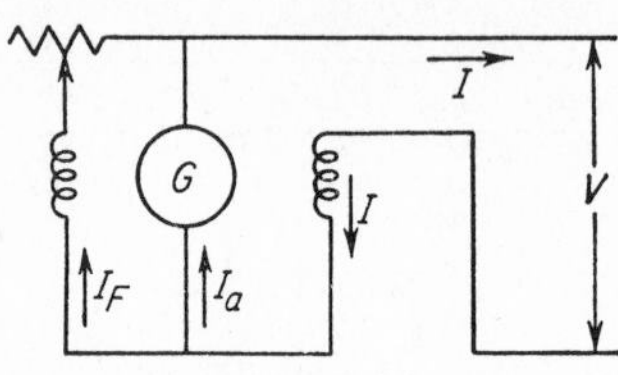

Fig. 10-24

ANSWER. Please refer to Fig. 10-24 and let

E_t = terminal voltage, volts
P = power output, watts
R_1 = series field resistance, ohms
R_f = shunt field resistance, ohms
I_f = shunt field current, amp
I_a = armature current, amp
I_L = load current, amp

(*a*)
$$I_L = P\frac{1}{E_t} = \frac{90{,}000}{125} = 720 \text{ amp}$$

Then the voltage drop in series field

$$I_L R_1 = 720 \times 0.001 = 0.72 \text{ volt}$$

Voltage at armature and shunt field is as follows:

$$E_t + 0.72 = 125 + 0.72 = 125.72 \text{ volts}$$
$$I_f = \frac{125.72}{R_f} = \frac{125.72}{7.1} = 17.72 \text{ amp}$$
$$I_a = I_L + I_f = 737.72 \text{ amp}$$

Total losses:

$I_L{}^2R_1$ = loss in series field = $720^2 \times 0.001$	=	519 watts
$I_f{}^2R_f$ = loss in shunt field = $17.72^2 \times 7.1$	=	2,220
$I_a{}^2R_a$ = loss in armature = $737.72^2 \times 0.0025$	=	1,352
Stray power loss	=	5,000
Total loss	=	9,091 watts

$$\text{Efficiency at full load} = \frac{90{,}000 \times 100}{90{,}000 + 9{,}091} = \textbf{91 per cent}$$

(*b*) $I_L = \dfrac{45{,}000}{125} = 360$ amp also

$$I_LR_1 = 360 \times 0.001 = 0.36 \text{ volt}$$

Voltage at armature and shunt field $E_t + 0.36 = 125.36$ volts

$$I_f = \frac{125.36}{R_f} = \frac{125.36}{7.1} = 17.65 \text{ amp}$$

$$I_a = I_L + I_f = 360 + 17.65 = 377.65 \text{ amp}$$

Total losses:

Series field = $360^2 \times 0.001$	=	130 watts
Shunt field = $17.65^2 \times 7.1$	=	2,220
Armature = $377.65^2 \times 0.0025$	=	356
Stray power	=	5,000
Total loss	=	7,706 watts

$$\text{Efficiency at half load} = \frac{45{,}000 \times 100}{45{,}000 + 7{,}706} = \textbf{85.6 per cent}$$

Note: In the solution to this problem the brush contact loss was considered negligible. Also note that shunt field and stray-power losses remain unchanged regardless of load conditions.

Q10-16. An induced voltage in generator *A* under load is 258 volts and in generator *B* 261 volts with the circuit constants as shown in Fig. 10-25. Calculate the current delivered by each generator, the terminal voltage of each generator, the voltage of the load, and the power taken by the load.

ANSWER. By Kirchhoff's laws

$$E_a = 258 - 0.1I_a = 0.3I_a + 3I_a + 3I_b \qquad (1)$$

$$E_b = 261 - 0.2I_b = 0.4I_b + 3I_a + 3I_b \qquad (2)$$

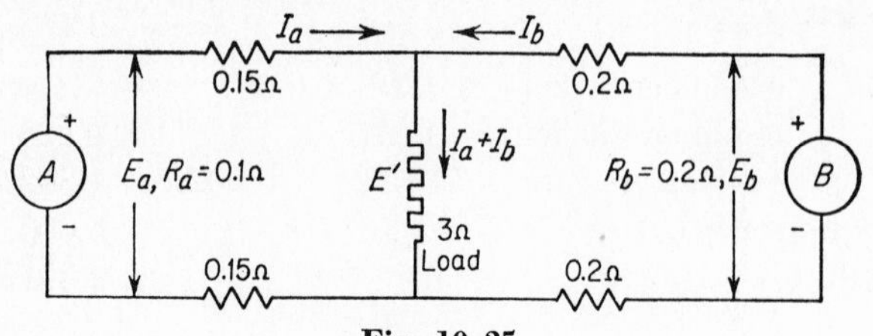

Fig. 10-25

After rearrangement we have

$$258 = 3.4I_a + 3I_b$$
$$261 = 3.6I_b + 3I_a$$

And

$$I_a = 48.6/1.08 = \mathbf{45\ amp}$$
$$I_b = [258 - (3.4 \times 45)]/3 = \mathbf{35\ amp}$$

Terminal voltage of each generator: From Eq. (1) in this problem

$$E_a = 258 - (0.1 \times 45) = \mathbf{253.5\ volts}$$
$$E_b = 261 - (0.2 \times 35) = \mathbf{254\ volts}$$

Voltage at load C: $E' = 3(I_a + I_b) = 3 \times 80 =$ **240 volts**
Power taken by load: $240 \times 80 \equiv$ **19.2 kw**

Q10-17. A d-c generator is capable of producing 120 volts on open circuit. A load of 30 amperes is connected, and the voltage at the terminals of the generator drops to 100 volts. Find the internal resistance of the generator.

ANSWER. Open-circuit voltage *minus* terminal voltage = voltage drop across internal resistance. Thus, $120 - 110 = IR = 30\ R$, from which $R = 10/30 =$ **1/3 ohm.**

Chapter 11

DIRECT-CURRENT MOTORS

11-1. D-C Motors. We saw previously that, whereas the generator converts mechanical energy into electric energy, the motor acts in the opposite manner. This process is reversible in that a motor may be used as a generator and vice versa. Furthermore, the speed of a generator is fixed by the speed of its prime mover, while that of the motor armature depends upon the electric and magnetic conditions within the armature as well as on the nature of the load.

11-2. Direction of Rotation. The direction of rotation of a motor may be determined by what is called the *left-hand rule.* This is as illustrated in Fig. 11-1. When the first finger of the left hand points in the direction of the lines of force, and the second finger points in the direction from which the current flows, the thumb, if extended level with the hand, will indicate the direction of rotation of the revolving armature of the motor.

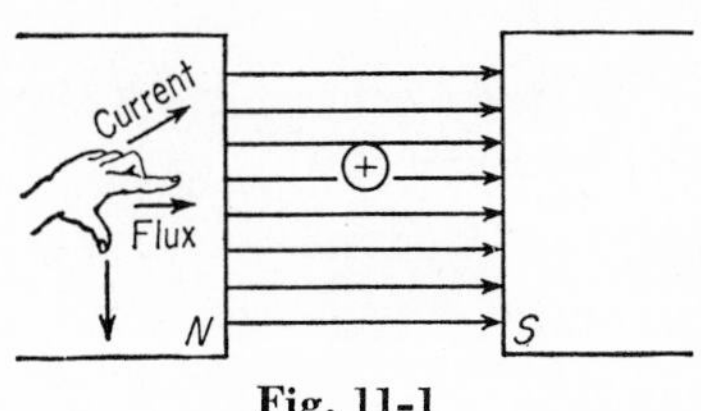

Fig. 11-1

This may be proved by well-known experiments showing the automatic twisting of two conductors to line up with a magnetic field. When this takes place within a motor, the combined effort of this pulling and twisting of the magnetic lines of force is called the *torque* of the motor, or its turning effort. Naturally, the strength of the torque depends on the strength of the magnetic flux of the field poles and of the strength of the magnetic lines of force surrounding the armature coils. When the load on the motor is increased, the current drawn from the supply lines is increased, and the torque also increases in proportion, and the motor continues to turn, taking care of the load.

11-3. Torque of a Motor. In a given motor its torque is proportional to the armature current and to the strength of the magnetic field. This torque is greater than that exerted by the armature on the external mechanical load, since it includes the torque required to overcome friction of the motor bearings, air friction, windage, etc.

$$T = KI_a\phi \tag{11-1}$$

where T = torque, lb-ft

K = constant involving number of poles, parallel paths through armature, choice of units, etc.

I_a = armature current, amp

Q11-1. When a certain motor armature is taking 50 amp from the line, it develops 60 lb-ft of torque. If the field strength is reduced to 75 per cent of its original value and the current increases to 80 amp, what is the new value of the torque developed?

ANSWER. If the current remained constant, the new value of torque due to the weakened field would be

$$60 \times 0.75 = 45 \text{ lb-ft}$$

Because of the increase in current, the final value of torque would be

$$45 \times 80/50 = \mathbf{72\ lb\text{-}ft}$$

11-4. Speed of a Motor. The emf induced in a generator will also apply to a motor as well. Thus, the counter emf is

$$E_a = K\phi N \tag{11-2}$$

Rearranging and solving for the motor speed

$$N = \frac{E_a}{K\phi} \tag{11-3}$$

With respect to the terminal voltage E_t Eq. (11-3) may also be written

$$N = \frac{E_t - I_aR_a}{K\phi} \tag{11-4}$$

Since the values of armature current and resistance (I_a and R_a) are small compared with the terminal voltage, it may be seen that the speed of the motor is nearly directly proportional to the impressed voltage and inversely proportional to the resultant flux per pole.

Q11-2. In a certain motor the armature resistance is 0.1 ohm. When connected to 110 volts, the armature current is 20 amp and its speed is 1,200 rpm. What is its speed when the armature takes 50 amp with the field increased 10 per cent?

ANSWER. $$\frac{N_2}{N_1} = \frac{K[110 - (50 \times 0.1)]/\phi_2}{K[110 - (20 \times 0.1)]/\phi_1} = \frac{105\phi_1}{108\phi_2}$$

With N_1 equal to 1,200 rpm

$$N_2 = 1{,}200 \times \frac{105\phi_1}{108\phi_2}$$

But $\phi_2 = 1.10 \times \phi_1$, so that finally

$$N_2 = 1{,}200 \times \frac{105}{108} \times \frac{\phi_1}{1.10\phi_1} = \mathbf{1{,}060\ rpm}$$

11-5. Counter Emf in Motors. When the motor armature is revolving through the flux of the field poles, producing mechanical power, it is simultaneously acting as a generator. The armature coils cutting through the field flux pick up energy which produces an emf in an opposite direction to the flow of current taken from the supply line. As in the generator, the strength of this emf is determined by the number of loops of wire being revolved, their manner of winding on the armature core, the speed of rotation, and the strength of the field through which they pass.

The current flowing through the armature, then, depends upon the difference between the applied voltage and the counter emf produced by the rotating of the armature. *If the counter emf were to equal the applied emf, no current would flow.*

Let us illustrate by referring to water flowing in a pipe. The effect of the counter emf on the armature current of a d-c motor would be similar to placing a back pressure at the outlet of the pipe. In the case of motors it should be apparent that the applied emf must always exceed the counter emf, and the difference between them is due to the armature resistance, which causes the "drop" in the generator armature.

The speed of the motor varies more or less under variation of load, and the counter emf developed is always in proportion to the applied emf. Then, as the motor accommodates itself to different loads, the current flowing in the armature varies with the developed counter emf because of the speed of the motor. Thus, only sufficient current to handle the load is taken from the line.

When the motor is run without load, its speed is increased, and the counter emf is increased, so that a small amount of current is flowing in the armature. When a load is applied to the pulley, or shaft, the speed is lowered, less counter emf is developed, and more current flows in the armature, increasing the torque, and thus the pulling power of the motor. It will be seen that the power which a motor draws from the supply lines is directly proportional to the work which it is required to do.

The resistance of the armature of a 10-hp 110-volt d-c motor is about 0.05 ohm. If this armature were connected directly to 110 volts, the current drawn, according to Ohm's law, would be $I_a = 110/0.05 = 2{,}200$ amp. The rated current is 90 amp approximately. Obviously, some other force is retarding the flow of current. This other force, as we saw, is the counter emf, or back voltage.

The net voltage acting in the armature circuit is $E_t - E_g$, where E_g is the generated voltage. The armature current I_a follows Ohm's law and

$$I_a = \frac{E_t - E_g}{R_a} \tag{11-5}$$

Transposing, we have the following

$$E_g = E_t - I_a R_a \tag{11-6}$$

This should be compared with the equation for the generator. In a generator E_g is equal to E_t minus the armature resistance drop. The back voltage must always be less than the terminal or impressed voltage if the current is to flow into the motor at the positive terminal.

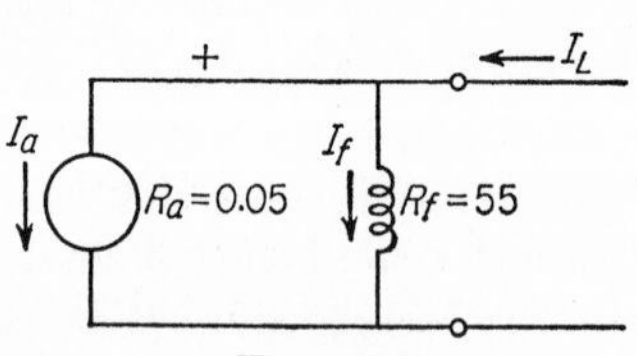

Fig. 11-2

Q11-3. Determine the back voltage of a 10-hp motor when the terminal voltage is 110 volts and its armature current is 90 amp, the armature resistance being 0.05 ohm.

ANSWER. $E_g = 110 - (90 \times 0.05) = 110 - 4.5 =$ **105.5 volts**

If the motor is of the shunt type and the field resistance is 55 ohms, what is the line current drawn by the motor? See Fig. 11-2.

$I_f = {}^{110}\!/_{55} = 2$ amp and $I_L = I_a + I_f = 90 + 2 =$ **92 amp**

Q11-4. The armature resistance of a motor is 0.24 ohm. If it has a counter emf of 108.8 volts when running on a 110-volt line, what current does it take?

ANSWER. Effective drop is 110 − 108.8, or 1.2 volts.

$$I_a = 1.2/0.24 = \mathbf{5\ amp}$$

If counter emf did not develop, the current flow through the armature coil would be 110/0.024 = 458 amp, and the armature would burn out.

11-6. How D-C Motors Function. Fundamentally, the d-c motor consists of a magnetic field and a rotating element (armature) which contains coils and a device for reversing the direction of the current in the coils. The first part consists of a frame, the poles, and the field coils around the poles.

When a current flows through the coils, the poles become magnetic. The coils are wound in such a manner that one pole will be a *north pole* and the next one a *south pole*, etc. The rotating part or armature consists of coils imbedded in a core and a commutator which reverses the current. The brushes work together with the commutator to accomplish this. Brushes are actually nothing more than devices to make and maintain electrical contact with a rotating element.

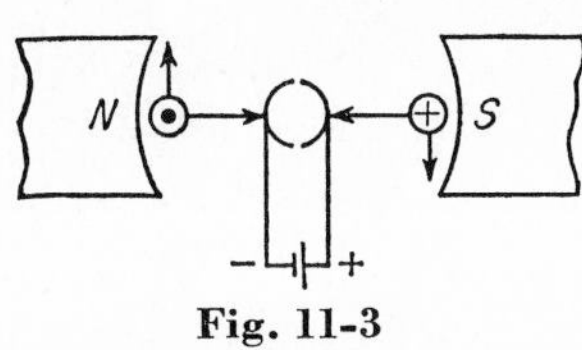

Fig. 11-3

Figure 11-3 illustrates this principle. The dot (.) is the current flow, normal to the paper, coming toward the observer and the plus (+) the current going away. The poles north N and south S produce a magnetic field, and the two sides of the loop of wire are located in the magnetic field. The battery is a source of power and current. We have already discussed the forces produced in the conductor.

These coils are located in slots in the core, and the core, in turn, is fastened to the shaft so that these two forces will produce a twisting moment, i.e., torque. Thus, the shaft will try to rotate. When the shaft is turned 90°, the sides of the coil will be located vertically above each other and the forces will no longer tend to produce a torque, but they will act in the same vertical plane.

Now suppose that the inertia of the rotating element will carry the wires past this vertical plane. Then the forces acting on the sides will attempt to return the coil to the vertical plane, if the currents in the respective sides of the coil still flow in the directions indicated in Fig. 11-3. However, if the current could be reversed at the instant that the coil passes this vertical plane, then the forces would also be reversed and torque would again be produced in the same direction of rotation. The two semicircles in the center of Fig. 11-3 make it possible to reverse this current. They rotate together with the wires of the coil while the brushes supply them with current from the battery. These two segments constitute the fundamental principles of the commutator.

In an actual motor, of course, there is not just one set of wires or one coil, but there are several such coils all acting in the same way. There are not merely two segments, as shown in Fig. 11-3, but there are a corresponding number of commutators or bars all insulated from each other by means of mica strips. Figure 11-3 also shows that the current in the coils must be reversed at a point halfway between the poles.

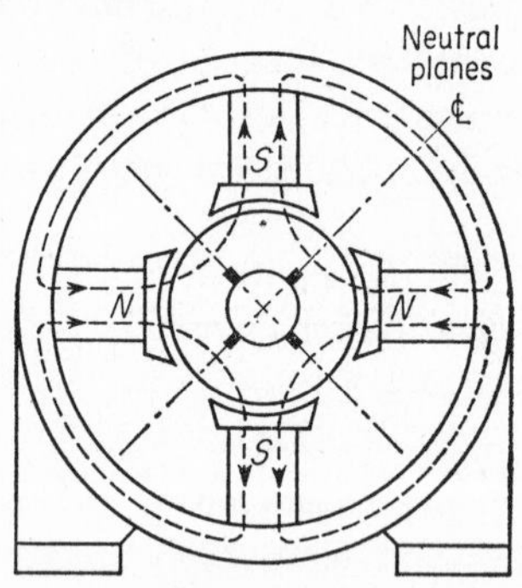

Fig. 11-4

Again in Fig. 11-3 there is one north pole and one south pole. However, it is possible to use four poles or more as long as they are alternately north and south poles. As previously stated, the coils of wire are located in the armature core which serves a double purpose. In the first place, it provides the coils with something on which to exert their turning force and, in the second place, it constitutes a part of the magnetic path. This latter is so because the iron core presents much less resistance to the lines of force than does the air.

Figure 11-4 shows a schematic cross section through a four-pole motor. The paths of the lines of force are shown as well as the positions where the current in the armature coils must be reversed. These are known as the neutral positions, and they must be determined by test. The brush assembly is made movable for the purpose of making adjustments. Close adjustment is necessary to prevent sparking between the brushes and the commutator.

It is necessary for the operation of the motor to set up a magnetic field between the poles, and in order to produce this field an electric current must be sent through the coils which surround the poles. This current may be supplied from an external source (Fig. 10-4), or it may be taken from the same source which sends a current through the armature. In the latter case, the coils may parallel the armature (Fig. 10-5), which results in a *shunt motor*. On the other hand, they may be connected in series with the armature circuit (Fig. 10-6), in which case the motor is called a *series motor*. Most generally, it is a motor in which both shunt winding and series winding are provided (Figs. 10-7 and 10-8). This is known as a *compound motor*. These different windings lend different characteristics to the motors; particularly, they influence the manner in which the speed changes when the load varies. By a proper selection of the portions of the field produced by each of the two windings (shunt and series), the desired characteristics of the motor can be obtained.

The difference between these motor types will be explained briefly. In a shunt motor the voltage across the field is always the network voltage, so that the field strength does not change. The current which flows back through the armature coils will vary with the load on the machine, and this will cause slight variations in the speed (usually 5 per cent approximately) between no load and full load. Since the speed is fairly constant regardless of load, the motor is suitable for lathes and other similar machine tools.

In a series motor the entire load current passes through the field coils, so that the field will become stronger when the load increases. More torque will be developed as a result. The speed of the motor will be decreased by the increase in load. The fact that the motor develops a large torque makes it suitable for cranes, street cars, elevators, and other uses where the starting load is heavy.

Compound-motor characteristics lie between the two. These motors are used where moderate changes in speed are not objectionable, while a fairly good torque is desired for starting purposes. The interactions of the different magnetic fields cause distortions and other electrical effects which make it difficult to obtain proper commutation, i.e., contact without sparking between the brushes and the commutator. Of great help in this respect is an *interpole winding*, which, in effect, is a small series winding located at one or more of the neutral positions. These coils are always connected in series with the armature (see Fig. 11-5).

The smaller sizes of the d-c motors are built with two field poles, and with either shunt coils, series coils, or compound coils. For the purpose of providing better commutation, one or two interpoles may be provided.

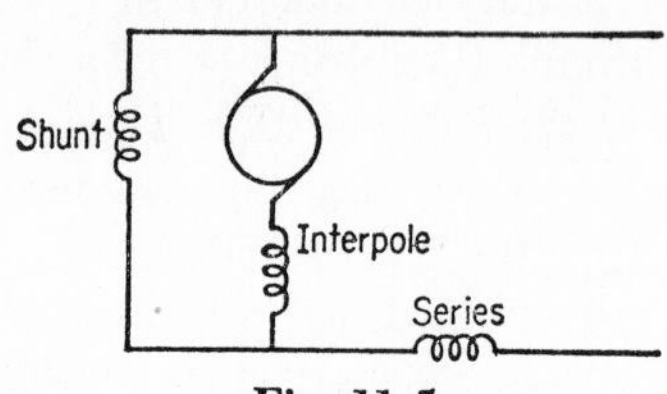

Fig. 11-5

The larger sizes have a larger number of poles and generally have the same number of interpoles. Most of these motors are compound wound, although series and shunt motors are used for specific purposes. The smaller sizes generally have cast-steel frames and laminated poles, while larger motors are preferably built with rolled frames, either in the form of a continuous ring or consisting of two halves.

Mechanically, there are many varieties depending on the purposes for which the motors are to be used. They may be equipped with sleeve bearings or with ball or roller bearings. They may be open, dripproof, or enclosed. They may be designed for horizontal or for vertical operation. They may be provided for ring mounting.

11-7. Mechanical Power of a Motor. The mechanical power of a motor is dependent upon the speed of the armature combined with the torque. It may be determined by finding the product of the counter emf and the current flowing in the armature, or the power developed is equal to volts × amperes. This includes the losses due to heat, friction, eddy currents, etc. Thus, the power output of a motor is equal to the power developed minus losses. The power input is the sum of the watts drawn by the armature and the watts drawn by the field. The watts drawn by the armature is equal to the applied emf × armature current, and the watts drawn by the field is equal to the applied emf × field current.

The *efficiency* of the motor is given by

$$\text{Efficiency} = \frac{\text{power output}}{\text{power input}} \tag{11-7}$$

Q11-5. Determine the mechanical power developed in the armature of the 10-hp motor in Q11-3.

ANSWER. The back emf is 105.5 volts and the armature current is 90 amp. The mechanical power developed is

$$P_m = 105.5 \times 90 = 9{,}495 \text{ watts, or } \mathbf{9.495 \text{ kw}}$$

This would be equivalent to **12.73 hp.**

11-8. Shunt Motor. The shunt motor is connected in the same manner as the shunt generator (see Fig. 11-6). The speed may be determined from

$$N = \frac{E_t - I_a R_a}{K\phi} \tag{11-8}$$

In the shunt motor K, E_t, R_a, and ϕ are all substantially constant for a given line voltage and excitation.

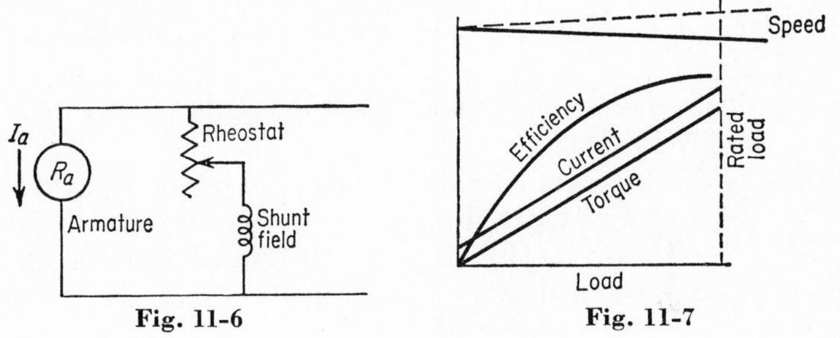

Fig. 11-6

Fig. 11-7

In so far as speed is concerned the shunt motor is used where substantially constant speed is required or where adjustable speed is required. Please refer to Fig. 11-7 where its characteristics of speed, torque, current drawn, and efficiency are shown.

The shunt motor has only one exciting coil connected across terminals. A field winding consists of a large number of turns of fine wire on each pole, and usually the windings on all poles are connected in series in one circuit. The current carried by the field winding is between 1 and 5 per cent of the full-load current. Since the armature flux is practically independent of the load, the motor has an approximately constant speed for all reasonable variations in load; the torque is directly proportional to the armature current irrespective of speed. The efficiency is high throughout the load range, and the load range is wide for a small speed range. The torque to supply additional load is produced by an increase in the armature current with only a slight change in speed.

Q11-6. The armature of a shunt motor contains 0.20 ohm resistance. The motor is to run on a 110-volt circuit. Suppose that it is thrown on the circuit suddenly (across-the-line starting), while the armature is standing still, what current will flow?

ANSWER. $$I_a = 110/0.20 = \mathbf{550\ amp}$$

With the motor running at normal speed on 110 volts, there is a back emf of 107 volts. What current does it then take?

$$I_a = (110 - 107)/0.20 = \mathbf{15\ amp}$$

11-9. Series Motor. The series motor has only one exciting coil in series with the armature, carrying full-load current. The field winding consists of a few turns of heavy wire on each pole, and the windings on all the poles are connected in series. The resistance of the coil is purposely made low so that losses there are small. The construction is more rugged than that of the shunt motor, and it is usually totally enclosed.

Since the flux in the series motor is produced by the load current, the flux increases with the load current. The torque increases faster than the increase in the load current. The speed varies inversely with the load. The motor has high efficiency throughout the wide range of speed as well as of load. The speed decreases much more rapidly than that of a shunt motor because both the armature current and the flux density increase as additional torque is required.

11-10. Compound Motor. This motor has both a series winding and a shunt winding on each pole. These are so wound and connected that the two windings assist each other in the production of magnetism. Thus, it is a combination of a shunt and a series motor designed to give the good starting qualities of the series motor and to avoid the danger of excessive speeds at light loads.

11-11. Differential Motor. In this motor the shunt and field windings oppose each other in the production of magnetism. Therefore, this motor has poor starting qualities, and increases in speed with an increase in load. However, it has no tendency to run at a dangerously high speed. Applications of this motor are very limited.

11-12. Caution in the Use of Series and Shunt Motors. Never open the field circuit of a shunt motor, for it will race when the field circuit is broken and the armature is left on the line.

Never start an unloaded series motor, and never remove the load from a series motor while it is running. It will race under these conditions.

Q11-7. A 110-volt shunt motor has an armature resistance of 0.8 ohm and a field resistance of 220 ohms. At full load the motor takes

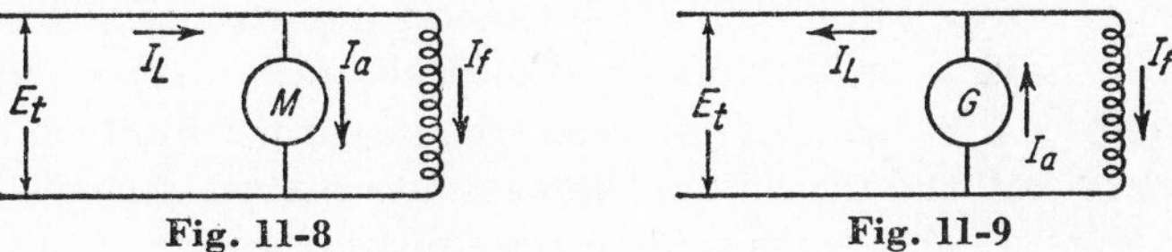

Fig. 11-8 **Fig. 11-9**

10 amp, and the speed is 1,200 rpm. At what speed must this motor run as a generator to deliver 10 amp to an external circuit at 110 volts? See Figs. 11-8 and 11-9.

ANSWER. *As a shunt motor*

$$I_f = \frac{E_t}{R_f} = \frac{110}{220} = 0.5 \text{ amp}$$

$$I_a = I_L - I_f = 10 - 0.5 = 9.5 \text{ amp}$$

$$\text{Counter emf} = E_t - I_aR_a = 110 - (9.5 \times 0.8) = 102.4 \text{ volts}$$

As a generator assume that the change in the effect of the armature reaction is negligible and that no change is made in the field rheostat. Then

$$I_a = I_L + I_f = 10 + 0.5 = 10.5 \text{ amp}$$

Generated voltage is

$$110 + (10.5 \times 0.8) = 118.4 \text{ volts}$$

This voltage must be generated by the armature. Therefore, as a generator the speed must be (voltage, rpm, and flux are constant)

$$1{,}200 \times (118.4/102.4) = \mathbf{1{,}387 \text{ rpm}}$$

The field current when the machine operates as a generator is the same as it was when operating as a motor, and if any change in the armature reaction is neglected, the effective flux ϕ would remain the same. Since the armature reaction varies with the armature current, and the armature current is increased by only 1 amp, the change in armature reaction may be neglected.

11-13. Speed Regulation of Motors. Speed regulation of motors, after they have been started, is accomplished with a rheostat. When starting d-c motors, care must be exercised. We have seen that the current which flows in the armature depends upon the difference between the impressed voltage and the back emf. The latter keeps the current from being excessive although the armature resistance is very low. We have also seen that the back emf depends upon the speed, since it is caused by the armature conductors cutting lines of force. When, therefore, the armature is not rotating, there is no back emf, and the current through the armature depends upon the impressed voltage and the armature resistance only.

Suppose then that full voltage were thrown on to an armature which has not started rotating. Since its resistance is small, the current through the armature would be excessive and would tend to burn out the windings.

In order to avoid this excessive current build-up at starting, a starting resistance is introduced into the armature circuit, which cuts down the voltage across the motor armature first and allows but a small current to flow through it. By slowly cutting out this resistance as the motor speeds up and sets up a back emf, it is possible to throw the full voltage on the motor. It is safe to have this high voltage across the motor only as long as the speed is high enough to oppose it by a high back emf. Figure 11-10 is a simple diagram of the starting resistance used with a shunt motor. The resistance prevents too large a current from entering the armature. As the motor picks up speed, the resistance is reduced in steps, as shown. Finally, the armature is put directly across the line by swinging the arm C to the point P. The handle should not be held on any of the intermediate points more than a

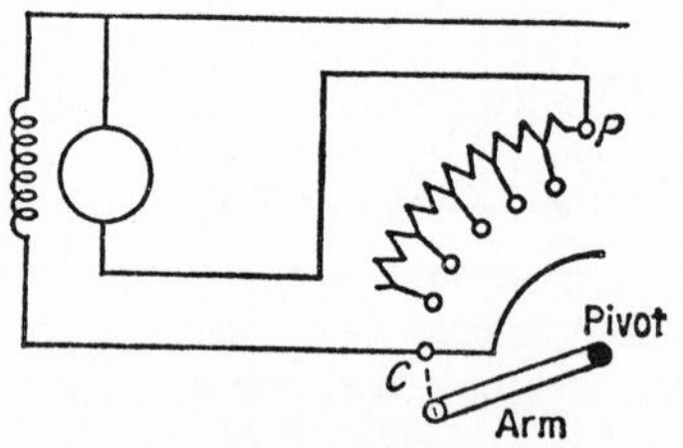

Fig. 11-10

few seconds, as the resistance is not designed to carry the necessary current for a longer period.

11-14. D-c Motor Speed Control. The speed of a simple series motor is easily varied by a resistance in series with it, as shown in Fig. 11-11. The current flowing through the resistor reduces the voltage impressed across the motor, thus causing it to slow down. After the speed is adjusted by this method, it is constant only as long as the load remains the same. If the load goes up, the motor will slow down. Reducing the load will cause the speed to increase.

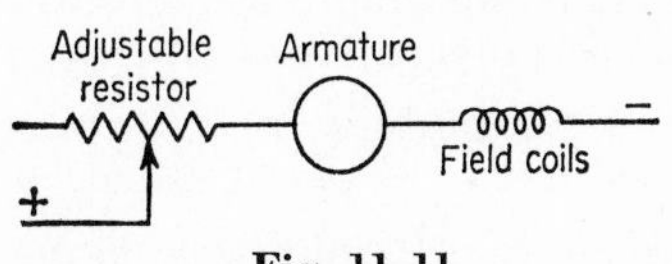

Fig. 11-11

Speed changes are caused by the variation in voltage impressed across the motor as a result of the difference in voltage drop across the series resistor. This type of control is suitable only for constant loads or where the motor is under the control of an operator. Figure 11-12 shows a method of increasing the speed of a loaded series motor above full-load value by shunting a resistor across the field windings. A combination of series and shunt resistances may also be used to control series motor speeds from zero values to those above normal full load.

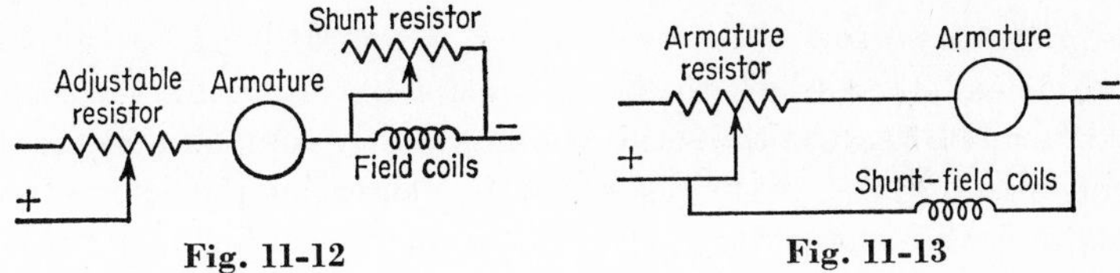

Fig. 11-12 **Fig. 11-13**

The speed of shunt motors can be decreased below normal value by the armature resistance, as shown in Fig. 11-13. This reduces the speed in the same way as the series motor above. A variable resistance in series with the field winding, Fig. 11-14, is used to control speed of shunt motors above normal value. The speed of general-purpose shunt motors can be increased about 25 per cent without trouble if the motor is not overloaded.

Speed adjustment of either shunt or compound motors is most economical and satisfactory by field control. The shunt field current is only about 5 per cent of the total current supplied to the motor; consequently the field rheostat is small, low in cost, and the power losses in it are also small. Once the speed is adjusted, it remains essentially constant for all changes in load.

Motors of this type are known as adjustable-speed motors because speed can be adjusted to some desired value, and more or less stays put. On the other hand, motor speeds controlled by regulating armature current vary with the load. Such motors are known as variable-speed machines.

When the shunt-motor speed is to be increased more than 25 per cent, a specially designed, adjustable-speed motor should be selected. Such motors are available for speed ranges of 6:1; that is, if normal speed is 250 rpm, the field can be *weakened* until speed is 1,500 rpm.

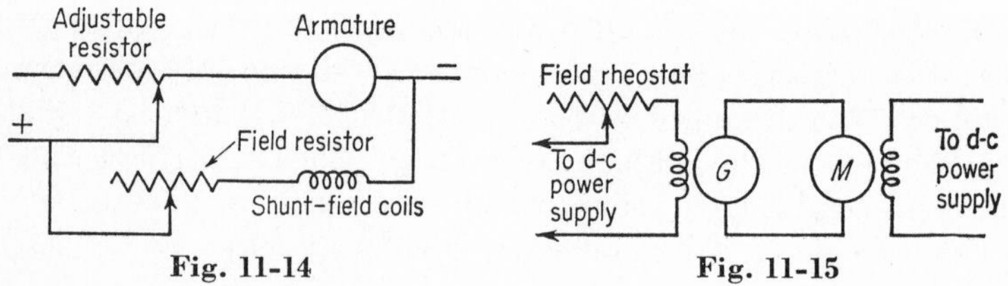

Fig. 11-14

Fig. 11-15

11-15. Compound Motors—Speed Control. These do not readily lend themselves to speed control by adjusting the shunt-field current, because the series winding partly compensates for any weakening of the shunt field strength by decreasing the field current. That is, if the magnetizing effect of the shunt winding is decreased, the series-winding effect becomes more noticeable. On certain applications, where a wide speed range is required, the series winding is used for starting only, and it is cut out after the motor comes up to speed. Then the motor speed can be adjusted by shunt field control as with a shunt motor.

When speed control over a wide range is required, say from zero up to a maximum, a combination of armature and field control or a variable-voltage control system, as in Fig. 11-15, is used. On a variable-voltage system, a separate generator is required for each motor. The armature of the generator is tied directly to the motor, and the field coils of both the motor and generator are supplied from a separate d-c source. The motor field coils are generally excited at full value most of the time, and the speed is controlled by regulating the generator field current through the resistor.

11-16. Dynamic Braking. When used on hoists, elevators, and other applications where the motor is started, stopped, and reversed frequently, it is often made to act as a brake to slow down the load. This is done by leaving the field coil energized, disconnecting the armature from the power source and connecting it to a resistance, as in Fig. 11-16. The motor acts as a generator, producing a braking action (dynamic braking) inversely proportional to the circuit resistance.

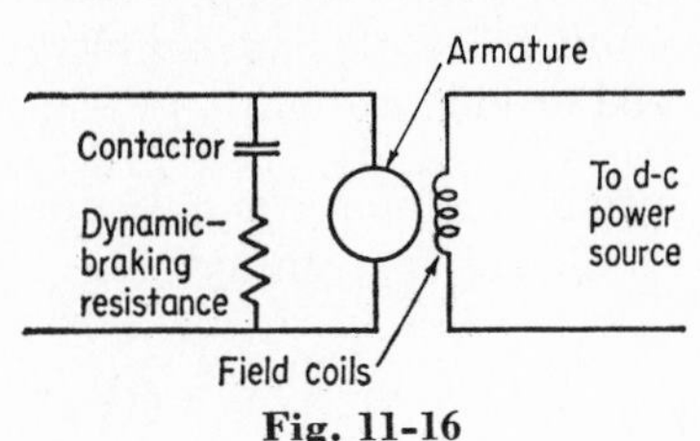

Fig. 11-16

11-17. Efficiency of Generators and Motors. This is known as the ratio of its useful output to its total input. It may be written

$$\text{Efficiency} = \frac{\text{input} - \text{losses}}{\text{input}}$$

$$\text{Efficiency} = \frac{\text{output}}{\text{output} + \text{losses}}$$

The first of the these general relations is used for determining the efficiency of motors, while the second is used for efficiency measurements on generators. If the losses in a motor are summarized, the first above may take on the form

$$\eta_m = \frac{E_tI_L - (E_fI_f + I_s^2R_s + I_a^2R_a + P_{sp})}{E_tI_L} \tag{11-9}$$

where E_t = terminal voltage

I_L = line current

E_fI_f = copper losses in shunt field

I_aR_a = copper losses in armature

P_{sp} = stray power losses

The second relation for generators takes on the form

$$\eta_g = \frac{E_tI_L}{E_tI_L + (E_fI_f + I_s^2R_s + I_a^2R_a + P_{sp})} \tag{11-10}$$

Terms in parentheses are the combined total losses in the motor and the generator, respectively.

The copper losses in a machine may readily be calculated in each case, the stray power losses represented by frictional losses of bearings and brushes, windage resistance, hysteresis, and eddy currents in the armature, and the pole faces cannot be calculated directly since they all are some function of speed or flux, or both.

11-18. The Prony Brake. If the output of a motor is measured by a Prony brake, then the horsepower output is given by

$$\text{Hp} = \frac{2\pi NT}{33{,}000} \tag{11-11}$$

where N = rpm

T = torque, lb-ft

Q11-8. A motor takes 20 amp at 220 volts. It is rigged with a Prony brake and runs at 1,200 rpm. With the net load of 48 lb (scale reading − tare) on the arm, and with the brake arm set at 5.4 in., find the efficiency (see Fig. 11-17).

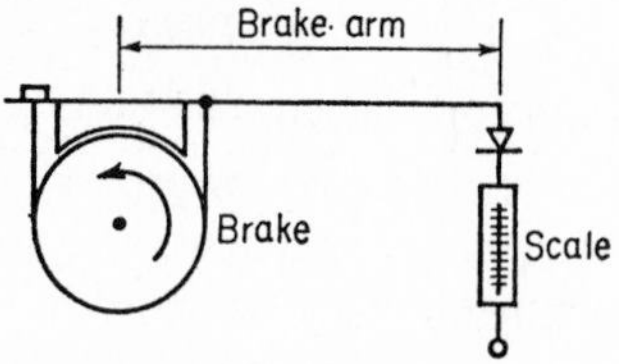

Fig. 11-17

ANSWER. Efficiency is the output divided by the input. Then

$$\frac{2\pi NT/33{,}000}{EI/746} \tag{11-12}$$

$$\frac{(2\pi \times 1{,}200 \times 5.4/12 \times 48)/33{,}000}{(220 \times 20)/746} = 0.835, \text{ or } \mathbf{83.5\ per\ cent}$$

11-19. Computing the Efficiency by Means of Losses. In any machine the output equals the input minus the losses, and the input equals the output plus the losses.

In some cases it is easier to measure the input and the losses. Thus, the output is found by subtracting the losses from the input.

Wherever the output and the losses can be measured easily, they are merely added to find the input. Certain losses in a generator or motor remain practically constant at all loads, providing the speed does not materially change. The constant losses in a shunt motor or generator are:

Field (I^2R) losses
Eddy-current losses
Hysteresis losses
Mechanical losses

These may be found as follows: Run the machine as a motor with no load. Then all the power put into it must go to supply the above constant losses, plus the I^2R loss in the armature. Knowing the current through the armature and its resistance, the I^2R loss may be computed from the armature. Then subtract it from the total input into the motor at no load. The remainder will be the constant losses, which, of course, will remain practically the same at all loads.

The motor is now run at a given load. The constant loss remains as computed at no load, and the new armature I^2R loss is computed from the new value of the armature current.

The sum of the constant losses and the armature I^2R losses equals the total losses. This subtracted from the input gives the output. The efficiency, therefore, equals the input minus the losses, all divided by the input.

Q11-9. A shunt motor takes 40 amp at 112 volts under full load. When running without load at the same speed, it takes 3 amp at

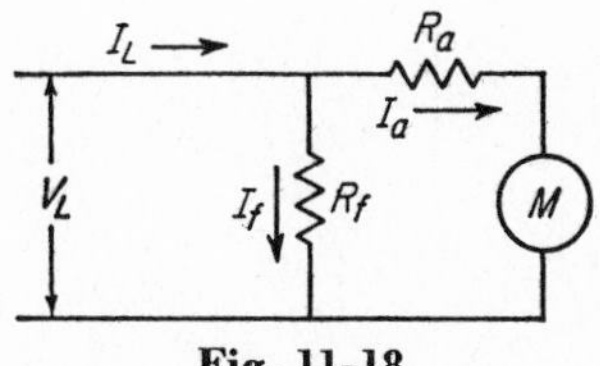

Fig. 11-18

106 volts. The field resistance is 100 ohms and the armature resistance is 0.125 ohm. What horsepower does it deliver at full load, and what is its commercial or over-all efficiency?

ANSWER. Please refer to Sec. 11-18 and Fig. 11-18.

At no load:
Field takes: $106/100 = 1.06$ amp $= I_f = E_L/R_f$
Armature current: I_a, $3 - 1.06 = 1.94$ amp
Armature loss: $I_a^2R_a = 1.94^2 \times 0.125 = 0.47$ watt
Power taken: $I_LE_L = 3 \times 106 = 318$ watts
Constant loss: $318 - 0.47 = 317.53$ watts

At full load:
Total input: $I_LE_L = 40 \times 112 = 4{,}480$ watts
Armature current: $I_a = 40 - 1.06 = 38.94$ amp
Armature loss: $I_a^2R_a = 38.94^2 \times 0.125 = 190$ watts
Total losses: $317.53 + 190 = 507.53$ watts
Output: input $-$ losses $= 4{,}480 - 507.53 = 3{,}972.47$ watts
Output in horsepower: $3{,}972.47/746 =$ **5.32 hp**
Efficiency: $3{,}972.47/4{,}480 = 0.89$, or **89 per cent**

Q11-10. An electric vehicle is driven by a d-c motor which receives power from a 100-volt battery. The vehicle is required to exert a constant tractive force of 200 lb at 7.5 rpm. The over-all efficiency is 74.6 per cent. Find current drawn from the battery.

ANSWER. The vehicle velocity in feet per second is found by $7.5 \times 88/60 = 11$ fps. Output horsepower required = (force $\times$ velocity)/550 $= (200 \times 11)/550 = 4.0$. Output watts required $= 4.0 \times 746 = 2{,}984$. Input watts required $= 2{,}984/0.746 = 4{,}000$. Finally, input current or current draw is 4,000 watts/100 volts = **40 amp.**

Chapter 12

CONDENSERS

12-1. General. A condenser is a device having capacity for holding an electrostatic charge. The earliest known condenser was the Leyden jar, invented in 1746 by Musschenbroek and named after the Dutch town where it was originally manufactured.

A peculiarity of condensers is that when one is connected in a circuit with a charging source, such as a battery, it will store up a charge from this battery while the circuit is closed and discharge it when the circuit is opened. In Fig. 12-1 the charge from the battery will create a positive strain on the upper plate of the condenser, and a negative condition on the lower plate. The plates are insulated from each other, so that *current* cannot pass through the condenser. A difference of potential between the two sets of plates is produced, and the pressure or strain set up by this condition is called an *electrostatic field.*

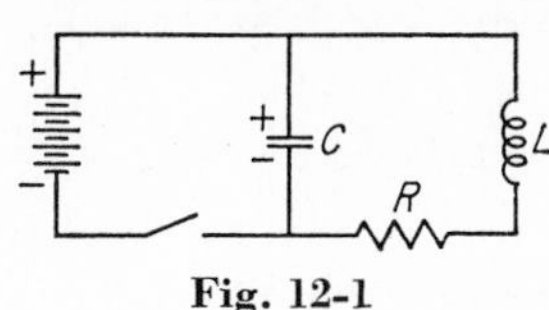

Fig. 12-1

An electric charge is there; but it is, for the instant, motionless or in a state of tension. Condensers operate on the law of attraction of unlike charges and repelling of like charges. A negative charge on the one plate will repel the negative electrons in the opposite plate, causing the latter to have a positive charge.

12-2. Back Pressure on Condensers. The above positive charge then has an attraction for the negative plate, reducing the potential of the latter. This can go on until the *back pressure* of the condenser equals the voltage of the charging current. Thus, at this stage, there is no further charge placed on the plates, and the condenser is said to be fully charged. In other words, it is said to be charged to the limit of its capacity.

As we have previously noted, if a continuous potential, direct current, is applied across a condenser, there will be no current flow. When the circuit is first closed, a charging current first flows into the condenser until the voltage between the plates has risen to the same value as that of the applied voltage. If the voltage source is then removed and the plates are shorted, a discharge current flows out of the condenser in the opposite direction to that of the charging current. The discharge ceases when the plates of the condenser have no potential difference. When an alternating current is applied to the condenser, it will alternately be charged and discharged to keep the voltage between the plates equal to the instantaneous value of the applied voltage.

Q12-1. Four capacitors are connected as shown in the diagram. Prior to closing the switch, the capacitors are charged to the potentials indicated. Calculate the voltage across each capacitor for the steady state after the switch has been closed (see Fig. 12-2).

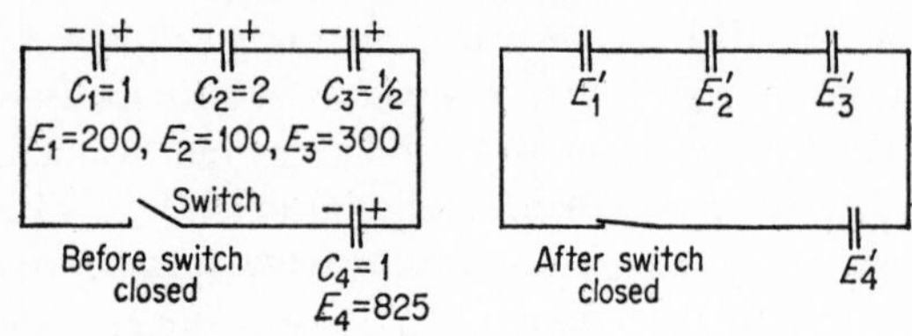

Fig. 12-2

ANSWER. Assume polarities as shown. Let Q be the circulated charge in coulombs. After a steady state has been reached at equilibrium:

$$E_1' + E_2' + E_3' + E_4' = 0$$

$$E_1 + \frac{Q}{C_1} + E_2 + \frac{Q}{C_2} + E_3 + \frac{Q}{C_3} + E_4 + \frac{Q}{C_4} = 0$$

Rearranging

$$Q\left(\frac{1}{C_1} + \frac{1}{C_2} + \frac{1}{C_3} + \frac{1}{C_4}\right) = -E_1 - E_2 - E_3 - E_4$$

$$Q\left(\frac{1}{1} + \frac{1}{2} + \frac{1}{0.5} + \frac{1}{1}\right) = -200 - 100 - 300 + 825$$

$$4.5 \times Q = 225 \qquad \text{or} \qquad Q = 50 \text{ microcoulombs}$$

Therefore

$$E_1' = 200 + 50/1 = \mathbf{250\ volts}$$
$$E_2' = 100 + 50/2 = \mathbf{125\ volts}$$
$$E_3' = 300 + 50/0.5 = \mathbf{400\ volts}$$
$$E_4' = -825 + 50/1 = \mathbf{-775\ volts}$$

Note: $E_1' = E_1 + \frac{Q}{C_1}$; $E_2' = E_2 + \frac{Q}{C_2}$; $E_3' = E_3 + \frac{Q}{C_3}$; $E_4' = E_4 + \frac{Q}{C_4}$.

12-3. Production of Oscillating Current. If the key in Fig. 12-1 is raised, the electric energy stored up in the condenser during the charging period will discharge around the circuit containing the coil (inductance) and the resistance. It will rush around and produce a negative charge on the *other side* of the condenser, causing a positive condition on the plate which has been negative. After the key has been opened and the released energy has charged the condenser from the opposite side, the process reverses. This process will repeat itself again and again until the energy is gradually dissipated in heat by the resistance of the circuit. Theoretically, were it not for the resistance of the circuit, this might go on forever. The movement of the current around the circuit during the discharge of the condenser is of an oscillating character, and is known as *oscillating current*.

The time period of each oscillation, and thus the frequency, is determined by the amount of inductance and capacitance in the circuit. The gradual decrease in amplitude of each successive oscillation is determined principally by the amount of resistance in the circuit. This is called *damping*, and the number of oscillations is determined by their rate of damping. They will continue until they have died down to zero.

12-4. Capacity of Condenser. The unit of capacity of a condenser is the farad. One farad is that capacity which will produce a back pressure of one volt when the charge is one coulomb. For convenience in measuring capacity the farad is divided into units of one millionth of a farad, or microfarads. Capacity may be defined as that property of a condenser, or a circuit, which enables it to hold a charge of electricity in electrostatic form. The capacity for holding an electrostatic charge is given as

$$C = \frac{Q}{E} \tag{12-1}$$

Then

$$Q = CE \tag{12-2}$$

12-5. Calculation of Capacitance. Calculation of capacitance can only be approximate because of the fact that the entire circuit in which they are connected has some capacitance. The turns of wire in a coil, with insulation and air between, combine to produce a capacity effect that will modify the characteristics of the circuit. It is impossible to add pure coil inductance to a circuit because of this.

The capacitance changes with the frequency of the charging current. The higher the frequency of the charging current, the smaller the charge that can be stored with lowered attendant capacity effectiveness.

The charge which a condenser can hold also depends upon the voltage of the charging current. That is, a certain amount of current at a certain voltage will cause a condenser of a certain capacity to reach a state of back pressure equal to the charging voltage. Now if the voltage were raised, it might be possible to place a still greater charge into that particular condenser.

12-6. Dielectric Constant. The medium which separates the plates of a condenser is known as the dielectric. The ability of the dielectric to allow the electrostatic field to permeate through it affects the capacity of the condenser. This ability is given the name *dielectric constant.* It signifies that ratio of capacity of the condenser using a given substance for a dielectric to the capacity of the same condenser with air for a dielectric. See Table 12-1 for dielectric constants for other substances.

TABLE 12-1. DIELECTRIC CONSTANTS FOR SOME SUBSTANCES

Substances	*Dielectric constant*	*Substances*	*Dielectric constant*
Air	1.0	Beeswax	3.2
Glass	4.0–10.0	Silk	4.6
Mica	4–8	Celluloid	7–10
Hard rubber	2–4	Wood, maple, dry	3.0–4.5
Paraffin	2–3	Wood, oak, dry	3.0–6.0
Paper, dry	1.5–3	Vulcanized fiber	5–8
Paper (treated as used in cables)	2.5–4.0	Castor oil	4.7
Porcelain, unglazed	5–7	Transformer oil	2.5
Sulfur	3.0–4.2	Water, distilled	81.0
Marble	9–12	Cottonseed oil	3.1
Shellac	3.0–3.7		

When a condenser is composed of several parallel circular plates, the following formula may be used to calculate capacitance

$$C = 0.0885K \frac{(N - 1)S}{t} \qquad (12\text{-}3)$$

where N = number of plates
S = surface area of one plate, sq cm
K = dielectric constant
t = thickness of dielectric

Q12-2. What is the capacitance in microfarads of a fixed condenser consisting of 43 circular plates, when each plate is 20 cm in diameter, and they are separated for a distance of 1 mm with transformer oil used as the dielectric?

ANSWER.

$$C = 0.0885 \times 2.5 \times (42 \times 314.16)/0.1 = 2{,}919.331 \ \mu\mu\text{f}$$

This is the equivalent of 0.00291 μf.

A useful rule for determining the number of plates, or sheets, of the dielectric to use in constructing a condenser of a desired capacity is

$$\text{Number of sheets} = \frac{tC \times 10^6}{0.0885K \times \text{area one plate}} \qquad (12\text{-}4)$$

Q12-3. How many sheets of a dielectric are necessary to make a condenser having a capacity of 0.0055 μf, when the sheets are 15×30 cm in area and 0.29 cm thick? K is equal to 8.

ANSWER. $N = \dfrac{0.29 \times 0.0055 \times 10^6}{0.0885 \times 8 \times 15 \times 30}$ = **5 sheets of dielectric**

12-7. Manner of Connecting Condensers. As in other electric devices, such as batteries and resistance coils, the manner of connecting the condensers in the circuit affects their total capacity in the circuit. When condensers are *connected in parallel*, the total capacity of the group is equal to the sum of all the capacities.

$$C_p = C_1 + C_2 + C_3 + \cdots + C_n \qquad (12\text{-}5)$$

When *connected in series*, the total capacity is less than the capacity of one condenser.

$$C_s = \frac{1}{1/C_1 + 1/C_2 + 1/C_3 + \cdots + 1/C_n} \tag{12-6}$$

When connected in series, the individual condensers remain of the same capacity and each is still capable of taking as much of a charge as when charged singly. However, as connecting them in series reduces the pressure, the result is to also reduce the capacity. In certain cases, this is done to protect condensers from breakdown, where the voltage is higher than one condenser would be likely to stand. For example, if we take a single condenser with a capacitance of 0.002 μf and connect it as in Fig. 12-3*a*, it will have an effective capacitance of 0.002 μf, neglecting the capacitance of the remainder of the circuit. If we connect two condensers of equal capacitance as in Fig. 12-3*d*, the resulting capacitance will be 0.004 μf.

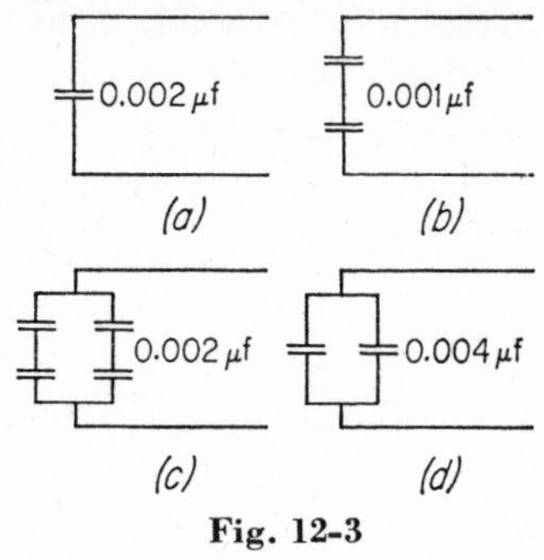

Fig. 12-3

However, if these condensers have insufficient dielectric strength to stand the voltage to which they will be subjected, we can still use them by sacrificing the capacity. In Fig. 12-3*b* it may be seen that the same two condensers in series with each other have a resulting capacity of only one-half that of one alone if connected as in Fig. 12-3*a*. Therefore, to obtain the capacity of the one condenser, but with protection given to them by series connections, we must connect four condensers, as in Fig. 12-3*c*.

Connecting them in parallel is mechanically the same as connecting all the plates on one side of the connection together, and all the plates on the other side together. Thus, we form one condenser of greater surface area. Connecting them in series has the opposite effect. This arrangement may be thought of as having greatly increased the dielectric thickness, thus reducing the total capacity and causing the group to discharge sooner under a given charge. This will also accelerate the frequency of the discharge current. *The voltage across capacitors in series varies inversely with the capacitance.* Condensers in any arrangement will permit alternating current to pass and prevent direct current from flowing.

Q12-4. Two condensers, one of 2 μf and one of 10 μf, are connected in parallel, and this combination is connected in series with a third condenser of 5 μf capacitance across a 120-volt d-c supply circuit. Find the charge of each condenser and the voltage across each. What is the total energy stored in the three condensers? See Fig. 12-4.

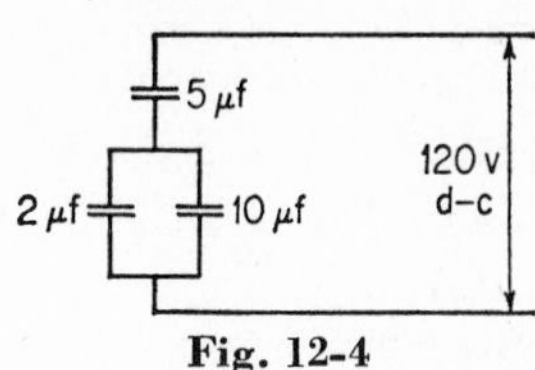

Fig. 12-4

ANSWER. For the condensers (capacitors) in parallel the effective value for this combination is $2 + 10 = 12$ μf. For the full combination including the 5 μf condenser the effective value is

$$\frac{12 \times 5}{12 + 5} = 3.52 \ \mu\text{f}$$

Voltage across 5 μf: $120 \times (3.52/5) =$ **84.5 volts**
Voltage across 2 μf and 10 μf: $120 \times (3.52/12) =$ **35.2 volts**
Charge in the 5 μf: $5 \times 10^{-6} \times 84.5 =$ **423×10^{-6} coulomb**
Charge in the 2 μf: $2 \times 10^{-6} \times 35.2 =$ **70.4×10^{-6} coulomb**
Charge in the 10 μf: $10 \times 10^{-6} \times 35.2 =$ **352×10^{-6} coulomb**

Energy stored is ½ charge × voltage
Thus, it follows

$$\tfrac{1}{2} \times 423 \times 10^{-6} \times 84.5 + \tfrac{1}{2} \times (70.4 + 352)10^{-6} \times 35.2$$
$$= 0.0179 + 0.00742 = \mathbf{0.02532\ joule}$$

Q12-5. Two condensers are connected in series across a 2,000-volt circuit. One has a capacitance of 0.1 μf, and the other has a capacitance of one-quarter as much. What is the voltage across the latter condenser? How much energy does it take or store, if the impressed voltage is unidirectional? See Fig. 12-5.

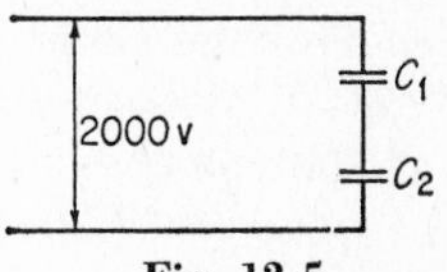

Fig. 12-5

ANSWER. For two condensers in series from Eq. (12-6)

$$C_s = \frac{C_1 C_2}{C_1 + C_2} = 1/10^6 \left(\frac{0.1 \times 0.025}{0.1 + 0.025}\right)$$

Also for condensers in series the following holds: $Q = Q_1 = Q_2$. Then

$$Q = CE = C \times 2{,}000 \qquad \text{coulombs}$$

Finally

$$E_2 = \frac{Q_2}{C_2} = \frac{Q}{C_2} = \frac{CE}{C_2} = \textbf{1,600 volts} \text{ across } C_2 \text{ condenser}$$

Energy stored in $C_2 = 0.5Q_2E_2$, joules. This same energy is also the same as or equal to QE_2. And by substitution in this equation the energy stored is found to be equal to **0.032 joule.**

Note: In this solution the following nomenclature was used:

C = circuit capacitance, farads
C_1 and C_2 = capacitance of each condenser, farads
E = impressed voltage, volts
E_1 and E_2 = potential difference across each condenser, volts
Q = charge on circuit, coulombs
Q_1 and Q_2 = charge on each condenser, coulombs

Q12-6. What is the combined capacity of four condensers of 10, 15, 25, and 30 μf, respectively? Assume these are connected in parallel.

ANSWER. $C_p = 10 + 15 + 25 + 30 =$ **80 µf**

Q12-7. What is the voltage of a condenser having a capacity of 0.016 μf when the charge is 0.000192 coulomb?

ANSWER. $E = \dfrac{Q}{C} = \dfrac{1.92 \times 10^{-4}}{1.6 \times 10^{-8}} =$ **12,000 volts**

Q12-8. A capacitor of 8 μf is to be charged by a voltage of 400 volts through a resistor of 100,000 ohms. How long will it take for the voltage across the capacitor from its initial zero value to 300 volts? What fraction of the final energy is stored in the condenser at 300 volts?

ANSWER. Voltage v at any time t sec is given by the equation

$$v = E(1 - e^{-(t/CR)})$$

where E = 400 volts
C = capacitance in farads equal to $8/10^6$
R = resistance, ohms, or 100,000

By substitution in above equation,

$$300 = 400(1 - e^{-(t/CR)}) = 400(1 - e^{-(t/0.8)})$$

Thus, $e^{t/0.8} = 4$, and

$$t = 0.8 \ln 4 = \frac{0.8 \log 4}{\log e} \quad \sec = \mathbf{1.10\ sec}$$

For final energy storage: $0.5CE^2 = 0.5\ (8/10^6)400^2 =$ **0.65 joule.** For the 300-volt stage similarly the energy stored is **0.36 joule.** Fraction of final energy stored is $0.36/0.65 =$ **0.56.**

Q12-9. An inductance coil with pure resistance of 10 ohms and inductance of 0.4 henry is connected to a 100-volt d-c source. Calculate: (*a*) Rate of change of current at the instant of closing the switch; (*b*) final steady value of current; (*c*) time constant of the circuit; (*d*) time taken for the current to rise to one-half its final value; (*e*) energy finally stored in the magnetic field in joules.

ANSWER. The formula for the flow of current after switching is

$$i = \frac{E}{R}(1 - e^{-Rt/L})$$

(*a*) Rate of change of current initially $= E/L = 100/0.4$, or **250 amp per sec.**

(*b*) Final current $I = E/R = 100/10 =$ **10 amp.**

(*c*) Time constant $= L/R = 0.4/10 =$ **0.04 sec.**

(*d*) $i = I(1 - e^{-Rt/L})$, $5 = 10(1 - e^{-25t})$, and $t =$ **0.0278 sec.**

(*e*) The energy stored is $0.5LI^2$ joules.

$$0.5 \times 0.4 \times 10^2 = \mathbf{20\ joules}$$

Q12-10. Refer to Fig. 12-6 and (*a*) find the polarity of E_{c1} and E_{c2} at $t = \infty$, (*b*) compute the total energy dissipated by resistor and that supplied by the battery.

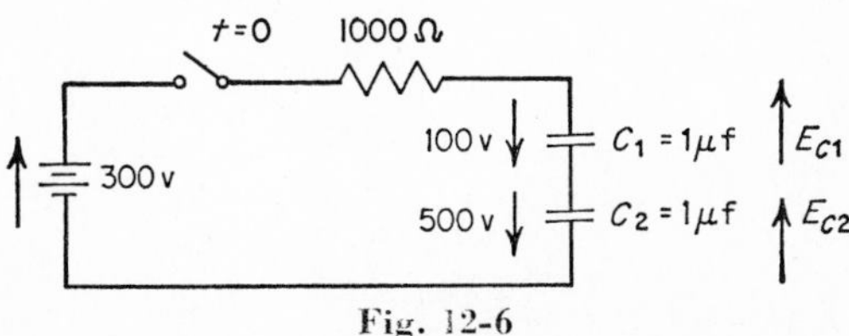

Fig. 12-6

ANSWER.

(a) $$i = i_o\epsilon^{-t/T} = 900/1000\ \epsilon^{-2,000t}$$

Then $$E_{c1} = -100 + 10^6 \int_0^t 900/1,000\ \epsilon^{-2,000t} = 350 - 450\ \epsilon^{-2,000t}$$

$$E_{c2} = -500 + 10^6 \int_0^t 900/1,000\ \epsilon^{-2,000t} = -50 - 450\ \epsilon^{-2,000t}$$

Now refer to Fig. 12-7.

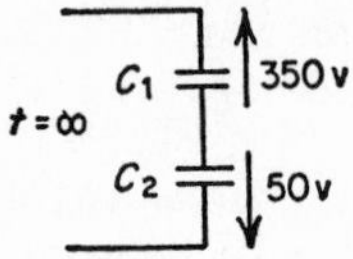

Fig. 12-7

(b) $$W_b = \int_0^t P\,dt = \int_0^t 300(900/1,000)\ \epsilon^{-2,000t}\,dt$$

$$= 0.135(1 - \epsilon^{-2,000t}) = \mathbf{0.135\ joule}$$

$$W_r = \int_0^t P\,dt = \int_0^t Ri^2\,dt = \int_0^t 1,000(900/1,000)^2\ \epsilon^{-4,000t}$$

$$= \mathbf{0.2025\ joule}$$

Q12-11. What capacitance in microfarads will give the same reactance as a choke coil of 0.150 henry, both at 100 cps?

ANSWER. $X_c = 1/(2\pi fc) = X_L = 2\pi fL$, from which and by rearrangement,

$$C = 1/(4\eta^2 f^2 L) = 1/(4\eta^2 \times 10^4 \times 0.150) = \mathbf{16.9\ \mu f}$$

Supplement **A**

DIRECT CURRENT

Questions and Answers

A-1. A sinusoidal current having a peak value of 1 amp is superimposed upon a direct current of 2 amp. The combined current passes through a d-c ammeter and a thermal ammeter. What values would each of these meters indicate?

ANSWER. Many types of a-c ammeters can be used to measure direct current where extreme accuracy is not required. It is, therefore, assumed that the d-c ammeter to which the problem refers is

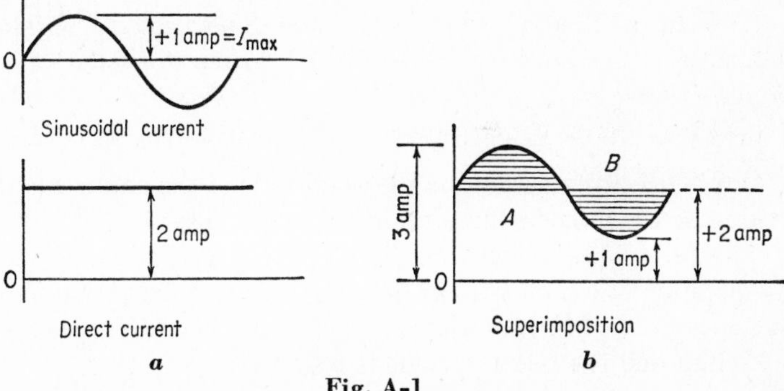

Fig. A-1

a moving-magnet type, which measures average values of current and is suitable for direct current only. Figures A-1*a* and *b* show the currents before and after the superimposition. If, as just stated, the d-c ammeter assumed measures only average current values, its reading will be 2 amp because the shaded additive area *A* offsets equal, shaded area *B*.

A thermal ammeter, on the other hand, measures the heating effect of the rms value of the current. When we speak of alternating current we always mean the rms or effective value. For a complex or superimposed wave, the resulting rms current is the square root of the summation of the squares of the effective values of all components.

Effective value of the direct current is **2 amp**

Effective value of the alternating current is 0.707 × maximum, or **0.707 × 1**

Effective value of the superimposed current is

$$\sqrt{0.707^2 + 2^2} = \mathbf{2.12\ amp}$$

A-2. Name four methods by which an electrical potential may be generated. Name additional methods if desired.

ANSWER.

1. Chemical (battery)
2. Mechanical (generator)
3. Heat (thermocouple)
4. Friction (static electricity)
5. Pressure and expansion (piezoelectric)
6. Light (photoelectric cell)

A-3. What method of connection should be used to obtain the maximum no-load output voltage from a group of similar cells in a storage battery?

ANSWER. Series connection.

A-4. What factors determine the amplitude of the emf induced in a conductor which is cutting lines of magnetic force?

ANSWER. Speed of cutting, flux density, relative angle between the direction of the motion of the conductor and the direction of the magnetic flux. Maximum emf is generated when the velocity is a maximum and the relative angle is 90°.

A-5. The charge is stored in what part of a condenser?

ANSWER. It is stored on the surface of the dielectric in the form of electrostatic lines of force.

A-6. How much heat is produced by a current of 20 amp flowing for one-half hour in a circuit having a resistance of 6 ohms?

ANSWER.

$$0.24I^2Rt\ (\text{sec}) = 0.24 \times 20^2 \times 6 \times (30 \times 60) = \mathbf{1{,}036{,}800\ calories}$$

A-7. How much greater is the heating effect of a current of 32 amp than that of 8 amp flowing through a wire of 10 ohms resistance?

ANSWER. I^2R is the heating effect. Then for the 32 amp flow

$$32 \times 32 \times 10 = 10{,}240 \text{ watts}$$

For the 8 amp flow circuit

$$8 \times 8 \times 10 = 640 \text{ watts}$$

Then $10{,}240/640 =$ **16 times as great**

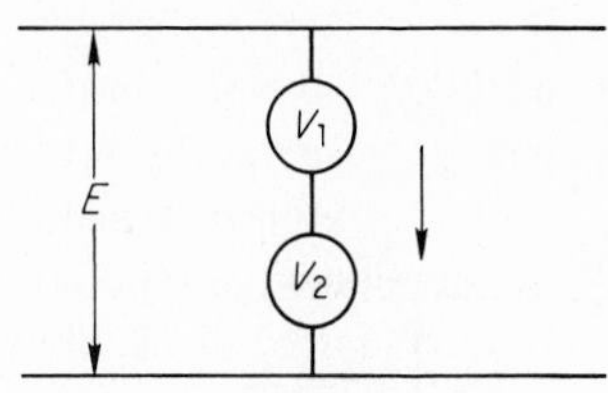

Fig. A-2

A-8. Voltmeters V_1 and V_2 are connected across a d-c line. V_1 reads 80 volts at full scale and has a per volt resistance of 200 ohms. V_2 has a total resistance of 32,000 ohms. What is the line voltage?

ANSWER. Refer to Fig. A-2. $E = IR_t = I(80 \times 200 + 32{,}000)$. Then $E = 48{,}000I$. Also $80 = I \times 16{,}000I = 1/200$. Finally, $E = 48{,}000/200 =$ **240 volts.**

Supplement B

BATTERIES

Questions and Answers

B-1. Three dry cells, each of 1.5 volts emf and 0.1-ohm internal resistance, are connected in series to send current through two external resistances of 0.5 ohm each, also connected in series. A resistance of 1 ohm is connected from a point between the two external resistances to a point between the first and second dry cells. What is the current through each resistance?

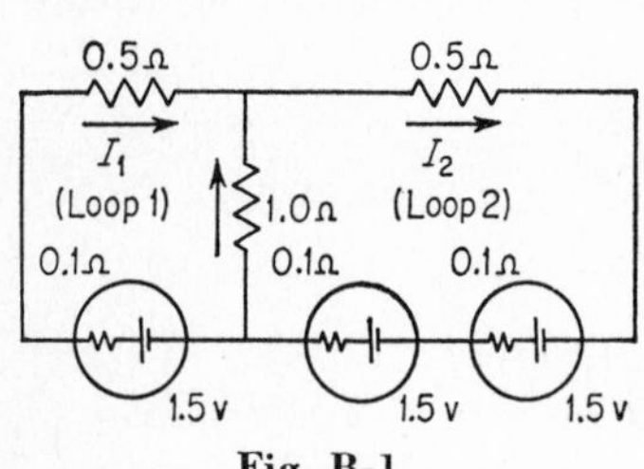

Fig. B-1

ANSWER. This problem may be solved by the application of Kirchhoff's laws. Now please refer to Fig. *B*-1.

Loop 1:

$$-1.5 + 0.1I_1 + 0.5I_1 + 1.0I_1 = 1.0I_2$$

or

$$1.6I_1 - 1.0I_2 = 1.5 \quad (1)$$

Loop 2:

$$1.0I_2 - 1.0I_1 + 0.5I_2 + 0.1I_2 - 1.5 + 0.1I_2 - 1.5 = 0$$

$$-1.0I_1 + 1.7I_2 = 3.0 \quad (2)$$

Now in order to cancel out the I_1's and solve for I_2, multiply Eq. (2) by 1.6 and add Eq. (1) to the result.

$$1.6(-1.0I_1 + 1.7I_2 = 3.0) = -1.6I_1 + 2.72I_2 = 4.8$$

$$1.6I_1 - 1.0I_2 = 1.5 \quad (1)$$

$$0 \quad + 1.72I_2 = 6.3$$

Thus, $I_2 = 6.3/1.72 = 3.67$

Now from Eq. (2),

$$I_1 = 1.7I_2 - 3 = (1.7 \times 3.67) - 3 = 6.23 - 3.0 = 3.23$$

Check:
$$I_1 = \frac{1.5 + I_2}{1.6} = \frac{1.5 + 3.67}{1.6} = \frac{5.17}{1.6} = 3.23$$

Therefore, **3.23 amp** through 0.5 ohm in loop 1
3.67 amp through 0.5 ohm in loop 2
0.44 amp through 1-ohm resistance, since

$$3.23 + 0.44 = 3.67$$

B-2. Two dry cells, each of 1.5 volts emf and internal resistance of 0.1 ohm, are connected in series with two resistances of 0.2 and 0.3 ohm each, also in series. The junction between the two cells

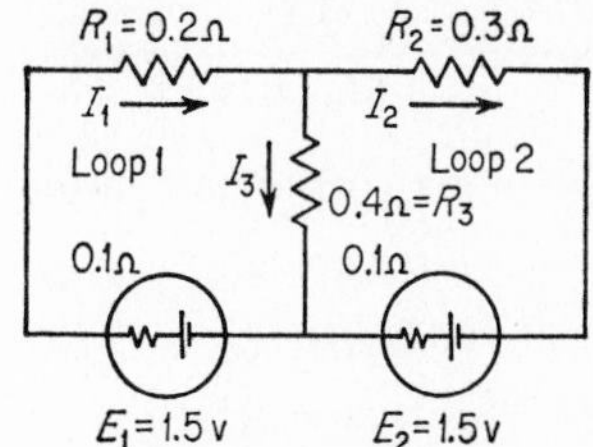

Fig. B-2

is connected to the junction between the two resistances by a third resistance of 0.4 ohm. What is the current in each resistance?

ANSWER. See Fig. B-2 and equate the voltages around each loop to zero.

Loop 1:

$$-1.5 + (I_1 \times 0.1) + (I_1 \times 0.2) + (I_3 \times 0.4) = 0 \qquad (1)$$

Loop 2:

$$-1.5 + (I_2 \times 0.1) - (I_3 \times 0.4) + (I_2 \times 0.3) = 0 \qquad (2)$$

Since $I_3 = I_1 - I_2$, Eqs. (1) and (2) above may be solved for I_1 and I_2. Therefore, by simple clearance,

$I_1 =$ **4.5 amp**
$I_2 =$ **4.126 amp**
$I_3 =$ **0.374 amp**

Note that our assumption of the directional arrow for I_3 was correct. If the current value came out a minus, we would have to merely reverse the arrowhead.

B-3. Two storage batteries, one measuring 50 volts and the other 51 volts on an open circuit, are to be connected in parallel. The respective internal resistances of these batteries are 0.23 ohm and 0.28 ohm, and the load resistance is 0.8 ohm. What will be the load current and what current will each battery supply? See Fig. B-3.

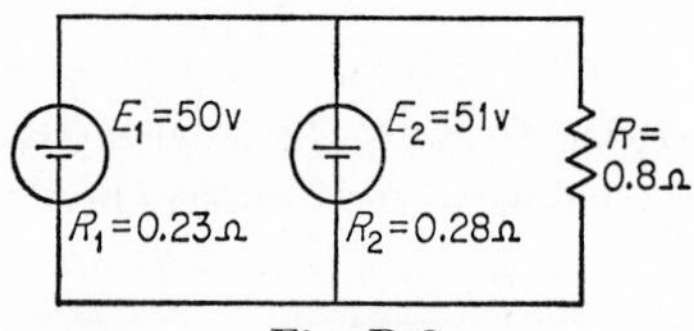

Fig. B-3

ANSWER. From previous problems under solution and the application of Kirchhoff's laws we write

$$E_1 - I_1R_1 = E_2 - I_2R_2 \tag{1}$$

$$I_1 + I_2 = I \tag{2}$$

From Eq. (1) we get

$$50 - 0.23I_1 = 51 - 0.28I_2 \tag{3}$$

$$I_2 = 3.57 + 0.82I_1 \tag{4}$$

Substituting Eq. (4) for Eq. (3) and solving, we have

$$I_1 = \mathbf{27.9\ amp}$$

$$I_2 = 3.57 + 0.82I_1 = 3.57 + (0.82 \times 27.9) = \mathbf{26.5\ amp}$$

$$I = I_1 + I_2 = 27.9 + 26.5 = \mathbf{54.4\ amp}$$

Supplement **C**

POWER, ENERGY, AND THERMAL UNITS

Questions and Answers

C-1. The use of electricity for melting snow in a driveway 10 by 50 ft long is being considered. At 2 cents per kwhr, what would be the cost of melting 6 in. of snow? Assume the following: the weight of the snow = 10 lb per cu ft, the temperature = 32°F, and the efficiency of the operation = 50 per cent.

ANSWER. The heat transferred during the melting process is equal to the weight of the snow × the heat of melting per pound of snow (144 Btu per lb). Thus

$$10 \times 50 \times 0.5 \times 10 \times 144 = 360{,}000 \text{ Btu}$$

For a 50 per cent operation this becomes 720,000 Btu, and the total cost

$$(720{,}000/3413) \times 0.02 = \$4.22 \qquad \text{Note 3413 Btu per kwhr}$$

C-2. A current of 15 amp flows through a resistance of 25 ohms within an electric boiler. If there is 10 gal water content and the thermal losses are neglected, to what temperature will the water rise in 20 min after closing the switch? The room temperature may be taken as 70°F.

ANSWER. Heat added is $0.057I^2Rt$, Btu. This is also equal to

$$10 \text{ gal} \times 8.33 \text{ lb per gal} \times 1 \text{ sp ht} \times \text{temp rise}$$

By rearrangement, the temperature rise may be made equal to

$$0.057 \times 15^2 \times 25 \times 20 \times 1/83.3 = 77°\text{F}$$

The final temperature is found to be **147°F.**

C-3. If a relay is designed to operate properly from a 6-volt d-c source and if the resistance of its windings is 120 ohms, what value of resistance should be connected in series with the winding if the relay is to be used with a 120-volt d-c source?

ANSWER. The resistor must drop the source voltage 120 − 6, or 114 volts.

Current through relay: $I = E/R = 6/120 = 0.05$ amp

New resistor: $R = E/I = 114/0.05 =$ **2,280 ohms**

C-4. If two voltmeters are connected in series to a power line and the first voltmeter, which has a resistance of 16,000 ohms, reads 60 volts, what is the voltage between the wires of the power lines provided the second voltmeter's resistance is 20,000 ohms? See Fig. C-1.

ANSWER. Second voltmeter reads 20,000/16,000 × first voltmeter

$$20{,}000/16{,}000 \times 60 = 1.25 \times 60 = 75 \text{ volts}$$

Total drop across the power lines: 60 + 75 = **135 volts.**

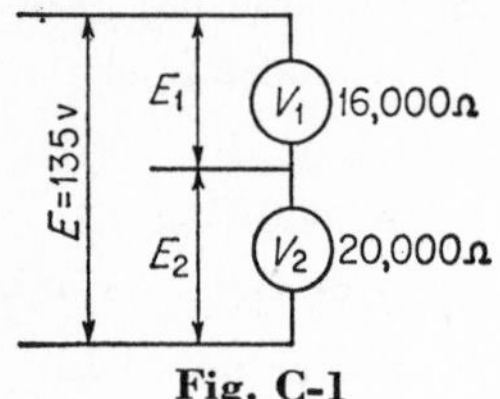

Fig. C-1

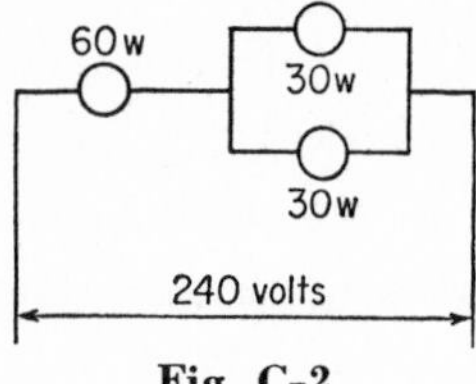

Fig. C-2

C-5. Could two 30-watt and one 60-watt 120-volt lamps be connected across a 240-volt circuit in such a way as to give normal illumination? If this can be done, show the circuit and explain (see Fig. C-2).

ANSWER. This can be done as shown by the diagram. The two 30-watt 120-volt lamps with a resistance of a value equal to the resistance of the 60-watt 120-volt lamp would give normal lighting. This is so because resistance equals volts²/watts.

For the 60-watt lamp: $R = 120^2/60 = 240$ ohms

For the two 30-watt lamps: $R = 120^2/(2 \times 30) = 240$ ohms

C-6. A 60,000-ohm voltmeter is used to measure the insulation resistance of a motor. A 40,000-ohm resistor is connected across the voltmeter terminals. This combination is then connected in series with the insulation resistance and a 550-volt battery. The voltmeter deflection is 12 volts. Determine the value of the insulation resistance in ohms.

ANSWER. Refer to Fig. C-3. We must first determine the equivalent parallel combination resistance of both the voltmeter and the 40,000-ohm resistor.

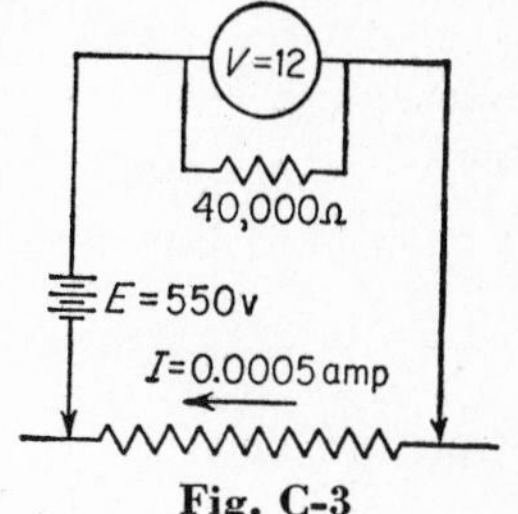

Fig. C-3

$$\frac{1}{R_p} = \frac{1}{40{,}000} + \frac{1}{60{,}000} = 0.0000417$$

$$R_p = \frac{1}{0.0000417} = 24{,}000 \text{ ohms}$$

Then

$$R_x = 24{,}000 \times \frac{550 - 12}{12} = 24{,}000 \times \frac{538}{12} = \mathbf{1{,}072{,}000\ ohms}$$

Note: It should be noted that the voltmeter is of sufficient range to take the full-scale deflection when connected across the source of potential. This is to protect the instrument against damage should the parallel resistor be left out of the circuit.

C-7. (*a*) Two storage cells connected in parallel supply jointly a current of 30 amp to an external circuit. One cell has an emf of 2.1 volts and the other an emf of 2.0 volts; each has an internal resistance of 0.06 ohm. Find the current flowing through each cell.

(*b*) A 20,000-ohm voltmeter is used in measuring the insulation resistance of a motor. When connected in series with the insulation across a 230-volt line, the instrument reads 1 volt. Find the value of the insulation resistance of the motor (see Fig. C-4*b*).

ANSWER.

(*a*) Refer to Fig. C-4*a*. Let I equal the current delivered to the external circuit; I_1 and I_2 equal the current flowing through cells 1 and 2, respectively; E_1 and E_2 equal emf of cell 1 and cell 2, respectively; R_1 and R_2 equal to the resistance in cell 1 and cell 2, respectively. Then by Kirchhoff's laws

$$E_1 - I_1R_1 = E_2 - I_2R_2$$

Also $$I_1 = I - I_2$$

Therefore $$2.1 - [(30 - I_2) \times 0.06] = 2.0 - (I_2 \times 0.06)$$

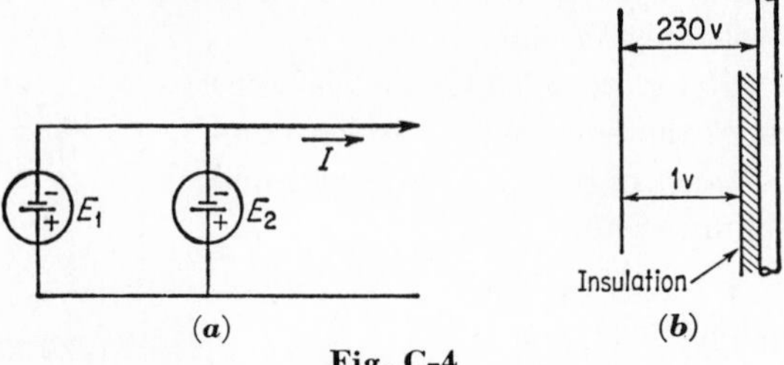

Fig. C-4

from which

$$I_2 = \mathbf{14.15\ amp}$$
$$I_1 = I - I_2 = 30 - 14.15 = \mathbf{15.85\ amp}$$

(b) Let I equal the current flowing through the series circuit; R_v be the resistance of the voltmeter; R_s be the resistance of the insulation. Then

$$IR_v + IR_s = 230 \text{ volts}$$

Since $IR_v = 1$ volt, $IR_c = 230 - 1 = 229$ volts.

Also
$$I = \frac{1}{R_v}$$

and
$$R_s = \frac{229}{I} = 229 \times 20{,}000 = \mathbf{4{,}580{,}000\ ohms}$$

C-8. Two 1,000-ohm per volt d-c voltmeters are used in series with a 50,000-ohm resistor to measure an unknown d-c voltage. The first voltmeter reads 125 volts when connected for its 150-volt scale. The second meter is connected for its 300-volt scale. What is the unknown voltage?

ANSWER. The effective resistance for the first voltmeter is $150 \times 1{,}000$. The current flowing through the voltmeter resistance is the same as that flowing through the remainder of the system (series circuit). This is

$$I = 125/150{,}000 = 0.883. \times 10^{-3} \text{ amp}$$

The reading on the second voltmeter with the same current flow is

$$E_2 = IR = 0.833 \times 10^{-3} \times 300 \times 10^3 = \mathbf{250\ volts}$$

The voltage drop across the 50,000-ohm resistor similarly is determined

$$E_r = 0.833 \times 10^{-3} \times 50{,}000 = 41.65 \text{ volts}$$

Supplement **D**

DIRECT-CURRENT GENERATORS

Questions and Answers

D-1. A 15-kw 110-volt shunt generator has a speed of 900 rpm at full load. The field resistance is 35 ohms and the armature resistance is 0.06 ohm. If this machine is run as a 115-volt motor, what must be its speed if the same amount of current is to flow through the armature?

ANSWER. Refer to Fig. 11-18.

As a generator:

Full-load current output $= \dfrac{15{,}000}{110} = 136.3$ amp $= I$

Field current $I_f = \dfrac{E}{R_f} = \dfrac{110}{35} = 3.14$ amp

Armature current $I_a = I + I_f = 136.3 + 3.14 = 139.44$ amp

Armature-resistance drop $E_a = I_aR_a = 139.44 \times 0.06 = 8.36$ volts

Line voltage $= 110$ volts

Voltage generated at 900 rpm $= 110 + 8.36 = 118.36$ volts

As a motor: Carrying the same armature current the armature-resistance drop is still 8.36 volts; but, since the terminal voltage is 115 volts, the armature emf is $115 - 8.36 = 106.64$ volts. This is the counter emf. If the magnetic field flux has not changed, the emf's are proportional to the speeds.

$$\frac{\text{Emf generator}}{\text{Emf motor}} = \frac{\text{speed of generator}}{\text{speed of motor}} = \frac{\text{emf}_1}{\text{emf}_2} = \frac{S_1}{S_2}$$

As a motor the rpm is S_2, or

$$S_2 = S_1 \times \frac{\text{emf}_2}{\text{emf}_1} = 900 \times \frac{106.64}{118.36} = \textbf{810 rpm}$$

Actually, the field current has increased when operating as a motor, and the magnetic flux will have increased, giving a still lower speed. However, shunt motors are operated above the "knee" of the saturation curve and the magnetic flux does not increase in proportion to the field current. This increase in the field current increases the flux density very little, and the armature speed will be reduced. Thus, the field magnetism is saturated and a slight change in the field current will have no effect on the generated voltages.

D-2. A shunt generator equipped with a 30-in. diameter pulley is driven by a belt at 450 rpm. When full load is taken from the machine, the armature current and the terminal voltage are 350 amp and 600 volts, respectively. The armature resistance is 0.04 ohm. The field resistance is 50 ohms. Assume the stray power losses to be 3 per cent of the output power and the brush drop to be 2 volts. Estimate the difference in tensions on the two sides of the driving belt.

ANSWER. Again refer to Fig. 11-18.

Losses:

Output $= 338 \times 600 = 202{,}800$ watts. Output $= I_L E_L$.
Armature $= I_a^2 R_a = 350^2 \times 0.04 = 4{,}900$ watts
Brush loss $= I_a E_b = 350 \times 2 = 700$ watts
Field $= I_f^2 R_f = 12^2 \times 50 = 7{,}200$ watts. Also $I_L = I_a - I_f$.

$$\text{And } I_f = \frac{E_L}{R_f} = \frac{600}{50} = 12 \text{ amp}$$

Stray power $= 0.03 \times 202{,}800 = 6{,}100$ watts
Electric input $=$ output $+$ losses $= 202{,}800 + [4{,}900 + 700 + 7{,}200 + 6{,}100]$
Electric input $= 221{,}700$ watts
Therefore, $I_L = 350 - 12 = 338$ amp

$$\text{Mechanical input} = \frac{221{,}700}{746} = 296 \text{ hp}$$

$$\text{Horsepower delivered by belt} = \frac{\text{belt velocity (fpm)} \times (T_2 - T_1)}{33{,}000}$$

$$\text{Difference in belt tensions} = T_2 - T_1 = \frac{33{,}000 \times \text{hp}}{V}$$

$$T_2 - T_1 = \frac{33{,}000 \times 296}{\pi DN} = \mathbf{2{,}760\ lb}$$

where D = wheel diameter $^{30}/_{12}$, ft
N = wheel rpm of 450

D-3. The field winding of a certain shunt generator has a resistance of 80 ohms and its field rheostat has a resistance of 15 ohms. At 1,000 rpm, with its separately excited field taking 1 amp, the open-circuit emf generated is 50 volts. What is the minimum speed at which this generator will operate satisfactorily as a shunt machine? Neglect the effect of magnetic saturation.

ANSWER. See Fig. D-1. To operate a machine as a straight shunt generator, the critical resistance curve must coincide or be less than the saturation curve of the generator. The relative position of the saturation curve will shift with speed.

It has been previously found that the voltage generated is proportional to flux and speed directly. Hence

$$\frac{\text{Min speed}}{1{,}000} = I_f \frac{R_f}{50} = \frac{1(80 + 15)}{50}$$

And by rearrangement

$$\text{Min speed} = \frac{95 \times 1{,}000}{50} = \mathbf{1{,}900\ rpm}$$

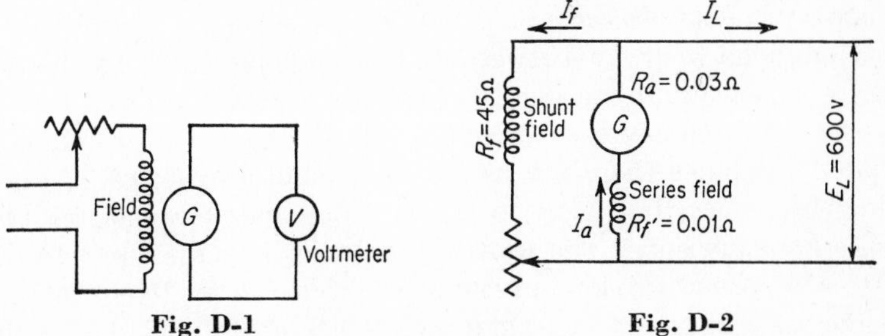

Fig. D-1 **Fig. D-2**

D-4. A 420-kw 600-volt 580-rpm long-shunt connected compound generator has the following circuit characteristics: armature resistance 0.03 ohm, series field resistance 0.01 ohm, shunt field resistance 45 ohms. The stray-power loss of the machine is 12 kw. Calculate the necessary driving power and torque at rated load.

ANSWER. See also question D-2. Now refer to Fig. D-2.

Losses: Neglect the brush loss.

$$\begin{aligned}
\text{Armature} &= I_a{}^2R_a = 713.3^2 \times 0.03 &&= 15{,}200 \text{ watts}\\
\text{Series field} &= I_f{}^2R_f = 713.3^2 \times 0.01 &&= 5{,}100\\
\text{Shunt field} &= I_f{}^2R_f = 13.3^2 \times 45 &&= 8{,}000\\
\text{Stray power} & &&= 12{,}000\\
\text{Output} & &&= 420{,}000\\
\text{Electric input (summation of above)} & && 460{,}300 \text{ watts}
\end{aligned}$$

Note: Line current $= I_L = \dfrac{420{,}000}{600} = 700$ amp

$$\text{Field current} = I_f = \frac{E_1}{R_f} = \frac{600}{45} = 13.3 \text{ amp}$$

$$\text{Armature current} = I_a = I_L + I_f = 713.3 \text{ amp}$$

Mechanical input: $\dfrac{460{,}300}{746} = \mathbf{617\ hp}$

$$\text{Torque} = \frac{33{,}000 \times \text{hp}}{2\pi N} = \frac{33{,}000 \times 617}{2\pi \times 580} = \mathbf{5{,}580\ lb\text{-}ft}$$

D-5. State the purpose, position, number, construction, and connection of interpole and compensating windings in d-c machines. Explain what compounding action results from the use of each of these types of windings and give reasons for your answers.

ANSWER. *Interpole windings.* These consist of coils placed at the neutral point midway between the main poles (see Fig. 10-13). These produce an opposing flux in the coils short-circuited by the brushes and thereby give sparkless commutation. With multiple-wound machines there are as many commutating poles as there are main poles, but with four-pole series-wound armatures two poles are sufficient. The width of the poles is made equal to the span of an armature slot and two teeth, while the length equals that of the main poles. The air gap is from 1.5 to 2.0 times that of the main poles. The interpole windings have about 40 per cent more ampere turns than the armature reaction and are made up of double-cotton-covered or enameled wires. These poles are connected in series with the armature, the brushes, and the series field.

Compensating windings. These are employed to neutralize the emf of the armature. They are distributed uniformly in slots in the faces of the main pole pieces with as many ampere turns as the armature reaction per pole. They are then connected in series with the armature so that the full-load current goes through them in a direction opposite to the armature current. These windings are ordinarily made of copper bars with from four to eight slots per pole equally spaced. Such windings are used on high-voltage high-speed high-capacity machines where the commutating poles are inadequately effective. With compensating windings and commutating poles a speed control of 6:1 can be obtained in shunt motors with sparkless commutation.

The armature reaction influences the commutation and regulation of all d-c machines. When current flows in the armature it twists or distorts the field flux, and the armature conductors under load constitute a source of magnetomotive force which acts along a plane which is at a leading angle with the direction of rotation. The armature reaction causes sparking when the brushes are left in the neutral position because the coil underneath each brush is cutting flux and generating a voltage so that the brush short-circuits.

It then appears that the action produced by the commutating poles consists in neutralizing the cross-emf under the faces of the commutating poles while forcing equal flux into the armature core to obtain sparkless commutation. They do not correct the field distortion under the main poles themselves. The action of the compensating windings, on the other hand, consists in neutralizing the magnetizing effect of the armature. It follows that they will alter both the distribution and the magnitude of the field winding flux. By correcting the twisting effect of the armature reaction they will have the greatest effect when the machine is running fully loaded or overloaded.

D-6. Explain how you would test for and localize the following faults in a simple lap-wound d-c armature: (*a*) open circuit, (*b*) ground, and (*c*) short circuit.

ANSWER. Please refer to Fig. D-3*a* adjacent.

(*a*) *Open circuit.* All the brushes but one pair are removed from the commutator. To this pair a dry cell is connected as shown. With a low-reading voltmeter the *IR* drop across the adjacent commutator segments is now read. If there is no open circuit and all the paths have the same number of coils, the drop for each path would be the same or greater depending on the number of coils in a path.

An open circuit is present when there is zero *IR* drop across all the coils of the same path except for the open-circuited coil where the voltmeter will read the voltage of the dry cell.

(*b*) *Ground.* To test for a ground, one side of a 110-volt line is connected through a voltmeter to the armature shaft while the other side of the line is connected to a commutator bar, as shown

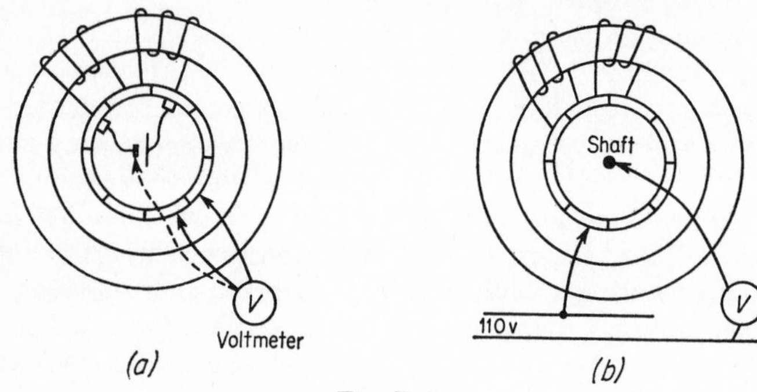

Fig. D-3

in Fig. D-3*b*. A ground in the winding will be present if the voltmeter has a considerable reading. If the winding is well insulated, the test will show a voltmeter reading of less than 1 volt.

To locate the ground, all brushes are removed from the commutator except one pair across which a dry cell is connected. The voltmeter is now connected between the shaft and each succeeding commutator segment. If a ground is present, the voltmeter will give varying results in readings, being zero at the segment connected to the grounded coil. It will be found that each path will indicate a grounded coil. Furthermore, on rotating the armature, a new grounded coil will be found for each path, called "phantom ground," except for the path with the real ground where the coil will coincide with the first test (see Fig. D-3*a*, dotted lines).

(*c*) *Short circuit.* A short-circuited coil is tested and located in exactly the same manner as an open-circuited coil, except that the *IR* drop read by the voltmeter will be zero for the short-circuited coil, or in any event much less than for the other coils.

D-7. A voltammeter has a moving element of 3-ohm resistance, which gives full deflection when traversed by a current of 25 ma. The instrument has two voltage ranges: 15 and 150 volts, and two current ranges: 1 and 10 amp. Show the internal connections of the instrument and calculate the resistances of the shunt and multiplier coils.

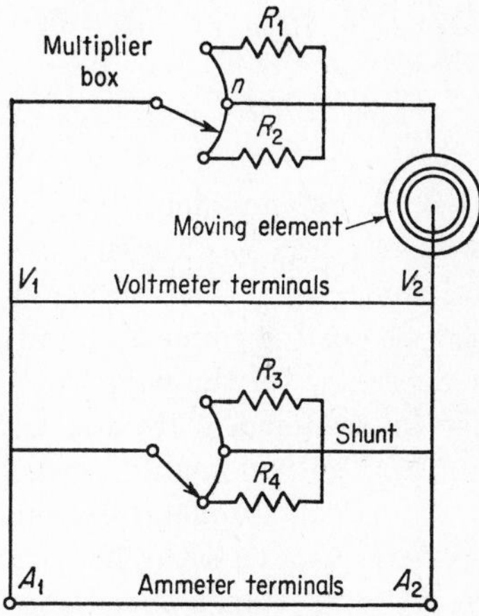

Fig. D-4

ANSWER. Refer to Fig. D-4.

Let R_1 = multiplier coil resistance when the meter measures up to 15 volts.

R_2 = multiplier coil resistance when the meter measures up to 150 volts.

R_3 = shunt resistance when the meter measures up to 1 amp.

R_4 = shunt resistance when the meter measures up to 10 amp.

E_1, E_2, E_3, and E_4 are voltage drops across respective resistances.

I_1, I_2, I_3, and I_4 are currents through respective resistances.

R, E, and I are resistance, voltage, and current respectively for the moving element.

Then, when measuring voltages with the meter, the source is connected to terminals E_1 and E_2. Thus

$$E = IR = 3 \times 0.025 = 0.075 \text{ volts}$$
$$E_1 = 15 - E = 15 - 0.075 = 14.925 \text{ volts}$$
$$I_1 = I \qquad \text{also} \qquad R_1 = \frac{E_1}{I_1} = \frac{14.925}{0.025} = 597 \text{ ohms}$$

It also follows that

$$E_2 = 150 - E = 150 - 0.075 = 149.925 \text{ volts}$$
$$I_2 = I \qquad \text{also} \qquad R_2 = \frac{E_2}{I_2} = \frac{149.925}{0.025} = 5{,}990 \text{ ohms}$$

When using the meter to measure currents, tap n in the multiplier box is connected to the circuit and the source is hooked up to terminals A_1 and A_2. Then

$$I_3 = 1 - 0.025 = 0.975 \text{ amp}$$

$$I_3R_3 = IR \qquad \text{and} \qquad R_3 = \frac{IR}{I_3} = \frac{0.075}{0.975} = \mathbf{0.0769\ ohm}$$

Also $I_4 = 10 - I = 10 - 0.025 = 9.975$ amp

$$I_4R_4 = IR \qquad \text{and} \qquad R_4 = \frac{IR}{I_4} = \frac{0.075}{9.975} = \mathbf{0.00752\ ohm}$$

D-8. A galvanometer coil produces a full-scale deflection with 0.10 amp and has a resistance of 5 ohms. It is desired to use this galvanometer as an ammeter having a range of 15 amp. Show how the coil is connected and determine the resistance of whatever is necessary for the purpose. Similarly, indicate how the same coil can be used as a voltmeter having a range of 150 volts.

Let R = galvanometer resistance
I = galvanometer-current full-scale deflection
R_1 = shunt resistance
R_2 = resistance of series resistor
I_1 = shunt current

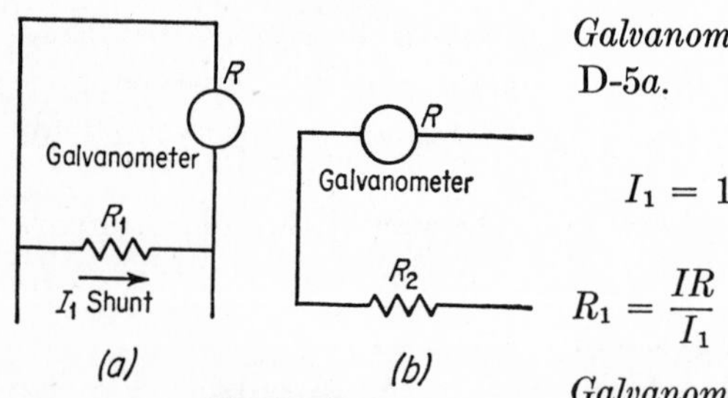

Fig. D-5

Galvanometer as ammeter: See Fig. D-5*a*.

$$I_1 + I = 15 \text{ amp}$$
$$I_1 = 15 - 0.10 = 14.9 \text{ amp}$$
$$I_1R_1 = IR$$
$$R_1 = \frac{IR}{I_1} = \frac{0.10 \times 5}{14.9} \ \mathbf{0.0335\ ohm}$$

Galvanometer as voltmeter: See Fig. D-5*b*. The galvanometer will have a series resistor instead of a shunt resistor.

$$I(R_2 + R) = 150 \text{ volts}$$

$$R_2 = \frac{150}{I} - R = \frac{150}{0.10} - 5 = \mathbf{1{,}495\ ohms}$$

D-9. Three points, A, B, and C, are connected by means of a "ring" distribution network. The total resistance of the system from A to B is 0.2 ohm, the total resistance from B to C is 0.4 ohm, and the total resistance from A to C is 0.6 ohm. If power is supplied by A to B and C, what must be the ratio of the currents drawn at B and at C so that the voltages at these two points are equal?

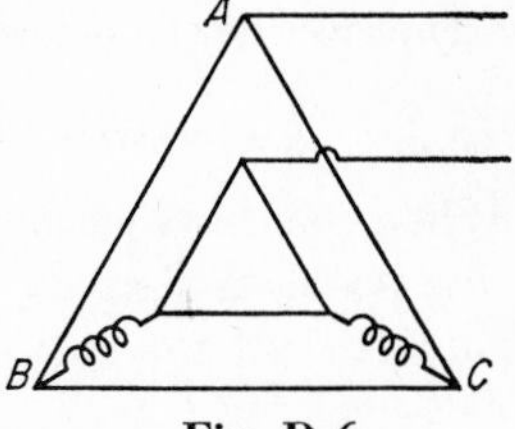

Fig. D-6

ANSWER. Refer to Fig. D-6 and let R_1 and R_2 be the resistances in the lines AB and AC, respectively (ohms).

I_1 and I_2 are the currents in the lines AB and AC, respectively (amperes).

$$I_1R_1 = I_2R_2$$

And the ratio of currents is equal to the following for the voltage to be equal at B and C. The voltage drops in the lines AB and AC must be equal.

$$\frac{I_1}{I_2} = \frac{R_2}{R_1} = \frac{0.6}{0.2} = \mathbf{3}$$

D-10. A 250-volt 10-kw d-c generator is separately excited. It has an effective armature circuit resistance of 0.5 ohm and an inductance of 0.1 henry when it is supplying rated current. Suddenly the terminals beyond its protective circuit breaker are short-circuited with a short-circuit resistance of 0.2 ohm. The breaker operates 0.02 sec after the fault occurs. Neglecting saturation of the magnetic circuit, determine the maximum current to which the generator is subjected.

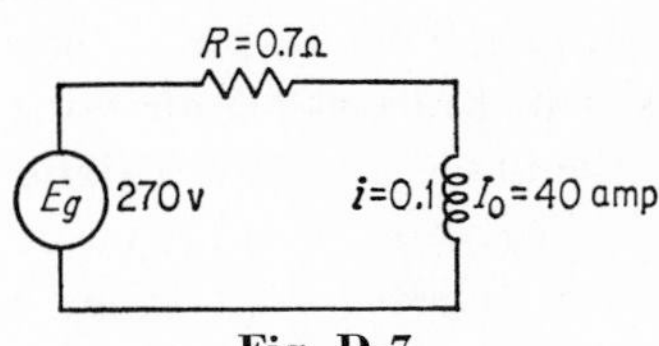

Fig. D-7

ANSWER. Refer to Fig. D-7. The transient equation for this situation is

$$i = I - (I - I_0)e^{-Rt/L} \text{ for } 0 \leq t \leq 0.02$$

The maximum surge of current that will be contained by the armature of the generator will take place at $t = 0.02$ sec. For the above equation, I is that current which will flow in the unprotected circuit if the circuit breaker did not operate; I_0 is that current which flows initially at $t = 0$; e is 2.71828; R is the circuit resistance at a short circuit; t is the time duration of the short; L is the circuit inductance.

Rated current flowing during normal operation is given by

$$I_L = 10{,}000/250 = 40 \text{ amp}$$

Generated emf during normal operation is

$$E_g = 250 + I_0 R_a = 250 + 40 \times 0.50 = 270 \text{ volts}$$

When the short circuit takes place, it is assumed for the purposes of this problem that E_g, L, and R_a remain constant. At time of the short circuit the entire circuit resistance becomes

$$R = R_a + R_{sc} = 0.5 + 0.2 = 0.7 \text{ ohm}$$

And

$$I = \frac{E_g}{R} = \frac{270}{0.7} = 386 \text{ amp}$$

$$I_0 = I_L = 40 \text{ amp}$$

Thus

$$I_{max} = 386 - (386 - 40)e^{-0.7\times(0.02/0.10)} = (386 - 346) \times 0.869$$

$$I_{max} = 386 - 301 = \mathbf{85\ amp}$$

Switching, sudden short circuits, and similar disturbances originating in a system produce surges which may assume the form of a traveling wave similar to that produced by lightning, and at times these surges reach magnitudes which are sufficient to damage apparatus. For a more complete discussion on transients see "Standard Handbook," 6th ed., Sec. 14, pp. 1400–1414. See also "The Engineers' Manual" by Ralph G. Hudson, 2d ed., pp. 211–214, John Wiley & Sons, Inc., New York, 1944.

D-11. Three d-c generators of negligible resistance are connected to a distribution system. The emf's of these generators are A of 120 volts, B of 121 volts, and C of 122 volts. The two lines connecting generators A and B have 0.1-ohm resistance each, while the

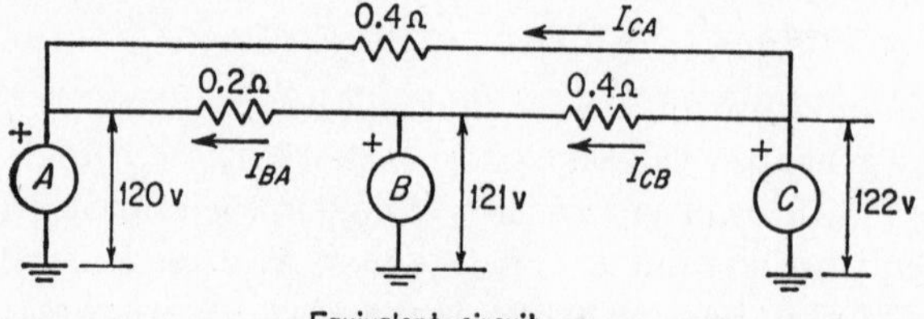

Equivalent circuit

Fig. D-8

other four lines have 0.2-ohm resistance each. Calculate the circulating currents in lines *AB*, *BC*, and *CA* and indicate the direction of these currents.

ANSWER. Refer to Fig. D-8 for equivalent circuit. Also refer back to the section on Kirchhoff's laws in Chap. 2. Now assume the direction of currents as indicated in the equivalent circuit diagram and apply Kirchhoff's law of voltages in a complete circuit.

$$E_B - E_A = 0.2I_{BA} \quad \text{or} \quad 121 - 120 = 0.2I_{BA}$$
$$I_{BA} = +5 \textbf{ amp.} \quad \text{Direction is correct.}$$
$$E_C - E_B = 0.4I_{CB} \quad \text{or} \quad 122 - 121 = 0.4I_{CB}$$
$$I_{CB} = +2.5 \textbf{ amp.} \quad \text{Direction is correct.}$$
$$E_C - E_A = 0.4I_{CA} \quad \text{or} \quad 122 - 120 = 0.4I_{CA}$$
$$I_{CA} = +5 \textbf{ amp.} \quad \text{Direction is correct.}$$

D-12. For the trolley line shown in, compute the voltage across each car and the efficiency of transmission.

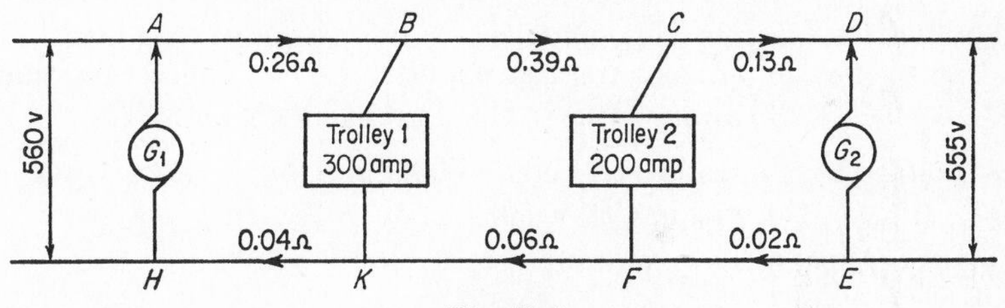

Fig. D-9

ANSWER. Refer to Fig. D-9 and assume current flowing in the directions shown. Should any current values come out negative, it only means that the direction of the current should have been the reverse of the direction originally indicated in the problem itself.

The solution to this problem is the application of Kirchhoff's laws. The sum of the *IR* drops around any one path of a circuit equates the emf's impressed along the same path. Now let us proceed as follows:

Let $$I_x = \text{current along } AB$$

Then the current passing through car 1 also passes along *KH*. Or

$$I_x = \text{current along } KH$$

Now by difference

$$I_x - 300 = \text{current along } BC \text{ and } FK$$

Also $$500 - I_x = \text{current along } DC \text{ and } FE$$

The assumptions we have now made are in accord with Kirchhoff's first law. The drop across trolley 2 is equal to

$$555 - 0.15(500 - I_x)$$

The drop across trolley 2 also is equal to

$$560 - 0.30I_x - 0.45(I_x - 300)$$

We then equate the two to obtain

$$555 - 0.15(500 - I_x) = 560 - 0.30I_x - 0.45(I_x - 300)$$

From the above considerations the voltage across a parallel group is the same for any other parallel path across the same points. This is a manifestation of Kirchhoff's second law. Accordingly, the voltage across trolley 2 is exactly the same whether we calculate it from G_1 or G_2.

Now solve for I_x from the last equation above, and we see that I_x is equal to 239 amp. This is the current in AB and KH.

$$\text{Voltage across trolley 1} = 560 - (239 \times 0.30) = \mathbf{488.3\ volts}$$
$$\text{Current in } CD = 500 - 239 = 261 \text{ amp}$$

and for trolley 2 we find

$$\text{Voltage across trolley 2} = 555 - (261 \times 0.15) = \mathbf{515.8\ volts}$$
$$\text{Current in } BC = 239 - 300 = -61 \text{ amp}$$

Note the minus sign. This tells us that actual current is flowing in a direction opposite to that shown on the figure.

Before we can determine the efficiency of transmission we must find the total power used by the trolleys as compared to the power delivered to the trolleys.

$$\text{Power delivered by } G_1 = 560 \times 239 = 134{,}000 \text{ watts}$$
$$\text{Power delivered by } G_2 = 555 \times 261 = \underline{145{,}000}$$
$$\text{Total power delivered} = 279{,}000 \text{ watts, or 279 kw}$$
$$\text{Power used by trolley 1} = 488.3 \times 300 = 146{,}500 \text{ watts}$$
$$\text{Power used by trolley 2} = 515.8 \times 200 = \underline{103{,}200}$$
$$\text{Total power used by trolleys} = 249{,}700 \text{ watts}$$

$$\text{Efficiency of transmission} = (250/279) \times 100 = \mathbf{89.6\ per\ cent.}$$

It is well to note at this point that the manner in which the above problem has been solved illustrates which current and voltage relations in "meshwork" may be worked.

D-13. Explain how a potentiometer is used to measure the voltage of a voltaic cell, and also how it is used to calibrate a high-range voltmeter.

ANSWER. Figure D-10 shows a potentiometer described as a slide-wire unit. To measure the voltage of a voltaic cell, its terminals are connected to V in such a manner that its emf opposes the emf of the cell, battery B. The sliding contact C is then moved

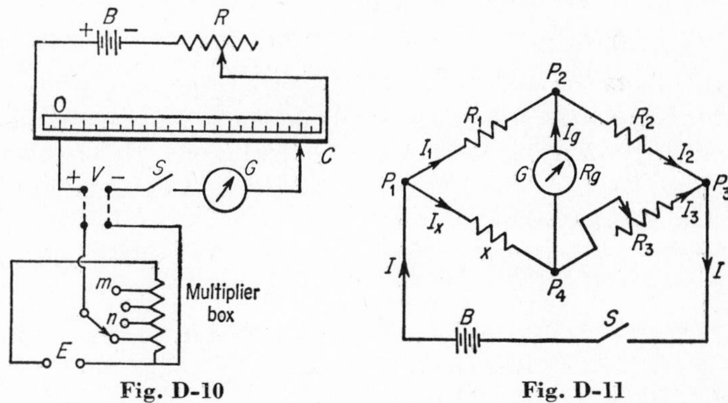

Fig. D-10 **Fig. D-11**

until the galvanometer G shows zero deflection, when the scale reading will indicate the voltaic-cell voltage.

To calibrate a high-range voltmeter, a *multiplier box* consisting of high resistances is connected to V in such a manner as to be in series with the voltmeter connected at E. After a suitable arrangement of the taps, such as m and n, so as to obtain a reading on the potentiometer scale, the contact C is moved and the scale reading recorded. This scale reading must be multiplied by 10 or 100, etc., depending on which contact of the multiplier box was used.

In any case, the resistance R must be high enough to prevent any appreciable amount of current from influencing the voltmeter potential difference.

D-14. Apply Kirchhoff's laws to an unbalanced bridge circuit by drawing the circuit, labeling each portion and writing all the current and voltage equations that must be satisfied.

ANSWER. Figure D-11 shows a Wheatstone bridge for measuring an unknown resistance, such as X. The bridge will be unbalanced whenever a current flows through the galvanometer G.

Referring to Fig. D-11, let

P_1, P_2, P_3, P_4 = junctions of bridge conductors
R_1, R_2, R_3 = known resistances
X = unknown resistance
B = battery supplying an emf
S = switch (assumed to have no appreciable resistance)
G = galvanometer
I_1, I_2, I_3, etc. = currents in respective branches of circuit
emf = terminal voltage of battery

Kirchhoff's first law states: The algebraic sum of the potential drops around every closed circuit is always equal to zero. In accordance with this law the following equations may be written for the unbalanced bridge shown:

$$\text{emf} - I_1R_1 - I_2R_2 = 0$$
$$\text{emf} - I_xX - I_3R_3 = 0$$
$$I_1R_1 + I_2R_2 - I_3R_3 - I_xX = 0$$
$$I_1R_1 - I_gR_g - I_xX = 0$$
$$I_2R_2 - I_3R_3 + I_gR_g = 0$$

Kirchhoff's second law states: The algebraic sum of the currents at any junction of conductors is always equal to zero. The following equations may be written in accordance with this law.

At junction P_1: $I - I_1 - I_x = 0$
At junction P_2: $I_1 + I_g - I_2 = 0$
At junction P_3: $I_2 + I_3 - I = 0$
At junction P_4: $I_x - I_g - I_3 = 0$

Supplement **E**

DIRECT-CURRENT MOTORS

Questions and Answers

E-1. A certain 230-volt shunt motor with an armature resistance of 0.07, including brushes, generates a torque of 250 lb-ft at 1,200 rpm and has an armature current of 210 amp. Calculate the torque developed and the speed when the armature current is 300 amp and the field is reduced 75 per cent of normal.

ANSWER.

Let T = motor torque, lb-ft

K_m = motor constant

I_a = armature current, amp

ϕ = flux per pole

$\phi_2 = 0.75 \times \phi_1$

(*a*) Using the following relation: $T = K_m I \phi$, we can establish the torque ratio with K_m canceling out

$$\frac{T_2}{T_1} = \frac{I_2\phi_2}{I_1\phi_1}$$

or

$$T_2 = T_1 \frac{I_2}{I_1} \frac{\phi_2}{\phi_1}$$

After insertion of the appropriate quantities

$$T_2 = 250 \times \frac{300}{210} \times \frac{0.75\phi_1}{\phi_1} = \textbf{268 lb-ft}$$

New torque developed.

(*b*) Let E_c = motor counter emf

E_L = line voltage

K_g = generator constant

$$E_c = E_L - I_a R_a = K_g N \phi$$

And by rearrangement we obtain for N

$$N = \frac{E_L - I_a R_a}{K_g \phi}$$

The ratio of new speed to old speed N_2/N_1 follows accordingly

$$\frac{N_2}{N_1} = \frac{E_L - I_{a2}R_{a2}}{E_L - I_{a1}R_{a1}}\left(\frac{\phi_1}{\phi_2}\right)$$

Solving for N_2 alone and inserting proper values in the above equation

$$N_2 = \frac{(230 - 300 \times 0.07)1}{(230 - 210 \times 0.07)0.75} \times 1{,}200 = \mathbf{1{,}550\ rpm}$$

E-2. A 15-hp, 220-volt d-c shunt motor draws an armature current of 60 amp at full load. The armature resistance, including brushes, is 0.26 ohm. Calculate the total resistance of a starting box which would permit this motor to develop a starting torque equal to twice that at full load.

ANSWER. Refer to Fig. 11-10. Assume that the motor is operated from no load to full load at constant field strength ϕ, or $\phi_1 = \phi_2$. Since for a shunt motor the torque developed is proportional to ϕ and the armature current I_a, and if we let E be the impressed voltage, I_{a1} will be the no-load armature current and I_{a2} the full-load armature current. Following quite closely the nomenclature of the above solution

$$T_1 = K_m \phi_1 I_{a1} \qquad \text{and also} \qquad T_2 = K_m I_{a2} \phi_2$$

from which we can set up the following ratio of armature currents

$$\frac{I_{a1}}{I_{a2}} = \frac{T_1/K_m\phi_1}{T_2/K_m\phi_2} = \frac{T_1}{T_2} = 2$$

The no-load current $I_{a1} = 2I_{a2} = 2 \times 60 = 120$ amp. Thus, the total resistance of the armature-resistance circuit at no load is

$$R_a + R_s = E/I_a$$

Therefore

$$R_s = \frac{E}{I_a} - R_a = \frac{220}{120} - 0.26 = \mathbf{1.57\ ohms}$$

E-3. When the field rheostat is cut out, a 230-volt shunt motor generates a counter emf of 220 volts at no load. The armature resistance is 2.3 ohms and that of the field is 115 ohms. Calculate (*a*) the current through the armature when sufficient external resistance is inserted in the field circuit to make the field current one-half as great, (*b*) the current through the armature when the field rheostat is cut out. Supply voltage at motor terminals = 230.

ANSWER.

(*a*) The field current without external resistance is

$$230/115 = 2 \text{ amp}$$

When the field current is made half as great by inserting external resistance, the field flux and, therefore, the counter emf will become half as great, or 110 volts. And the armature current I_a becomes

$$I_a = \frac{230 - 110}{2.3} = \mathbf{52.2\ amp}$$

(*b*) When the field rheostat is cut out, I_a becomes

$$I_a = \frac{E_t - E_a}{R_a} = \frac{230 - 220}{2.3} = \mathbf{4.35\ amp}$$

E-4. Enumerate the reasons why a 230-volt 5-hp 4-point d-c starting box cannot be employed to start a 115-volt 5-hp shunt motor.

ANSWER. Refer to Fig. E-1. Then let us assume that in either case the motor efficiency at full load is 85 per cent and the armature resistances are equal. Then the full-load armature current for the 230-volt motor would be

$$(5 \times 746)/(0.85 \times 230) = 19 \text{ amp}$$

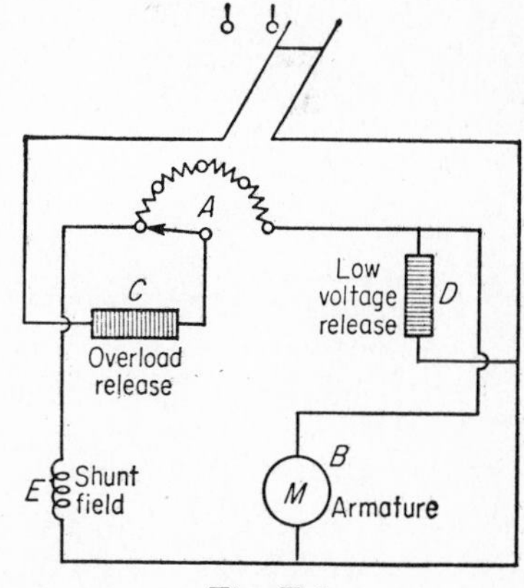

Fig. E-1

and its starting box would be designed for a current one and one-half times the full-load current of 19 amp, or 19 × 1.5 = 28.5 amp. Thus, the combined resistance of the armature and the starter would be

$$230/28.5 = 8.07 \text{ ohms}$$

Similarly, for the 115-volt motor the full-load armature current is

$$(5 \times 746)/(0.85 \times 115) = 38 \text{ amp}$$

By using the starting box, the maximum current that would flow through this motor, when starting it at full load, would be

$$115/8.07 = 14.25 \text{ amp}$$

only 37½ per cent of the full-load current. Remembering that the torque developed by the motor is directly proportional to the armature current, it follows that only 37½ per cent of the required torque will be developed. *Therefore, the motor will not start.* If the rheostat arm is moved to subsequent points, thus reducing the armature current resistance, the current may increase enough over 38 amp to start the motor, but the box would probably burn out because it is only good for 28.5 amp. If it should last until contact 4 is reached, with all the series resistance cut out, the low-voltage release would send the rheostat arm back to its "off-position," breaking the circuit.

E-5. What is the approximate resistance and current-carrying capacity of a starter for a 15-hp 220-volt shunt motor. Assume a motor efficiency of 85 per cent, an armature resistance of 0.15 ohm, and the maximum allowable current as 150 per cent of its rated value. Explain how you arrive at your answer.

ANSWER.

$$\text{Output} = 15 \times 746 = 11{,}190 \text{ watts}$$
$$\text{Input} = 11{,}190/0.85 = 13{,}160 \text{ watts}$$
$$I_L \text{ (line current)} = 13{,}160/220 = 60 \text{ amp, rated value}$$

Neglecting I_f, it being small by comparison,

$$I_{\max} = 1.5 \times 60 = \mathbf{90\ amp}$$

the maximum allowable current. Resistance of the starter is

$$R = \frac{E_L}{90} - R_a = \frac{220}{90} - 0.15 = 2.44 - 0.15 = \mathbf{2.29\ ohms}$$

E-6. A 25-hp shunt-wound 230-volt d-c motor is operated at constant field strength and used to drive a "frictional" type of load where the torque is substantially constant at all speeds. If the motor draws an armature current of 80 amp at normal operating speeds with 230-volt supply, what ohmic value of the armature circuit resistor should be used to allow the motor to develop a torque at starting 25 per cent in excess of the torque required at operating speed? The armature resistance may be taken as 0.1 ohm.

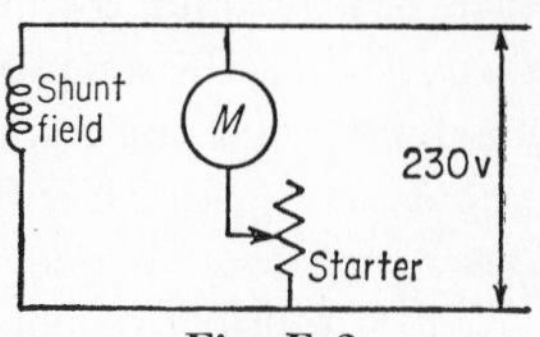

Fig. E-2

ANSWER. Refer to Fig. E-2 and let the following be the basis:

T_1 and T_2 = no-load and full-load torques, respectively
ϕ_1 and ϕ_2 = no-load and full-load field strengths, respectively
K_m = motor design constant
E_m = impressed voltage on motor
I_{a1} and I_{a2} = no-load and full-load armature currents, respectively
R_a = armature resistance, ohms
R_s = resistor ohmic value, ohms

Then, from previously established relations we remember that

$$T_1 = K_m\phi_1 I_1 \qquad \text{also for} \qquad T_2 = K_m\phi_2 I_2$$

From the problem itself (constant field strength) $\phi_1 = \phi_2$ and $T_1/T_2 = 1.25$. Thus, it follows that for

$$I_1 = T_1/T_2 I_2 = 1.25 \times 80 = 100 \text{ amp}$$

From the well-known relation for counter emf $= K_m\phi N$, where is motor and ϕ is its field strength, we conclude that on starting, the counter emf is zero because N is zero. Also,

$$E_m = (\text{counter emf}) + [I_1(R_a + R_s)] = I_1(R_a + R_s)$$

Therefore $$R_a + R_s = \frac{E_m}{I_1} = \frac{230}{100} = 2.3 \text{ ohms}$$

And in conclusion

$$R_s = 2.3 - R_a = 2.3 - 0.1 = \mathbf{2.2\ ohms}$$

E-7. A four-pole lap-wound shunt motor has 100 armature conductors, the effective length of each being 12 in. The radius of the armature is 6 in. and the flux per pole is 10^7 lines. Calculate (*a*) the torque when the armature current is 100 amp; (*b*) the speed of the machine when the applied voltage is 230 volts. Assume that the armature resistance is 0.1 ohm.

Let T = motor torque, lb-ft
p = number of poles
m = number of armature paths between brushes
ϕ = flux per pole in armature
Z = number of active conductors in armature
I_a = armature current, amp
R_a = armature resistance, ohms
E = motor counter emf, volts
E_m = applied voltage, volts
N = motor speed, rpm

(*a*) In a lap-wound motor there are as many paths in the armature as there are poles. Therefore, m is equal to 4. Thus

$$T = \frac{p\phi ZI_a}{852m \times 10^6}$$

Substituting in the above equation

$$T = \frac{4 \times 10^7 \times 100 \times 100}{852 \times 4 \times 10^6} = \mathbf{117.5\ lb\text{-}ft}$$

This assumes that all armature conductors are effective, which may not be true in a particular case.

(*b*) $E_m = E + I_aR_a$

And therefore

$$E = E_m - I_aR_a = 230 - (100 \times 0.1) = 220 \text{ volts}$$

Also since $E = K_m\phi N$, where $K_m = pZ/m \times 60 \times 10^8$, it remains

$$N = \frac{E}{K_m\phi} = \frac{220 \times 4 \times 60 \times 10^8}{4 \times 10^7 \times 100} = \mathbf{1{,}320\ rpm}$$

E-8. Explain how the shunt motor, the induction motor, and the synchronous motor accommodate themselves to changes in load, stating just how the input current is caused to change in each.

ANSWER. *Shunt motor:* When the motor load increases, the torque delivered by the motor must increase for the operation to be maintained. An increase in torque T must carry with it an increase in armature current I_a; this is very nearly the input current, in accordance with the relation

$$I_a = \frac{T}{K_m \phi}$$

where ϕ is flux per pole. This may also be derived by considering that as the load on the motor increases, the tendency for it to be slowed down with an attendant decrease in the motor counter emf in accordance with counter emf $= K_m \phi N$, where N is speed of motor in rpm and K_m and ϕ remain constant as before. The terminal voltage E_m is ordinarily constant as is the armature resistance R_a from which

$$I_a = \frac{E_m - E(\text{counter emf})}{R_a}$$

We can readily see that a decrease in E means an increase in input current. The converse also holds true; i.e., a decrease in load will result in a decrease in input current.

Induction motor: With this type of a-c motor the armature consists of the rotor. So with a shunt motor, an increase in load tends to decrease the motor speed, but so long as the rotor turns at a speed less than the magnetic field set up by the stator, it will be cutting more lines of force per unit of time. The rotor current must, therefore, increase. Furthermore, the motor torque depends on the product of the rotor current and the stator flux. And since the stator flux remains substantially constant, an increase in torque must be followed by an increase in the rotor, or input, current.

Synchronous motor: The speed of this motor remains constant regardless of the load, i.e., until the pull-out load is reached. The field is excited by a d-c line, while the armature draws current from an a-c source. The counter emf generated at no load overcomes the frictional losses when the motor runs at synchronous speed at no load. Because the motor takes no load, it will, as in the case of other motors, slow down for an instant so that the counter emf will lag for an instant, producing a voltage which will immediately send a current through the winding. This current will be proportional to the lag and since that load is what produces the lag, the greater the load the greater will be the current. Since the lag is measured in electrical degrees, a very small shift of the armature (less than one-half the pitch of a pole) is all that is needed for a large increase in its current to take place.

E-9. A 440-volt 70-hp shunt motor has an armature resistance of 0.185 ohm and a field resistance of 350 ohms. The current drawn by this machine is 135 amp at full load. If this machine is to deliver a torque equal to 175 per cent of that at full load, what would be the resistance of its starter? What is the maximum power dissipated in this starter? See Fig. E-3.

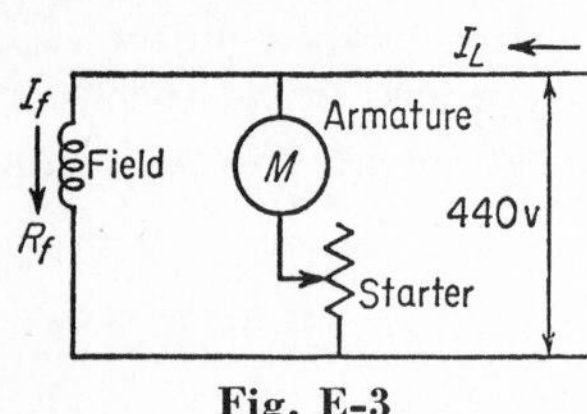

Fig. E-3

ANSWER. Since a shunt motor is considered to be a constant-speed machine and its field strength constant, then the counter emf at all loads is the same, but zero at the start. The problem states that

$$T_2 = 1.75 \times \text{full-load torque} = 1.75T_1 \text{ (full-load torque)}$$

Also it follows that $I_{a2} = 1.75I_{a1}$. But in order to determine I_{a1} (armature current at full load) we must first find the field current

$$I_f = {}^{440}\!/_{350} = 1.26 \text{ amp}$$

And since $I_{a1} = I_L - I_f = 135 - 1.26 = 133.74$ amp

then $I_{a2} = 1.75 \times 133.74 = 235$ amp

Counter emf is zero and full starting torque is utilized at this instant. The resistance of the starter is found by

$$\frac{440 - [0 + (235 \times 0.185)]}{235} = \frac{440 - 43.5}{235} \mathbf{1.68 \text{ ohms}}$$

Maximum power dissipated by the starter is

$$I_{a2}^2R_a = 235^2 \times 1.68 = \mathbf{90{,}000 \text{ watts}}$$

E-10. A 25-hp shunt motor is used to drive a pump. The armature current drawn by this machine from 230-volt mains is 80.6 amp. The armature resistance, including brushes, is 0.086 ohm. The iron, windage, and friction losses, as determined from another test, were found to be 754 watts for the speed and field excitation used in driving the pump. Calculate the power (mechanical) delivered by this motor to its load.

ANSWER. Refer to the section on efficiency of motors. The power required to drive the load is the difference between input and losses. First let us find the input to the motor, i.e., the input to the armature:

$$I_aE_L = 80.6 \times 230 = 18{,}538 \text{ watts}$$

Now for the losses:

Stray-power losses (754 watts, given) $=$ 754 watts
Brush and armature losses: $I_a{}^2R_a$,
$80.6^2 \times 0.086 = 559$ watts $=$ 559 watts
Input minus losses $= 18{,}538 - (754 + 559)$ $= 17{,}225$ watts

$$\text{Horsepower delivered to load, } 17{,}225/746 = \mathbf{23.1\ hp}$$

E-11. The operation of a d-c 220-volt shunt motor driving its normal load is expressed by the following data: armature current 130 amp, field current 3.5 amp, armature resistance 0.05 ohm, speed 1,200 rpm, and brush drop 2 volts. It is required to determine the stray-power losses for this machine under normal load conditions.

(*a*) Explain briefly but clearly how the test should be conducted, give the circuit diagram, and state the range of all meters used.

(*b*) Specify the voltage to be applied to the armature, the current in the field, and the speed at which the test is to be conducted. Give reasons for your specifications.

(*c*) Explain how you would determine the efficiency of the machine.

ANSWER. In the solution let us work out (*a*) and (*b*) together. Three tests are required: full-load, no-load, and armature resistance.

Full-load circuit diagram: See Fig. E-4*a*. Rated voltage is to be applied and the field current is to be adjusted to produce normal load speed. This will give the armature current for normal test load together with the field current for rated speed.

$$\begin{aligned} \text{Armature current} &= 150 \text{ amp} \\ \text{Field current} &= 3.5 \text{ amp} \\ \text{Speed} &= 1{,}200 \text{ rpm} \end{aligned}$$

No-load circuit diagram: See Fig. E-4*b*. Rated voltage is to be applied. Normal load speed is to be produced by varying the field. This will give the armature current to overcome friction, windage, and brush loss, etc.

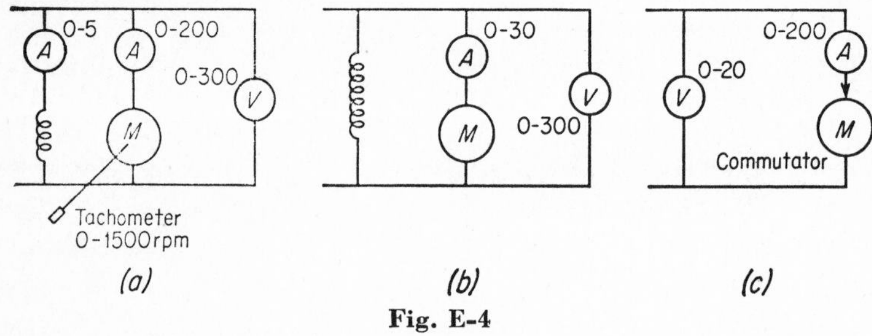

Fig. E-4

Armature-resistance measurement: See Fig. E-4*c*. Apply voltage of sufficient intensity to opposite commutator bars to produce rated current. From this calculate field armature resistance (R_a equal to 0.05 ohm). Assume brush drop to be 1 volt per brush as shown in AIEE standards.

(*c*) *The efficiency* may be determined from the ratio of power input minus stray-power losses, all divided by the power input. The power input is the sum of all volt-amperes of the armature and the field. The stray-power losses are

$$\begin{aligned} \text{Shunt field loss: } EI_f &= 220 \times 3.5 = 770 \text{ watts} \\ \text{Armature loss: } I_a{}^2R_a &= 130^2 \times 0.05 = 845 \text{ watts} \\ \text{Brush contact loss: } E_bI_a &= 2 \times 130 = 260 \text{ watts} \end{aligned}$$

(*d*) *The rotational losses:* armature no-load loss minus (no-load brush-contact loss plus copper loss)

$$EI_a - (E_bI_a + I_a{}^2R)$$

The I_a's are at *no load.* Total losses = summation of $a + b + c + d$. And the efficiency may be found by

$$\frac{E(I_a + I_f) - \text{total losses}}{E(I_a + I_f)} \times 100$$

E-12. The power input to a 30-hp 220-volt shunt motor is 2,100 watts, taken at no load. The field and armature resistances are 73 and 0.11 ohms respectively. Calculate the efficiency and horsepower output for an input current of 100 amp.

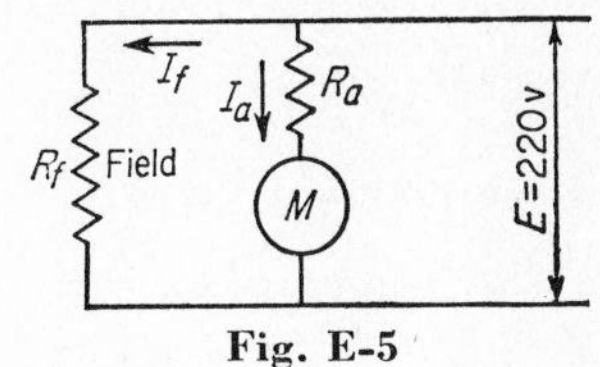

Fig. E-5

ANSWER. Refer to Fig. E-5. At no load the current through the armature of the motor is zero, and we may write

$$I_f = \frac{E}{R_f} = \frac{220}{73} = 3.01 \text{ amp}$$

Therefore, at no load

$$I_a^2R_a \text{ negligible}$$
$$I_f^2R_f = 3.01^2 \times 73 = 662 \text{ watts}$$

And the stray-power loss is 2,100 − 662 = 1,438 watts. At 100 amp input, remembering that stray-power and shunt field losses remain constant, we have

$$I_a = 100 - I_f = 100 - 3.01 = 96.99 \text{ amp}$$

The armature and field losses are as follows:

$$\begin{aligned} I_a^2R_a &= 96.99^2 \times 0.11 = 1{,}035 \text{ watts} \\ I_f^2R_f &= 3.01^2 \times 73 = 662 \\ \text{The stray-power loss} &= \underline{1{,}438} \\ \text{Total losses} &= 3{,}135 \text{ watts} \end{aligned}$$

$$\text{Motor efficiency} = \frac{\text{input} - \text{losses}}{\text{input}} = \frac{EI - \text{losses}}{EI}$$

Motor efficiency = (220 × 100) − 3,135/(220 × 100) = 0.857, or **85.7 per cent**

Hp output = [(200 × 100) − 3,135]/746 = **25.3 hp**

E-13. The armature and shunt field resistances of a 30-hp 240 volt 1,200-rpm shunt motor are 0.102 and 150 ohms respectively. This machine has an efficiency of 88 per cent at full load. Calculate efficiency and speed at 125 per cent of full load.

ANSWER. Let subscript 1 refer to full-load conditions and subscript 2 refer to 125 per cent of full load.

At full load:

Motor output $= 30 \times 746 = 22{,}400$ watts

Motor input $= \dfrac{\text{output}}{\text{efficiency}} = \dfrac{22{,}400}{0.88} = 25{,}400$ watts

Total losses = input − output $= 25{,}400 - 22{,}400 = 3{,}000$ watts

Field current $= I_{f_1} = \dfrac{E}{R} = \dfrac{240}{150} = 1.6$ amp

Input current $= \dfrac{\text{input}}{E} = \dfrac{25{,}400}{240} = 106$ amp

Armature current $= I_{a_1} = I_1 - I_{f_1} = 106 - 1.6 = 104.4$ amp

Armature loss $= I_{a_1}{}^2R_{a_1} = 104.4^2 \times 0.102 = 1{,}113$ watts

Shunt field loss $= I_{f_1}{}^2R_{f_1} = 1.6^2 \times 150 = 384$ watts

Stray-power loss = total losses − $I_1{}^2R_1$ losses $= 3{,}000 - (1{,}113 + 384) = 1{,}503$ watts

The stray-power and shunt field losses remain constant irrespective of load. Thus, at 125 *per cent of full load:*

$I_2 = 1.25 \times 106 = 132.6$ amp

$I_{a_2} = I_2 - I_{f_1} = 132.6 - 1.6 = 131$ amp

Armature loss $= I_{a_2}{}^2R_{a_1} = 131^2 \times 0.102 = 1{,}750$ watts

Shunt field loss $= I_{f_1}{}^2R_{f_1} = 384$ watts (same as for full load)

Stray-power loss $= 1{,}503$ watts (same as for full load)

Total losses $= 1{,}750 + 384 + 1{,}503 = 3{,}637$ watts

Motor efficiency $= \dfrac{\text{output}}{\text{output} + \text{losses}}$

or

$$\text{Motor efficiency} = \frac{22{,}400 \times 1.25}{(22{,}400 \times 1.25) + 3637} = 0.886$$

or **88.6 per cent**

In order to find rpm:

At full load:

Counter emf $= E - I_{a_1}R_{a_1} = 240 - (104.4 \times 0.102) = 229.4$ volts

At 125 *per cent of full load:*

$$\text{Counter emf} = E - I_{a_2}R_{a_1} = 240 - (131 \times 0.102) = 226.6 \text{ volts}$$

From the general equation for counter emf, $K_m\phi N$, where K_m is the motor constant and ϕ is the constant field flux, we find that motor rpm at 125 per cent of full load may result in the form

$$1{,}200 \times (226.6/229.4) = \mathbf{1{,}186\ rpm}$$

E-14. A 25-hp d-c motor has been water-soaked by seepage following a fire in a building in which it was in operation. Explain how you would recondition this machine and how you would test it, using only an ammeter and a voltmeter.

ANSWER. Using direct current to dry out the windings or baking the machine in an oven under controlled conditions are two methods in common practice today.

Using direct current. Open the shunt field and series field, if any, and cut out of circuit while the armature is short-circuited by means of a low-resistance cable capable of carrying two or three times the normal armature current. By driving the motor at normal speed, a current produced by the residual magnetism will flow through the armature. This may be sufficient to dry it out. If this is not sufficient, the shunt field may be slightly excited from a low-voltage external source. The hot air blown off by the rotating armature is usually enough to dry out the field. However, if this is not the case, direct current up to normal voltage may be passed through it while the machine is at rest.

Throughout this drying process the insulation resistance must be checked. This may be accomplished by first connecting the winding to be tested to one side of a d-c supply, while the other side is hooked up to a voltmeter in such a manner that the voltmeter and windings are in series. In this position the voltmeter reading E is taken. Connection is then made to the frame of the motor and the voltmeter reading E_1 taken again. Knowing the voltmeter resistance R, the insulation resistance may be obtained.

$$\text{Resistance} = R\,\frac{E}{E_1 - 1} \qquad \text{ohms}$$

The rate of change of the insulation resistance is the best indication of how long the drying process should continue. When the curve of resistance values plotted against time flattens out, after first increasing, the drying operation should be stopped. A common rule to remember is that the insulation resistance with respect to rated voltage and kilowatt is

$$\frac{\text{Rated voltage}}{\text{Rated kw} + 1{,}000} \quad \text{megohms}$$

Using a baking oven. Where a baking oven of adequate size is available, the motor may be dried out by baking it at a temperature from 100 to 150°C (212 to 302°F) until the insulation resistance measures approximately 500,000 ohms. Sometimes this value cannot be reached because moisture entrapped in the clamping ring of the commutator cannot be dried out. In this event, holes must be drilled, care being taken not to destroy the mica insulation. Insulation tests for resistance are made as before described.

E-15. A shunt motor takes 40 amp at 112 volts under full load. When running without load at the same speed, it takes 3 amp at 106 volts. The field resistance is 100 ohms and the armature resistance is 0.125 ohm. What horsepower does the motor deliver at full load and what is its commercial or over-all efficiency? See Fig. E-6.

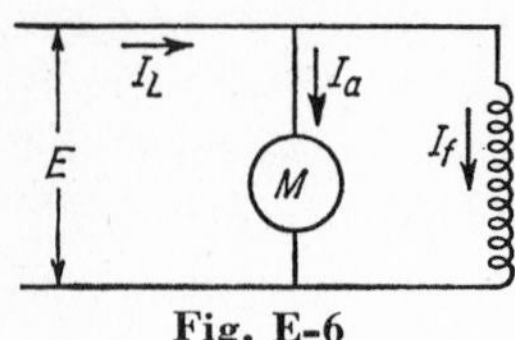

Fig. E-6

ANSWER. Since the stray-power losses vary with the generated voltage and the speed, the first step is to calculate the generated voltage for the load condition.

$$I_f = \frac{E}{R_f} = \frac{112}{100} = 1.12 \text{ amp}$$

$$I_a = I_L - I_f = 40 - 1.12 = 38.88 \text{ amp}$$

The generated voltage may be determined in accordance with

$$112 - (38.88 \times 0.125) = 107.14 \text{ volts}$$

When the motor is operated at no load, the generated voltage as calculated above, 107.14 volts, and the speed are approximately the same as for the load conditions. Therefore, the input to the armature at no-load conditions is approximately equal to the stray-power losses under the load conditions specified.

From no-load data the field current is found

$$I_f = {}^{106}\!/_{100} = 1.06 \text{ amp}$$
$$I_a = I_L - I_f = 3 - 1.06 = 1.94 \text{ amp}$$

Then stray-power losses are

$$106 \times 1.94 = 200 \text{ watts plus}$$

Full-load conditions:

Output = input − losses

Output = $(112 \times 40) - (1.12 \times 112 + 38.88^2 \times 0.125 + 200)$

Output = $4{,}480 - (125.5 + 188 + 200) = 3{,}967$ watts

$$\text{Hp output} = \frac{\text{watts output}}{746} = \frac{3{,}967}{746} = \mathbf{5.32\ hp}$$

$$\text{Efficiency} = \frac{\text{output}}{\text{input}} = \frac{3{,}967}{4{,}480} = 0.885, \text{ or } \mathbf{88.5\ per\ cent}$$

E-16. A 230-volt shunt motor, employed to drive a load up to 15 hp at 1,000 rpm with a permissible speed regulation of 20 per cent, is damaged and requires replacement. A 15-hp 230-volt compound motor is available. This machine is used to replace the damaged one and has series field turns equal to 1 per cent of the shunt turns. May this machine be used to replace the damaged one? Explain your assumptions and justify your conclusion.

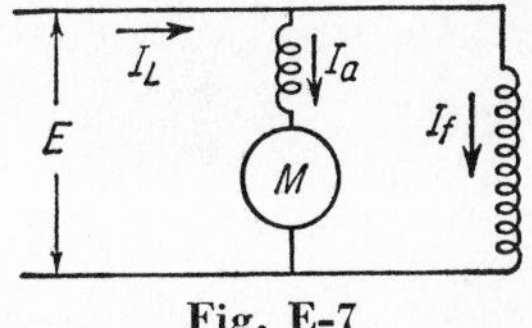

Fig. E-7

ANSWER. Refer to Fig. E-7. With a permissible speed regulation of 20 per cent, the allowable speed would be from 1,200 to 1,000 rpm, i.e., from no load to full load respectively. From the general speed equation for d-c motors, an analysis can be made of the factors which affect the speed and how these factors can be modified.

$$\text{Speed} = \frac{\text{terminal voltage} - \text{armature current} \times \text{total resistance}}{\text{measure of effective field flux}}$$

For a compound motor operating on a constant voltage E_t there are two factors that affect the speed as the load is increased from no load to full load.

(*a*) The $I_aR'_a$ drop in the armature circuit which for a compound motor would be $I_a(R_a - R_s)$ and amounts to about 3 per cent of rated voltage E_t and would cause an equal per cent decrease in speed.

(*b*) As the load increases, the ampere turns in the series field will increase the effective flux ϕ in the field and cause a corresponding drop in speed. The amount the flux is increased depends on the point on the saturation curve at which the motor is operating at no load. For compound motors it is generally about the knee of the curve.

To analyze the effect of the series field turns of 7 per cent, it is necessary to know the relative number of ampere turns in both series and shunt fields. Assume 90 per cent efficiency at full load for the 15 hp motor. Then

$$I_L = \frac{\text{hp} \times 746}{\text{eff} \times E_t} = \frac{15 \times 746}{0.9 \times 230} = 54 \text{ amp}$$

The shunt field current I_f for a 15-hp motor is approximately 2 per cent of full-load current. Then

$$I_f = 0.02 \times 54 = 1.00 \text{ amp}$$

Also

$$I_a = I_L - I_f = 53 \text{ amp}$$

Then, assuming 100 turns in the shunt field, there will be one turn in the series field and the ampere turns will be in

$$\text{Shunt field} = 1 \times 100 = 100 \text{ amp turns}$$
$$\text{Series field} = 53 \times 1 = 53 \text{ amp turns}$$

From no load to full load the total field ampere turns are increased about 50 per cent. This would probably cause an increase in the effective field flux of about 25 per cent and a corresponding per cent decrease in speed. With no physical changes in the motor, the speed regulation would exceed the permissible 20 per cent, but the motor could be used as a replacement by inserting a resistance shunt around the series field which would by-pass some of the series field current and thus keep the speed regulation within the permissible 20 per cent.

E-17. From the general speed equation for d-c motors indicate the various methods available for speed control of motors. By what four criteria should methods of speed control be adjudged?

ANSWER. The general speed equation for d-c motors is given by the following:

$$N = \frac{E_t - I_a R'_a}{\phi Z'}$$

where E_t = terminal voltage
R'_a = total resistance in armature circuit
ϕ = effective field flux per pole linking armature circuit
Z' = design constant given by

$$Z' = \frac{pZ}{a \times 60 \times 10^8}$$

where p = total poles
a = parallel paths in armature windings
Z = total number of conductors in armature windings

Refer to the section on d-c motor control (p. 90). It may be seen that there are basically two methods of changing the speed of d-c motors: (*a*) changing the voltage on the armature circuit (Fig. 11-11) and (*b*) changing the effect of the field flux (Fig. 11-12). In order to change the voltage on the armature circuit, add resistance in series with the armature or provide the armature with an external source of voltage which may be adjusted. In order to change the field flux, the ampere turns may be changed or the reluctance of the field magnetic circuit may be changed.

Criteria for speed control should be judged on the basis of regulation, efficiency, simplicity of construction and operation, effect on commutation, and cost.

E-18. The no-load armature current of a 220-volt shunt motor is 3 amp. The no-load speed is 950 rpm and the total resistance of the armature circuit R'_a including brushes is 0.2 ohm. If the motor is loaded, the field flux reduced 15 per cent, and a resistance of 2 ohms inserted in the armature, the speed is 609 rpm. What is the armature current and the total electromagnetic torque in pound-feet?

ANSWER. From the general equation for motor speed we can calculate $\phi Z'$ for no-load conditions. This will remain the same for all load conditions.

$$\phi Z' = \frac{E_t - I_a R'_a}{N} = \frac{220 - (3 \times 0.2)}{950} = 0.231$$

At the loaded condition, 609 rpm, general equation considerations, and under loaded conditions

$$609 = \frac{220 - (I_a \times 2.2)}{(1 - 0.15) \times 0.231}$$

from which I_a is found to be 45.6 amp. The equation for electromagnetic torque is

$$T = 7.04\phi Z' I_a$$
$$T = 7.04 \times 0.231 \times 45.6 \times 0.85 = \mathbf{63\ lb\text{-}ft}$$

E-19. Explain the construction and operation of that type of d-c motor which has minimum commutation difficulties when heavily overloaded.

ANSWER. Such motors are usually provided with commutating, or interpoles. As may be seen from Fig. 10-13 interpoles are placed between the main poles in the axis of commutation and the interpole windings are in series with the armature circuit (Fig. 11-5). Because overheating and sparking reduce the output of a d-c motor, the insertion of interpoles improves the operating range of loading and speed. To minimize the voltage of self-inductance in the armature, it is necessary to keep the number of ampere turns in each coil as low as possible and to cause the reversal during commutation to take place as slowly as possible. Interpoles neutralize the voltage of self-inductance and improve the performance of the machine, reducing sparking at the brushes.

E-20. A four-pole shunt motor has the following characteristics:

Flux density under poles	=	6,000 gausses
Active length of armature conductors	=	15 cm
Rated current in armature conductors	=	12 amp
Diameter of armature	=	28 cm
Number of armature slots	=	34
Conductors per armature slot	=	12
Slots covered by each pole	=	6

Neglecting the fringing at the edges of the poles, calculate the torque in pound-foot of this machine with rated current.

ANSWER. Developed torque in this machine is given by

$$T = 7.04\phi Z' I_a$$

where $$Z' = \frac{pZ}{a \times 60 \times 10^8}$$

In this problem $p = 4$, number of poles
$Z =$ conductors $34 \times 12 = 408$
$a =$ parallel paths in armature winding, etc.
See also E-17.

$$\phi = 6{,}000 \times 15 \times 6/34 \times \pi \times 28 = 1.4 \times 10^6 \text{ lines}$$

Solving for torque

$$T = \frac{7.04 \times 1.4 \times 10^6 \times 4 \times 408}{60 \times 10^8} \times 12 = \mathbf{32.2\ lb\text{-}ft}$$

E-21. A 10-hp d-c shunt motor has armature and field resistances of 0.25 and 100 ohms. The full-load efficiency is 83 per cent. Determine the value of the starting resistance in order that the armature starting current will not exceed 200 per cent of the full-load value (see Fig. E-8).

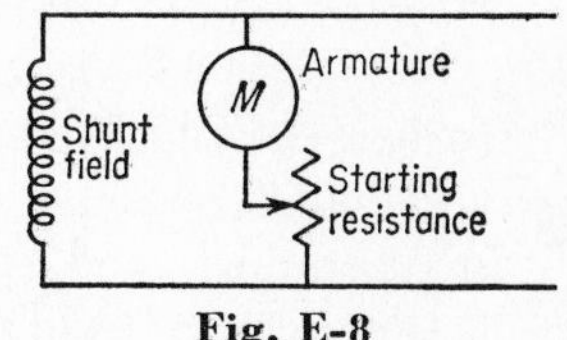

Fig. E-8

ANSWER.

Motor output $= 746 \times 10 = 7{,}460$ watts

Motor input $= \dfrac{\text{output}}{\text{efficiency}} = \dfrac{7{,}460}{0.83} = 8{,}990$ watts

$I = \dfrac{\text{motor input}}{E} = \dfrac{8{,}990}{220} = 40.8$ amp at full load

I_f (starting field current) $\dfrac{E}{R_f} = \dfrac{220}{100} = 2.20$ amp

I_a (full-load armature current) $= I - I_f = 40.8 - 2.2 = 38.6$ amp
I_s (starting armature current) $= 2I_a = 2 \times 38.6 = 77.2$ amp

Since, when starting, the motor emf is zero, we may write

$$E = I_s(R_a + R)$$

From which we obtain

$$R = \frac{E}{I_s} - R_a = \frac{220}{77.2} - 0.25 = 2.85 - 0.25 = \mathbf{2.60\ ohms}$$

E-22. A client planned to use a 230-volt hoisting motor on a construction job, but it is found that the voltage will be only 200 volts. Advise your client as to how the low voltage will affect the operating characteristics of the motor and whether or not it should be used.

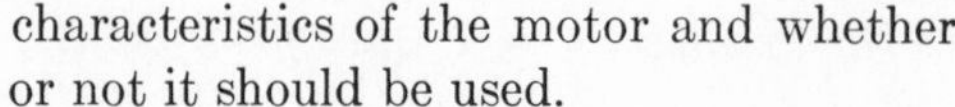

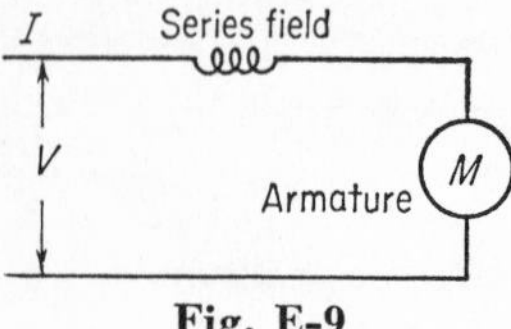

Fig. E-9

ANSWER. For hoists requiring a large starting torque and high speed at low speeds, a d-c motor is generally employed. In this type of motor, the field coils are in series with the armature (series motor), as shown in Fig. E-9. The relation between the applied voltage E and the motor speed is

$$N = \text{speed} = \frac{E - I_a R_a}{K\phi}$$

Since K, I, ϕ, and R_a are not affected by a change in E, the motor speed will drop when E is decreased from 230 to 200 volts.

On the other hand, the starting torque, which is proportional to the product, ϕI, will remain unchanged. The same motor can, therefore, be used on either 230 or 200 volts.

E-23. A 200-hp 60-cycle three-phase 440-volt synchronous motor is used to drive a pump. When starting as an induction motor with full line voltage applied, the per unit I is 5.50 and the per unit torque developed is 1.60. To start the pump unloaded, 50 per cent of the normal motor torque is required. What line current in amperes will the motor take when supplied by autotransformers delivering the minimum voltage necessary, if the motor efficiency at full load is 92 per cent? Neglect starter losses and the d-c power input.

ANSWER. The torque varies as the square of the voltage.

$$\frac{E_s^2}{E_L^2} = \frac{0.5 \text{ torque}}{1.6 \text{ torque}}$$

E_s = starting voltage for 50 per cent normal torque
E_L = line voltage for 160 per cent normal torque

$$E_s^2 = 440 \times 440 \times 0.5/1.6 = 60{,}500 \text{ volts}$$
$$E_s = 246 \text{ volts}$$

I'_s = starting current at 440 volts

$$I'_s = \frac{\text{hp} \times 746 \times 5.5}{1.73 \times E_L \times \text{eff} \times pf} = \frac{200 \times 746 \times 550 \text{ per cent}}{1.73 \times 440 \times 92 \text{ per cent} \times 1} = 1{,}170 \text{ amp}$$

$$\frac{I_s}{I'_s} = \frac{E_s}{E_L}$$

where I_s = motor starting current at 246 volts
I'_s = motor starting current at 440 volts

$$I_s = \frac{246}{440} \times 1{,}170 = 655 \text{ amp}$$

E line $\times$ I line $= E_s \times I_s$
I line = starting current at 50 per cent torque

$$I \text{ line} = \frac{246}{440} \times 655 = \mathbf{366\ amp}$$

E-24. A 2,200-volt 60-cycle three-phase 100-hp synchronous motor has a synchronous impedance at starting by action of the damper grids of 0.15 per unit. If a starting compensator is to be supplied, which reduces the current from the line to 2.0 per unit, what would be the voltage rating of the compensator?

ANSWER.

$$\text{Starting current} = I_s = \frac{E}{Z} = \frac{1.0}{0.15} = 6.67 \text{ per unit}$$

Since the line current is proportional to the square of the voltage, then

$$\frac{\text{Starting current}}{\text{Full voltage inrush}} = \frac{I_s}{I_i} = \left(\frac{E_s}{E}\right)^2$$

Then
$$E_s^2 = \frac{I_s}{I_i} E^2$$

from which
$$E_s = E\sqrt{\frac{I_s}{I_i}}$$

or finally

$$E_s = 2{,}200 \times \sqrt{\frac{2.0}{6.67}} = 2{,}200 \times 0.547 = \mathbf{1{,}200\ volts}$$

E-25. A 300-hp 2,300-volt 60 cps synchronous motor has a synchronous impedance at starting by action of the amortisseur grids, so that 6.00 per unit current will be drawn if rated voltage is applied. It is desired to reduce the starting current drawn from the line

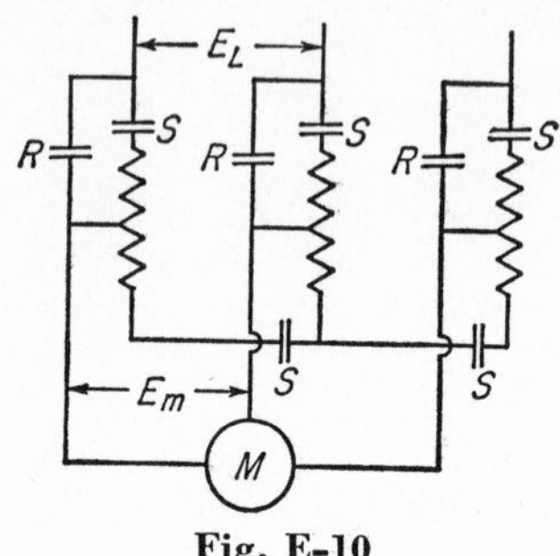

Fig. E-10

to 2.5 per unit. (*a*) Indicate by sketch how this would be done. (*b*) What voltage would be applied to the motor at starting?

ANSWER.

(*a*) See Fig. E-10. *E* has same meaning as *V*.

Autotransformer

Starting, *S* closes	S = starting contacts
R opens	R = running contacts
Run, *S* opens	E_m = voltage at motor
R closes	V_L = voltage on line

(*b*) The 6.00 per unit (600 per cent) inrush current is at starting. It is desired to limit the starting or inrush current to 2.5 per unit, or 250 per cent of rated. In autotransformers the starter (compensator) line current is proportional to the square of the voltage applied to the motor. Then $(V_m/V_L)^2 = 2.5/6.0$.

And $$V_m = \sqrt{(2.5/6.0)}\,V_L = \sqrt{0.416}\,V_L = 0.647V_L$$

or use the 65 per cent tap on the autotransformer. Finally, $V_m = 0.65 \times 2{,}300 =$ **1,495 volts.** *Note:* See also ways to start induction motors (p. 225).

E-26. A d-c series motor, used for railway purposes, is rated at 550 volts and 60 amp. The field resistance is 0.3 ohm and the armature resistance is 0.2 ohm. A starter is desired that will maintain the motor torque within 100 to 400 per cent of its rated value. Assume that the saturation curve is a straight line and determine the necessary resistance of the starter.

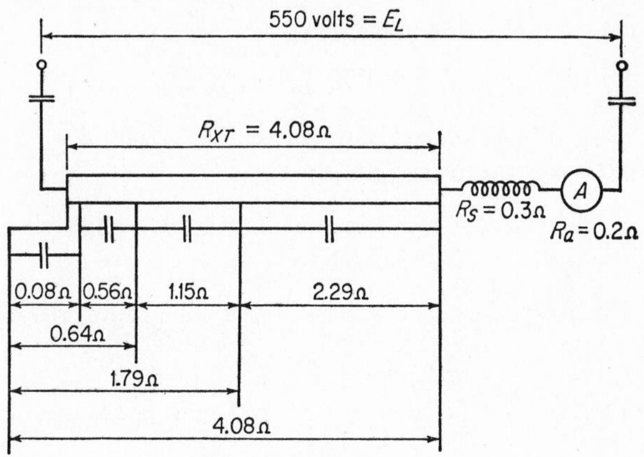

Fig. E-11

ANSWER. Refer to Fig. E-11. For a d-c motor torque $= KI_a\phi$. For a series motor $\phi = K_1I_a$. Therefore, for a series motor

$$\text{Torque} = K_tI_a^2$$

For 400 per cent torque I_a is equal to 200 per cent. Now in order to limit the starting torque to 200 per cent, the total resistance must be

$$R_t = \frac{E}{2I} = \frac{550}{2} \div 60 = 4.58 \text{ ohms}$$

The series or net external resistance is found to be

$$R_x = R_t - (R_s + R_a) = 4.58 - (0.3 + 0.2) = 4.08 \text{ ohms}$$

The motor will accelerate, and the current will decrease to 100 per cent or 60 amp. At this point, the voltage across the motor terminals is

$$E_m = I_aR_m = I_a(R_s + R_a) = 60 \times 0.5 = 30 \text{ volts}$$

The voltage across the external resistance is

$$E_x = E_L - I_aR_x = 550 - (60 \times 4.08) = 305.2 \text{ volts}$$
$$E_g = E_x - E_m = 305.2 - 30 = 275.2 \text{ volts}$$

Cutting out the resistance for the first step, the total resistance for the first step is

$$R_t = \frac{E_L - E_g}{2I_a} = \frac{550 - 275.2}{2 \times 60} = 2.29 \text{ ohms}$$

The balance of the external resistance

$$R_{x1} = R_t - (R_s + R_a) = 2.29 - 0.5 = 1.79 \text{ ohms}$$

The motor will accelerate, and the current will decrease. The voltage across the motor resistance

$$E_{rx1} = E_L - I_a R_{x1} = 550 - (60 \times 1.79) = 443 \text{ volts}$$
$$E_g = 443 - 30 = 413 \text{ volts}$$

Cutting out the resistance for the second step. The total resistance for the second step

$$R_{tx2} = \frac{E_L - E_g}{2I_a} = \frac{550 - 413}{2 \times 60} = 1.14 \text{ ohms}$$

The external resistance of the second step

$$R_{x2} = R_{tx2} - (R_s + R_a) = 1.14 - 0.5 = 0.64 \text{ ohm}$$

The motor will continue to accelerate, and the current will decrease. The voltage across the motor resistance is shorted.

$$E_{gx2} = E_L - I_a R_{x2} = 550 - (60 \times 0.64) = 511.6 \text{ volts}$$

If at this point all external resistance is shorted,

$$E_g = 511.6 - 30 = 481.6 \text{ volts}$$
$$I_t = \frac{E_L - E_g}{R_s + R_a} = \frac{550 - 481.6}{0.5} = 136.8 \text{ amp}$$

This is too high and another step is indicated. The balance of the total resistance

$$R_{tx3} = \frac{550 - 481.6}{2 \times 60} = 0.58 \text{ ohm}$$

The balance of the external resistance

$$R_{x3} = R_{tx3} - (R_s + R_a) = 0.58 - 0.5 = 0.08 \text{ ohm}$$

Part II
ALTERNATING CURRENT

Chapter 13

ALTERNATING CURRENT

13-1. A-C Circuits. Alternating current is so called because the current in the circuit, upon which an alternating current is impressed, changes direction periodically. For example, in a 60-cycle circuit the current reverses direction completely 60 times per second (60 cps) as in Fig. 13-1. Taking a direct current and changing its magnitude and direction periodically results in an alternating current. In $\frac{1}{60}$ of 1 sec the current in this 60-cycle circuit increases from zero to a maximum value in one direction, then decreases to zero; after reversing direction it increases to a maximum in the opposite direction, then decreases again to zero, completing one cycle. Sixty such cycles occur in 1 sec. Likewise for 50 and 25-cycle alternating current.

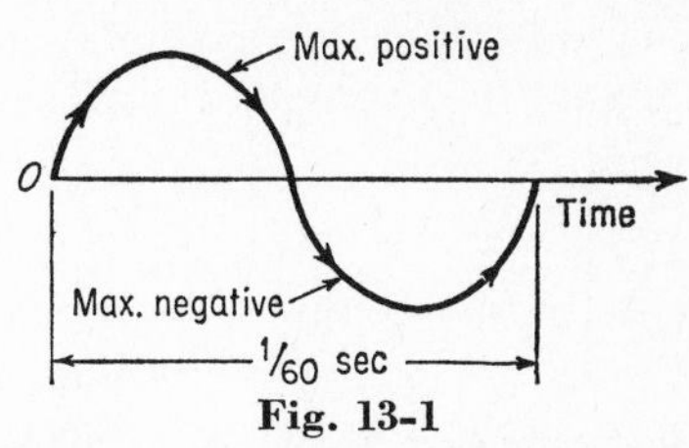

Fig. 13-1

Alternating current varies regularly from positive maximum to zero to negative maximum, as we have seen. The *average current* is therefore zero for each cycle. This is what we would read if we used a d-c meter measuring current (an ammeter) to measure alternating current. The varying alternating current does have an *effective* value of current I, and a-c ammeters measure this value. An *a-c ampere* is the current that produces the same amount of heat in a given resistor as one ampere of direct current.

It can be shown mathematically that for a sine-wave voltage or current the effective value equals $1/\sqrt{2}$, or 0.707, times the maximum value. Thus

$$E = 0.707E_{\max} \tag{13-1}$$

$$I = 0.707I_{\max} \tag{13-2}$$

The voltage goes through a similar sequence of changes in each cycle. But voltage and current cycles may not occur at the same time. The current cycle can start ahead, behind, or at the same time as the voltage cycle; i.e., the current leads the voltage, the current lags the voltage, or the current is in phase with the voltage. The type of circuit through which the current flows determines the relative positions of the voltage and current cycles.

An a-c volt is the emf required to cause one a-c ampere to flow through one ohm resistance. A-c voltmeters read the effective voltage E. Note that an alternating voltage subjects insulation to higher instantaneous stress than an equivalent direct voltage. Effective values of a-c waves are called *root-mean-square* (*rms*) values because they are found by taking the square root of the average (mean) value of the squared current or voltage wave. It can be found by squaring a number of the instantaneous values in a cycle, finding the average of these squares, and then extracting the square root.

13-2. Circuit Components. There are three types of electric-circuit elements: resistor, inductor, and capacitor. These are diagrammatically shown in Fig. 13-2. Typical apparatus using resistors are electric irons, heaters, grids, and incandescent lamp filaments. Typical electric apparatus using the coil (inductor) are transformers, motor and generator windings, and relays. Capacitors or condensers are the third and last electric-circuit element. In this group are radio circuit condensers, power-factor correction condensers, and condensers used with fractional horsepower motors.

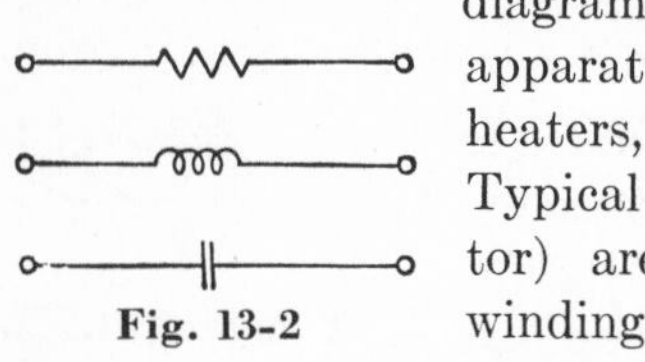

Fig. 13-2

13-3. Current and Voltage Relations. Now let us assume we have a pure resistor, one having no coil or condenser characteristics. Current flowing through such a device is always in step with the voltage across it. Figure 13-3 shows the diagram of a resistor in such an a-c circuit with the current and voltage across it in phase or resonance. Figure 13-4 shows the sine-wave forms. Here it is represented by the coincidence of the zero points of both current and voltage waves. As in direct current voltages and currents can be represented by weighted arrows, each pointing in the same direction (see Fig. 13-5). Calculation of power is the same as for direct current; namely, active power equals volts × amperes equals watts. A power factor does not come into the picture at this point since for a pure resistor it mathematically equals unity.

Now let us assume a pure coil or inductor, one having *no* resistor or condenser characteristics. In a circuit containing such a device

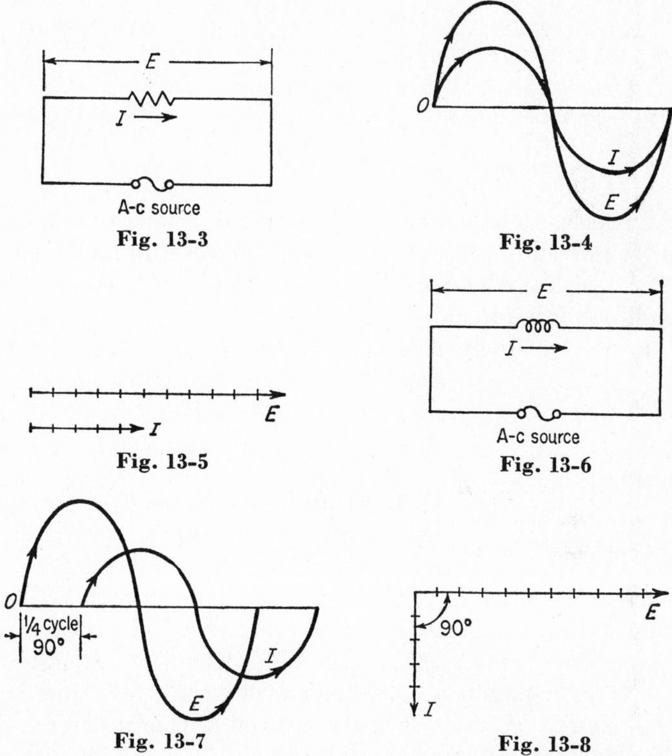

Fig. 13-3

Fig. 13-4

Fig. 13-5

Fig. 13-6

Fig. 13-7

Fig. 13-8

it is important to remember that the current *lags* behind the voltage by one-quarter cycle or 90°, one complete cycle being 360°. Figures 13-6 and 13-7 show the circuit diagram and sine-wave forms, with the current *I* lagging the voltage *E*. Note that the current passes through zero one-quarter of a cycle after the voltage has passed through the zero axis. This condition can also be represented by our current and voltage arrows, if we assume that the arrows pivot at their junction point and that they are capable of moving in a counterclockwise direction only (see Fig. 13-8). With this assumption the current arrow can be drawn 90° away from and lagging the voltage arrow.

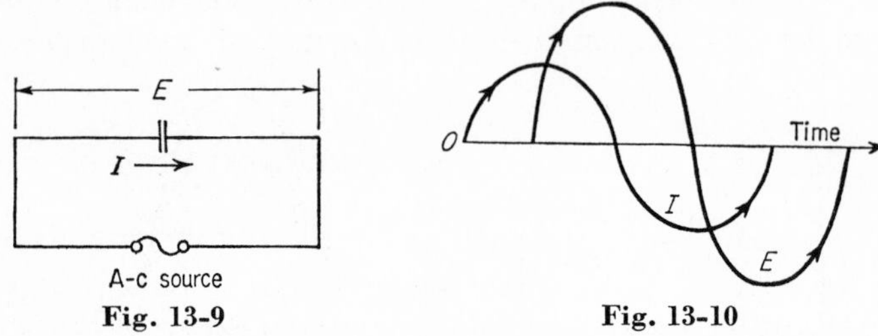

Fig. 13-9 **Fig. 13-10**

Finally, let us assume a pure capacitance or condenser having no resistor or inductor. Putting this pure condenser in an a-c circuit results in the current leading the voltage by 90°. Arrows show how this condition is represented. Assuming the same arrow rotation as for the coil, generally accepted, the current is now drawn 90° in a counterclockwise direction from the voltage arrow. Refer to Figs. 13-9, 13-10, and 13-11.

I

90°

E

Fig. 13-11

13-4. Combined Circuit Components. Up to this point we have talked about pure resistance, pure conductance, and pure capacitance circuits. Of course, it is impossible to find such pure circuits in practical work. A coil is made up of wire; hence it must have some resistance. It also has negligible capacitance in addition to its inductance. A condenser has some resistance in its leads and plates. A resistor wire, wound on a cylindrical form, also has some coil and condenser characteristics.

This should not throw us out of step if we remember that currents in pure resistors are in phase with their voltages; currents in coils are out of phase, lagging, whereas currents in condensers are out of phase, leading, with their respective voltages.

Adding resistance to a pure coil has the effect of rotating the current arrow in a counterclockwise direction toward the stationary voltage arrow. This results in reducing the angle between the arrows. If the resistance becomes large compared to the inductance, the arrows (vectors) practically coincide. Now at this point, if capacitance is inserted in the circuit, the arrow moves away from the voltage in a counterclockwise direction. Adding more capacitance widens this angle. If the capacitance becomes great compared to the resistance and the inductance, the current arrow will end up leading the voltage by almost 90°.

13-5. Simple A-C Calculations. Now let us assume a relay coil has an a-c voltage E applied across the terminals. This coil has little resistance, large inductance, and very small condenser characteristics. The relay coil is represented by the circuit in Fig. 13-12. Since the relay coil has a large inductance, the current lags almost

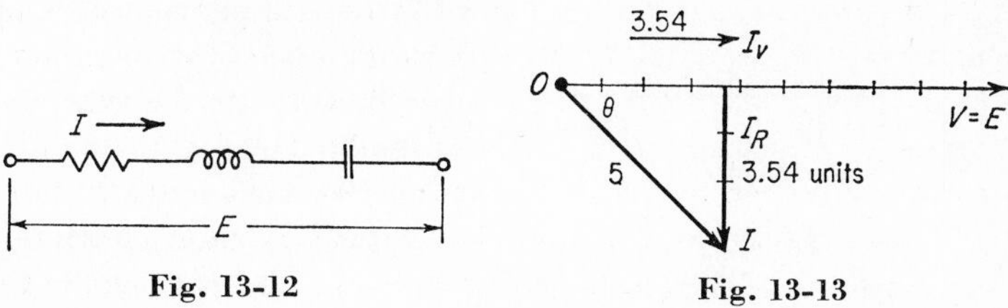

Fig. 13-12

Fig. 13-13

90° behind the voltage. If the voltage across the terminals is 100 volts and the current flowing is 50 amp, the total va power flowing to the circuit is 100 × 50, or 5,000 va, or 5 kva.

The coil now consumes active power since there is a component of the total current which is in phase with the voltage because of the resistance present in the coil. This component is obtained by drawing a line from the tip of the current arrow I_1 perpendicular to the voltage arrow V. The component of current in phase with the voltage is that length on the voltage arrow between the origin 0 and the intersection of the perpendicular with the voltage line (see Fig. 13-13).

Active power in the coil is the product of the voltage V and the component of the current I_v in phase with V. This component measured on the arrow diagram is 3.54 units, or 35.4 amp.

The active power in the circuit is then $VI_v = P$, or

$$100 \times 35.4 = 3{,}540 \text{ watts}$$

or 3.54 kw. Now the power factor of the coil may be calculated as the ratio of the active to the total va power.

$$\text{Power factor} = 3{,}540/5{,}000 = 0.707, \text{ or } 70.7 \text{ per cent lagging}$$

If we had assumed a condenser circuit, it could have been represented as in Fig. 13-14. Since a condenser has almost negligible resistance and inductance, or coil, characteristics, the current leads the voltage by almost 90°. Here you can readily see that the total va is 5,000, and from the arrow diagram that the total active power is 860 watts and the power factor is 0.172, or 17.2 per cent leading.

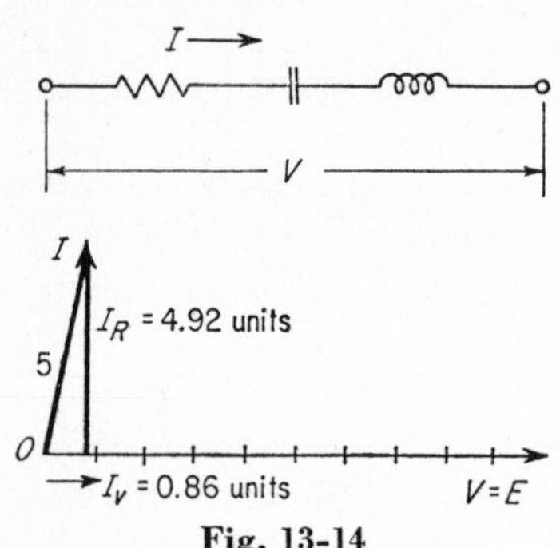

Fig. 13-14

If we had been talking about an electric heater, the diagram would be represented by Fig. 13-15. Since a heater has more inductance than capacitance, the current I lags the voltage as shown. Here the power factor is 4,750/5,000, or equivalent to 95 per cent lagging. Also refer to Fig. 13-16.

13-6. Inductive Reactance. We have learned that if an electric circuit contains inductance, an emf is induced in it whenever there is a change in the current flowing in the circuit. This induced emf always opposes the change in the current. Thus, the current lags behind the voltage.

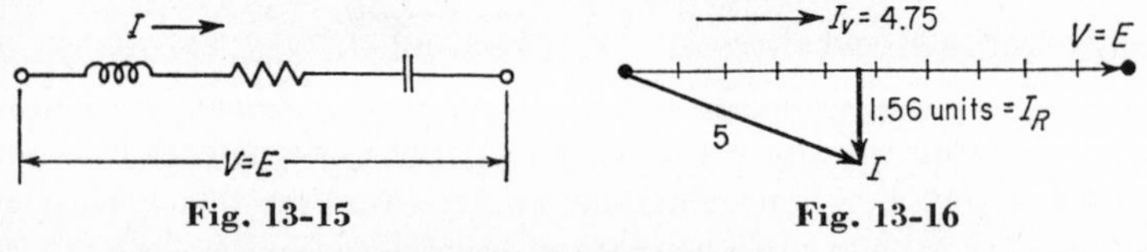

Fig. 13-15 Fig. 13-16

Since a simple resistance in a circuit does not cause a current to "lag" or "lead," we can say that inductance has an effect which is at 90°, or at right angles, to the effect of resistance. The result of this we will see later when the combined effect of inductance and resistance is considered. If there is also a resistance in series with the inductance, the angle of lag will be less than 90°. This opposition of self-inductance to the flow of the current is called *inductive reactance*, and is measured in ohms just as pure resistance is. As we already know, the current flowing through a resistance is found by the equation

$$\text{Current} = \frac{\text{voltage}}{\text{resistance}}$$

Hence the current flowing through an inductive reactance is found by the equation

$$\text{Current} = \frac{\text{voltage}}{\text{reactance}}$$

or

$$I = \frac{E}{X_L} \tag{13-3}$$

where I = current, amp

X_L = inductive reactance, ohms

E = voltage necessary to force current through inductive reactance

We must remember that if we use a maximum value of voltage we get a maximum value of current; an effective value of voltage gives an effective value of current; an average value of voltage gives us an average value of current.

Q13-1. An inductance coil contains resistance of 10 ohms and an inductance of 0.1 henry. If the frequency of the alternating current is 60 cps, what is the voltage required to cause a current of 3 amp to flow through the circuit?

ANSWER. The inductive reactance X_L is found as follows:

$$X_L = 2\pi fL = 2\pi \times 60 \times 0.1 = 37.7 \text{ ohms}$$

For pure resistance the voltage requirement is

$$E_r = IR = 3 \times 10 = 30 \text{ volts}$$

For the inductive reactance

$$E_L = IX_L = 3 \times 37.7 = 113.1 \text{ volts}$$

The applied voltage is therefore

$$E = \sqrt{E_r^2 + E_L^2} = \sqrt{20^2 + 113.1^2} = \mathbf{117\ volts}$$

Q13-2. A circuit containing pure inductance of 0.2 henry is connected across 110-volt 60-cycle mains. What current flows?

ANSWER. The reactive effect is found and then inserted into the standard formula.

$$X_L = 2\pi fL = 2\pi \times 60 \times 0.2 = 75.4 \text{ ohms}$$

$$I = \frac{E}{X_L} = \frac{110}{75.4} = \mathbf{1.46\ amp}$$

Q13-3. An electric device operates on 120 volts and 60 cycles, taking 5 amp with a power factor of 90 per cent lagging. What is the inductance of the device?

ANSWER. The impedance of the device is found in the usual manner.

$$Z = \frac{E}{I} = \frac{120}{5} = 24 \text{ ohms}$$

To obtain the pure resistance we must consider the effect of the power factor

$$R = Z \cos \phi = 24 \times 0.90 = 21.6 \text{ ohms}$$

$$X_L = \sqrt{Z^2 - R^2} = \sqrt{24^2 - 21.6^2} = 10.49 \text{ ohms}$$

$$L = \frac{X_L}{2\pi f} = \frac{10.49}{2\pi \times 60} = \mathbf{0.0278\ henry}$$

Q13-4. A coil of wire is of such inductance that a current changing at the rate of 1 amp per sec induces a reverse pressure of 0.025 volt. An alternating current having a frequency of 100 passes through it. Neglecting the ohmic resistance, what is the ohmic equivalent of inductance?

ANSWER. $X_L = 2\pi fL = 6.28 \times 100 \times 0.025 = \mathbf{15.7\ ohms}$

13-7. Capacitive Reactance. As we reviewed before, when pure capacitance is in a circuit, the current leads the voltage by 90°. The opposition which the capacitance offers to the flow of the current is called *capacitive reactance*, and is measured in ohms, just as resistance. The symbol for capacitive reactance is X_C. The equation for current and voltage relations in a circuit containing capacitive reactance of X_C is similar to Ohm's law for the current-voltage relations in a circuit containing a resistance of R. Thus,

$$I = \frac{E}{X_C} \tag{13-4}$$

where the nomenclature is similar to that for Eq. (13-3).

The relation between capacitive reactance, frequency f, and capacitance C is given by

$$X_C = \frac{1}{2\pi fC} \tag{13-5}$$

Q13-5. What is the capacitive reactance of an a-c circuit containing 35 μf capacitance if the frequency is 60 cycles?

ANSWER. $X_C = \frac{1}{2\pi fC} = \frac{1}{6.28 \times 60 \times 35 \times 10^{-6}} =$ **75.8 ohms**

13-8. Combinations of Inductive and Capacitive Reactance. As we have seen, inductive and capacitive reactance have exactly opposite effects upon the phase relation of the current curve to the voltage curve. For the sake of repetition:

Inductive reactance causes the current to "lag."

Capacitive reactance causes the current to "lead."

Accordingly, when an a-c circuit has a combination of inductance and capacitance in series, they tend to neutralize each other. This combined effect, being the difference between them, is called the reactance.

$$X = X_L - X_C \tag{13-6}$$

where all are in ohms. When the result is positive, the reactance is the result of more inductive reactance than capacitive reactance; when negative, there is a greater preponderance of capacitance over inductance.

Q13-6. What is the reactance of an a-c circuit having 20 ohms inductive reactance in series with 10 ohms capacitive reactance?

ANSWER. $20 - 10 =$ **10 ohms reactance** (inductive)

Q13-7. What is the reactance of a 60-cycle a-c circuit containing an inductance of 1.5 henrys in series with a capacitance of 40 μf?

ANSWER. Proceed by first obtaining the inductive reactance, then the capacitive reactance, and finally subtracting X_L from X_C to determine the reactance.

$$X_L = 2\pi fL = 6.28 \times 60 \times 1.5 = 565 \text{ ohms}$$

$$X_C = \frac{1}{2\pi fC} = \frac{1}{6.28 \times 60 \times 40 \times 10^{-6}} = 66.3 \text{ ohms}$$

$$X = X_L - X_C = 565 - 66 = \textbf{499 ohms reactance} \text{ (inductive)}$$

13-9. Impedance—Combination of Resistance and Reactance. The resultant effect of a series combination of resistance and reactance cannot be found by a simple addition of the two. It may be found by the graphical method.

Let us assume that we are required to find the end result of a series combination of 4 ohms pure resistance and 3 ohms pure inductive reactance. Draw line R to scale to represent 4 ohms pure resistance (Fig. 13-17). Now since we already have learned that the effect of reactance is at right angles (90°) to pure resistance, we can draw line X_L to scale at right angles from the end of R, representing 3 ohms reactance. The line Z then is the resultant and represents impedance. If we proceed to scale off line Z, we find it to be equal to 5 ohms impedance. Note this is not the sum of 3 and 4, nor the difference between the two.

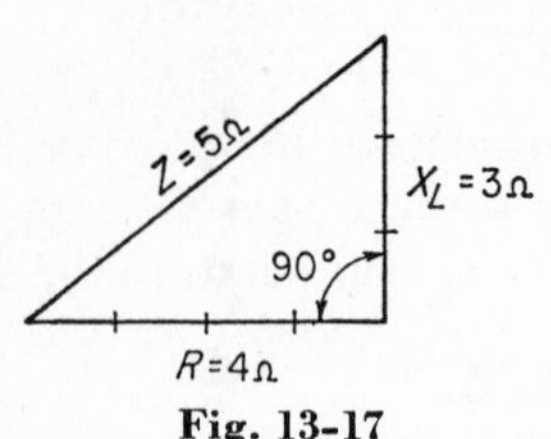

Fig. 13-17

We see that the lines R, X_L, and Z always form a right triangle with Z as the hypotenuse. And with the application of the Pythagorean proposition we can find the length of line Z by use of the simple formula: "In a right-angled triangle the square of the hypotenuse equals the sum of the squares of the other sides." Thus,

$$Z^2 = R^2 + X_L^2 \tag{13-7}$$

and

$$Z = \sqrt{R^2 + X_L^2} \tag{13-8}$$

In the example above (Fig. 13-17),

$$Z = \sqrt{4^2 + 3^2} = \sqrt{16 + 9} = \sqrt{25} = 5 \text{ ohms} \tag{13-9}$$

If the reactance is replaced with a capacitive reactance, we can draw the diagram as in Fig. 13-18, with the line representing the capacitive reactance X_C drawn downward, showing that it acts in a direction opposite to the inductive reactance. Mathematically, the result and the equation are the same as we obtained in the case of the inductive reactance.

$$Z = \sqrt{R^2 + X_C^2} \tag{13-10}$$

If a circuit contains both inductance and reactance in series with resistance, we can easily determine the impedance by first subtracting the capacitive reactance X_C from the inductive reactance X_L and then com-

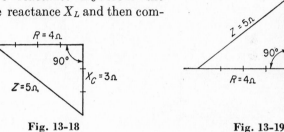

Fig. 13-18

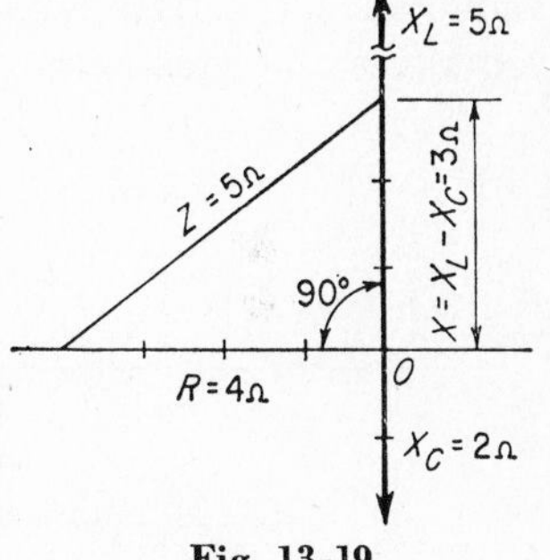

Fig. 13-19

bining the result X with the resistance R as in our previous discussion. The equation would then be

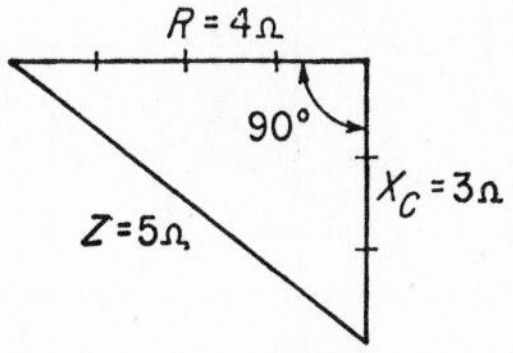

$$Z = \sqrt{R^2 + (X_L - X_C)^2} \tag{13-11}$$
$$Z = \sqrt{R^2 + X^2} \tag{13-12}$$

We can represent the graphical method of obtaining the impedance in a series circuit consisting of resistance, inductance, and capacitance, as in Fig. 13-19.

Let us suppose that 5 ohms inductive reactance and 3 ohms capacitive reactance are in series with 4 ohms resistance.

The resistance R is shown by the horizontal line R (Fig. 13-19). Then the inductive reactance X_L is shown by the line X_L drawn *up* from the end of line R and at right angles to line R. The capacitive reactance is then shown by the line X_C drawn *down* from the end of line R. The resultant of X_L and X_C is determined by subtracting X_C from X_L equal to X. Thus,

$$X = X_L - X_C \tag{13-13}$$
$$= 5 - 3 = 2 \text{ ohms}$$

Line Z is then drawn as the resultant of R and X. By scaling off the result is 4.47 ohms.

Using the purely mathematical equation method

$$
\begin{aligned}
Z &= \sqrt{R^2 + (X_L - X_C)^2} \\
&= \sqrt{4^2 + (5 - 2)^2} = \sqrt{16 + 4} = \sqrt{20} = 4.47 \text{ ohms}
\end{aligned}
$$

As an example of inductance and resistance in series we have the induction coil. Here the same current must be forced through both reactance and resistance.

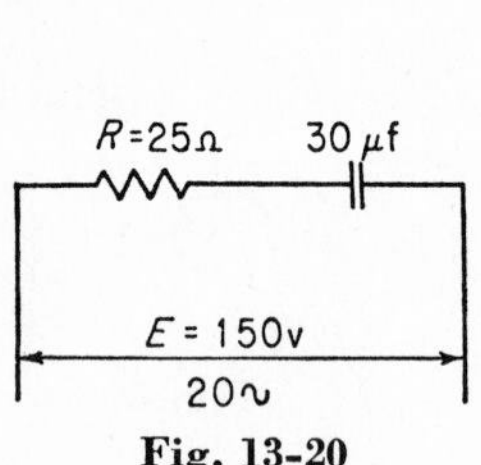

Fig. 13-20

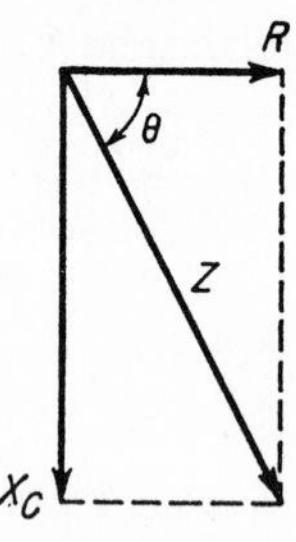

Fig. 13-21

Q13-8. What current would a 25-cycle a-c emf of 150 volts produce in a series circuit having 25 ohms resistance and 30 μf capacitance? What would be the power factor of this circuit? What effect on the current in the circuit would the addition of a one-half henry inductance produce if also connected in series?

ANSWER. Refer to Fig. 13-20. Note that 1 farad is equal to 10^6 μf. Let

E = impressed voltage of 150 volts
I = circuit current, amp
Z = circuit impedance, ohms
X_L = inductive reactance, ohms
X_C = capacitive reactance, ohms
$\cos\theta$ = power factor
f = current frequency, cps
L = inductance, henrys
C = capacitance, farads
R = circuit resistance, ohms

Now refer to Fig. 13-21.

$$X_C = \frac{1}{2\pi fC} = \frac{10^6}{6.28 \times 25 \times 30} = 212 \text{ ohms}$$

$$Z = \sqrt{R^2 + X_C{}^2} = \sqrt{25^2 + 212^2} = 214 \text{ ohms}$$

$$I = \frac{E}{Z} = \frac{150}{214} = \mathbf{0.7\ amp}$$

$$\text{Power factor} = \cos\theta = \frac{R}{Z} = \frac{25}{214} = 0.117, \text{ or } \mathbf{11.7\ per\ cent}$$

If a coil of one-half henry is introduced into the circuit in series (Figs. 13-21, 13-22, and 13-23)

$$X_L = 2\pi fL = 6.28 \times 25 \times 0.5 = 78.5 \text{ ohms}$$

$$X_C - X_L = 214 - 78.5 = 133.5 \text{ ohms} = X$$

$$Z = \sqrt{R^2 + (X_C - X_L)^2} = 136 \text{ ohms}$$

$$I = \frac{E}{Z} = \frac{150}{136} = \mathbf{1.10\ amp}$$

$$\cos\theta = \frac{R}{Z} = \frac{25}{136} = 0.184, \text{ or } \mathbf{18.4\ per\ cent}$$

Thus, the addition of one-half henry inductance will increase the circuit current by $1.10 - 0.7 = 0.4$ amp and the power factor from 11.7 per cent to 18.4 per cent. This problem gives us an introduction to the power factor and the important matter of its improvement.

Q13-9. A series *LCR* circuit with an inductance of 100 henrys has a resonant frequency of 5 cps. (*a*) What is the period, (*b*) the capacity? Neglect the effect of *R* on the frequency.

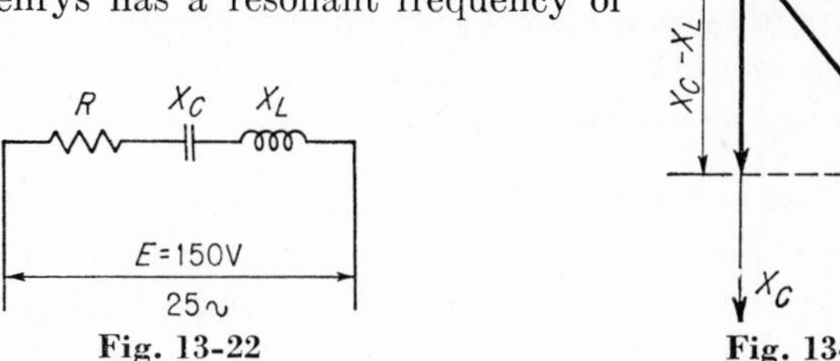

Fig. 13-22

Fig. 13-23

ANSWER.

(*a*) The period of any oscillation is the time required per cycle. Hence, the period T

$$T = \frac{1}{f} = \frac{1}{5} = \mathbf{0.2\ sec}$$

(*b*) In the absence of resistance, the natural undamped frequency of transient oscillations is the same as the a-c resonant frequency. Thus, the angular frequency is

$$\omega_0 = \frac{1}{\sqrt{LC}}$$

whence

$$C = \frac{1}{\omega_0^2 L} = \frac{1}{(5 \times 2\pi)^2 \times 100}$$

$$C = 10.1 \times 10^{-6} \text{ farad, or } \mathbf{10.1\ \mu f}$$

13-10. Reactive Power. Both condenser and coil circuits are said to be reactive. It should be apparent now that in a pure reactive circuit the current and the voltage are displaced from each other by 90°. How about power in a reactive circuit? Remember that active power is obtained by multiplying the current and the voltage that are in phase, i.e., when they can be represented by arrows pointing in the same direction. In a pure reactive circuit, the current and the voltage are not in phase; hence there can be no active power consumed in the circuit. But since current flows in the circuit, there must be flow of power, which is not consumed in the circuit but flows between the coil or condenser and the source of voltage applied to the circuit.

This power, which is not consumed in the circuit, is measured in volt-amperes, VA, and is found by multiplying the voltage by the current just as in direct current, i.e., volts × amperes.

13-11. Power Factor. Entering into practically all considerations of a-c machinery and circuits is the term "power factor." We have used it in the solution of problems but merely as an introductory measure.

The power factor (pf) is simply a multiplier. It indicates what part of the total "apparent" power (measured in d-c fashion) flowing in an a-c circuit is real or consumed power. The power factor has also to do with the *reactive* or so-called "idle" power in an a-c system.

The use of the power-factor multiplier is not difficult. However, it immediately brings up the question: Why is not all the apparent power flowing to a load consumed by it? Why is it necessary to generate more than apparently required for the actual kilowatts used by the loads in the plant? What is "idle" power in an a-c system having low power factor, and how can it be removed? The answers to these questions will be made in due course.

These considerations can be discussed conveniently in terms of the *induction motor*. Induction motors, although they are the simplest and most effective general purpose a-c drive, are, nevertheless, the most prevalent cause of low power-factor conditions. When a system has a lagging current, it frequently is because the load is composed of a number of induction motors. By using rotors in motors composed of special windings, an induction motor can be made to take current in phase with or even leading the voltage.

Electric motors rotate and do work by means of forces produced by the reaction of magnetic fields. To establish the magnetic fields all types of motors must be magnetized from some source.

D-c motors and *synchronous motors* are magnetized from a source of direct current. For d-c motors this source is the d-c line. For synchronous motors the magnetization is supplied to field coils on the rotor of the motor from a d-c source: a generator often called an "exciter."

Induction motors, however, have no d-c exciter, nor do they utilize any direct current for their magnetization. Induction motors must be magnetized by alternating current from the a-c line to which they are connected. It is this fact that inherently makes induction motors the chief offender from the viewpoint of power factor.

The current of a *squirrel-cage motor* lags particularly far behind the voltage on starting or when running under light load. At full load the angle of lag is usually between 25 and 40°.

13-12. Input to Induction Motors Has Two Components. When an induction motor is connected to the line, alternating current flows into it. This current is made up of two components.

One of these components establishes the magnetic fields in the motor. The other component, acting through the magnetic fields, is consumed in the motor. That is, it is converted from electric energy to mechanical energy to do work.

The magnetic field in the stator, or stationary part of the motor, whirls around in the stator at a speed slightly faster than the operating speed of the motor. This magnetic field sweeping across the air gap between the stator and the rotor generates currents by induction in the rotor bars. These rotor currents then create a rotor magnetic field that embraces the revolving magnetic field in the stator causing the rotor to turn the shaft.

The component of current that flows into the motor to establish the magnetic fields is called the *magnetizing-current component.* The component of current that flows into the motor to be consumed in doing work is called the *power-current component.* The relative proportions of these two current components flowing into the motor determine the *power factor* of the motor.

13-13. Active and Reactive Power. In working with power-factor problems it is usually preferable and more convenient to deal in terms of power rather than current. In other words, it is more convenient to deal with *current* × *voltage.*

The product of the power-current component and the voltage is measured in kilowatts. This is the power component from the line to the induction motor, which is consumed in the motor doing work and in taking care of the motor losses.

The product of the magnetizing-current component and the voltage is measured in reactive kilovolt-amperes (reactive kva). This is the magnetizing component from the line to the induction motor. Reactive kva surges back and forth to the line from the motor to the generator with each alternation of current to establish the magnetic fields of the motor. It is not consumed in the motor.

Thus, we see that the total input to an induction motor is made up of a kilowatt component and a reactive kva component. These two components make up the "apparent" power, or kilovolt-ampere (kva), to the motor. The kva is not the simple addition of these two components, but is their right-angle or "vectorial" sum.

13-14. Idle Power. Reactive kva is idle power. It does no real work. Magnetic fields have a property much like the inertia of a flywheel. It takes power to establish magnetic fields just as it takes power to bring a flywheel up to speed. But when a magnetic field collapses, the power that was put into it is returned to the circuit. Similarly, a flywheel returns power when brought to rest.

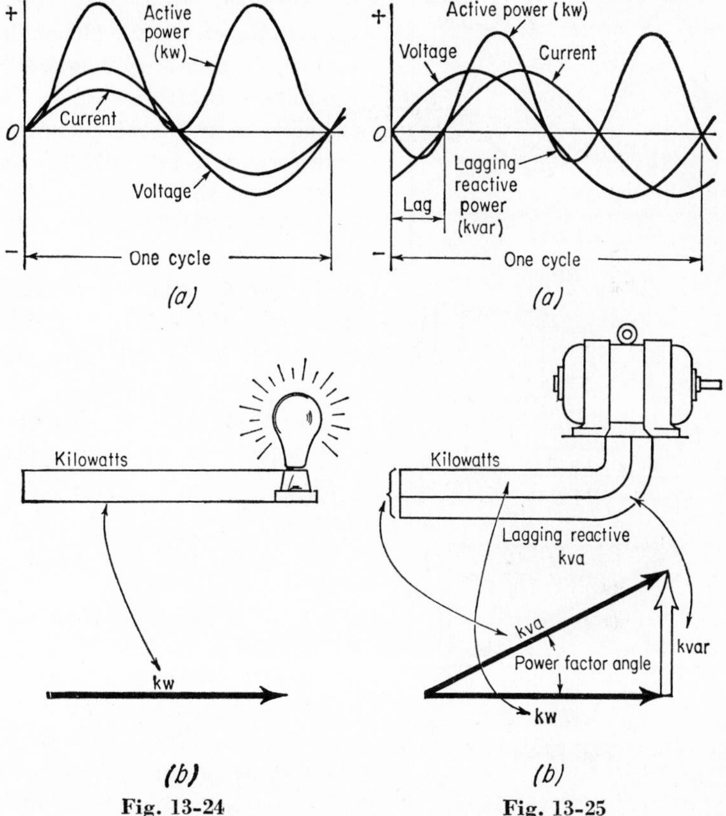

Fig. 13-24 Fig. 13-25

Thus, the magnetic fields of an induction motor give rise to a pulsating magnetizing power, called reactive power because it is not permanently consumed power. It merely surges back and forth between the generator and the induction motor, twice for each cycle of the alternating current to provide the magnetic fields of the induction motor.

13-15. Reactive Kva Not Metered. Because it is not permanently consumed power, reactive kva does not register on the regular kilowatthour meter, and for that reason it is often called "wattless," "idle," or just "deadhead" power. Also, because it is not permanently consumed power, reactive kva is drawn at right angles to the kilowatts in any graphical representation of a-c loads. Reactive kva, although it performs a necessary magnetizing function, nevertheless takes up valuable room in the generators, transformers, and distribution lines—room that could be utilized to carry useful kilowatts. It is the purpose of *power-factor improvement* to relieve as large a portion of the a-c power system as possible of this reactive kva.

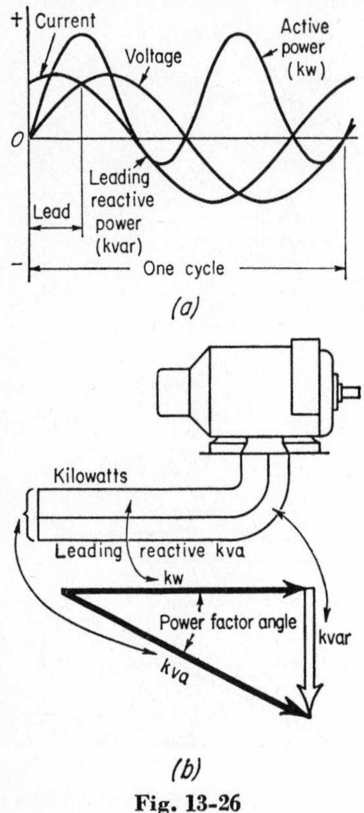

Fig. 13-26

13-16. Power-factor Terms. Power factor is spoken of as being "unity," "lagging," or "leading." The significance of these terms is illustrated in Figs. 13-24*a*, 13-24*b*, 13-25*a*, 13-25*b*, 13-26*a*, and 13-26*b*. They have to do with the effect of the particular type of load on the relationship of voltage and current. Whenever the voltage and current shift in relation to each other, reactive kva is produced. The power factor becomes less than unity—and lagging or leading—as shown by the chart, Table 13-1, for various a-c loads.

13-17. Low, Zero, and High Power Factor. We often hear of the power factor as being *low* or *high*. These are relative terms.

By low power factor is generally meant a relatively large, lagging reactive kva component compared to the kilowatt component. Low, lagging power factor will usually be found in plants where the load consists largely of induction motors. Low power factor can,

TABLE 13-1. POWER FACTOR OF TYPICAL A-C LOADS

Unity (or near unity) power factor		Lagging power factor		Leading power factor	
Load	Approximate power factor	Load	Approximate power factor	Load	Approximate power factor
Incandescent lamps (Power factor of lamp circuits operating off step-down transformers will be somewhat below unity)	1.0	Induction motors (at rated load):		Synchronous motors	0.9, 0.8, 0.7, 0.6, etc., leading power factor depending on the rated leading power factor for which they are built
		Split-phase below 1 hp	0.55–0.75		
		Split-phase 1–10 hp	0.75–0.85		
		Polyphase, squirrel-cage:			
		High-speed, 1–10 hp	0.75–0.90		
Fluorescent lamps (With built-in capacitor)	0.95–0.97	High-speed, 10 hp and larger	0.85–0.92	Synchronous condensers	Nearly zero leading power factor (Output practically all leading reactive kva)
Resistor heating apparatus	1.0	Low-speed	0.70–0.85		
Synchronous motors (Operate at leading power factor at part loads; also built for leading power factor operation)	1.0	Wound-rotor	0.80–0.90		
		Groups of induction motors	0.50–0.85	Capacitors	Zero leading power factor (Output practically all leading reactive kva)
		Welders, motor-generator type	0.50–0.60		
		transformer type	0.50–0.70		
		Arc furnaces	0.80–0.90		
Rotary converters	1.0	Induction furnaces	0.60–0.70		

however, also be leading. This is often the case with a generator connected to a long, lightly loaded transmission line.

When we speak of zero power factor, we mean that *all* the kva in an a-c circuit is reactive kva. This condition holds for either zero power factor *leading* or zero power factor *lagging*. An electromagnet is nearly zero power factor lagging, since it draws practically only magnetizing current from the a-c circuit. A capacitor is nearly zero power factor leading, since it draws practically only capacity or leading current from the a-c circuit.

When we speak of high power factor, we generally mean a relatively small amount of reactive kva in proportion to the kilowatts. Achieving high power factor is the objective of power-factor improvement.

13-18. Power-factor Improvement. Whenever the power factor of an industrial plant is *below unity*, which is usually the case, the kva will be larger than the kilowatts, and there will be reactive kva burden on the power system. Now, because generators, transformers, and distribution lines are limited in their capacity by the total kva they must carry, it is desirable to relieve the power system wherever possible of reactive kva, that is, to keep the power factor high.

13-19. The Power Bill. Now with the terms defined, we are ready to tackle the most involved rate schedules. Remember first that most industrial schedules of power are made up of two major components, the demand charge and the energy charge.

The demand charge is based on kilowatt or kva demand. When the kilowatt demand is used, there is probably a power-factor clause that adds a charge if the power factor is less than a predetermined amount. When the kva demand is used, there is no need for a power-factor clause since such a charge is included in the demand. This is evident on noting the relation between power factor and kva, kva being equal to kw divided by power factor.

If the kilowatt load were 100, the kva would be 125 at 0.80 power factor and only 100 kva at unity power factor. This shows it is good business sense to maintain as high a power factor as possible. The power-factor clause is introduced by most utilities as a means of distributing cost of larger lines and equipment on an equitable basis. By improving the power factor, transmission lines supplying the load can carry additional power with the same voltage drop as before under lower power-factor conditions.

13-20. Economics of High Power Factor. If the power factor can be corrected to 90 per cent or better, operating costs may be reduced in addition to I^2R losses. The power triangle can be used to determine the amount of reactive kva needed to correct the power factor to improved conditions. When speaking of power factor in industrial plants, we automatically imply that the power factor is lagging, since most loads consist of induction motors, transformers, and lighting. These all have large inductance or coil characteristics. Hence, an uncorrected power factor is bound to be lagging.

To correct the power factor means adding enough devices with leading power-factor characteristics to overcome some of the lagging characteristics of the load. In condensers, you will recall, the current always leads the voltage, and so these are suitable leading power-factor devices. Synchronous motors may also be used both to supply a mechanical load and to improve the power factor.

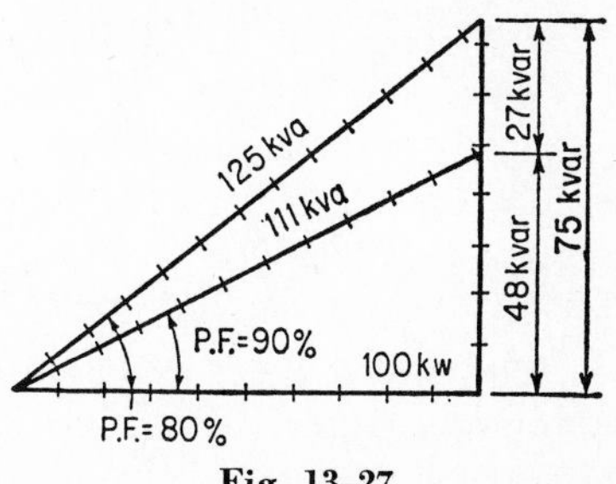

Fig. 13-27

Let us analyze a case showing all the steps in determining the amount of corrective reaction kva required. At 90 per cent power factor the ~~reactive~~ kva would be kw/pf = 100/0.9 = 111. Reactive kva would then be equal to

$$\sqrt{111^2 - 100^2} = 48$$

Now let us draw the power triangle for 90 per cent power factor superimposed on the triangle for 80 per cent power factor (Fig. 13-27). The amount of leading condenser capacity required to correct to 90 per cent is evidently

$$75 - 48 = 27 \text{ kva}$$

If the cost of 27 kva of condensers is \$270, it would take three and one-third years to return the investment cost by eliminating the power-factor charge, assuming certain unit costs.

Q13-10. An industrial plant has a rate schedule that requires a demand charge of \$1.50 per kva per month. A group of induction motors operate at an average power factor of 0.60 and draw 50 kw from the 440-volt supply. (*a*) What kva of capacitor is needed to raise the power factor to 0.80? (*b*) What will be the annual savings

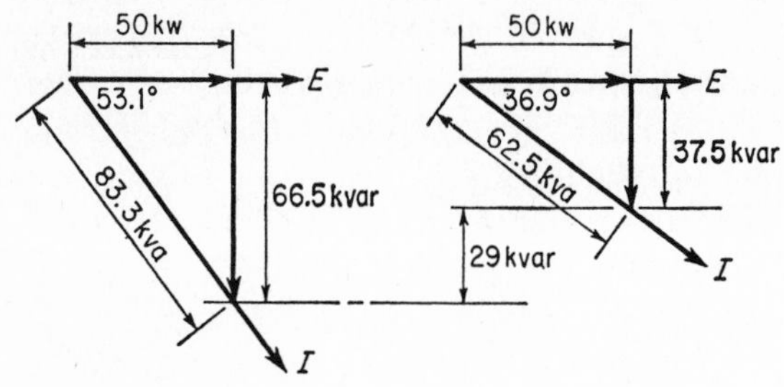

Fig. 13-28

in the power bill for this group of motors from such an increase in the power factor? See Fig. 13-28.

ANSWER.

$$\cos^{-1} 0.60 = 53.1^\circ \qquad \sin 53.1^\circ = 0.8 \qquad \tan 53.1^\circ = 1.33$$
$$\cos^{-1} 0.80 = 36.9^\circ \qquad \sin 36.9^\circ = 0.6 \qquad \tan 36.9^\circ = 0.75$$

Refer to figures.

$$\text{Reactive kva @ 60\% pf} = 50 \times 1.33 = 66.5 \text{ reactive kva}$$
$$\text{Reactive kva @ 80\% pf} = 50 \times 0.75 = 37.5 \text{ reactive kva}$$

(*a*) To raise 60 per cent pf to 80 per cent pf the capacitor reactive kva should be

$$\text{Reactive kva @ 60\%} - \text{reactive kva @ 80\%} = 66.5 - 37.5$$
$$= \mathbf{29\ reactive\ kva} \text{ to be added}$$

(*b*)
$$\text{Kva @ 60\% pf} = \text{kw}/0.6 = 83.3 \text{ kva}$$
$$\text{Kva @ 80\% pf} = \text{kw}/0.8 = 62.5 \text{ kva}$$

Annual savings in demand charge

$$\text{Months} \times (\text{kva } \% \text{ @ } 60\% - \text{kva @ } 80\%)(\text{cost/kva})$$
$$12(83.3 - 62.5)(1.50) = \mathbf{\$374}$$

Q13-11. Calculate the capacitor rating necessary to improve the power factor of a 10-hp 900-rpm squirrel-cage motor. The full-load efficiency of the motor is 84.5 per cent, and the motor power factor is 0.793.

ANSWER. Power input to motor is $\text{kw} = \dfrac{\text{hp} \times 0.746}{\text{efficiency}}$

$$\frac{10 \times 0.746}{0.845} = 8.73 \text{ useful kw}$$

Kva taken from lines is kva = kw/pf = 8.73/0.793 = 11 kva. If we refer back to Fig. 13-27

$$\text{Reactive kva} = \sqrt{11^2 - 8.73^2} = 6.7 \text{ reactive kva}$$

However, it is seldom economical to do this. Let us assume an improvement to 95 per cent. At this power factor the load kva is

$$\frac{\text{kw}}{\text{Power factor}} = \frac{8.73}{0.95} = 9.19$$

The reactive kva is found to be

$$\text{Reactive kva} = \sqrt{9.19^2 - 8.73^2} = 2.9 \text{ reactive kva component}$$

And the kva rating of the new capacitor is 6.7 − 2.9 = 3.8 kva. Economically, this would be sound.

13-21. Why Low Power Factor Is Detrimental. Low power factor is detrimental for three important reasons:

(*a*) The power factor cuts down the loadability; that is, it reduces the capacity of the power system to carry useful power as kilowatts. As a result larger generators, transmission lines, transformers, feeders, and switches must be provided for each kilowatt of the load when the power factor is low than when it is high. Thus the capital investment per kilowatt of the load is higher.

(*b*) A low power factor makes each kilowatt of power carry a higher burden of line losses. That is, it makes it cost more to transport each kilowatt of power.

(*c*) A low power factor may depress the voltage, and even cause some production slump by making the motors sluggish, dimming the lights, and slowing up the heaters.

13-22. Effect of Low Power Factor on the Generator. In order to accommodate some low power-factor load, generators are usually rated at 0.8 power factor. The full-load kilowatt capacity is thus 80 per cent of the kva, provided the rated power factor or better is maintained. When the power factor is lower than 0.8, then the resulting increase in total kva because of a large reactive kva component reduces the generator capacity, as shown in Fig. 13-29, if the rated voltage is to be maintained.

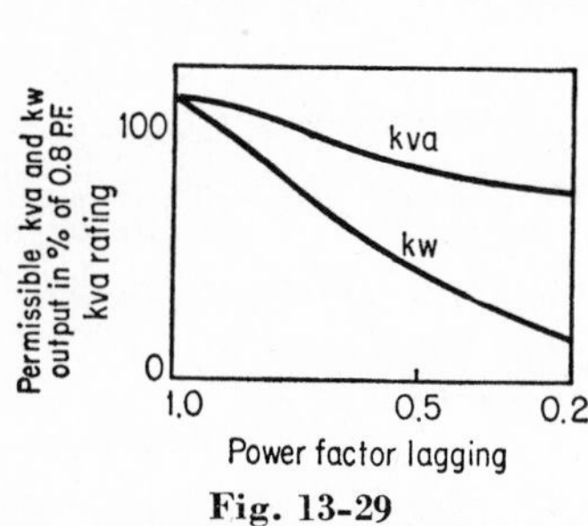

Fig. 13-29

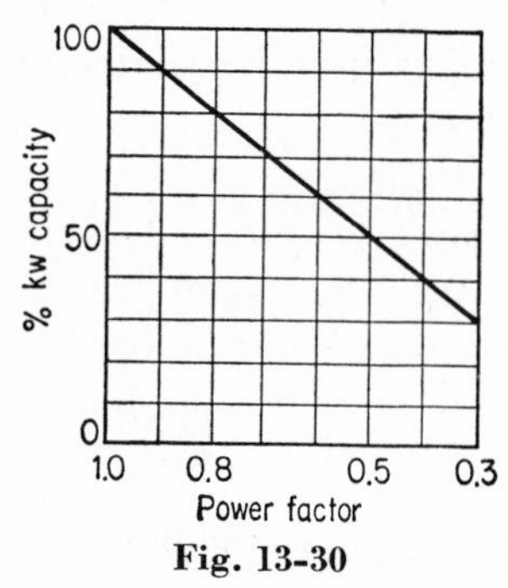

Fig. 13-30

At a low power factor the efficiency of the generator is also appreciably reduced. The following table shows approximately those multipliers by which the full-load efficiency of the engine-type generators is changed at various power factors.

TABLE 13-2

Power factor	0.9	0.8	0.7	0.6	0.5
Efficiency factor	1.02	1.00	0.98	0.95	0.91

13-23. Effect on Transformers. The kilowatt-carrying capacity of transformers is affected by the power factor of the load, as shown in Fig. 13-30. Then, too, the drop in voltage of a transformer with an increase in load through it is greater at a low power factor than at a high power factor. As a result, the voltage regulation of the transformer is impaired at low power factors.

13-24. Effect on Distribution. The size of conductors in electric circuits must be selected to carry safely the required current. Since at a low power factor the circuit must carry reactive kva in addition to useful kilowatts, it should be apparent that the distribution wires must be larger for a given kilowatt load at a low power factor than they would be at a high power factor.

The I^2R or resistance losses of the conductors are proportional to the square of the current passing through them. See Fig. 13-31 on the effect of the power factor on the losses of a 400-ft distribution line.

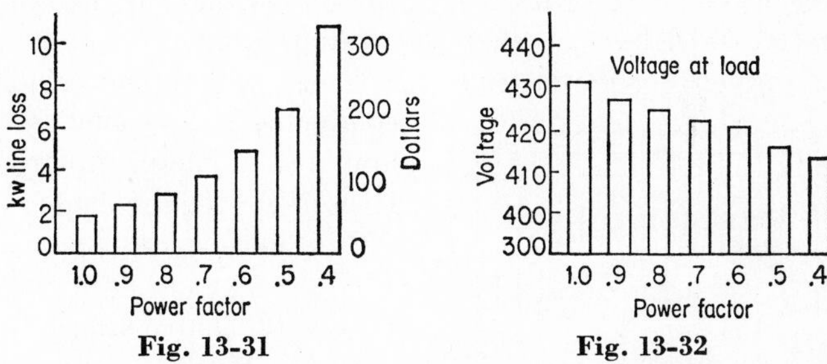

Fig. 13-31 **Fig. 13-32**

13-25. Effect on Voltage. As we know, the voltage drop increases in conductors with a decrease in the power factor. Figure 13-32 gives a pictorial representation of this effect on a distribution circuit. The voltage drop at the load ends of the distribution lines can be easily the result of a low power factor.

A low power factor may lower the production by causing a low voltage. The effects of undervoltage of several important pieces of equipment are shown in Table 13-3 below.

TABLE 13-3

Lights:	
5 per cent undervoltage means	16 per cent less light from incandescent lights
	7 per cent less light from fluorescent lights
10 per cent undervoltage means	30 per cent less light from incandescent lights
	12 per cent less light from fluorescent lights
Induction motors:	
10 per cent undervoltage means	19 per cent less maximum running torque
	1½ per cent less full-load speed
	11 per cent more full-load current
Heaters (resistor type):	
10 per cent undervoltage means	19 per cent less normal heat
	19 per cent lower heating speed

TABLE 13-4

Power factor	Amp per conductor	Wire size
1.0	131	No. 00
0.9	146	No. 00
0.8	164	No. 000
0.7	187	No. 000
0.6	220	No. 0000
0.5	263	300 MCM

The National Electrical Code (NEC) states that the size of feeder conductors shall be such that the voltage drop will not be more than 3 per cent for power loads. Table 13-4 above shows how the conductor size is affected by a low power factor in order to meet NEC requirements for the 400-ft circuit shown in Fig. 13-31.

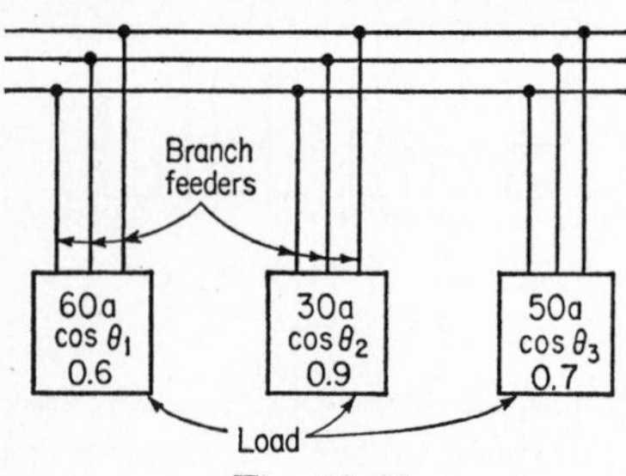

Fig. 13-33

Q13-12. Three branch feeders shown in Fig. 13-33 draw 60, 30, and 50 amp at 60, 90, and 70 per cent lagging power factor, respectively. How may the current in the main feed line feeding these units be calculated?

ANSWER. This is a simple problem in vector calculation. Since each current is known as to size and direction, a problem of this kind may be solved either graphically or by direct calculation.

$$I = \sqrt{[(60 \times 0.6) + (30 \times 0.9) + (50 \times 0.7)]^2 + [(60 \times 0.8) + (30 \times 0.435) + (50 \times 0.715)]^2}$$

$$= \mathbf{137.5\ amp}$$

The angle of lag of the resultant vector is found by use of $\cos \theta$.

$$\cos \theta = \text{power factor} = 98/137.5 = 0.7127$$

From which $\theta = 44.5°$. Thus the current lags the voltage at a power factor of 71 per cent.

13-26. How to Get a High Power Factor. A high power factor is obtained by making the ratio between the kilowatts and kva of the a-c circuits as nearly 1:1 as possible. This is accomplished, as seen from the power triangle (Fig. 13-34).

The quantities discussed above can be related to each other through the use of vector diagrams. We remember that active power is always in phase with the circuit voltage, while kva power either lags or leads it by less than 90°. We also have seen that the amount of lag or lead depends on whether or not the circuit load has largerly coil or condenser characteristics, respectively. Var power also lags or leads the circuit voltage by 90°.

In any given circuit the power quantities may be related by a right-angled triangle of arrows, as in Fig. 13-35. It is now possible with the basic facts at hand, a right-hand triangle, a suitable scale, and a knowledge of simple arithmetic, to make the following a-c

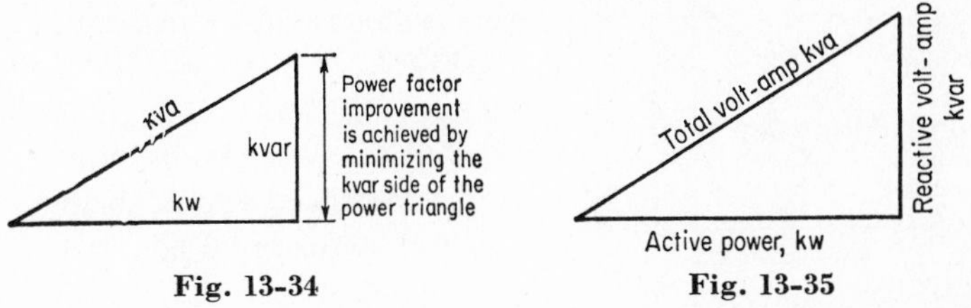

Fig. 13-34 **Fig. 13-35**

power calculations: determine the power factor, reactive kva, total kva, and kilowatts, provided any two are known for any circuit.

The reactive kva side of the power triangle may be minimized by minimizing the reactive kva component. This is a preventive method, such as by avoiding oversize or low-speed induction motors and keeping the induction motors as fully loaded as possible.

A high power factor may be achieved by eliminating the reactive kva, that is, where a particular motor drive permits, using unity-power-factor synchronous motors. These require no reactive kva from the line. Also a high power factor may be achieved by offsetting lagging reactive kva with leading-power-factor synchronous motors and capacitors.

It is a known fact that the magnetizing kva of induction motors remains almost constant from full load to no load, while the kilowatts vary with the load. Thus at light loading the motor input is largely reactive kva; hence the power factor suffers and is low.

It becomes apparent that any practical steps that will result in maintaining as nearly as possible full-load conditions will prove beneficial to power factor. There are four general rules that will help attain the goal of a high power factor:

(*a*) Select the motors for their jobs with care. Use high-slip motors for punch presses. This will enable the application of smaller horsepower rating than where normal-torque motors are used.

(*b*) Rearrange the induction motors in the plant where practical so as to secure an over-all average higher loading.

(*c*) Use wound-rotor induction motors where a high starting torque requires oversize squirrel-cage motor.

(*d*) Avoid the use of slow-speed induction motors having a low power factor even at full load. Use synchronous motors instead.

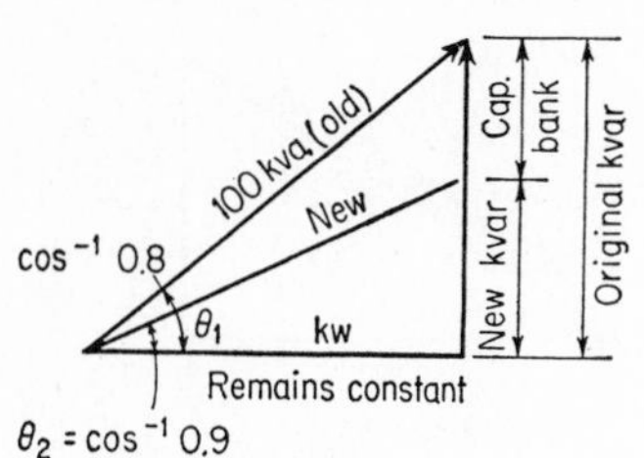

Fig. 13-36

Q13-13. A load of 100 kva of induction motors on the three-phase 208-volt 60-cycle power supply to a shop is balanced and is operating at a power factor of 80 per cent. The proposition is to bring the power factor up to 90 per cent, lagging, by means of a Δ-connected bank of capacitors. Calculate the total kva of capacitors required in this bank.

ANSWER. Please refer to the diagram in Fig. 13-36.

$$\tan \theta_1 = 0.7508 \qquad \text{also} \qquad \tan \theta_2 = 0.4845$$

$$\text{kw} = \text{pf} \times \text{kva} = 0.8 \times 100 = 80 \text{ kw}$$

$$\text{Old reactive kva} = \tan \phi_1 \times \text{kw} = 0.7508 \times 80$$
$$= 60.06 \text{ reactive kva}$$

$$\text{New reactive kva} = \tan \phi_2 \times \text{kw} = 0.4845 \times 80$$
$$= 38.76 \text{ reactive kva}$$

$$\text{Rating of capacitor bank} = \text{old} - \text{new}$$
$$= 60.06 - 38.76 = \mathbf{21.3 \text{ reactive kva}}$$

Q13-14. Show how condensers may be connected to improve the power factor. It is desired to install a condenser to obtain

200 kva (leading) on a 600-volt system at a frequency of 50 cycles. If each element has a capacity of 1 μf how many will be needed?

ANSWER. Capacitors are stationary devices, the sole function of which is to deliver leading reactive kva for the magnetizing current requirements of induction motors and other inductive equipment on the power system.

The advantages of capacitors include the following: They are a static device requiring little attention for maintenance or operation; they can be installed in practically any location and without disturbance to production; installation can be made in a wide variety of capacities to fit the power-factor requirements of individual motors or circuits; losses are low.

The power factor of a distribution or transmission system may be improved by connecting the capacitors in parallel with induction motors or other inductive equipment. A single large capacitor unit connected at the power entrance corrects the power factor at meters for power-bill reduction purposes, but it does not improve the power factor of plant distribution circuits.

$$\text{Condenser current } I_c = \frac{\text{va}}{\text{system voltage}}$$

$$I_c = \frac{200 \times 1{,}000}{600} = 333.3 \text{ amp}$$

$$\text{System voltage } E_t = I_c X_c = 600$$

Then by rearrangement and equating

$$X_c = \frac{10^6}{2\pi f C} = \frac{E_t}{I_c} = \frac{600}{333.3} = 1.8 \text{ ohms reactance}$$

Total capacitance is then obtained as follows:

$$C = \frac{10^6}{2\pi f X_c} = \frac{10^6}{2 \times 3.1416 \times 50 \times 1.8} = \mathbf{1{,}770\ \mu f}$$

Since each condenser has a capacity of 1 μf, and since condensers connected in parallel have an additive effect, we would have to connect 1,770 1-μf condensers in parallel.

Q13-15. What ten points should be investigated in improving the power factor?

ANSWER.

(*a*) What are the power rates and power-factor clause?

(*b*) How is the plant power factor determined for power billing purposes?

(*c*) What are the variations in load and power factor? Is the lowest power factor at average or peak load?

(*d*) Is the power metered on the low or the high side of the transformers?

(*e*) What are the sizes and loads on transformers and feeders?

(*f*) What drives can use synchronous motors?

(*g*) How high is it economically possible and practical to raise the power factor?

(*h*) What will be the investment required in the power-factor corrective equipment?

(*i*) What will be the returns from the power-factor improvement in power bill savings, released capacity, reduced line losses, better voltage?

Q13-16. A synchronous motor has the ability to operate with either a leading or a lagging power factor. In what ways is this utilized in plant practice and transmission lines?

ANSWER. This is utilized in two ways: (a) In industrial plants to deliver leading reactive va for power-factor correction, (*b*) for voltage control at the receiving end of the transmission lines to deliver leading reactive va at times when the line is heavily loaded.

13-27. Methods for Determining the Power Factor. Where power is purchased, and is billed on a power-factor basis, the installed power company meters will provide a means of determining the over-all plant power factor. This may be the average power factor for the billing period or the power factor at the time of the maximum kilowatt demand, or other values depending on the kind of power-factor clause and the metering method used. When power company meters are not available, or for purposes of surveys and checks on the power factor of distribution lines or individual loads within the plant, the following power-factor and load-measuring methods can be used.

Power-factor-meter method. An indicating or recording power-factor meter will provide (*a*) the power factor of the circuit to which it is connected, and (*b*) whether the power factor is leading or lagging. The methods listed below are, however, generally more useful, because at the same time that they measure the power factor they also show the power data necessary in making power-factor calculations.

Voltmeter, ammeter, wattmeter method. When loads are quite steady, simultaneous readings of volts, amperes, and kilowatts can be made from indicating or recording meters, connected for three-phase circuits. If the voltage is known and is steady, a voltmeter may not be necessary.

$$\text{Power factor} = \frac{\text{kw} \times 1{,}000}{1.73 \times \text{volts} \times \text{amp}} \tag{13-14}$$

Two single-phase watt-hour-meter method. In some plants the power consumption on three-phase circuits is measured by two single-phase watt-hour meters. The readings of these meters can be utilized to measure the average power factor over a short or longer period of time. To obtain the power factor over a short period of time, the number of revolutions of the meter disks are read simultaneously over the given period of time. The ratio of the two readings (smaller over a larger reading) determines the power factor according to the curve in Fig. 13-37. If one of the disks runs backwards, a minus ratio is obtained, and the power factor is below 0.5.

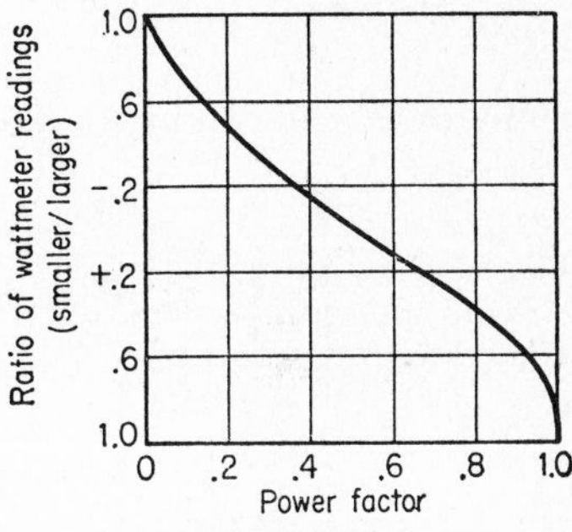

Fig. 13-37

Three-phase wattmeter method. If the load is reasonably steady, the power factor may be obtained fairly accurately by using a three-phase indicating or recording wattmeter. The meter is connected up in the proper manner for measuring three-phase kilowatts. Then, the potential lead in one phase is disconnected and a reading taken. This lead is reconnected and the potential lead in the other phase is disconnected and the second reading taken. In this manner two single-phase readings are obtained, the same as in the previous manner, and the curve will show the power factor.

Watt-hour meter, split-core ammeter method. This method is probably the most readily available of any, for the only requirement, in addition to the power company's watt-hour meter, is a split-core ammeter. It also has a very real advantage that it is not necessary to disturb any connections. The procedure is as follows:

(*a*) *Determine kilowatts*

$$\text{Average kw demand} = \frac{DMN \times 3{,}600}{\text{sec} \times 1{,}000} \qquad (13\text{-}15)$$

In the above, D = kwhr per revolution of meter disk, as given by meter manufacturer

M = meter constant as determined by ratios of instrument transformers (usually marked on meter by power company)

N = number of turns of disk in any chosen time

sec = number of seconds in time chosen for counting of disk revolutions

(*b*) *Determine amperes*

Simultaneously with the counting of the disk revolutions, the amperes in one conductor should be read by means of a split-core

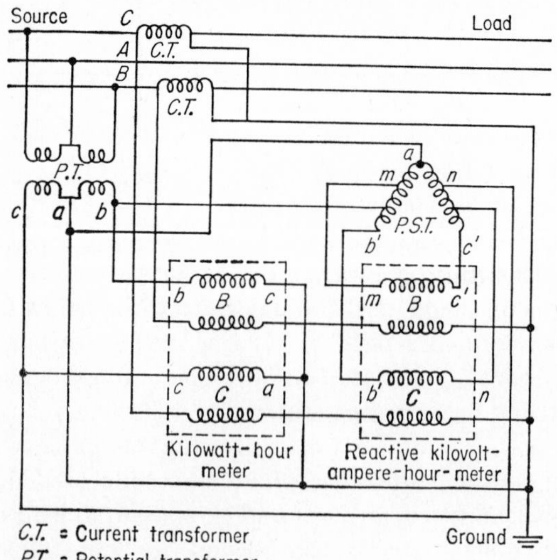

Fig. 13-38

ammeter. It is assumed that the three-phase load is fairly well balanced. Then, knowing the voltage

$$\text{Kva} = \frac{1.73 \times \text{volts} \times \text{amp}}{1{,}000} \tag{13-16}$$

and

$$\text{Power factor} = \frac{\text{kw}}{\text{kva}} \tag{13-17}$$

Watt-hour meter, reactive kva-hour meter method. A common method of metering power is by means of two three-phase integrating watt-hour meters. One meter is connected in the regular way to measure the kilowatt hour consumption. The other meter, as shown in Fig. 13-38, has a phase-shifting transformer in the potential circuit so that the meter registers reactive kva-hours.

TABLE 13-5. THE REACTIVE KVA TABLE METHOD

Power factor	Ratio reactive kva/kw	Power factor	Ratio reactive kva/kw	Power factor	Ratio reactive kva/kw
1.00	000	0.80	0.750	0.60	1.333
0.99	0.143	0.79	0.776	0.59	1.369
0.98	0.203	0.78	0.802	0.58	1.405
0.97	0.251	0.77	0.829	0.57	1.442
0.96	0.292	0.76	0.855	0.56	1.480
0.95	0.329	0.75	0.882	0.55	1.518
0.94	0.363	0.74	0.909	0.54	1.559
0.93	0.395	0.73	0.936	0.53	1.600
0.92	0.426	0.72	0.964	0.52	1.643
0.91	0.456	0.71	0.992	0.51	1.687
0.90	0.484	0.70	1.020	0.50	1.732
0.89	0.512	0.69	1.049	0.49	1.779
0.88	0.540	0.68	1.078	0.48	1.828
0.87	0.567	0.67	1.108	0.47	1.878
0.86	0.593	0.66	1.138	0.46	1.930
0.85	0.620	0.65	1.169	0.45	1.985
0.84	0.646	0.64	1.201	0.44	2.041
0.83	0.672	0.63	1.233	0.43	2.100
0.82	0.698	0.62	1.266	0.42	2.161
0.81	0.724	0.61	1.299	0.41	2.225

The ratio of the reading of the reactive kva meter to that of the kilowatt hour meter, over a given period of time, used with Table 13-5, gives the average power factor for that period.

13-28. Parallel Circuits. We found in our previous considerations of series a-c circuits that the current throughout the circuit was equal, while the voltage varied in each part of the circuit in accordance with the characteristics of the opposing elements.

In a parallel circuit, as shown in Fig. 13-39*a*, voltage across each branch is the same. It is then convenient to choose the voltage vector as a reference, by laying it along the horizontal (Fig. 13-39*b*).

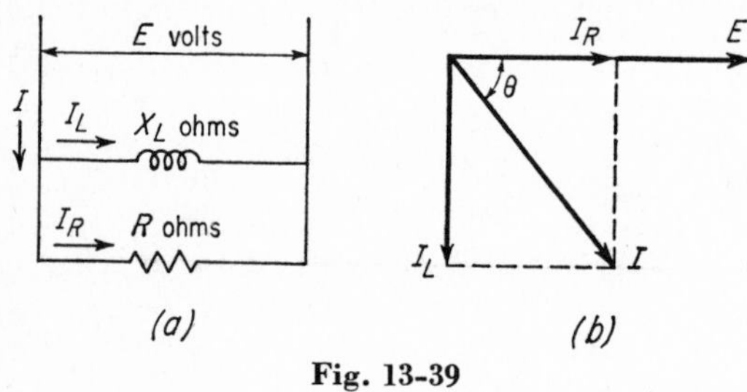

Fig. 13-39

The current through each branch may be represented with reference to this common voltage vector, and the total current may then be determined by adding vectorially the individual currents.

With reference to the circuit and vector diagram, let us write the following equations.

$$E = I_r R \text{ (through resistor)} \tag{13-18}$$

$$E = I_L X_L \text{ (through induction coil)} \tag{13-19}$$

$$I = \sqrt{I_r^2 + I_L^2} \tag{13-20}$$

$$\cos \theta \text{ (pf)} = \frac{I_r}{I} = \frac{I_r}{\sqrt{I_r^2 + I_L^2}} \tag{13-21}$$

Application of these formulas will be illustrated in the following examples.

Q13-17. Referring to Fig. 13-40*a* find, (*a*) the current through each branch, (*b*) the line current, (*c*) the power factor, (*d*) the power consumed (see Fig. 13-40*b*).

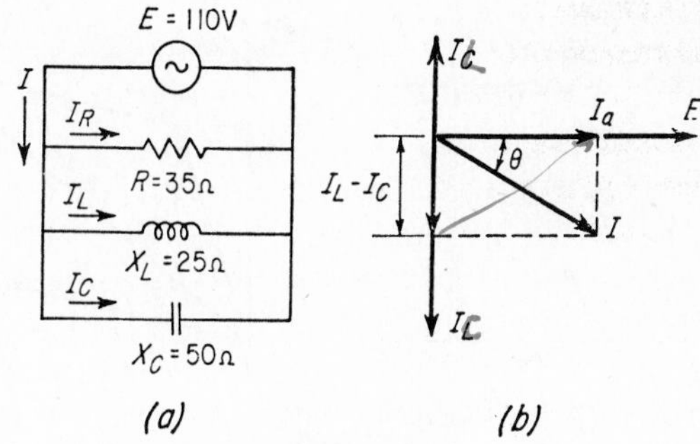

Fig. 13-40

ANSWER.

(a)
$$I_r = \frac{E}{R} = \frac{110}{35} = \mathbf{3.14\ amp}$$
$$I_L = \frac{E}{X_L} = \frac{110}{25} = \mathbf{4.4\ amp}$$
$$I_c = \frac{E}{X_c} = \frac{110}{50} = \mathbf{2.2\ amp}$$

(b) $$I = \sqrt{I_r^2 + (L_L - I_c)^2} = \sqrt{3.14^2 + 2.2^2} = \mathbf{3.83\ amp}$$

(c) $$\cos\theta = \frac{I_r}{I} = \frac{3.14}{3.83} = 0.82, \text{ or } \mathbf{82\ per\ cent}$$

(d) $$\text{Power} = I_r^2R = 3.14^2 \times 35 = \mathbf{345\ watts}$$

Q13-18. A transmission line having a resistance of 0.1 ohm and an inductance reactance of 0.6 ohm supplies a load of 100 amp at a lagging power factor of 80 per cent. The voltage at the input end of the line is 440 volts. Calculate the voltage at the end.

ANSWER. Please refer to Figs. 13-41*a* and *b* and let

I = current in circuit, amp
Z = circuit impedance, ohms
E_1 = voltage at input end, volts
E_2 = voltage at output end, volts
X_{L1} = line inductive reactance, ohms
X_{L2} = load inductive reactance, ohms
R_1 and R_2 = line and load resistance, respectively, ohms
Z_1 and Z_2 = line and load impedance, respectively, ohms
$\cos\theta$ = power factor

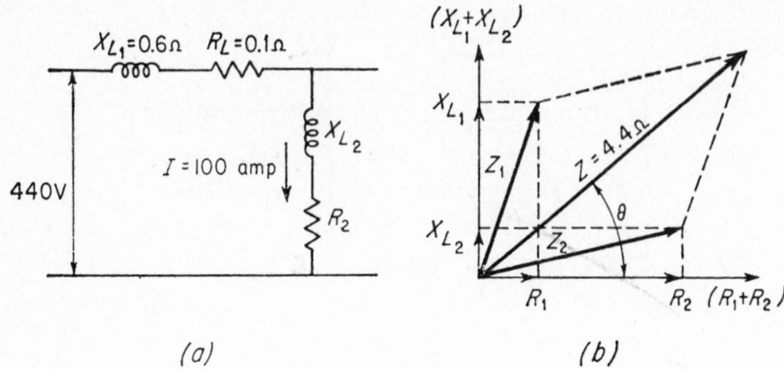

Fig. 13-41

Then $E = IZ$, or $Z = E/I = {}^{440}\!/_{110} = 4.4$ ohms. Now for the power factor

$$\theta = \cos^{-1} 0.8 = 36°52'$$

$$R_2 = Z \cos \theta - R_1 = (4.4 \times 0.8) - 0.1 = 3.42 \text{ ohms}$$

$$X_{L2} = Z \sin \theta - X_L = (4.4 \times 0.6) - 0.6 = 2.04 \text{ ohms}$$

$$Z_2 = \sqrt{R_2{}^2 + X_{L2}{}^2} = \sqrt{3.42^2 + 2.04^2} = 3.98 \text{ ohms}$$

$$\text{Voltage at load end} = E_2 = IZ_2 = 100 \times 3.98 = \mathbf{398 \text{ volts}}$$

13-29. Impedance of a Parallel Circuit. We were a little premature in working the problem immediately above. But let us list the various equations that are used in circuits involving impedance in a parallel circuit.

$$Z = \frac{E}{I} \tag{13-22}$$

$$Z = \frac{E}{\sqrt{I_r{}^2 + I_L{}^2}} \tag{13-23}$$

If the current and potential are not given, then for R and X_L in parallel

$$Z = \frac{RX_L}{\sqrt{R^2 + X_L{}^2}} \tag{13-24}$$

For R and X_c in parallel

$$Z = \frac{RX_c}{\sqrt{R^2 + X_c{}^2}} \tag{13-25}$$

For R, X_L, and X_c in parallel

$$Z = \frac{RX_LX_c}{\sqrt{X_L{}^2X_c{}^2 + R^2(X_L - X_c)^2}} \tag{13-26}$$

For all practical intents and purposes a practical circuit is more likely to have the appearance of that shown in Fig. 13-42. The associated vector diagram is as shown in Fig. 13-43. Here one branch is considered to be made up of pure resistance and pure inductance only. Treat each brand as a separate series circuit with

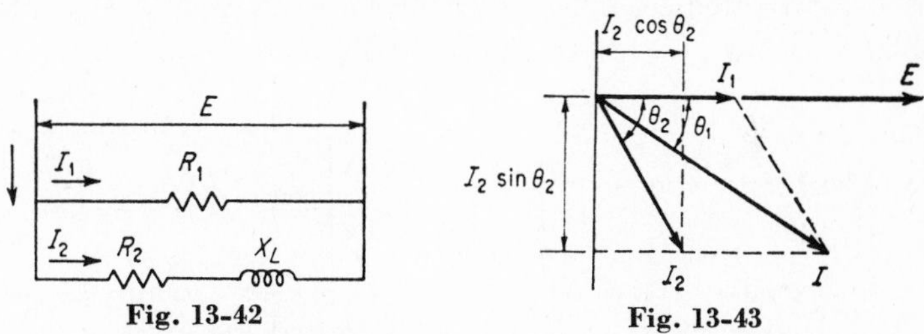

Fig. 13-42

Fig. 13-43

the common voltage impressed across it. Then find the impedance of the inductive branch of the circuit.

13-30. A-C Calculations Using Complex Quantities. When an a-c circuit becomes rather complicated and does not lend itself to a simple solution as previously indicated, a time-saving approach to solution called complex algebra is used.

In this method vector quantities are resolved into two components at right angles to each other. The component with direction along the X or horizontal axis is called the *axis of reals.* The other component at right angles along the Y axis is called the *axis of imaginaries.* An operator j is used in electrical engineering to denote direction along the X axis or the Y axis.

A quantity preceded by $(+j)$ has its direction along the X axis of the imaginaries in an *upward* direction. A quantity preceded by a $(-j)$ has its direction along the axis of the imaginaries in a downward direction.

13-31. Series Circuit Using Complex Quantities. In a series circuit having resistance R_1 and inductive reactance X_1

$$Z_1 = R_1 + jX_1 \tag{13-27}$$

The current

$$I_1 = E\left[\frac{R_1}{R_1^2 + X_1^2} - \left(j\frac{X_1}{R_1^2 + X_1^2}\right)\right] = E(g_1 - jb_1) \quad (13\text{-}28)$$

where g_1 = conductance of circuit, mhos
b_1 = susceptance of circuit, mhos
Refer to Fig. 13-44.

13-32. Parallel Circuit Using Complex Quantities. Each branch is treated separately and solved for I_1 and I_2 (Fig. 13-45). Then the total current flow is the summation of both currents.

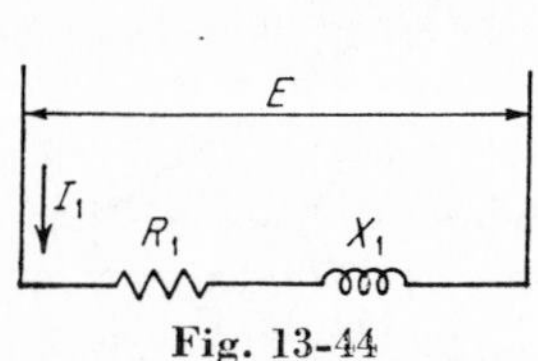

Fig. 13-44

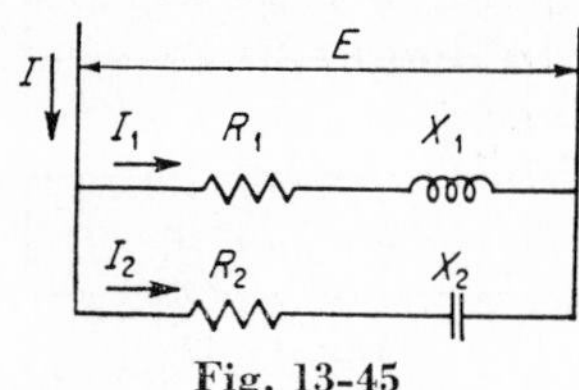

Fig. 13-45

Q13-19. What will be the current in Fig. 13-44 when E is 100 volts, R_1 is 4 ohms, and X_1 is 3 ohms?

ANSWER.
$$I = \frac{E}{Z} = \frac{100}{4 + j3} = \frac{100}{4 + j3} \times \frac{4 - j3}{4 - j3}$$
$$= \frac{400}{25} - j\frac{300}{25} = \mathbf{16 - j12\ amp}$$

The numerical value would be

$$I = \sqrt{16^2 + 12^2} = \mathbf{20\ amp}$$

Q13-20. The two circuits of Fig. 13-45 have the following values: $R_1 = 5$ ohms, $X_1 = 3$ ohms, $R_2 = 7$ ohms, and $X_2 = 4$ ohms. If the terminal voltage is 120 volts, find (*a*) the current through each branch, (*b*) the total current, (*c*) the power factor of the circuit.

ANSWER. The current through the upper branch of the circuit is given by

$$I_1 = \frac{E}{R_1 + jX_1} = \frac{120(5 - j3)}{34} = 17.7 - j10.6$$

And
$$I_1 = \sqrt{17.7^2 + 10.6^2} = \mathbf{20.6\ amp}$$

For the current through the lower branch

$$I_2 = \frac{E}{R_2 - jX_2} = \frac{120(7 + j4)}{65} = 12.9 + j7.4$$

From which $\quad I_2 = \sqrt{12.9^2 + 7.4^2} = \mathbf{14.9\ amp}$

The total current flowing is the summation of the two.

$$I_1 + I_2 = 17.7 - j10.6 + 12.9 + j7.4 = 30.6 - j3.2$$

The power factor is $\cos \theta = 30.6/30.8 =$ **0.993,** or nearly unity

13-33. Resonance in Parallel Circuits. In a series circuit resonance exists when the *inductive reactance equals the capacitive reactance,* or in another way of saying the same thing, when the lagging component of the voltage across the combination equals the leading component.

As for a parallel circuit resonance exists when the lagging component of current through the combination equals the leading component.

When a series circuit with given resistance is resonant, the impedance is a minimum and the current for a given voltage is a maximum.

With a given parallel circuit with inductive reactance in one branch and capacitive reactance in the other, resonance exists when the impedance of the combination is a maximum and the current through the combination is a minimum.

13-34. Phase. Phase, as we know, is a term applied to designate the circuits of an a-c system. In the *single-phase* system the voltages are in the same time phase in all parts of the system. In the *two-phase* system, the two voltages are 90° apart and in the *three-phase* system they are 120° apart.

In another sense, it may be considered that there are, in effect, one, two, or three circuits in the single-, the two-, or the three-phase systems. Of course, the load and power factor in the different phases may differ; hence the computation of power in polyphase circuits becomes difficult.

13-35. Circuits. The single-phase two-wire circuit shown in Fig. 13-46 is the simplest and most used connection. It is the connection used to supply lamps and small motors and appliances in even the largest systems.

The single-phase three-wire circuit shown in Fig. 13-47 is really two single-phase circuits with one wire in common. It is used where more power is required than can economically be supplied by the two-wire system. It also permits 220-volt units to be connected to the two fused wires, and yet the system has no more than

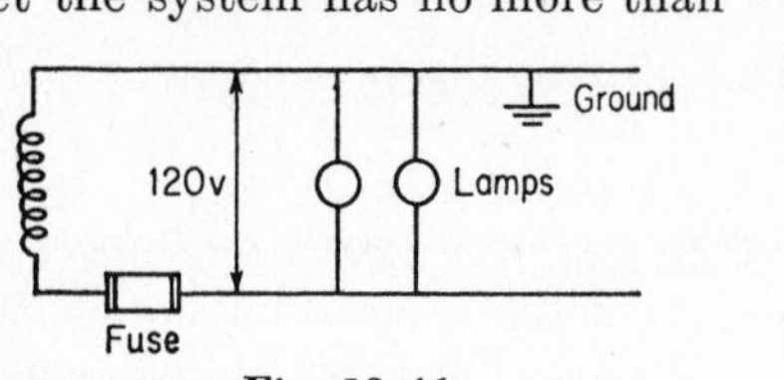

Fig. 13-46

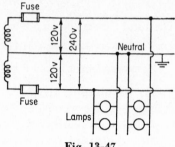

Fuse
120v
240v
Neutral
120v
Fuse
Lamps

Fig. 13-47

110 volts to ground. When the loads are identical, the neutral wire carries no current, but since one fuse may blow, the neutral must carry the same as the phase wires, and thus it must be of the same size. Yet even then this system saves one of the four wires that would be required for two two-wire circuits.

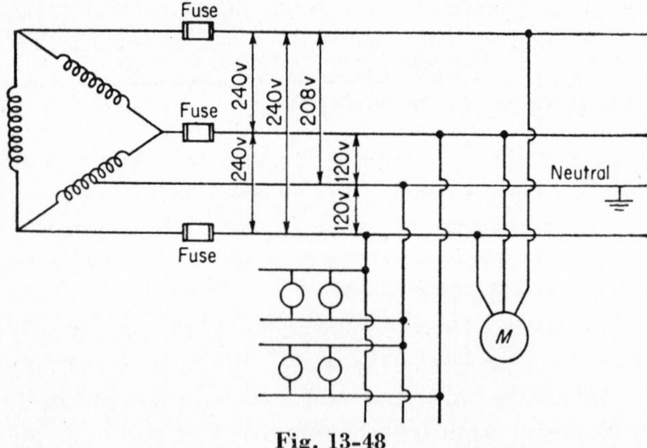

Fig. 13-48

The three-phase circuits shown in Figs. 13-48 and 13-49 are really three single-phase circuits connected together, with the important difference that the three phases are equally spaced in time relationship (120° apart). There are two connections generally used for three-phase circuits, the Y and the Δ, each of which has certain advantages. In general, the Δ connection is more reliable, when used for power only, in that if one transformer is lost, the remaining two can supply 87 per cent of their combined capacity. But to offset this advantage when lighting is needed, unless the motor load is small enough to be operated at 110 volts, one of the neutrals of one of the transformers must be grounded, thus increasing the chances for power failure.

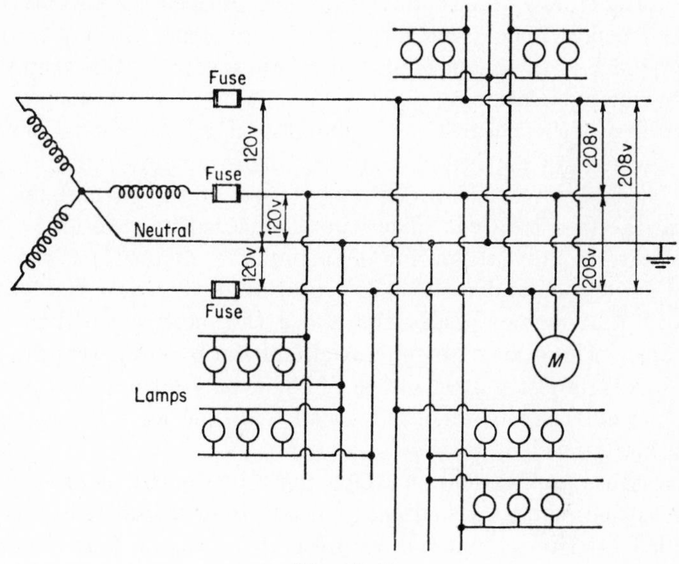

Fig. 13-49

Then, too, many power companies will limit the lighting load that may be placed on the one phase that supplies it (Fig. 13-48); and the lighting is badly affected with "flickering" particularly if the motors are frequently started and stopped as for elevators.

The Y connection has the advantage that the lighting load can be divided between the three phases, but the three-phase voltage for motor use is less than that for which motors are usually designed.

Hence, it may be said that for large loads three-phase systems are always used. In plant practice it may be found that motors less than 5 hp are supplied by three Y-connected transformers; the large motors and elevators are supplied from three Δ-connected 480-volt transformers.

13-36. Fuses and Circuit Breakers. Fuses and all electric circuit breakers are similar in function to the safety valve on a boiler; and therefore they should be seriously considered. They are all intended to open the circuit, thus stopping the flow of current, should the current exceed a predetermined or dangerous rate.

While the problem of circuit protection becomes at times quite involved, still there is much that can here be discussed to advantage. The NEC is now almost universally accepted as the authority for the safe utilization of electricity in all except central station systems. Consequently, the engineer should obtain a copy of this useful and important publication.

Fuses are the cheapest and simplest of all protective devices. They consist of a piece of relatively high-resistance alloy wire or tape, designed to melt, and thereby part, at the current flow for which the fuse is rated. The wire is enclosed in a suitable flame-proof case to prevent ignition of surrounding materials at the time the fuse flashes and melts. The use of fuses is now confined to circuits that do not justify the use of the more expensive circuit breaker. There may be some slight deviation from this practice but this is almost universally so. The type known as the "refill" type permits the renewal of the fuse link and the use of the expensive part (the cartridge) over and over again.

For those applications in which the replacement of fuses is frequently necessary, thermal relays or air circuit breakers are recommended. Motor circuits fall in this classification. Thermal relays and circuit breakers can permit a greater than rated current to pass for a short time, as during the starting period of a motor, and therefore they are of especial advantage for this service. The current capacity as well as the voltage ratings of this class of protective devices are limited, and when greater limits of protection are needed, the oil circuit breaker may be used.

The oil used in circuit breakers must be especially selected for this service; and no other oil should ever be used. This precaution is essential for the simple reason that the oil must have high insulation value, high flash point, low carbon content, and wide temperature viscosity constant—all in order to quickly quench the arc that is formed with each opening of the contacts.

13-37. Relays. The overload operation of a circuit breaker is usually actuated by an auxiliary device known as a relay. Relays may be of several types: the instantaneous, the inverse-time limit, and others. The modern relays are truly marvelous devices; however, the types needed are usually covered by three.

(*a*) *The instantaneous type*, dashpot relays with direct mechanical trip. Here a coil which carries all or a definite part of the load current raises a solenoid when the current flow exceeds a predetermined value. The rising solenoid mechanically releases a trigger, which in turn permits the circuit breaker to open. An oil-filled dashpot retards the solenoid movement. The dashpot opening is adjustable, and thus the time for opening for a given overcurrent can be selected as desired. The oil in the dashpot should be inspected at intervals and replaced if dirt or sludge has accumulated. While rather crude, this device is sufficiently accurate for the protection of most small motor circuits.

(*b*) *The air bellows type*, an air bellows with an adjustable escape valve is used in place of the oil dashpot described above. This type is essentially obsolete.

(*c*) *The induction type*, which is more expensive than those mentioned above, but its use is justified in the protection of all important circuits. The advantage of this type is that the time of opening the main circuit breaker can be determined to a fraction of a second; also, the limits of control are so wide that these relays, with suitable current transformers, become almost universally applicable and highly accurate protective devices.

In the face of being repetitious, too much care cannot be given the subject of circuit protection. Overfusing (use of fuses with too great capacity) of circuits is dangerous and uneconomical. An all too frequent operation of the overload protection of an appliance warrants prompt investigation. The source of the trouble should be remedied by correction, rather than by following the easiest path of enlarging the circuit protection.

Chapter 14

ALTERNATING-CURRENT MOTORS

14-1. Motors—General. A-c motors are of two general types, synchronous and induction. The main difference between these two types is that, while the synchronous motor can run only at a definite constant speed (synchronous speed), the induction motor must operate at a speed dependent upon the load applied, but *never* at synchronous speed. There are other important differences which will be mentioned later.

14-2. Synchronous Motors. At a given frequency the speed of a synchronous motor may be determined by the equation

$$N = \frac{120f}{P} \qquad (14\text{-}1)$$

where N = revolutions of rotor per minute, rpm
f = frequency, cps
P = number of poles

Since the standard frequency is 60 cps, Eq. (14-1) is generally expressed in the form

$$N = \frac{7{,}200}{P} \qquad (14\text{-}2)$$

From these considerations we can see that the maximum speed for a 60-cycle motor is 7,200/2, or 3,600 rpm, and the speed N from Eqs. (14-1) and (14-2) is known as the synchronous speed of the motor. This, for 60-cycle motors, is the constant 7,200 divided by the number of poles. Now, since the number of poles must be an even number, synchronous speeds become 3,600, 1,800, 1,200, 900, etc., rpm. One of the largest limitations of the standard synchronous motor is that it cannot be built to operate at other than synchronous speeds.

14-3. Characteristics of Synchronous Motors. By varying the field current, the power factor of this motor may be varied so that the motor can be operated at unity power factor; that is, the armature current is in phase with the voltage and, under this condition, the armature current is the minimum for any given load. When the strength of the field is made weaker, the power factor is less, and the armature current lags behind the voltage. With a stronger field the power factor is also less, but the armature current leads the voltage and acts as a condenser. In either case the current drawn from the line will be greater than when operated at unity power factor, because with overexcited field strengths the motor acts as a condenser with leading current. The motor can be used to counteract other lagging currents on the circuit and still carry some mechanical load at the same time.

The efficiency of the synchronous motor is somewhat higher than the corresponding induction motor of the same rating. This is especially true of the lower-speed motors.

The synchronous motor has one unfavorable characteristic. The necessity for an auxiliary direct current is unfortunate, because synchronous motors are difficult to maintain. Hence, greater skill and maintenance are required for synchronous than for induction motor operation.

On the advantage side, the synchronous motor does possess the ability to operate with a leading power factor, contrasted to the rather decidedly lagging power factor of the standard induction motor. As indicated before, many power companies offer a lower rate to consumers whose requirements constitute a desirable power factor. Where this occurs, the ability of the synchronous motor to operate at the leading power factor may have influence on the selection of the type of motor adopted.

14-4. Starting Torque of Synchronous Motors. This is the turning effort which the motor develops when full voltage is applied to the armature winding. Earlier motors developed very little starting torque, but present designs can develop almost any reasonable starting torque with the use of the damper winding, different values of torque being obtained by changing the resistance and size of this winding. As a general rule starting inrush current increases with an increase in the starting torque.

14-5. Pull-in Torque of Synchronous Motors. When a synchronous motor has been started as an induction motor, it will run somewhat (2 to 5 per cent) below synchronous speed until such time as excitation is applied. Then the rotor pulls into step. The amount of torque or load which the motor will pull into step is called the pull-in torque. Sync motors are usually designed for a definite application. For proper use, then, the designer should know the nature of the load which the motor is required to start. In this way he can effectively take into consideration the necessary starting and pull-in torques.

14-6. Pull-out Torque of Synchronous Motors. During normal operation of a synchronous motor, when it is running in sync with its load, the individual pole pieces of the rotor have a fixed position with respect to the revolving magnetic field of the armature. When being loaded or when a resisting torque is being applied to the shaft of the motor, it develops a torque to balance the requirements of the load. This increased torque requirement is produced by the backward shift or lag in the position of the field poles with respect to the revolving magnetic field. During all this, however, the shaft or rotor will still maintain its sync speed. Too much of a shift of the field poles (about one-half the distance between adjacent poles) due to increased torque requirements will cause the motor to pull out of step and stop. The maximum torque which the motor will develop without pulling out of step is called the pull-out torque.

14-7. Application of Synchronous Motors. There are several classes of service to which synchronous motors are quite well adapted.

(*a*) *Constant-speed service.* Because of their higher efficiency synchronous motors can be advantageously applied to most loads where constant speeds are desirable.

(*b*) *Reciprocating compressor drives.* Synchronous motors can be built in low speeds, with high efficiency, for direct connection to reciprocating compressors. Because of their pulsating characteristics such loads produce pulsations of power. With proper design care the addition of flywheel effect to the motor will limit the pulsation effects to values that would be acceptable to industry.

(*c*) *Power-factor correction.* If the power factor of a plant is low because of a large induction-motor load, a synchronous motor can be used to raise the power factor. The overexcitation of the field of a synchronous motor causes the current to lead the voltage. This leading current counteracts the lagging current of the inductive load and the total power factor is raised.

(*d*) *Voltage regulation.* On the ends of long transmission lines the voltage tends to vary greatly, especially if a large inductive load is present. If the inductive load is thrown off, then because of the capacity effect or condenser action of the long transmission line the voltage may rise above normal. By installing a synchronous motor with a voltage regulator to control its field, the voltage change can be compensated for. The action of the voltage regulator is to increase the field strength of the synchronous motor when the voltage drops due to the inductive load. This raises the power factor and maintains the voltage at the same level. If the voltage tends to rise because of capacity effects of the line, the regulator weakens the field and causes the motor to have a lagging or inductive load so as to hold the voltage normal.

14-8. Auxiliary Equipment Required with a Synchronous Motor. For speeds below 500 rpm synchronous motors may be started using a full-voltage starter. On the other hand, higher-speed motors may be started by first applying reduced voltage and then switching over to line voltage when the motor has accelerated to almost sync speed. For this purpose of starting, a reduced voltage compensator or autotransformer is generally used. After full voltage is applied the fields are excited.

Additional equipment consists of meters, an ammeter to measure the motor line current and an ammeter to measure the current in the exciter circuit. These meters are usually mounted on a panel with the main line circuit breakers, an exciter field rheostat, means of automatically applying field excitation, and various protective relays.

Q14-1. A single-phase synchronous motor for use on a 500-volt circuit gives, when running as a generator, a short-circuit current of 150 amp with normal excitation applied. The armature resistance is 0.2 ohm. What must be the excitation voltage in order to develop 40 hp at unity power factor? What will be the armature current? The mechanical loss is 5 hp at the terminal voltage of 500.

ANSWER. It will be remembered that the d-c generator operates satisfactorily as a motor. Similarly, an alternator will operate as a motor without any changes being made in its construction. When so operated, the machine is called a synchronous motor.

The excitation voltage E' may be determined vectorially by the following relationship

$$E' = \sqrt{(E \cos \theta - I_aR_a)^2 + (E \sin \theta + I_aX)^2}$$

or E' may be determined by the complex notation

$$E' = E - I_a(\cos \theta + j \sin \theta)(R_a + jX)$$

where E = terminal voltage
I_a = armature current
R_a = armature resistance
X = leakage reactance

See Dawes: "A Course in Electrical Engineering," 4th ed., McGraw-Hill Book Company, Inc., vol. II, pp. 383 and 384. Under short-circuit conditions for normal excitation the motor operates at unity power factor at no load. Thus, the generated voltage E'_1 is equal to the terminal voltage E. Then the synchronous impedance Z_s is found by $Z_s = E'_1/I_a = 500/150 = 3.33$ ohms. The leakage reactance is found

$$X = \sqrt{Z_s^2 - R_a^2} = 3.33 \text{ ohms}$$

From our review of generators and motors (Chaps. 10 and 11) we noted that the motor input was equal to the output plus the mechanical losses plus the copper losses. At unity power factor

$$E'I_a \cos \theta = 500I_a \times 1 = 40 \times 746 + 5 \times 746 + 0.2I_a^2$$

After rearrangement

$$0.2I_a^2 - 500I_a + 33{,}600 = 0$$

from which $I_a = 70$ amp. Then by complex notation

$$E' = 500 - 70(1 + j \times 0)(0.2 + j \times 3.33) = 486 - j \times 233$$

and

$$E' = \sqrt{486^2 + 233^2} = \textbf{540 volts}$$

Q14-2. The input to a 600-volt 100-hp synchronous motor is measured by the two-wattmeter method. One wattmeter reads plus (+) 62.5 kw, and the other reads plus (+) 30.5 kw. The motor is known to be taking a leading current. Determine the line current and the output if the efficiency of the motor at this loading is 0.90 exclusive of the d-c field loss.

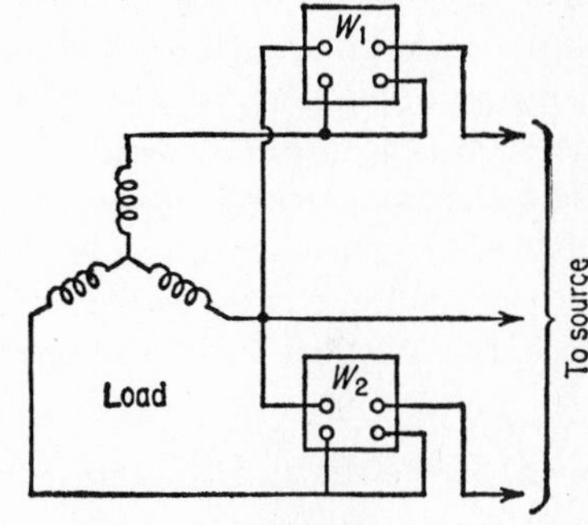

Fig. 14-1

ANSWER. There are two principal methods whereby power may be measured in a three-phase system. They are the three-wattmeter method and the two-wattmeter method. This problem involves the application of the two-wattmeter method. In this method the wattmeters will measure the total amount of power if connected, as shown in Fig. 14-1. The three-phase power is $P = W_1 + W_2 = EI \cos (30 - \theta) + EI \cos (30 + \theta)$. Or more simply, $P = \sqrt{3}\, EI \cos \theta$ watts. When the phase angle is less than 60° and the power factor is less than 0.5, both wattmeters will read "on scale" and the total power is the sum of the two wattmeter readings. The relation between the phase angle and the two meter readings is given by

$$\tan \theta = \sqrt{3} \times (W_1 - W_2)/(W_1 + W_2)$$

In this formula θ is the angle of lag or lead of the current and W_1 and W_2 are the readings of the wattmeters. Then

$$\tan \theta \text{ (pf angle)} = \sqrt{3} \times \frac{62.5 - 30.5}{62.5 + 30.5} = 0.596$$

From trigonometry tables the angle value of 0.596 is tan **30.8°**, and the power factor is 0.859. The power input is given by

$$\sqrt{3}\, EI \cos \theta = \sqrt{3} \times 600I \times 0.859 = 93{,}000 \text{ watts}$$

Thus, I for each line is **104.3 amp.** And the power output is

$$(93{,}000/746) \times 0.90 = \mathbf{112.2\ hp}$$

Q14-3. A 400-hp 0.8 power factor 600 rpm 440-volt three-phase 60-cycle synchronous motor is selected to drive a band saw in a lumber mill. The full-voltage starting current is 550 per cent of rated current, while the corresponding torque is 150 per cent of that at full load. Some means of limiting the current at starting is desired. (*a*) If 50 per cent of normal torque is adequate to start and accelerate the unloaded saw, will 65 per cent of the normal voltage applied at starting be sufficient to cause the motor to develop this torque? Neglect the effects of saturation. (*b*) Compare the line current, taken at the instant of starting, when 65 per cent of the normal voltage is applied by means of reactors, with the corresponding current, when it is applied by means of an autotransformer. Neglect losses in the reactors and in the autotransformer.

ANSWER.

(*a*) Since the starting torque is said to be 150 per cent of full load

$$T_s = 1.5T_f \text{ (for full voltage, or 100 per cent voltage)}$$

For the 65 per cent voltage condition and letting X be per cent full-load torque

$$T_s = XT_f$$

Since the torque is proportional to the voltage squared by direct proportion

$$\frac{100^2}{65^2} = \frac{1.5T_f}{XT_f}$$

from which we obtain $X = 0.634$, or **63.4 per cent.** Thus, it is apparent that 65 per cent of normal voltage will be adequate to start the motor and develop the required torque.

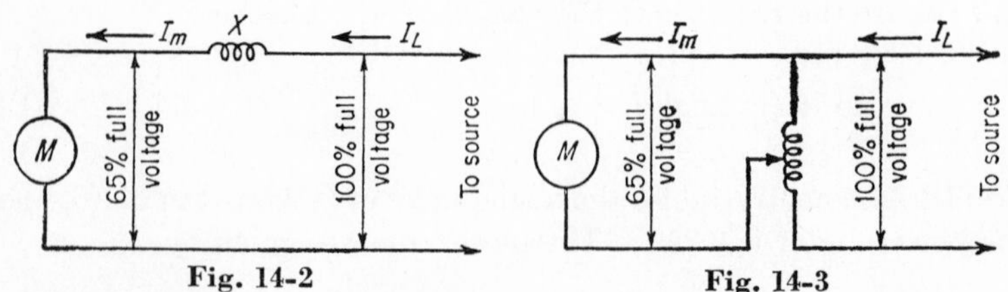

Fig. 14-2

Fig. 14-3

(*b*) Refer to Figs. 14-2 and 14-3. At 65 per cent of the normal voltage the motor current I_m is

$$I_m = {}^{65}\!/_{100} \times 5.5I \text{ (rated current)} = 3.57I$$

From inspection of Fig. 14-2 for the reactor hookup

$$I_L = I_m = \mathbf{3.57I}$$

From inspection of Fig. 14-3 for the autotransformer hookup

$$I_L = {}^{65}\!/_{100} \times I_m = 0.65 \times 3.57I = \mathbf{2.32I}$$

Q14-4. An air compressor using a 100-hp motor is to be added to a plant. What will be the effect on the plant load and power factor with the motor operating at nearly full load using an 0.8 leading power-factor synchronous motor? Existing average full-load plant conditions are as follows:

Voltage at main board...	440
Amperes (three-phase)...	470
Kilowatts..............	250
Motor efficiency	91 per cent

The existing load is as follows:

$$\text{kva} = \frac{1.73 \times 440 \times 470}{1{,}000} = 357$$

$$\text{Power factor} = \frac{\text{kw}}{\text{kva}} = \frac{250}{357} = 0.7$$

Using the reactive kva table method (Table 13-5), we must first obtain for the synchronous motor with an efficiency of 91 per cent and unity power factor

$$\text{kw input} = \frac{(100 \times 0.746 \times 100)}{91} = 82$$

$$\text{kva at unity power factor} = \text{kw} = 82$$

$$\text{Reactive kva} = 0$$

With the synchronous motor added, we will determine the over-all kilowatts, reactive kva, power factor, and kva.

Kilowatts:

Existing plant load.........	250 kw
1.0 pf synchronous motor...	82 kw (add)
	332 kw

Reactive kva:

Existing plant load.........	255 reactive kva
1.0 pf synchronous motor...	0 reactive kva (add)
	255 reactive kva

The existing plant reactive kva was obtained by use of Table 13-5. In the table the ratio for the power factor of 0.7 = 1.02, and the reactive kva = 1.02 × 250 = 255.

Power factor:

$$\text{Reactive kva ratio} = {}^{255}\!/_{332} = 0.768$$
$$\text{Power factor (see Table 13-5)} = 0.793$$
$$kva = 332\text{ kw}/0.793 = 418$$

On the basis of a 0.8 power-factor synchronous motor
Kilowatts = 84
Power factor = 0.8 leading
Reactive kva ratio for 0.8 pf = 0.75
Reactive kva of 0.8 pf synchronous motor = 84 kw × 0.75 = 63

Then

Kilowatts:

Existing plant load.........	250 kw
0.8 pf synchronous motor...	84 kw (add)
Total kw................	334 kw

Reactive kva:

Existing plant load.........	255 reactive kva
0.8 pf synchronous motor...	63 (subtract)
Total reactive kva........	192 reactive kva

Power factor:

$$\text{Reactive kva ratio} = {}^{192}\!/_{334} = 0.575$$
$$\text{Power factor (see Table 13-5)} = 0.867$$
$$\text{kva} = 334/0.867 = 385$$

Q14-5. A small industrial plant operates with an average load of 400 kw, consisting of small induction motors with some electric heating. The average power factor is 70 per cent lagging. A

200-hp synchronous motor driving a compressor is added to the plant. Neglecting the effect of the motor losses, calculate the power factor at which this motor must be operated in order to raise the power factor of the entire plant to 80 per cent.

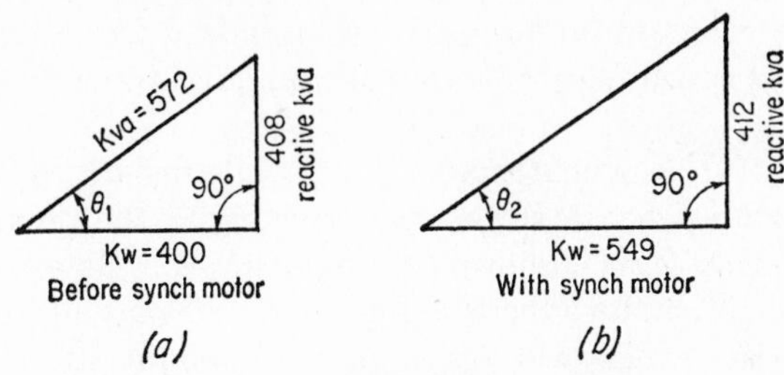

Fig. 14-4

ANSWER. For the existing plant installation, where $\cos \theta = 0.70$ (power factor),

$$\text{kva} = 400/0.7 = 572 \qquad \text{kvar} = 572 \times \sin \theta = 572 \times 0.714 = 408$$

See Fig. 14-4*a*. Assuming 100 per cent efficiency (neglecting the motor losses) the synchronous motor kilowatts is found to be $200 \times 0.746 = 149$ kw. This is the added kilowatt load because of the sync motor. Then the new plant load becomes

$$\text{kw (total)} = 400 + 149 = 549$$

Since we are looking for a new power factor to be 0.8 ($\cos \theta$), it follows

$$\text{Reactive kva} = 549 \times 0.75 \ (\tan \theta) = 412$$

See Fig. 14-4*b*. We see that the new reactive kva is greater than the original reactive kva by 4 reactive kva. This indicates that the sync motor is operating at a lagging power factor because it is taking reactive kva. Thus, in order for the plant to operate at 80 per cent power factor

$$\tan \theta = (412 - 408)/(200 \times 0.746) = 4/149 = 0.0259$$

From tables $\cos \theta$ so determined is close to unity. And we see that it is required that the sync motor must have a *power factor* of close to *unity* in actual operation.

Q14-6. Draw a family of V curves for a three-phase synchronous motor, indicating and explaining clearly: (*a*) the variables and parameters involved; (*b*) how the kilowatt input at unity power factor may be determined for each curve shown; (*c*) how a line may be drawn through these points at which operation is at: 80 per cent lagging power factor, 80 per cent leading power factor, the reason why the lines called for are located as you have drawn them on the V curves.

ANSWER. (*a*) A synchronous motor, while driving its load, can have the a-c input into its stator varied by varying the strength of the field excitation. This is shown by the V curves. Given a particular synchronous motor: there is for each load one adjustment of field excitation which will result in a minimum stator current. Weaker excitation and stronger excitation both produce an increase of stator current. If the power P delivered to a three-phase synchronous motor is kept constant and the field current I_f varied, the power factor of the motor will change. The power for the three-phase motor is

$$P = \sqrt{3}\,EI \cos \theta$$

where E = terminal voltage
I = line current
$\cos \theta$ = power factor of motor

As both E and P are constant, any decrease in the power factor $\cos \theta$ must be accompanied by a corresponding increase in the current I. Likewise, an increase in the power factor must be accompanied by a decrease in the current I.

Therefore, a change in the field current I_f at constant load changes the line or armature current I. In order to determine the relation between the field current and the armature current and also the characteristics of a synchronous motor as regards its ability to correct the power factor of a system, the so-called V curves of the motor are drawn. These V curves show the relation which exists between the armature current and the field current for different constant power inputs. Usually several curves are obtained, each curve representing a constant value of power input (see Fig. 14-5).

Figure 14-5 shows a set of typical curves. The curve AB is obtained when the motor is running at a very light load. At very low values of the field current the armature current is large and lagging. As the field current is increased, the power factor increases, and the armature current decreases until it reaches its minimum

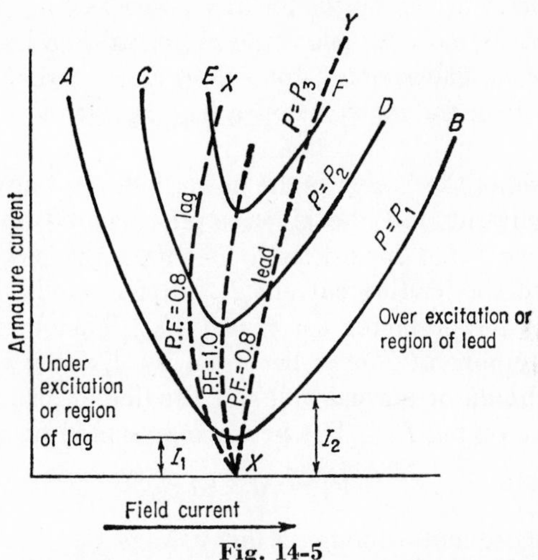

Fig. 14-5

value I_1. If the field current is still further increased, the armature current begins to increase and becomes leading. In other words, the motor passes from *underexcitation to overexcitation* when the field current is increased from low to high values.

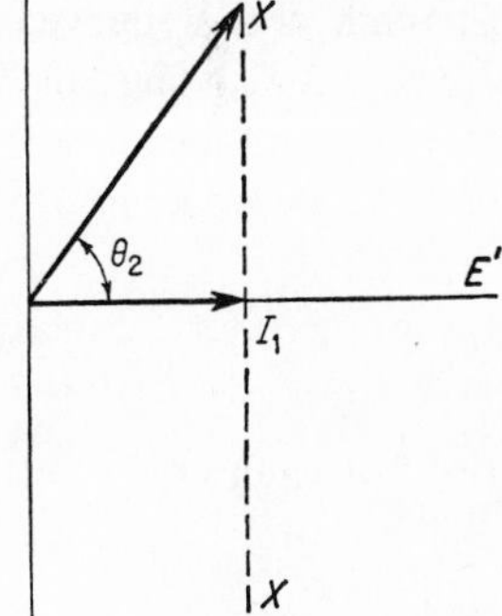

Fig. 14-6

The current I_1 is the value of the current at unity power factor. This is illustrated in Fig. 14-6. If we let I_2 be the value of the line current at some power factor $\cos \theta_2$, the power for one phase is

$$P_1 = E'I_2 \cos \theta_2$$

where E' is the phase voltage; but

$$I_2 \cos \theta_2 = I_1$$

for all values of θ_2. I_2 is line at angle θ_2.

In other words, for the constant power, P_1, I_1 is always the energy component of the current regardless of the power factor. Therefore, the current vector will always terminate on the line XX perpendicular to E'. The current is a minimum at I_1, where the current is in phase with E'. The power factor is then unity. The excitation corresponding to the armature current I_1 is called the *normal excitation* of the motor for any given load. For an excitation less than the normal value, the motor takes a lagging current and is said to be *underexcited;* for values of excitation greater than the normal value, the motor takes a leading current and is said to be *overexcited.*

With the aid of the V curves the power factor for any other value of the line current and the given input may be obtained. For example, assume that it is desired to obtain the power factor for some value of the leading current I_2. Refer to Fig. 14-5. From Fig. 14-6 the power factor $\cos \theta_2 = I_1/I_2$. Therefore, the power factor for any current I may be found by dividing the current I into the minimum or normal value of the line or armature current I_1 for the given input P_1. The power represented by curve AB is

$$P_1 = \sqrt{3}\,EI$$

for a three-phase motor having a line voltage E.

CD (Fig. 14-5) is a V curve taken for a value of power P_2 which is greater than P_1. EF is a third curve taken for a still greater value of power P_3. A curve drawn through the lowest points of the V curves is a unity power-factor curve. Curves XX and XY, drawn through the V curves at the proper points, are 0.8 power-factor curves, XX being for *lagging* current and XY for *leading* current. Curves for other power factors may also be found in a similar manner. These curves are called *compounding curves.*

(*b*)
$$\text{kw input} = \frac{\sqrt{3}\,EI\cos\theta}{1{,}000}$$

(*c*) For each curve (constant power input) the current is found by

$$I = \frac{\text{power input}}{\sqrt{3}\,E \times 0.8}$$

The points where the constant power-factor lines cross the power input line (one set for the leading power factor and one set for the lagging power factor) are indications of the constant armature current. Thus the power-factor curves may be drawn, all meeting at the one point at the bottom of the graph at zero armature current.

14-9. Induction Motors. The induction motor differs from the synchronous type in that it requires no separate excitation, and it operates at a variable speed, as contrasted to the constant-speed synchronous motor.

At no load an induction motor will run almost up to its synchronous speed. It should be understood that while the induction motor has no salient poles the stator is wound to produce two, four, or any even number of poles, just as is the stator of a synchronous motor; hence, it has a synchronous speed of 3,600, 1,800, or 900 rpm, just as does the synchronous motor.

14-10. Characteristics. Standard induction motors have very favorable characteristics, requiring a relatively small starting current and producing 200 or more per cent of full-load torque. They are the most rugged and reliable of all motors. But, while they will operate successfully under the most adverse conditions, reasonable care should be taken to keep their air ducts and windings free from lint and dirt. The oil in the bearings must be flushed and changed at least once each year.

14-11. Operation of Induction Motor. An induction motor is one in which the magnetic field in the rotor is induced by currents flowing in the stator. The rotor has no connections whatever to the line. This motor is by far the most commonly used of any of the types because it is the most simple in construction and has the broadest application. The most common form is the squirrel-cage motor.

The conventional type of induction motor consists of a stator, which is in every way similar to the armature of a synchronous motor. It also has a squirrel-cage rotor with bearings to support it. The stator, because it receives the power from the line, is often called the primary; the rotor is often called the secondary.

In the induction motor, the squirrel-cage winding takes the place of the field of the synchronous motor. The current flowing in the cage makes a loop which establishes magnetic fields in the rotor core with north and south poles. There is, however, one very interesting and important difference between the synchronous motor and the induction motor; the rotor of the latter does not rotate as fast as the rotating field in the armature. If the squirrel cage were to go as fast as the rotating field, the conductors in it would be standing still with respect to the rotating field rather than cutting across it. As a result there would be no voltage induced in the squirrel cage, no currents in it, no magnetic poles set up in the rotor, and no attraction between it and the rotating field in the stator. The rotor revolves just enough slower than the revolving field as it slips by, and thus induces the necessary currents in the rotor windings. Hence the motor can never go quite as fast as the revolving field, but it is always slipping back, and this difference in speed is called the *slip*. The greater the load on the motor the greater will be the slip; that is, the slower the motor will run. But even at full load the slip is not every great; in fact, this motor is commonly considered to be a constant-speed motor. When the load is applied, the motor slows down just enough to produce the necessary torque.

14-12. Slip. The slip is usually expressed as a percentage of the synchronous speed. For instance, a motor rated at 1,800 rpm operates at 1,738 rpm when loaded. Then the slip is said to be

$$(1{,}800 - 1{,}738)/1{,}800 = 0.035, \text{ or } 3\tfrac{1}{2} \text{ per cent slip}$$

Slip is also a measure of the rotor winding losses. From the per cent slip above calculated, 3½ per cent of the total power input was lost in the heating of the rotor windings. Slip is also connected with motor efficiency; the higher the slip, the lower the efficiency. Slip is held between no load and full load for a maximum of 5 per cent for best results.

When operating at light loads the power factor and efficiency of induction motors are lower than when operating at full load. It is, therefore, apparent that when a motor is selected it should be run at as close to full load as possible. Low-speed motors have lower power factors than high-speed motors. Here, too, care should be exercised to select a motor that will give the most satisfactory and economical arrangement with the driven unit. Table 14-1 shows the effect of these variables on efficiency, power factor, and loading.

TABLE 14-1. INDUCTION MOTORS*
Approximate, average data on normal-torque normal starting kva squirrel-cage induction motors
440 volts, 60 cycles 3, phase

Motor hp	Synchronous rpm	Efficiency		Power factor		Full-load kilowatts	Full-load amperes	Locked-rotor amperes
		Full load	½ load	Full load	½ load			
5	3,600	83.5	81.0	0.86	0.71	4.5	6.8	43
	1,800	83.5	81.0	0.86	0.73	4.5	6.8	43
	1,200	82.5	80.0	0.82	0.66	4.5	7.2	43
	900	82.0	79.0	0.76	0.58	4.5	7.9	42
10	3,600	85.0	82.5	0.88	0.75	8.8	13.0	93
	1,800	85.0	82.5	0.88	0.77	8.8	13.0	77
	1,200	84.5	82.0	0.84	0.72	8.8	13.8	77
	900	84.0	81.0	0.80	0.65	8.8	14.6	77
20	1,800	86.5	84.0	0.89	0.78	17.2	25.4	150
	1,200	86.5	84.0	0.86	0.75	17.2	26.2	150
	900	85.0	82.5	0.84	0.70	17.5	27.3	150
	600	84.5	82.0	0.74	0.60	17.6	31.2	140
30	1,800	87.5	85.0	0.90	0.80	25.5	37.2	215
	1,200	87.5	85.0	0.88	0.77	25.5	38.0	215
	900	86.5	84.0	0.86	0.72	25.9	39.6	215
	600	86.0	83.5	0.78	0.64	26.0	43.8	200
50	1,800	88.5	86.0	0.91	0.79	42.2	60.8	400
	1,200	88.5	86.0	0.88	0.76	42.2	62.8	400
	900	88.0	85.5	0.85	0.69	42.4	65.5	400
	600	87.5	85.0	0.80	0.67	42.6	70.0	360
75	1,800	90.0	87.5	0.91	0.79	62.1	89.9	600
	1,200	89.5	87.0	0.88	0.76	62.5	93.3	600
	900	89.0	86.5	0.87	0.71	62.8	94.7	600
	600	88.5	86.0	0.82	0.69	63.2	101	550
100	1,800	90.5	88.0	0.91	0.79	82.4	119	800
	1,200	90.0	87.5	0.89	0.77	82.8	122	800
	900	89.5	87.0	0.87	0.71	83.2	126	800
	600	89.0	86.5	0.83	0.69	83.8	132	700
150	1,800	91.0	88.5	0.91	0.79	123.0	177	1,200
	1,200	91.0	88.5	0.89	0.77	123.0	181	1,200
	900	90.5	88.0	0.88	0.72	123.5	185	1,200
	600	90.0	87.5	0.85	0.69	124.5	193	1,000
200	1,800	91.5	89.0	0.91	0.79	163.0	235	1,600
	1,200	91.5	89.0	0.89	0.77	163.0	240	1,600
	900	91.0	88.5	0.88	0.72	164.0	245	1,500
	600	91.0	88.0	0.86	0.69	164.0	250	1,500
	450	90.5	87.0	0.82	0.65	165.0	264	1,300

Note: For 220 volts the full-load amperes and starting amperes will be twice those given above; for 2,200 volts approximately one-fifth those given above.

* Courtesy of Electrical Machinery Manufacturing Co., Minneapolis.

14-13. Torque. The starting torque of the motor is the turning effort, or torque, which the motor exerts when full voltage is applied to the motor terminals at starting. The amount of starting torque which a given motor develops depends in a way on the resistance of the rotor winding. Starting torque is usually expressed as a percentage of full-load torque. Increased rotor resistance gives an increase in starting torque, with a corresponding increase in slip and a decrease in efficiency. Manufacturers can build motors with high torque alone or high efficiency alone, but there are commercial and industrial limitations which prevent the manufacture of motors which excel in all characteristics.

Induction motors are classified by starting torque, locked rotor current, and full-load speed. The NEMA Standards classify squirrel-cage induction motors as follows:

Class A. Normal starting torque—normal starting current
Class B. Normal starting torque—low starting current
Class C. High starting torque—low starting current
Class D. High slip

Although there are six NEMA classifications of a-c squirrel-cage induction motors by starting torque and starting current basis, the above classes B, C, and D are used to the greatest extent.

Normal starting torque, listed in the NEMA Standards as being more than 150 per cent for four poles, is usually in the 180 to the 225 per cent range for the smaller motors, and it is sufficient to start the majority of loads. High-starting-torque motors usually have a minimum of about 225 per cent starting torque. Figure 14-7 shows typical speed-torque curves of the four common classes A, B, C, D of induction motors.

Class A motors are not commonly found at the present time in practice, and class B motors are used unless the extra starting torque of the class C motor is required by the application.

Class B motors usually have in excess of 200 per cent breakdown torque, while the class C motors have slightly less. *Breakdown torque* is the greatest torque that the motor will carry at full voltage without a sudden change in speed. However, the maximum peak of the load torque should not exceed two-thirds or three-quarters of the value of the breakdown torque.

Standard induction motors may, at the discretion of the manufacturer, carry a service factor which defines the additional load that the motor will carry without reaching a temperature that will cause undue shortness of insulation life. For open or dripproof motors this factor is 1.15. This means that the motor will carry 15 per cent more horsepower than stamped on the name plate. This 15 per cent can be used as a safety factor to take care of the unknown things that may cause overloading of the motor, or if the load is known with considerable certainty, this extra output of

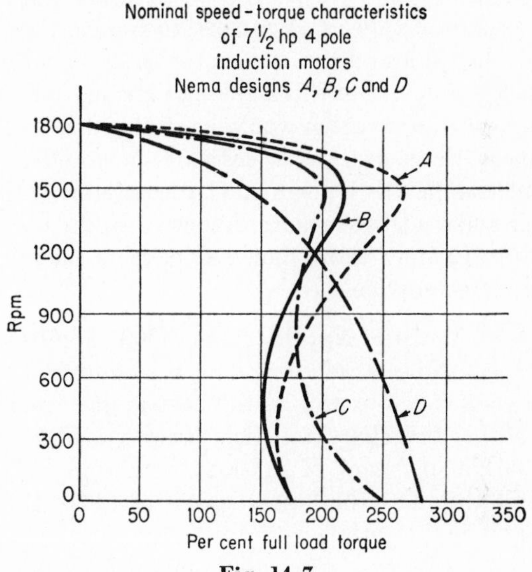

Fig. 14-7

the machine can be used under normal conditions. Splashproof and totally enclosed motors normally do not carry a service factor greater than 1.00, so when one of these motors is used on an application that normally uses an open motor, care must be exercised to see that the motor is not overloaded to any dangerous extent.

In this connection, it must be remembered that the overload device will remove the motor from the line when conditions for which it is set are reached. The NEC specifies what these conditions should be, and most wiring inspectors abide by the code rules. The usual overload device mounted in the controller case must remove the motor from the line if the current exceeds 125 per cent of full-load current in a 40°C (104°F) ambient temperature. These devices are usually made so that the motor will carry a greater load in a lower ambient temperature, and will allow the motor to maintain approximately the same total temperature.

Since 40°C is not uncommon in many locations, especially in the summer, and since there must always be some leeway in the effective capacity of the motor, these rules rather effectively limit the maximum expected load on the motor to the service factor rating of the motor. In certain cases where the load is accurately known, both in horsepower and in time, it is possible to trade on the time lag of the overload device and allow the motor to carry an overload for a short period of time. But this is taking a chance; and if something should go wrong, the machine would be shut down.

When the overload protection device is mounted on the motor itself the time-temperature overload characteristics of the combination of motor and control are better known. Under such conditions it is much safer to allow short time overloads on the motor without fear of interrupted service.

14-14. Higher Torque Applications. For applications where higher starting torque is required than is obtainable with class B motors, the class C motor will be the next most economical to use in order to obtain this characteristic. This type of motor has generally 200 to 250 per cent of full-load torque available at start, and is generally used at the higher speeds of 1,800, 1,200, and 900 rpm. For still higher starting torques to approximately 300 per cent of the full load torque the class D motor is used, although the lower efficiency during running may well justify the use of a larger horsepower rating of a class B or class C motor. The class D motor is primarily used for pulsating load applications where such motor and flywheel are used to reduce current peaks.

14-15. Wound-rotor Induction Motors. Wound-rotor induction motors are similar to squirrel-cage induction motors except that the rotor has an insulated winding with collector rings, between which resistance can be inserted to vary the characteristics. This resistance in the rotor circuit develops high starting torque with low values of starting current. As the motor accelerates up to speed, the resistance is gradually cut out until at full speed the rotor windings are short-circuited. The motor will then operate with high efficiency and low slip.

Wound-rotor induction motors are used for large motor drives requiring high slip or high starting torque. By inserting various values of resistance in the rotor circuit as above, a variety of speed torque values is obtainable. High or low starting torque and high or low speeds under load are possible. The higher cost of the wound-rotor motor and control generally is not warranted for many applications unless the characteristics obtainable can be economically justified compared with standard or semistandard general-purpose motor designs.

It should be remembered that the starting torque (and maximum torque) varies as the second power of the applied voltage. For example, a motor designed to have 100 lb-ft starting torque on a 440-volt power line will have only 82.5 lb-ft torque if the voltage drops to 400 volts. Starting torque requirements also may vary at different times because of frequency starting, temperature changes, type and amount of lubricant, and other conditions. The motor torque available at the motor shaft must at all times be greater than the torque required by the driven machine under the worst conditions.

Q14-7. (*a*) The efficiency of a 550-volt three-phase induction motor is 90 per cent when the line current is 100 amp per wire at a power factor of 92 per cent. Calculate the motor output. (*b*) Explain why an induction motor having a very low rotor resistance with respect to its reactance gives a poor starting torque.

ANSWER.

(*a*) Let E be the impressed voltage on the line, I be the line current, and $\cos\theta$ be the power factor. Then the motor input is

$$EI\sqrt{3} \times \cos\theta \qquad \text{for a three-phase system}$$

$$550 \times 100\sqrt{3} \times 0.92 = 87{,}700 \text{ watts}$$

$$\text{Motor ouput} = \frac{\text{input} \times \text{efficiency}}{100} = \frac{87{,}700 \times 90}{100} = \mathbf{79{,}000\ watts}$$

(*b*) With a very low rotor resistance with respect to the reactance, the current in the bars will lag considerably behind the impressed voltage. The result is that the maximum current at any given time will not coincide with the maximum voltage. Now the torque developed is proportional to the product of current and flux according to the expression

$$\text{Torque} = K\phi I$$

Since the flux density is greatest when the emf is greatest, its maximum will not coincide with the current maximum and the torque will be proportionally less. A lagging also means that motor conductors carrying current in a positive direction will be situated in a negatively magnetic field with further lessening of the developed torque.

Q14-8. A 20-hp three-phase induction motor operates at full load with 440 volts across its terminals. It takes 25 amp per line, lagging 30° behind the phase voltage. Determine the efficiency of this motor.

ANSWER.

$$\begin{aligned} \text{Motor input} &= EI\sqrt{3} \times \cos\theta/746 \\ &= 440 \times 25\sqrt{3}\cos 30/746 = 22.1 \text{ hp} \\ \text{Efficiency} &= 20.0/22.1 = 0.905, \text{ or } \mathbf{90.5\ per\ cent} \end{aligned}$$

Q14-9. A three-phase Y-connected induction motor operating at 220 volts has a power factor of 70 per cent and an efficiency of 82 per cent when delivering 8 hp. What is the current in the phase windings, and what is the voltage across each?

ANSWER.

$$\begin{aligned} \text{Motor output} &= 8 \times 746 = 5{,}970 \text{ watts} \\ \text{Motor input} &= 5{,}970/0.82 = 7{,}280 \text{ watts} \end{aligned}$$

For a Y-connected induction motor

$$\text{Motor input} = EI\sqrt{3} \times \cos\theta = 7{,}280 \text{ watts}$$

We rearrange and solve for

$$I = 7{,}280/(220\sqrt{3} \times 0.70) = \mathbf{27.3\ amp} \text{ (phase current)}$$

$$\text{Voltage across each phase is: } e = 220/1.732 = \mathbf{127\ volts}$$

Q14-10. A 150-hp three-phase 440-volt wound-rotor six-pole motor with a 220-volt secondary coupled to a jackshaft, driving an exhaust fan, has burnt out in a mine. The speed control on this motor was obtained by changing the secondary resistance. It is proposed to use two 75 hp three-phase 440-volt wound-rotor motors having 220-volt secondaries until the burnt-out motor can be repaired. These motors would be coupled to the jackshaft, both primaries would be connected to the line switch, and both secondaries would be joined up in parallel in order to use the secondary resistance control. Point out what difficulties may be incurred and what precautions might be taken.

ANSWER. Refer to Fig. 14-8. If the motors are to be connected in parallel, it is to be assumed that they will have to share the load

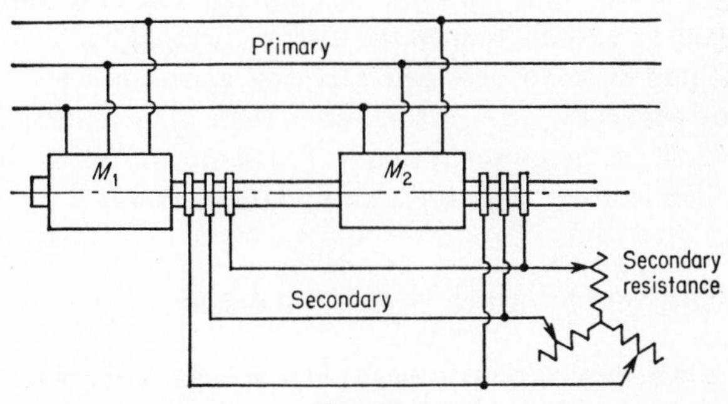

Fig. 14-8

equally. For this to happen the motors must have the same speed-load characteristics; that is, at a given speed common to both by virtue of the jackshaft connection they must give the same torque. If their characteristics vary slightly, bring it in line with the other. However, if the characteristics are far enough apart, it may be necessary to supply each secondary with its own resistance, discarding the available 150-hp resistance. Of course, the primaries must be connected so as to give the same rotation for both motors. Since both motors have the same secondary nominal voltage, no difficulty is expected in this respect.

14-16. Polyphase and Single-phase Motors. Polyphase motors have two (for two-phase) or three (for three-phase) separate windings, one for each phase. These windings are usually distributed uniformly around the inner circumference of the stator. The currents in these windings alternate progressively and continually so as to produce a uniformly revolving field that drags the rotor around with it.

In a single-phase induction motor there is but one winding, the alternating current in which produces a magnetic field that also alternates in polarity but does not tend to revolve. In other words, it has no starting torque. If a slight auxiliary torque is provided, the motor will start in either direction, depending on the initial direction of this torque. Then it soon accelerates up to normal speed. In very small motors this initial torque was provided by hand. Today, practically all single-phase motors of the induction type have some auxiliary means for starting. After a single-phase motor is up to speed, it operates as a polyphase machine. It is, however, limited as to overload capacity and is not built in large sizes. Its efficiency is a little lower than that of a polyphase machine of the same power rating. The same applies to the power factor, which is lower than for a polyphase machine.

14-17. Single-phase Motors. Single-phase motors may be divided into two principal classes, split-phase and commutator motors.

The earliest form of single-phase motor was probably the *split-phase motor*. Its rotor is the same as that of the squirrel-cage motor in many respects. Many manufacturers who offer squirrel-cage and split-phase motors in the same sizes and capacities make the rotors of the two types interchangeable. From the viewpoint of maintenance service, this is of some advantage, as not infrequently a temporary split-phase stator is slipped over a squirrel-cage rotor to keep a machine running while the regular stator is being rewound.

The stator of the split-phase motor consists of two windings, starting and running. The starting winding is very light and heats up quickly. It is thus able to carry current for a short time only. It must be opened by a centrifugal device the instant the proper speed is attained. Figure 14-9 schematically shows this motor. *AA* are the main poles; *BB* are the starting poles; *C* is the centrifugal switch. *D* is the resistance in the starting winding.

14-18. Repulsion-induction Motors. These motors are of the commutator type. They start by electrical repulsion and then run by induction, just as all other types of a-c motors, except the synchronous type.

Starting is accomplished by placing a winding on the rotor, in which a current is induced from the current applied to the stator. The nature of the motor requires that after it reaches speed, a short-circuiting device must be brought into contact with all segments of the commutator. In Fig. 14-10, as the motor starts, the brushes make successive circuits around one of the rotor conductors. When the rotor turns fast enough to operate a centrifugal device, a short-circuiting mechanism makes a single contact between all rotor conductors. The line voltage across L_1 and L_2 is 220; by

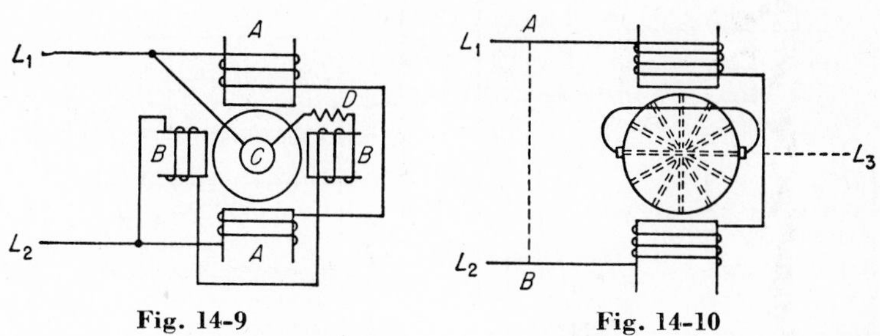

Fig. 14-9 **Fig. 14-10**

connecting A to B and changing L_1 to become L_3, the motor will run on 110 volts.

14-19. Capacitor Motors. This is a split-phase type. However, it overcomes most of the objectionable features inherent in the split-phase and repulsion-induction types; the high starting current of the former and the wound rotor and brushes of the latter are not found in the capacitor type. In addition, the rotor can be made identical to and interchangeable with the polyphase rotor. In Fig. 14-11 a capacitor start-and-run motor is shown diagrammatically. When the motor is at rest, L_2 connects to the "run" position in the current relay. At the instant of starting, a heavy inrush current activates the current relay, connecting L_2 to the "start" position. When the rotor reaches the three-quarter speed, the diminishing current drops out the relay, reconnecting L_2 to the "run" position. The autotransformer, besides serving during the starting period, may also be equipped with a manual setting device to induce different voltages, which, in turn, will result in different motor speeds.

Figure 14-12 shows a capacitor-start induction-run motor. When the relay pulls in, caused by changes in voltage due to the motor approaching speed, the starting winding is opened. As shown, the connections are for a line voltage of 220 volts. By merely making the connections indicated by the broken lines and removing the connections between *A* and *B*, the motor will run on 110 volts.

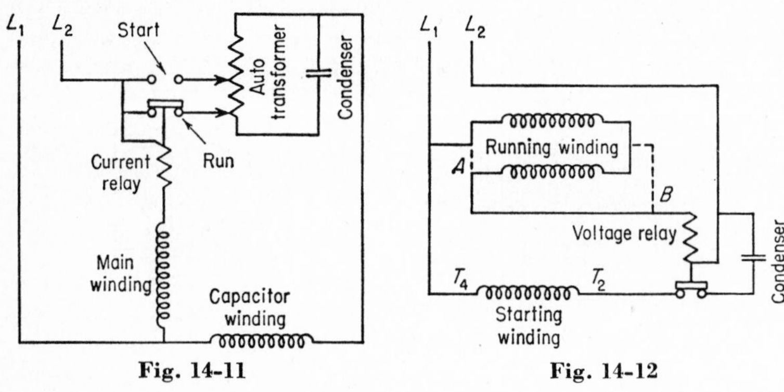

Fig. 14-11 **Fig. 14-12**

Q14-11. The circuit of a capacitor motor is as shown in Fig. 14-11. Explain why the autotransformer is used.

ANSWER. See Sec. 14-19 above.

14-20. Motor Speed–Torque Curves. Figure 14-13 shows a typical speed-torque curve for single-phase commutator motors. Figure 14-14 shows a typical speed-torque curve for single-phase capacitor motors.

14-21. Determination of Motor Rating. When current flows through the stator and rotor coils of a motor, heat is generated in overcoming the resistance. Additional heating effect is produced in the core laminations. The machine rating depends on its ability to dissipate this heat so that the machine will operate continuously within safe temperature limits. General-purpose motors carry a safety factor to permit operation at higher temperatures for short periods of time, without damaging the insulation on the windings. This service factor we discussed previously on p. 217. The 40°C rating mentioned is called a continuous rating and refers to the load the machine can carry steadily without a time limit. This load capacity is called *rated* load or *full* load. Just in passing it might

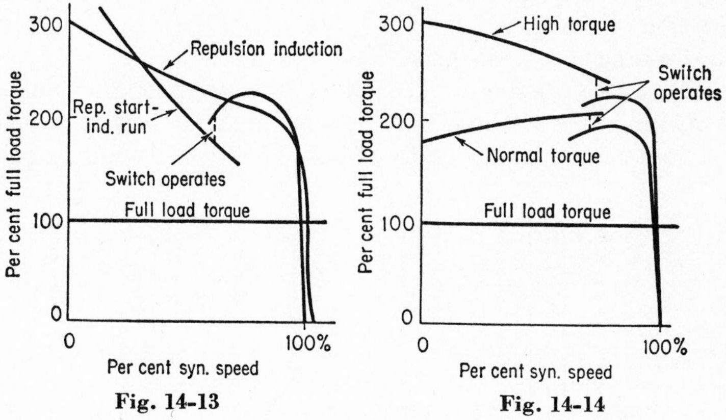

Fig. 14-13

Fig. 14-14

be well to note that d-c generators are rated in kilowatts, alternators in kilovolt-amperes, and motors in horsepower.

14-22. Induction-motor Starting. Induction motors up to 10 hp can generally be started *across-the-line.* In this method, the main line contacts to the motor can be closed and full voltage applied to the motor while it is at rest. Such starting is often done by having a solenoid coil pull in an iron bar which will in turn carry with it the motor-starting contacts. The typical across-the-line starter will have a push-button station if starting manually, with one starting and one stopping button, the contact being in the coil circuit. For automatic starting the coil circuit is closed through some automatic control device.

The comparatively high inrush current when starting above 10 hp prevents throwing them across the line. There are two methods of starting large squirrel-cage motors, both serving to reduce the starting voltage. The first uses a resistance to reduce the voltage, the second an autotransformer. There are also other methods which will be outlined next.

14-23. Ways to Start Induction Motors. There are five methods for starting induction motors in the United States. These are in common use today and are as follows:

(*a*) Across-the-line
(*b*) Autotransformer (compensator)
(*c*) Series-resistance
(*d*) Part-winding
(*e*) Wound-rotor resistance

Across-the-line method. The simplest way to start induction motors up through 10 hp is at full voltage. Larger motors may also be started across the line under certain conditions, because modern

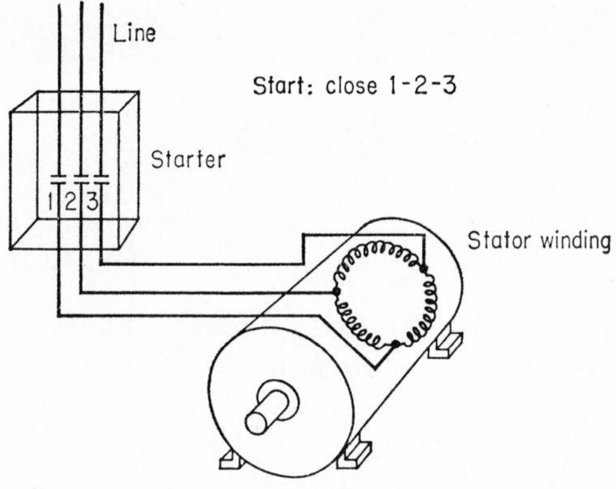

Fig. 14-15

motors are designed to withstand shock. Full voltage across-the-line starting is impractical if the utility company will not permit large starting-current inrush or surge, or under conditions where surge causes a troublesome voltage dip in a plant. In order to avoid the shock of across-the-line starting torque effects, reduced voltage starting is used most often for the larger motors to get smooth acceleration of the load. Figure 14-15 (*courtesy of Power Magazine*) schematically shows how this is accomplished.

Autotransformer (compensator) method. Manual or automatic operation connects the motor to the autotransformer taps and then the autotransformer to the line when starting. The motor is switched over to the line at about 70 or 80 per cent of the synchronous speed. Then the autotransformer is disconnected.

Usually autotransformers have two or more taps available so you can set the compensator for the starting requirements of the connected load. Voltage taps provide for a range from 65 per cent to 80 per cent of the line voltage. There are several advantages to autotransformer starting: lower power losses, lower line current required, and less radiated heat. Figure 14-16 (*courtesy of Power Magazine*) gives the hookup.

Series-resistance method. If a series-resistance-type starter is used to reduce the voltage, say to half value, the current drawn

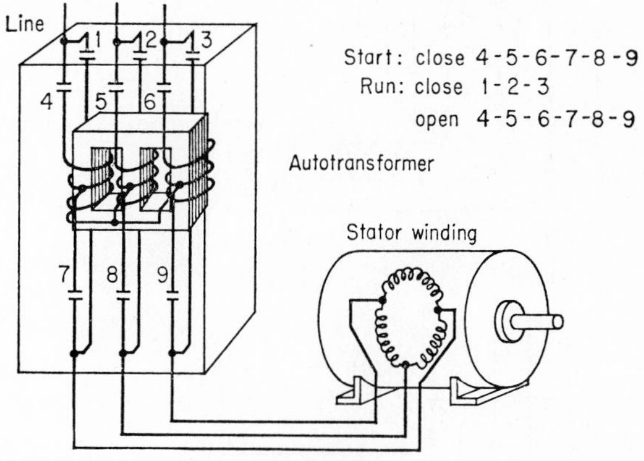

Fig. 14-16

from the line will also be at half value of the full voltage. Whereas, with the autotransformer type, the current drawn from the line varies as the square of the voltage ratio of the transformer. Thus, if half voltage is applied to the motor terminals from the secondary of the autotransformer, the current drawn from the line will only be one-fourth the full voltage value.

The circuit (see Fig. 14-17) is closed throughout the starting period, the contactors closing in sequence to short out the resistance. Such arrangements reduce the sharpness of the steps and produce smoother speed-torque curves. However, the attendant power loss and initial expense of equipment make this method less popular than compensator starting except where series-resistance characteristics are desired for a particular reason. (*Figure courtesy of Power Magazine*).

Part-winding method. See Fig. 14-18 (*courtesy of Power Magazine*). This method is gaining popularity for large motors because it is low in first cost, but the motor must be arranged for part-winding starting. In this design the motor is set up with two windings or more, so that the motor will start up with one section or winding.

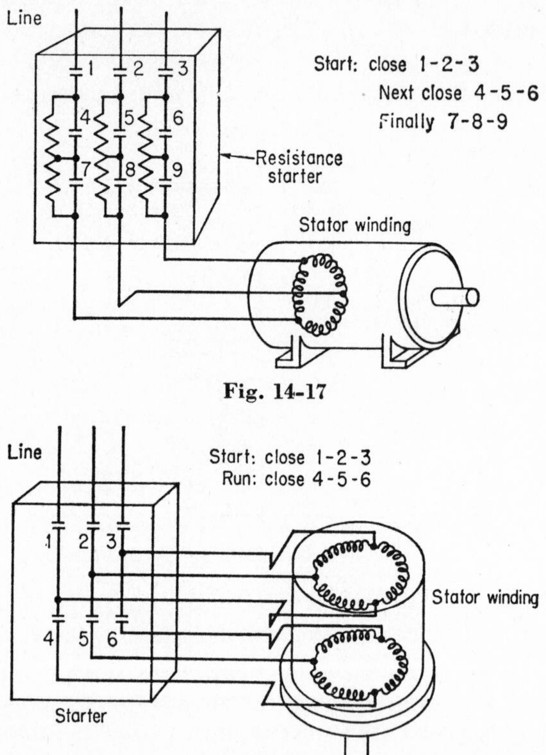

Fig. 14-17

Fig. 14-18

As motor speeds up, each winding will be connected in parallel until all are operating when the motor is up to speed. The impedance of a single section or winding is of such magnitude that it limits the starting current when full voltage is applied.

Wound-rotor resistance method. By inserting degrees of resistance in the rotor circuit the starting current is cut while the starting torque is kept high. Such a system of starting is useful when starting high inertia or flywheel loads. A drum controller is used to gain the advantage of many steps. Under desirable conditions where resistance is employed for starting only, the rotor winding is shorted out in the run position. If speed control is incorporated, part of the resistance remains in the running circuit and power is lost (I^2R loss). Resistors are designed for short-time service during starting operation and will burn out if subjected to continuous

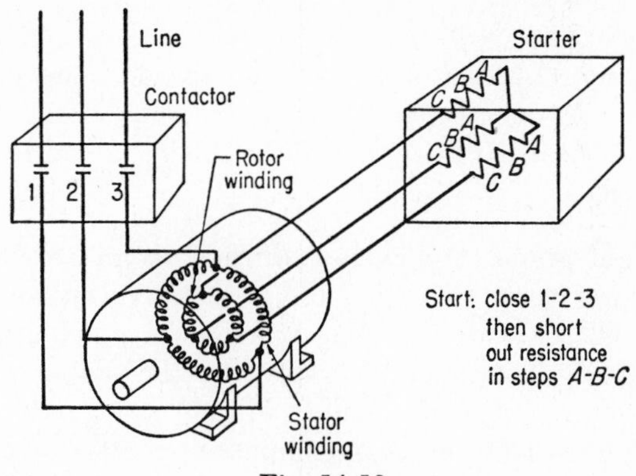

Fig. 14-19

speed-control service current requirements [see Fig. 14-19 (*courtesy of Power Magazine*)].

14-24. Motor Speeds. The relation by which the synchronous speed of an induction motor is determined is given by

$$\text{rpm} = \frac{120f}{p} \tag{14-3}$$

where f = frequency, cps

p = number of poles

Thus the common condition of 60 cps and a four-pole winding would give the customary speed of 1,800 rpm.

Since the speed is fixed, such motors are sometimes wound with a method for reconnecting the stator windings so as to change the number of poles. By arranging two windings each for two sets of poles it is possible to have four speeds. For example, with one winding for four and eight poles and the other for six and twelve poles, the motor would have synchronous speeds of 600, 900, 1,200, and 1,800 rpm on 60 cps. In such a case, the torque will vary as the square of the speed. Hence, if a given motor will develop 10 hp at 1,200 rpm, it will develop only 2.5 hp at 600 rpm.

In order to show the effect of slip in Eq. (14-3) it may be written in the following form

$$\text{rpm} = \left(\frac{120f}{p}\right)(1 - s) \tag{14-4}$$

where s = slip as a decimal

Thus a change of slip will control the speed. The slip may be changed by introducing resistance in the rotor circuit. This has already been discussed in connection with the wound-rotor type of motor, p. 219.

14-25. Single-phase Operation of a Polyphase Motor. As mentioned previously the single-phase induction motor is inferior to the polyphase motor. If one phase of a polyphase motor is opened, possibly because of overheating, the motor will operate as a single-phase motor, although it will not start under these conditions. Under conditions of single-phase operation of a polyphase motor the rating and breakdown torque of the polyphase motor are quite reduced. If run under these conditions at rated polyphase motor loading, the motor may overheat. One leg can be made open through maloperation of the compensator starter. The best way to tell is by testing each leg with an ammeter.

14-26. Tests for Output and Efficiency. The mechanical output of a motor is commonly measured by means of a Prony brake test. The power may then be calculated by means of the following formula

$$\text{hp} = \frac{2\pi NT}{33{,}000} \tag{14-5}$$

where N = motor speed, rpm

T = torque developed by motor, measured along moment arm in lb-ft

To measure the efficiency we must measure by means of a wattmeter or other suitable method the input to the motor at the same time as the output is determined. Both the output and the input are then converted to the same units, preferably watts, and the efficiency determined by applying the formula

$$\text{Efficiency} = \frac{\text{output}}{\text{input}} \times 100 \tag{14-6}$$

A more accurate method is to use either a dynamometer for measuring the output or to calculate the efficiency by the loss method as established by the American Institute of Electrical Engineers (AIEE).

Q14-12. What is the efficiency of a motor which takes 43 kw when it is delivering 50 hp?

ANSWER.

Efficiency = (50 × 746)/(43 × 1,000) × 100 = **86.6 per cent**

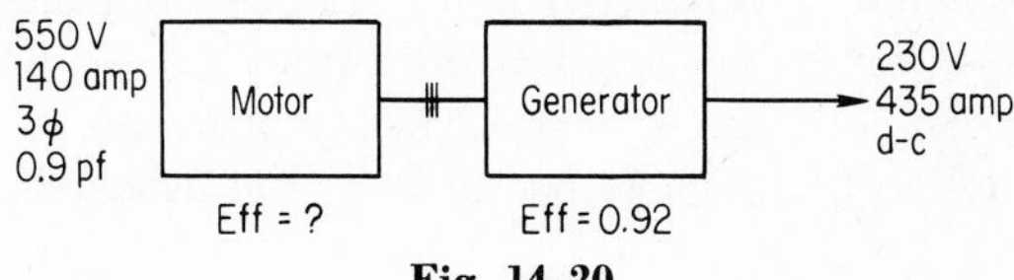

Fig. 14-20

Q14-13. A motor-generator set delivers direct current of 435 amp at a 550-volt, 3-ϕ, 3-wire line which delivers 140 amp at each terminal of the machine at a power factor of 90 per cent. The efficiency of the d-c generator under these conditions is 92 per cent. Calculate efficiency of the motor.

ANSWER. Refer to Fig. 14-20. Then, input = output/efficiency, where input power = $\sqrt{3}\, EI \cos \phi$. Power = $\sqrt{3} \times 550 \times 140 \times 0.9 = 120{,}000$ watts. Output = $P = EI = 230 \times 435 = 100{,}000$ watts.

Motor efficiency = 100,000/(120,000 × 0.92) = **90.7 per cent**

Chapter 15

POWER CIRCUITS

15-1. Polyphase Systems. If instead of a single coil (single phase) we use three coils in the armature of an alternator, each coil is called a *phase* of the generator. Power may be taken from each of the three armature coils, or phases, separately when a conductor is attached to each. Normally the three phases are joined together and only three wires (conductors) are connected to them. There are two ways of joining the phase coils, as we shall see later; one way is the delta (Δ) or mesh connection, the other the Y or star connection.

In a three-phase system the voltage across each phase is at an angle of 120° to the voltage across the other phases. When Y connected and balanced, the voltage across any phase equals $\sqrt{3}$ times the voltage between one line wire and the neutral. The current in each path in the armature is the same as the current in one phase.

When connected in Δ and balanced, the voltage across any phase equals the voltage across any one path in the armature. The current in each line wire equals $\sqrt{3}$ times the current in each armature path. As we have noted before, the power in a balanced three-phase circuit is found by the equation $P = 1.732EI \cos \theta$. This holds whether or not the circuit is Δ or Y connected.

The rating of a given motor (or generator) may be shown to increase with the number of phases. This is a very important con-

TABLE 15-1

Single-phase machine	100
Two-phase machine	140
Three-phase machine	148
Six-phase machine	148
Direct-current machine	154

sideration. On the basis of a single-phase machine giving 100 rating, Table 15-1 gives approximate power ratings.

Q15-1. A three-phase alternator has three coils each rated at 1,330 volts and 160 amp. What is the voltage, kva, and current rating of this alternator if the three coils are connected in Y?

ANSWER.

$$\text{Line voltage } E = \sqrt{3} \times 1{,}330 = \mathbf{2{,}300 \text{ volts}}$$
$$\text{Alternator rating} = \sqrt{3} \times 2{,}300 \times 160 = \mathbf{640 \text{ kva}}$$
$$\text{Current rating} = \mathbf{160 \text{ amp}}$$

15-2. Grounding. Grounding is the connection of one wire of an electric system or any metallic part of the system to the earth.

All alternating currents should be grounded and the grounded wire is called the *ground*, or neutral; the wires not grounded are known as the *hot*, or *phase*, wires. In practice, however, many large industrial plants operate with *ungrounded* systems, usually Δ. A neutral may be (rarely) ungrounded; a system without a neutral may be grounded on one leg or corner of a Δ. The neutral is the point of equal potential to the phase wires (see Figs. 13-47 and 13-49). The neutral of Fig. 13-48 is neutral to the two bottom but not to the upper phase wires. If there were no ground of the transformer neutral or of the phase wires in Fig. 13-48, the neutral would be an imaginary point.

The ground must be equal in size or larger than the phase wires. In addition, it must not be fused or connected to any switch. For identification the wire used for the neutral must have white covering. The neutral must also be connected to the white screw in all receptacles. In the single-phase two-wire system one of the wires must be grounded, and the same rules as given for the neutral conductor must be observed.

The approved maximum voltage to ground for any wire connected to standard "house-wiring" switches and receptacles is 150 volts. This is the reason why only one phase of a 220-volt Δ system may be used for the lighting circuits. Where greater voltages exist for motors, etc., the wires must be encased in a metal conduit, and the frames of the motors and conduit systems electrically connected and grounded. Grounding should be carefully inspected at sufficiently frequent intervals to make certain that the essential protection it provides is in perfect condition.

Grounds should be carefully made, preferably using not less than No. 2 B & S gauge copper wire. Several grounds are to be preferred to a single ground connection so that the possibility of broken connections would be a rarity. Use the water system, and the larger the pipe, the better the results. Do not use gas, gasoline, or oil pipelines.

It is necessary to ground the electric circuit so as to prevent dangerous high voltages (because of mistakes or accidents) from increasing the normally low potential circuits to a dangerous value. As we previously noted, the wiring for lighting systems of buildings is designed for a maximum of 150 volts to ground. What sort of an accident could occur? Let's assume that a street series circuit operating at 5,000 volts should fall across the service wires from the pole to your home. Without a ground to the neutral of the secondary wiring, the voltage would be raised from its low normal potential to a maximum of 5,000 volts. Definitely, this would be a dangerous condition. But if the neutral were grounded, the high voltage would safely pass to the ground.

Q15-2. Two coils *A* and *B* with resistances and inductances as shown in Fig. 15-1 are connected in series. If an alternating current of 100 amp and 50 cycles flows through the coils, find:

(*a*) Impedance of coil *A*
(*b*) Impedance of coil *B*
(*c*) Combined impedances of both coils
(*d*) Potential drop across coil *A*
(*e*) Potential drop across coil *B*
(*f*) Potential drop across both coils
(*g*) Phase difference between current and "drop" across *A*
(*h*) Phase difference between current and "drop" across *B*
(*i*) Phase difference between current and total "drop"
(*j*) Power consumed in coil *A*
(*k*) Power consumed in coil *B*
(*l*) Power consumed in both coils

ANSWER. Refer to Fig. 15-1.

(*a*)

$$Z_a = \sqrt{R_a^2 + X_{La}^2} = \sqrt{5^2 + (2\pi \times 50 \times 0.0107)^2} = \mathbf{6\ ohms}$$

(*b*)

$$Z_b = \sqrt{R_b^2 + X_{Lb}^2} = \sqrt{20^2 + (2\pi \times 50 \times 0.5)^2} = \mathbf{158\ ohms}$$

(*c*)

$$Z_{ab} = \sqrt{(R_a + R_b)^2 + [2\pi f(L_A + L_B)]^2}$$
$$= \sqrt{(5 + 20)^2 + [(2\pi \times 50)(0.0107 + 0.5)]^2} = \mathbf{162\ ohms}$$

(*d*) $E_a = IZ_a = 100 \times 6 = \mathbf{600\ volts}$

(*e*) $E_b = IZ_b = 100 \times 158 = \mathbf{15{,}800\ volts}$

(*f*) $E = IZ_{ab} = 100 \times 162 = \mathbf{16{,}200\ volts}$

(*g*) $\cos \theta_a = \frac{R_a}{Z_a} = \frac{5}{6} = 0.833$; from which $\theta_a = \mathbf{33.6°}$

(*h*) $\cos \theta_b = \frac{R_b}{Z_b} = \frac{20}{158} = 0.126$; from which $\theta_b = \mathbf{82.7°}$

(*i*) $\cos \theta_{ab} = \frac{R_{ab}}{Z_{ab}} = \frac{5 + 20}{162} = 0.154$; from which $\theta_{ab} = \mathbf{81.1°}$

(*j*)

$$P_a = I^2R_a = 100^2 \times 5 = \mathbf{50{,}000\ watts,}\ \text{or 50 kw}$$

(*k*)

$$P_b = I^2R_b = 100^2 \times 20 = \mathbf{200{,}000\ watts,}\ \text{or 200 kw}$$

(*l*)

$$P_{ab} = I^2R_{ab} = 100^2 \times (20 + 5) = \mathbf{250{,}000\ watts,}\ \text{or 250 kw}$$

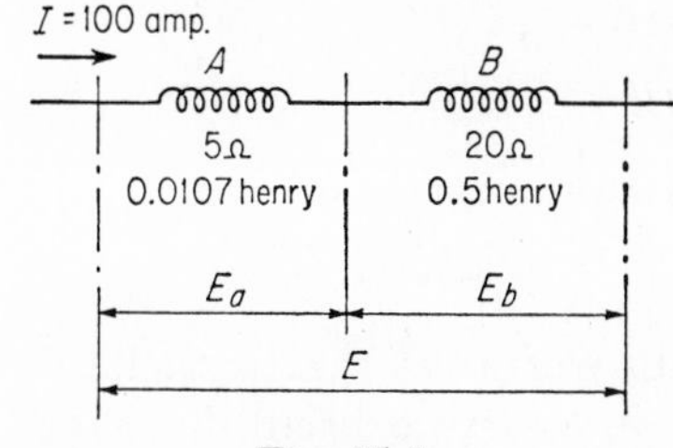

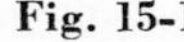

Fig. 15-1

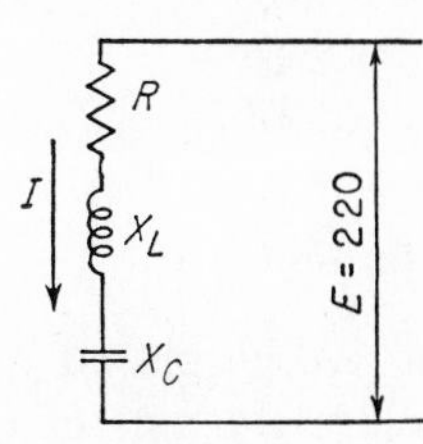

Fig. 15-2

Q15-3. A series circuit comprises a pure resistance of 6 ohms, an inductive reactance of 10 ohms, and a condenser of 25 ohms capacitive reactance. If the power source is a 200-volt alternating current, what is the current flow in the circuit and what power is being consumed?

ANSWER. Refer to Fig. 15-2. The impedance of the circuit is found

$$Z = \sqrt{R^2 + (X_L - X_C)^2} = \sqrt{6^2 + (10 - 25)^2} = 16.2 \text{ ohms}$$

$$I = \frac{E}{Z} = \frac{220}{16.2} = \mathbf{13.6\ amp}$$

$$\text{Power factor} = \cos\theta = \frac{R}{Z} = \frac{6}{16.2} = 0.37 \text{ (see Fig. 15-3)}$$

$$\text{Power consumed} = EI\cos\theta = 220 \times 13.6 \times 0.37 = \mathbf{1{,}105\ watts}$$

Q15-4. A circuit taking 20 amp from a 60-cycle 120-volt main operates at a lagging power factor of 80 per cent. Show how a

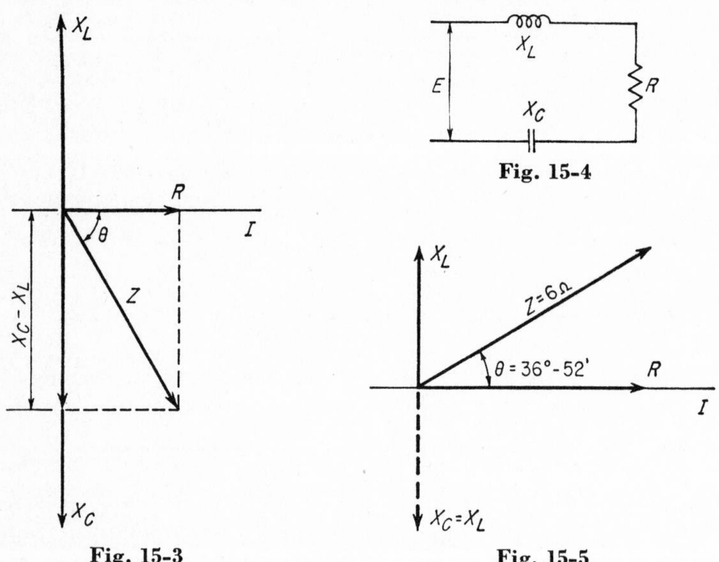

Fig. 15-3

Fig. 15-4

Fig. 15-5

condenser can be placed in the circuit to correct the power factor and calculate the size of condenser necessary to bring the power factor to 100 per cent.

ANSWER. Refer to Fig. 15-4 for the circuit and Fig. 15-5 for the vector diagram. In any a-c circuit in which the inductive reactance is greater than the capacitive reactance, the current lags behind the voltage, and the power factor is less than 100 per cent. Therefore, in order to bring the power factor to 100 per cent, the original circuit may be corrected by installing a condenser in series with the inductance and resistance. The condenser must be of such a size that the inductive reactance equals the capacitive reactance. Now let us consider the original without the addition of the condenser.

$$Z = \frac{E}{I} = \frac{120}{20} = 6 \text{ ohms total impedance of circuit}$$

$$\theta^{-1}0.80 = 36°52'$$

$$X_L = Z \sin \theta = 6 \times 0.6 = 3.6 \text{ ohms}$$

The condenser reactance to be added, therefore, is $X_C = X_L = 3.6$

$$X_C = \frac{1}{2\pi fC}$$

from which $$C = \frac{1}{2\pi \times 60 \times 3.6} = \mathbf{737 \times 10^{-6}\ farads}$$

Q15-5. A coil of 0.1 henry inductance and 18 ohms resistance is connected to a 220-volt 60-cycle power supply. Find the current flowing in the coil and the angle by which the current lags behind the voltage.

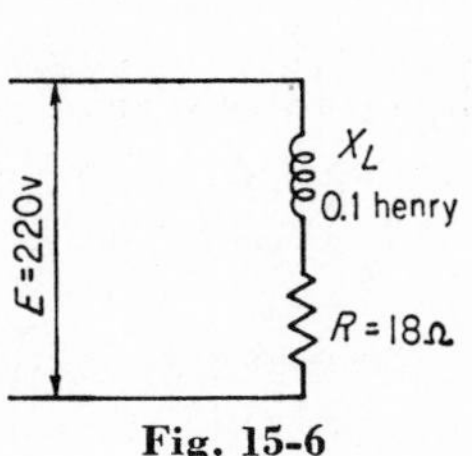

Fig. 15-6

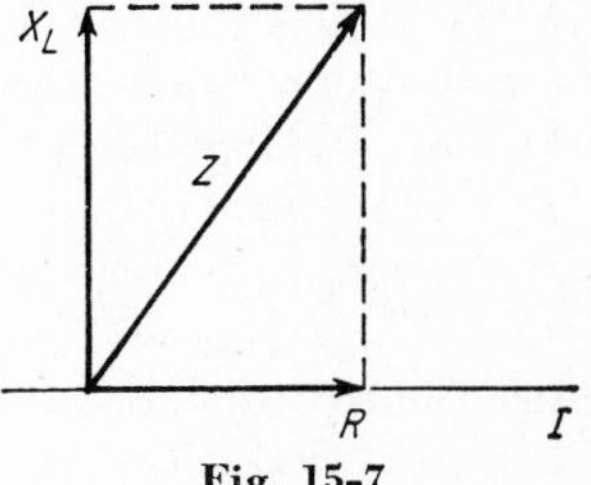

Fig. 15-7

ANSWER. Refer to the diagram of circuit (Fig. 15-6) and its vector diagram (Fig. 15-7). Then

$$\text{Load current} = I = \frac{E}{Z} \qquad \text{also} \qquad Z = \sqrt{R^2 + X_L{}^2}$$

Now find $X_L = 2\pi fL = 6.28 \times 60 \times 0.1 = 37.7$ ohms

$$Z = \sqrt{18^2 + 37.7^2} = 41.7 \text{ ohms impedance}$$

Finally, current flowing in $I = 220/41.7 =$ **5.26 amp**

Now refer to the vector diagram with the power factor

$$\frac{18}{41.7} = 0.43$$

from which $\theta = 64.5°$. Thus the current lags behind the voltage by the angle θ. This type of circuit effect is generally found where the electric load consists primarily of incandescent lighting and squirrel-cage motors. Improvement of the power factor would require the insertion of a condenser or addition of a synchronous motor into the circuit.

Q15-6. If an alternating current of 5 amp flows in a series circuit composed of 12 ohms pure resistance, 15 ohms inductive reactance, and 40 ohms capacitive reactance, what is the voltage across the circuit?

ANSWER. Drop across pure resistance = $5 \times 12 = 60$ volts

Drop across condenser = $5 \times 40 = 200$ volts

Drop across inductance = $5 \times 15 = 75$ volts

Drop across the entire circuit = $\sqrt{60^2 + (200 - 75)^2} =$ **138.7 volts**

Q15-7. Explain why it is desirable or necessary to correct the power factor of low power-factor loads and describe a practical way in which this correction may be accomplished.

ANSWER. The power transmitted by a three-phase circuit is $EI \cos \theta \sqrt{3}$. From this equation we can see that the higher the power factor the greater will the value of power become, i.e., useful power.

Since the equation also represents the power station output, it is also apparent that generator capacity supplying the load is increased with an increase in power factor.

By using overexcited synchronous motors to carry some of the load, the power factor of the system is improved because these motors can be operated with a leading power factor to offset the lagging current effect from the other equipment. When available, rotary converters can also be used to correct the power factor in the line. Many industrial plants in attempting to increase production have added electric loads to electric facilities that are inadequate. We have already learned that the power factor of an electric system depends on the character of the loads. If load consists entirely of transformers, electromagnets, and induction motors that require magnetizing current, then the power factor may be as low as 60 per cent lagging, particularly if the motors are only partially loaded.

On the other hand, when a considerable portion of the load consists of resistance devices and synchronous motors, the power factor may be nearly unity or leading. This condition is ideal but seldom evidenced in practice.

Low lagging power factor is objectionable for several reasons:

(*a*) Generators, transformers, and power lines must be larger (more copper) to supply a given kilowatt load. Nearly 35 per cent more generator, transformer, and power-line capacity is required to supply a given kilowatt load at a power factor of 75 per cent than at unity power factor, assuming the same voltage in both cases.

(*b*) A low power factor increases the power losses in generators, transformers, and transmission lines, nearly 75 per cent greater at 75 per cent power factor than at unity power factor.

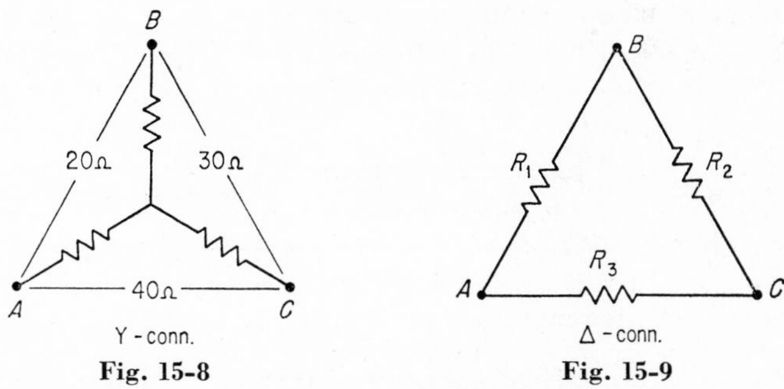

Fig. 15-8 **Fig. 15-9**

(*c*) The voltage regulation is impaired by a low power factor. This may cause poor operation of equipment on the line. A low voltage may reduce the motor speed with attendant overheating and reduced engine output.

Q15-8. A resistance box has three terminals, *A*, *B*, and *C*, and the resistances between them are: $AB = 20$ ohms, $BC = 30$ ohms, and $CA = 40$ ohms. There are three resistors within the box, and they are Y connected. (*a*) Determine the resistance of each of these resistors; (*b*) if these resistors are connected in Δ, what would be the resistances between the terminals?

ANSWER. Refer to Figs. 15-8 and 15-9 for Y and Δ connections respectively. Use delta-star transformation.

(*a*) For Y connection.
$$R_1 + R_2 = 20 \text{ ohms} \quad (1)$$
$$R_2 + R_3 = 30 \text{ ohms} \quad (2)$$
$$R_3 + R_1 = 40 \text{ ohms} \quad (3)$$

Therefore
$$R_1 = 40 - R_3 \quad (4)$$

Substituting in Eq. (1)

$$40 - R_3 + R_2 = 20 \tag{5}$$

Rearranging we obtain: $R_3 = 40 - 20 + R_2 = 20 + 30 - R_3$. From which we get

$$R_3 = (20 + 30)/2 = \mathbf{25\ ohms}$$
$$R_1 = 40 - R_3 = 40 - 25 = \mathbf{15\ ohms}$$
$$R_2 = 20 - R_1 = 20 - 15 = \mathbf{5\ ohms}$$

(b)
$$R_1 = 5 \times 15(1/5 + 1/25 + 1/15) = \mathbf{18.5\ ohms}$$
$$R_2 = 5 \times 25(1/5 + 1/25 + 1/15) = \mathbf{30.9\ ohms}$$
$$R_3 = 25 \times 15(1/5 + 1/25 + 1/15) = \mathbf{92.6\ ohms}$$

Q15-9. A group of lamps is to be wired so that they may be turned "on" and "off" by either of two switches that are located

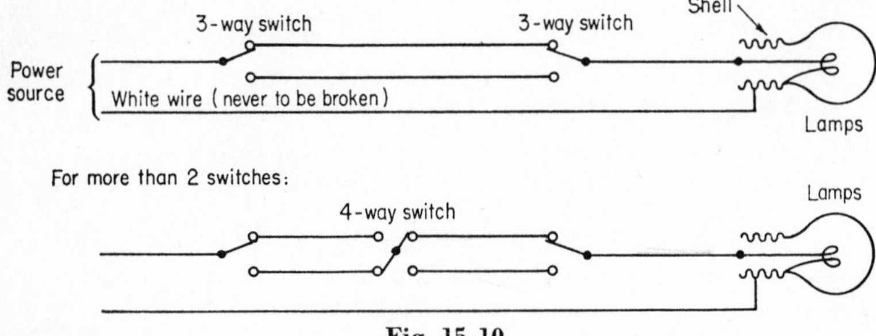

Fig. 15-10

some distance apart. Show by means of a diagram how the circuit and the switches should be wired.

ANSWER. Refer to Fig. 15-10.

Q15-10. A three-phase alternator is connected to a balanced receiving circuit of 80 per cent power factor lagging. Determine the power supplied by the alternator when the voltage across the line wires is 440 volts and the current in each line is 15 amp. Make a vector diagram to show the voltages, currents, and their phase relations. If the alternator is Δ connected, what current will flow in each of the windings?

ANSWER. Refer to Fig. 15-11. This shows the vector diagram for the three phases of the alternator with the line currents lagging at an angle θ behind the line voltages. The figure also shows the relation between the generated voltage V_g and the terminal or line voltage for one phase of the alternator. Now let us make use of the following nomenclature. Let

V_t = line voltage, volts
V_g = generated voltage, volts
I = line current, amp
P = power supplied, watts
$\cos\theta$ = power factor
i = phase current, amp
R = alternator resistance, ohms
X = alternator reactance, ohms
Z = alternator impedance, ohms

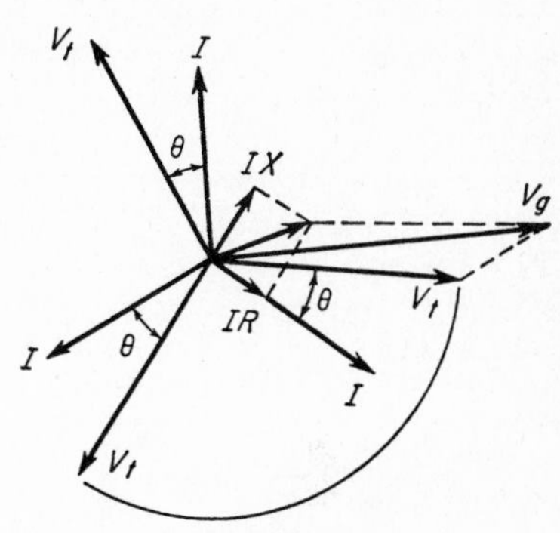

Fig. 15-11

Then, for power supplied, $P = VI\sqrt{3} \times \cos\theta$ from which we obtain

$$440 \times 15 \times 1.732 \times 0.80 = \textbf{9,150 watts}$$

For a Δ-connected alternator, the current in each winding (phase current) will be

$$i = I/\sqrt{3} = 15/1.732 = \textbf{8.66 amp}$$

Q15-11. A three-phase 208-volt system supplies power to the following loads: (*a*) a 30-hp motor of the induction type at an efficiency of 85 per cent and a power factor of 80 per cent; (*b*) a 12-kw incandescent lighting load which may be considered as balanced; (*c*) a synchronous motor drawing 20 kw at a leading power factor of 90 per cent. Calculate the total current supplied and the power factor of the entire load.

ANSWER. Refer to Fig. 15-12 showing diagrammatically the circuit indicated. Let us first determine the current values for each part of the circuit. For the *induction motor* the power is

$$P = (30 \times 746 \times 100)/85 = 26{,}300 \text{ watts}$$

Then the equation relating power, voltage, current, power factor, and $\sqrt{3}$ and the rearrangement to solve for I_1

$$I_1 = 26{,}300/(208 \times 0.80 \times 1.732) = 91.2 \text{ amp}$$

For the *lighting circuit,* because it is balanced and considered to be entirely a resistive load,

$$I_2 = (12 \times 1{,}000)/(208 \times 1 \times 1.732) = 33.3 \text{ amp}$$

For the *synchronous motor* with leading power factor

$$I_3 = (20 \times 1{,}000)/(208 \times 0.9 \times 1.732) = 61.7 \text{ amp}$$

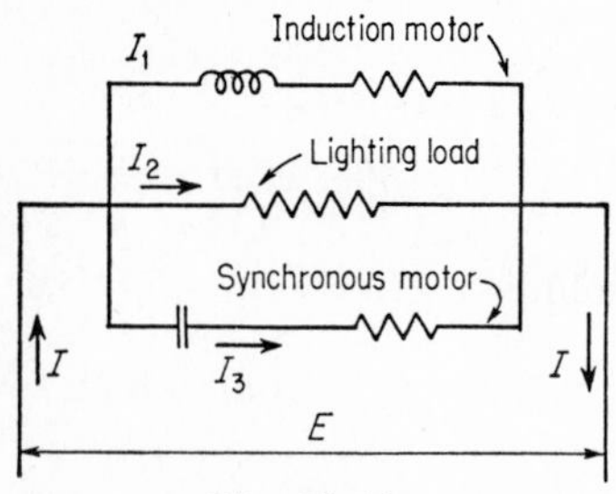

Fig. 15-12

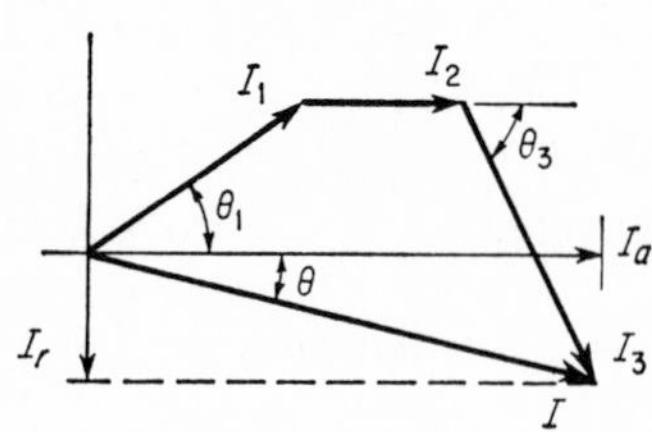

Fig. 15-13

Now let us refer to the vector diagram in Fig. 15-13.

$$\begin{aligned}
I_a &= I_1 \cos \theta_1 + I_1 \cos \theta_2 + I_3 \cos \theta_3 \\
I_a &= 91.2 \times 0.8 + 33.3 \times 1 + 61.7 \times 0.9 = 161.8 \text{ amp} \\
I_r &= -I_1 \sin \theta_1 + I_2 \sin \theta_2 + I_3 \cos \theta_3 \\
I_r &= -91.2 \times 0.6 + 0 + 61.7 \times 0.436 = -27.8 \text{ amp}
\end{aligned}$$

$$\text{Current supplied} = I = (I_a^2 + I_r^2)^{1/2} = [161.8^2 + (-27.8^2)]^{1/2}$$

$$= \mathbf{164.2 \text{ amp}}$$

$$\text{Power factor} = \cos \theta = 161.8/154.2 = 0.985, \text{ or } \mathbf{98.5 \text{ per cent}}$$

The total current and the current per line are the same in value.

Q15-12. A three-phase four-wire 208/120-volt system delivers power to an unbalanced load. Show by means of a diagram the connection necessary to measure, with a single-phase wattmeter, the total power drawn by the load. Indicate the positions of the plus and minus terminals of the current and voltage coils in each measurement. Explain how the total power is calculated.

ANSWER. Refer to Fig. 15-14. To measure the power of an unbalanced three-phase four-wire circuit with a single-phase wattmeter, it is necessary to measure the power of each phase separately. The total power of the circuit is, then, the sum of the separate circuits.

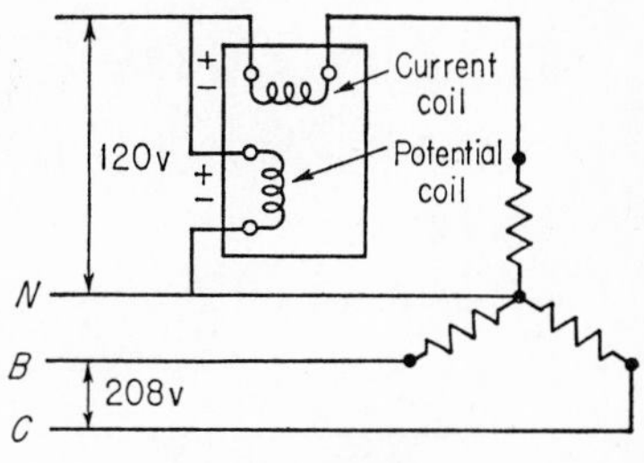

Fig. 15-14

The current coil is connected in series with the phase wire, such as in the diagram, with the (±) terminal on the line side. The potential coil is connected between the phase wire and the neutral (*N* in the figure) with its (±) terminal joined to the phase wire. In this discussion it has been assumed that the current in each phase is small enough not to require a current transformer to protect the meter.

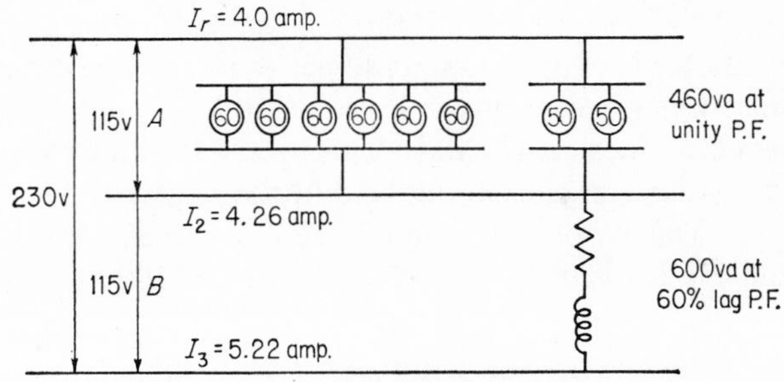

Fig. 15-15

Q15-13. A three-wire 115/230-volt single-phase circuit supplies the following load:

On side *A*: six 60-watt and two 50-watt lamps

On side *B*: one oil-burner motor drawing 600 va at 60 per cent power factor.

Calculate the current in each line and the neutral.

ANSWER. Refer to Fig. 15-15 for the circuit and to Fig. 15-16 for its vector diagram.

$$I_1 = 460/115 = \mathbf{4\ amp}$$
$$I_3 = 600/115 = 5.22 \text{ amp at unity power factor}$$

However, at a power factor of 60 per cent $I_3 = 5.22 \times 0.6 = 3.13$ amp. Now see vector diagram (Fig. 15-16), and adding in-phase and out-of-phase quantities vectorially

$$\sqrt{3.13^2 + x^2} = I_3 = 5.22$$

from which we obtain $x = 4.176$. Now calculate for I_2 current in the neutral.

$$I_2 = \sqrt{4.176^2 + (4.0 - 3.13)^2} = \mathbf{4.26\ amp}$$

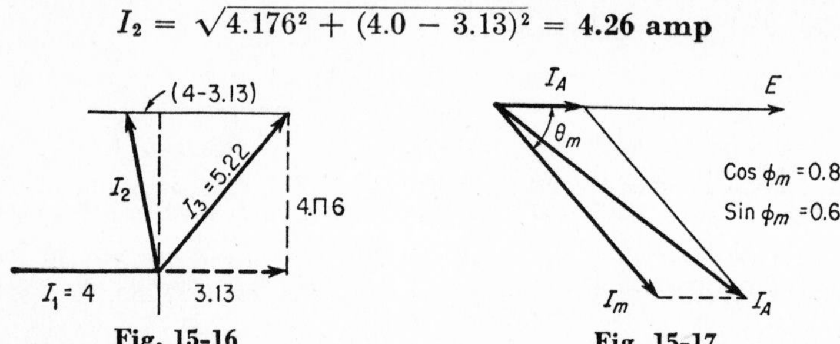

Fig. 15-16

Fig. 15-17

Q15-14. A four-wire 60-cycle three-phase 208-volt line supplies a balanced motor load of 250 kw at 80 per cent lagging power factor. In addition, unity power-factor loads of 50, 100, and 100 kw are drawn between lines A, B, and C, respectively, and the neutral. Calculate the current in each line and in the neutral.

ANSWER. The motor load is equal to 250 kw at 80 per cent power factor; lighting loads of $P_a = 50$ kw, $P_b = P_c = 100$ kw. Now for the motor load

$$I_m = 250{,}000/(208 \times 1.732 \times 0.8) = 866 \text{ amp}$$

For lighting loads

$$I_a = 50{,}000/120 = 417 \text{ amp}$$
$$I_b = I_c = 100{,}000/120 = 833 \text{ amp}$$

Now refer to the vector diagrams, Figs. 15-17, 15-18, 15-19. The summation Σ of horizontal and vertical components (see Fig. 15-17) gives

$$I_A = \mathbf{1{,}220\ amp}$$

In Fig. 15-18 the summation of horizontal and vertical components gives

$$I_B = I_C = \mathbf{1{,}640\ amp}$$

In Fig. 15-19 the summation of horizontal and vertical components gives

$$I_N = \mathbf{417\ amp}$$

Q15-15. A 60-cycle 220-volt single-phase source supplies to a load 100 amp at a lagging power factor of 80 per cent. It is necessary to raise the voltage from 220 to 240 volts. The load is one of constant impedance and power factor. (*a*) From the standpoint of low cost and high efficiency, what is the best device for effecting the change? (*b*) State the specifications for the device you recommend.

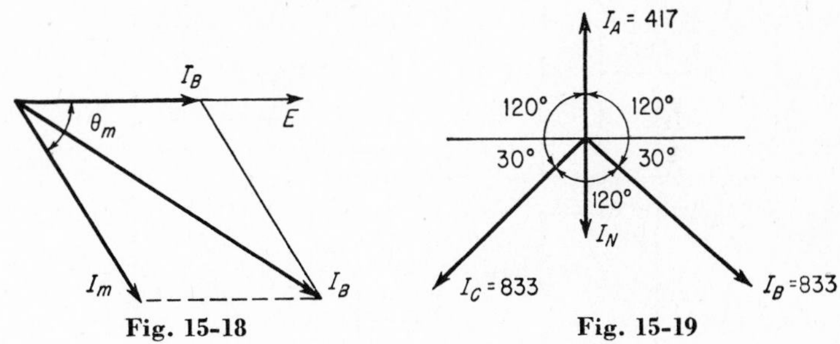

Fig. 15-18

Fig. 15-19

ANSWER.

(*a*) It is best to use a step-up or autotransformer.

(*b*) The transformer should be wound for 240 volts on the secondary and 220 volts on the primary. Since the impedance and power factor of the load are constant, the higher voltage will send more current through the load. The new load current will then be

$$100 \times (240/220) = 109 \text{ amp}$$

The transformer should have a minimum rating of

$$109 \times (240/1{,}000) = 26.2 \text{ kva}$$

Q15-16. A 208-volt three-phase four-wire line delivers power to a group of three-phase motors and a number of incandescent lamps. The motor load of 100 kva at 80 per cent lagging power factor is connected to the three main lines of the system marked *A*, *B*, and *C*. The lighting load is connected as follows: 12 kw from line *A* to the neutral, 18 kw from line *B* to the neutral, and 6 kw from line

C to the neutral. Draw the circuit and indicate the connection of three wattmeters that are arranged to measure the output. Calculate the power that would be indicated by each meter, and calculate the current in the neutral wire. Phase rotation is A-B-C.

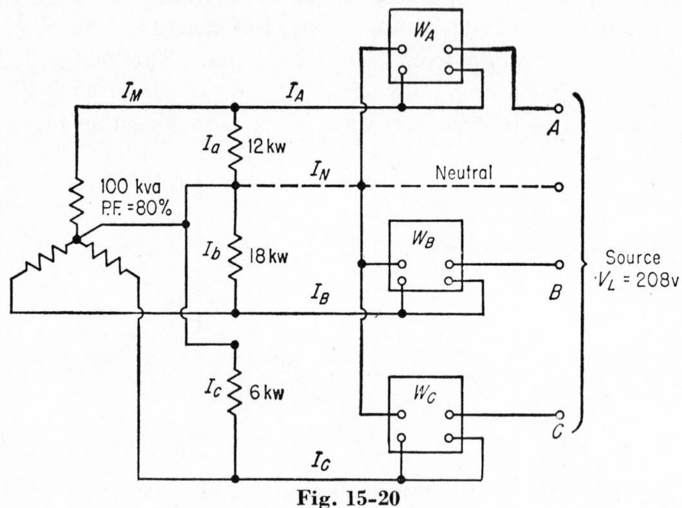

Fig. 15-20

ANSWER. Refer to Fig. 15-20 as drawn to represent circuit. The total power is equal to the sum of the individual meter readings or equal to the sum of the power in each phase.

The voltage across the neutral and each main is determined as follows:

$$208/1.732 = 120 \text{ volts}$$

$$\text{Power to motor load} = \text{kva} \times \text{power factor}$$
$$= 100 \times 0.80 = 80 \text{ kw}$$

$$\text{Motor power per phase} = 80/3 = 26.7 \text{ kw}$$

Therefore, the meter readings are

$$W_A = 26.7 + 12 = \mathbf{38.7\ kw}$$
$$W_B = 26.7 + 18 = \mathbf{44.7\ kw}$$
$$W_C = 26.7 + 6 = \mathbf{32.7\ kw}$$

Current in the neutral: For lighting loads $I_a = 12{,}000/120 = 100$ amp

$$I_b = 18{,}000/120 = 150 \text{ amp}$$
$$I_c = 6{,}000/120 = 50 \text{ amp}$$

Refer to the vector diagram for current relationships (see Fig. 15-21). The current through the neutral conductor is equal to the square root of the sum of the squares of the vertical components of I_N and the horizontal components of I_H.

I_N vertical $100 - 150 \times \frac{1}{2} - 50 \times \frac{1}{2} = 0$
I_N horizontal $0 + 150 \times 0.866 - 50 \times 0.866 = 86.6$ amp

$$I_N = \sqrt{(I_N \text{ vertical})^2 + (I_N \text{ horizontal})^2}$$
$$= \sqrt{0^2 + 86.6^2} = \mathbf{86.6\ amp}$$

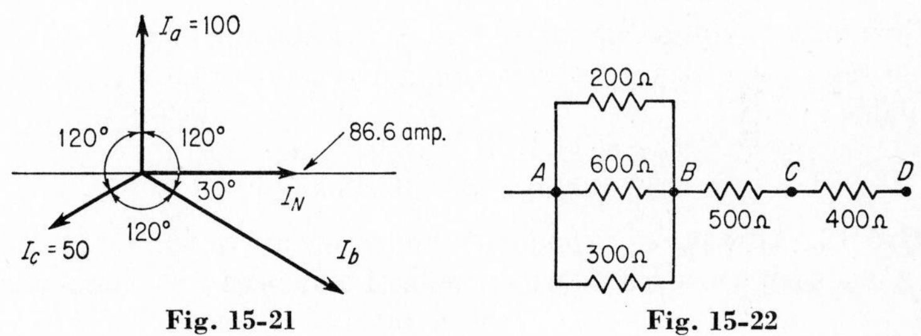

Fig. 15-21

Fig. 15-22

Q15-17. Referring to the sketch in Fig. 15-22, there is a difference in potential of 110 volts between *A* and *D*. Determine

(*a*) The drop in potential between *B* and *C*

(*b*) The current in the resistance *CD*

(*c*) The current flowing through the 200-ohm resistance between *A* and *B*

ANSWER. We must first determine the equivalent parallel resistance between *A* and *B*. We will then find the current flowing through the series of resistances. Knowing this current and the resistance between *B* and *C*, we can find the drop between *B* and *C* by the product of current and resistance (slide-rule computation).

Now determine the equivalent parallel resistance between *A* and *B*.

$$\frac{1}{R_e} = \frac{1}{200} + \frac{1}{600} + \frac{1}{300} = 0.005 + 0.0016 + 0.0032 = 0.0098$$

$$R_e = \frac{1}{0.0098} = 102 \text{ ohms, the equivalent parallel resistance}$$

$$\text{Current flowing: } I = \frac{E}{\Sigma R} = \frac{110}{102 + 500 + 400} = 0.1098 \text{ amp}$$

(*a*)

$$\text{Drop across } BC = E_2 = I \times 500 = 0.1098 \times 500 = \mathbf{54.8\ volts}$$

(*b*) Current through *CD* is the same as through other parts of circuit, or **0.1098 amp**

(*c*) First determine the drop across *AB*. Now since the total drop is equal to the sum of the drops across each section

$$E = E_1 + E_2 + E_3 = E_1 + 54.8 + 0.1098 \times 400 = 110$$
$$E_1 = 110 - 54.8 - 43.8 = 11.4 \text{ volts}$$

Current through the 200-ohm resistor is $11.4/200 =$ **0.057 amp**

Although not required in the solution, the current in each of the other resistors in this parallel grouping may be determined likewise as follows:

$$I_{600} = 11.4/600 = 0.019 \text{ amp}$$
$$I_{300} = 11.4/300 = 0.038 \text{ amp}$$

Q15-18. An outgoing conductor and a return conductor of No. 10 AWG each has a resistance of one-half ohm, and each conductor has a maximum power dissipation rating of 450 watts. The combined outgoing and return conductor dissipation is 900 watts when supplying current to the combination of R_1, R_2, and R_3. Determine the resistance of R_3 in ohms and determine the power dissipated for R_3 in watts (refer to Fig. 15-23). E_1 equals 120 volts rms, R_1 equals 15 ohms, and R_2 equals 175 ohms.

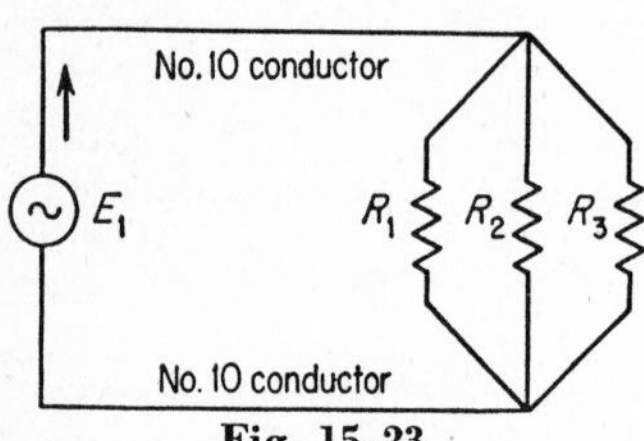

Fig. 15-23

ANSWER. The power dissipated through the circuit is I^2R total. Thus

$$P = 900 = I^2(0.5 + 0.5)$$
$$I^2 = 900/1 = 900 \qquad I = \sqrt{900} = 30 \text{ amp through circuit}$$

The total resistance of the circuit is equal to the total voltage impressed, divided by the current flowing through the entire circuit, or $R_t = E_1/I = 120/30 = 4$ ohms. The resistance of the parallel combination is $4 - (0.5 + 0.5) = 3$ ohms.

Then
$$\frac{1}{R_p} = \frac{1}{15} + \frac{1}{175} + \frac{1}{R_3} = \frac{1}{3}$$

Solving for R_3, we find this to be equal to **3.84 ohms,** and the power dissipated is

$$\text{Power} = 30^2 \times 3.84 = \mathbf{3{,}450\ watts}$$

Q15-19. It is known that the resistances of R_1, R_2, and R_3 in Fig. 15-24 are respectively 5, 10, and 3 ohms. The currents I_1, I_2, and I_3 are respectively 25, 10, and 15 amp. Determine the necessary supply voltage and the value of the resistance R_4 for the network given. Both must meet the above requirements.

ANSWER. By application of Kirchhoff's laws we can see that the IR drops must first be calculated.

$$E_1 = I_1R_1 = 25 \times 5 = 125 \text{ volts}$$
$$E_2 = I_2R_2 = 10 \times 10 = 100 \text{ volts}$$
$$E_3 = I_3R_3 = 15 \times 3 = 45 \text{ volts}$$

Now let us consider the right-hand loop involving the resistance R_4. The sum of the IR drops is equal to the drop across R_2.

$$45 + I_3R_4 = 45 + 15R_4 = 100$$
$$R_4 = (100 - 45)/15 = 55/15 = \mathbf{3.67\ ohms}$$

Supply voltage is

$$E = \text{drop across } R_1 + \text{drop across } R_2 = 125 + 100 = \mathbf{225\ volts}$$

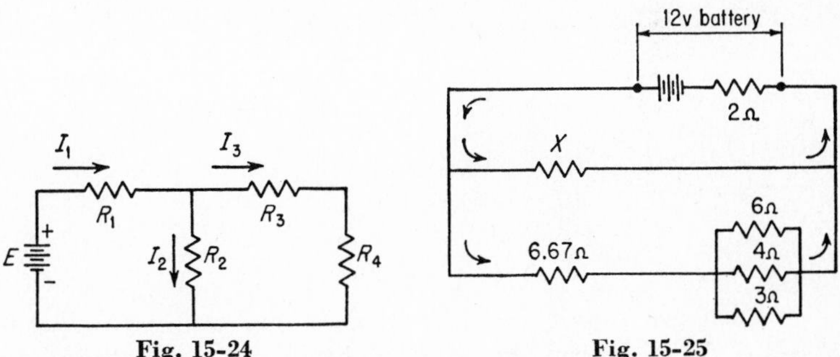

Fig. 15-24

Fig. 15-25

Q15-20. The battery in Fig. 15-25 supplies a current of 2 amp to the network shown. Determine the resistance X in ohms if the other resistances are as indicated.

ANSWER. Note carefully that the 2-ohm resistor is included within the battery terminals. Thus, the 12 volts is the net available voltage. Now we must first determine the equivalent parallel combination resistances in the lower loop.

$$R_t = 1/6 + 1/4 + 1/3 = 0.167 + 0.25 + 0.33 = 0.747$$
$$= 1/0.747 = 1.34 \text{ ohms}$$

In order to find the value of X, knowing the voltage impressed across that part of the circuit, we must first determine I_1, for $I_1 = 2 - I_2$. But we do not know the value of I_2 as yet. Then

$$I_2 = 12/(6.67 + 1.34) = 1.5 \text{ amp}$$

from which $I_1 = 2 - 1.5 = 0.5$ amp. Now determine the value of X,

$$X = 12/0.5 = \mathbf{24\ ohms}$$

Q15-21. Given the electrical circuit shown in Fig. 15-26 with associated constants: $E_1 = 240$ volts, $R_1 = 40$ ohms, $R_3 = 42.9$ ohms, $R_4 = 28$ ohms. R_2 is an incandescent lamp with a rating of 200 watts and 120 volts. Neglect the change in resistance caused by temperature effects. Determine the power in watts used by the lamp in the above circuit.

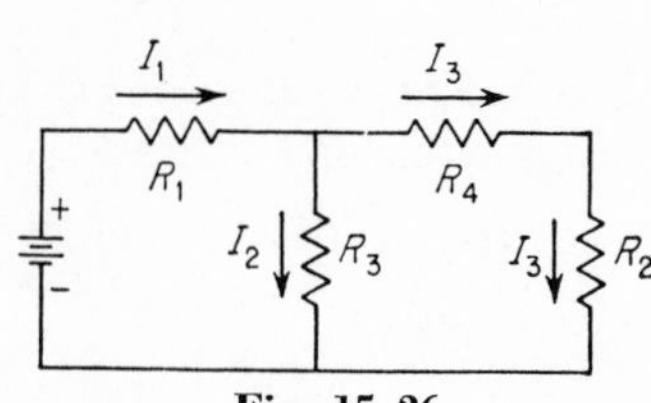

Fig. 15-26

ANSWER. Now by application of Kirchhoff's laws we can set up the following:

Loop 1: $$240 = I_1 \times 40 + I_2 \times 42.9 \qquad (1)$$
Loop 2: $$240 = I_1 \times 40 + I_3 \times 28 + I_3 \times 72 \qquad (2)$$

Subtract Eq. (2) from Eq. (1).

$$\begin{aligned} 240 &= 40I_1 + 42.9I_2 && (1)\\ -240 &= 40I_1 + 100I_3 && (2)\\ \hline 0 &= 0 + 42.9I_2 - 100I_3 \end{aligned}$$

from which we obtain $42.9I_2 = 100I_3$. But we can see that $I_3 = I_1 - I_2$. Substituting for I_3 in the above, we rearrange the equation as follows:

$$42.9I_2 = 100(I_1 - I_2) = 100I_1 - 100I_2$$
$$I_1 = \frac{(42.9 + 100) \times I_2}{100} = \frac{142.9I_2}{100} = 1.43I_2$$

If we substitute in Eq. (1) the value $1.43I_2$ for I_1, it follows

$$240 = 1.43I_2 \times 40 + 42.9I_2$$

Solving for I_2 $\qquad I_2 = 240/106.4 = 2.26$ amp

Using loop 1 again we can set up the following equation with $I_2 = 2.26$.

$$240 = 40I_1 + 2.26 \times 42.9 = 40I_1 + 97$$

$$I_1 = (240 - 97)/40 = 143/40 = 3.58 \text{ amp}$$

Thus $\qquad I_3 = I_1 - I_2 = 3.58 - 2.26 = 1.32$ amp

Power consumed by lamp $= I_3{}^2R$ of lamp

$= 1.32^2 \times 72 =$ **125 watts**

Note that

$$\text{Resistance of lamp} = \frac{E^2}{\text{rating}} = \frac{120^2}{200} = 72 \text{ ohms}$$

Q15-22. An a-c 60-cycle single-phase 240-volt rms motor delivers a mechanical full load of 25 hp with an efficiency of 85 per cent for 0.80 power factor for a speed of 1,130 rpm. Determine the rms starting current in amperes if it is known to be six times the full-load current.

ANSWER. The full-load current is found by the use of the following relationship for a single-phase system.

$$I = \frac{\text{hp} \times 746}{E \times \text{eff} \times \text{pf}} = \frac{25 \times 746}{240 \times 0.85 \times 0.8} = 114.2 \text{ amp}$$

The starting current is, therefore, $6 \times 114.2 =$ **685.2 amp**

Q15-23. Assume that incandescent lamps will cease the conduction of current if subjected to a voltage in excess of rated voltage and also assume that the resistance of the lamp is not changed unless a voltage is applied in excess of rated voltage. Lamp A is rated as 300 watts, 110 volts, and lamp B as 100 watts, 110 volts. It is desired to connect lamp A and lamp B in series across a 220-volt supply. Determine the combined wattage output (see Fig. 15-27).

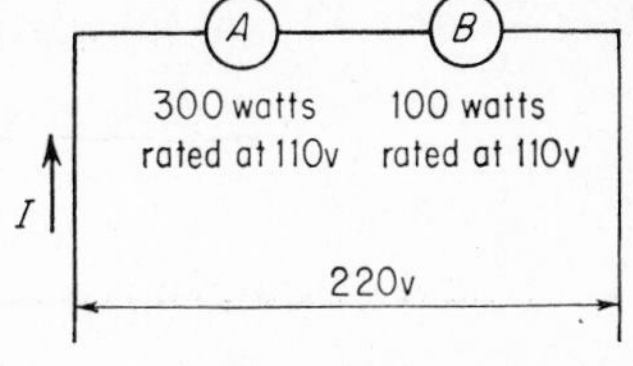

Fig. 15-27

ANSWER. The resistance of the lamps must first be determined so as to know the current flow under the conditions of the problem.

This is reasonable because it is assumed from the wording of the problem that the resistance of the lamps does not change unless overheated. The resistance of lamp A is

$$R_A = \frac{E^2}{W} = \frac{110^2}{300} = 40.4 \text{ ohms}$$

$$R_B = 110^2/100 = 121 \text{ ohms}$$

Current flowing is $I = 220/(40.4 + 121) = 1.36$ amp. Lamp A will deliver

$$I^2R_A = 1.36^2 \times 40.4 = 1.85 \times 40.4 = 74.5 \text{ watts}$$

Lamp B will deliver

$$I^2R_B = 1.36^2 \times 121 = 224 \text{ watts}$$

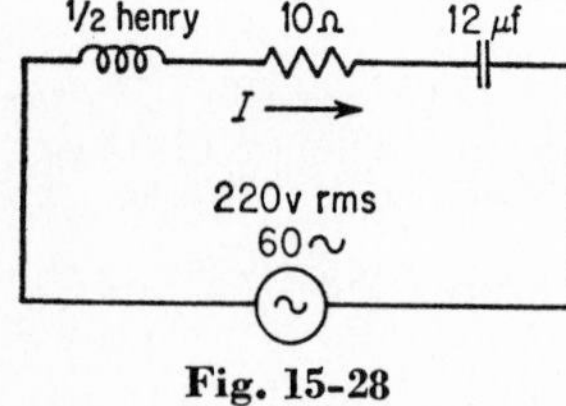

Fig. 15-28

Lamp B will burn out, and the **combined wattage will be zero.**

Q15-24. A series circuit consists of an inductor of one-half henry, a resistor of 10 ohms, a capacitor of 12 μf, and an a-c generator of 220 volts (rms) with a frequency of 60 cps. Determine:

(*a*) The average value of energy dissipated per second in the form of heat from the circuit. Give your answer in watts.

(*b*) The rms voltage across the capacitor.

ANSWER.

(*a*) Refer to Fig. 15-28. Knowing the voltage impressed together with the circuit impedance we can find the current flowing in this series circuit. Now since $Z = \sqrt{R^2 + (X_L - X_C)^2}$, and not knowing X_L and X_C, we calculate for the latter two first.

$$X_L = 2\pi fL = 6.28 \times 60 \times 0.5 = 189 \text{ ohms}$$

$$X_C = \frac{1}{2\pi fC} = \frac{1}{6.28 \times 60 \times 0.000012} = 232 \text{ ohms}$$

$$Z = \sqrt{10^2 + (189 - 232)^2} = \sqrt{10^2 + 43^2} = 44.1 \text{ ohms}$$

$$I = \frac{E}{Z} = \frac{220}{44.1} = 5 \text{ amp}$$

$$\text{Power loss} = I^2R = 5^2 \times 10 = 250 \text{ watts}$$

There is no heat dissipation in either the inductor or the capacitor.

(*b*) The voltage across the capacitor E_C is

$$IX_C = 5 \times 232 = \mathbf{1{,}160\ volts}$$

The power factor is given by

$$\cos\theta = \frac{R}{\sqrt{R^2 + (X_L - X_C)^2}}$$

If a vector diagram is drawn, it may be seen that the voltage across the inductance and that across the capacitor are *in opposition*, so that the resulting voltage of these two is their arithmetical difference. In this case IX_L is greater than IX_C. Thus, IX_C is subtracted directly from IX_L. The line voltage E is the vector sum of the three voltages and the hypotenuse of a right triangle of which IR and $(IX_L - IX_C)$ are the other sides. Therefore

$$E = \sqrt{(IR)^2 + (IX_L - IX_C)^2}$$
$$= I\sqrt{R^2 + (X_L - X_C)^2}$$

Solving for I

$$I = \frac{E}{\sqrt{R^2 + (X_L - X_C)^2}} = \frac{E}{Z}$$

It must be observed that the voltage across the capacitance is considerably greater than the line voltage impressed across the entire series circuit. This would be impossible in a d-c circuit, for under those conditions the voltage across any one part of the circuit cannot exceed the impressed total voltage. This condition can exist in an a-c circuit, because the capacitance voltage and the inductance voltage are in direct opposition. Both may be large, provided their difference is less than the line voltage.

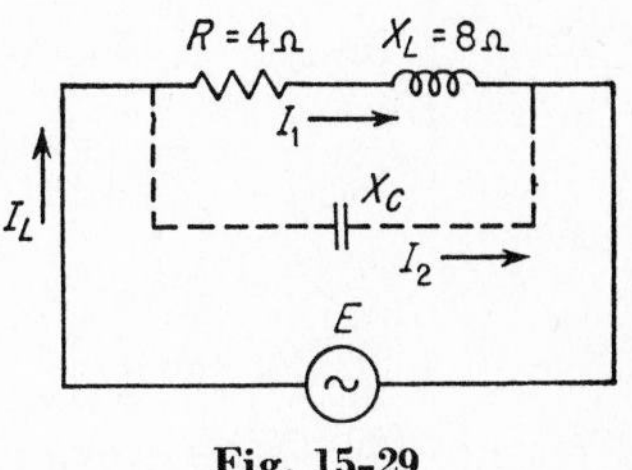

Fig. 15-29

Q15-25. A generator of 60 cps and 220 volts rms delivers power to a load consisting of a resistor of 4 ohms and an inductor of 8 ohms inductive reactance connected in series. Determine the capacitive reactance in ohms to be connected in parallel across the load to give a total power factor of 0.90 with the total current lagging the load voltage.

ANSWER. Refer to Fig. 15-29. We must first determine the original load current *before* the application of the condenser. Then for the sake of comparison we will determine the capacitive reactance for correction to 100 per cent power factor. Now

$$I_{L_1} = \frac{E}{Z_1}$$

but first $\quad Z_1 = \sqrt{R^2 + X_L^2} = \sqrt{4^2 + 8^2} = 8.85 \text{ ohms}$

Then $\quad I_{L_1} = \dfrac{220}{8.85} = 24.9 \text{ amp}$

The power factor before the addition of the condenser is $\cos \theta_1$.

$$\cos \theta_1 = \frac{R}{Z_1} = \frac{4}{8.85} = 0.45$$

from which $\quad \theta_1 = 63° \text{ and } \sin \theta_1 = 0.893$

Now refer to the vector diagram in Fig. 15-30. For correction to 100 per cent power factor we see that I_2, the current through the bypass capacitor, is equal to I_1, the current through the load, times $\sin \theta_1$, or

$$I_2 = I_1 \sin \theta_1 = 24.9 \times 0.893 = 22.2 \text{ amp}$$

Therefore, for 100 per cent correction

$$X_C = \frac{E}{I_2} = \frac{220}{22.2} = 9.9 \text{ ohms}$$

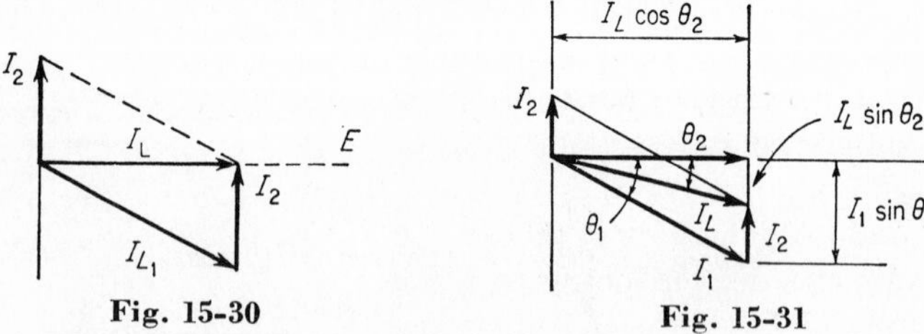

Fig. 15-30 Fig. 15-31

For correction to 90 per cent power factor: $\cos \theta_2 = 0.90$, $\sin \theta_2 = 0.437$. Refer to the vector diagram in Fig. 15-31.

$$I_2 = I_1 \sin \theta_1 - I_L \sin \theta_2$$
$$I_2 = 24.9 \times 0.893 - 5.2 = 17.1 \text{ amp}$$

Therefore, for correction to 90 per cent

$$X_C = \frac{E}{I_2} = \frac{220}{17.1} = \mathbf{12.8\ ohms}$$

In practice, it usually does not pay to increase the power factor over 90 or 95 per cent lagging. Not much more is gained by approaching unity power factor beyond these figures. The last few per cent improvement require a much greater proportionate increase in condenser capacity.

Q15-26. A simplified equivalent circuit for a single-phase a-c motor is a series circuit consisting of a resistance R_2 and an inductive reactance X_L. The power factor for the motor is 0.5. Determine the capacitive reactance in ohms to be connected across the motor, which results in a power-factor improvement to 0.85 (the voltage leading the current) across the terminals *ab*. *E* is equal to 117 volts rms, R_1 is equal to 3 ohms, and R_2 is 7 ohms.

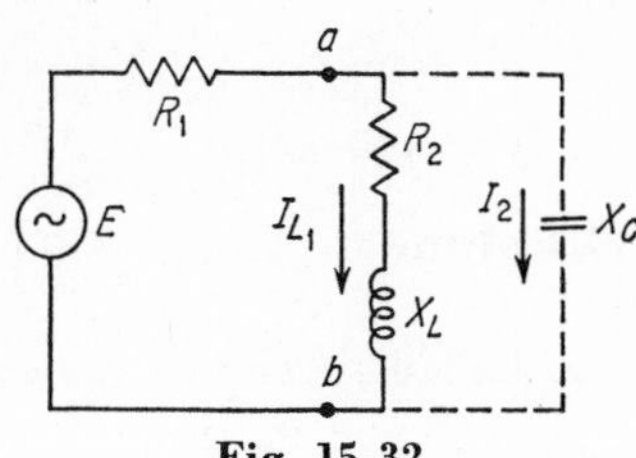

Fig. 15-32

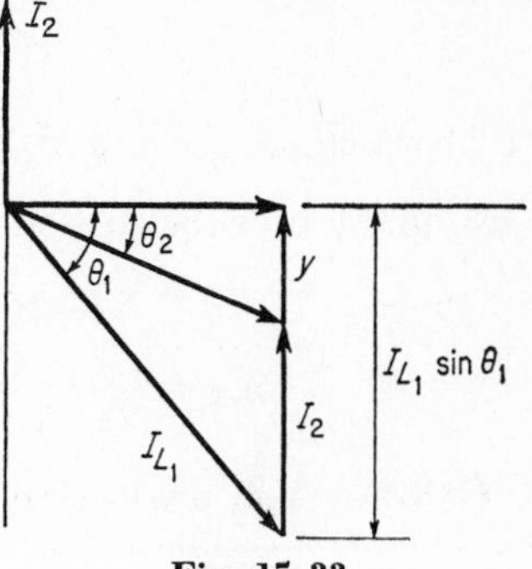

Fig. 15-33

ANSWER. Refer to the circuit in Fig. 15-32 and its vector diagram in Fig. 15-33. The latter accounts for the power-factor correction. We will first find the motor load current which will remain intact in value throughout the solution. Proceed as follows: Find the impedance *Z* knowing the total resistance and the power factor; thus

$$\cos \theta_1 = 0.5 = \frac{R}{Z}$$

By rearrangement

$$Z = \frac{R_1 + R_2}{0.5} = \frac{3 + 7}{0.5} = \frac{10}{0.5} = 20 \text{ ohms}$$

Load current $\qquad I_{L_1} = 117/20 = 5.85 \text{ amp}$

The voltage across the motor terminals *ab* is by Kirchhoff's law

$$117 - I_{L_1}R_1 = 117 - 5.85 \times 3 = 117 - 17.5 = 99.45 \text{ volts}$$

Refer to Q15-25 for the procedure to find X_C for 0.85 per cent power factor. The current through the condenser is I_2. From the vector diagram in Fig. 15-33

$$I_2 = I_{L_1} \sin \theta_1 - y$$

But we must find the value of y. This is done by first assuming a correction to 100 per cent power factor as in Q15-25.

$$I_{L_1} \cos \theta_1 = 5.85 \times 0.5 = 2.93 \text{ amp}$$

Now since $\cos \theta_2 = 0.85$, we can find θ_2 to be 31°30′. Back to the vector diagram

$$\tan \theta_2 = \frac{0.522}{0.85} = \frac{y}{2.93}$$

from which

$$y = 2.93 \times \frac{0.522}{0.85} = 1.8$$

and finally by substitution

$$I_2 = 5.85 \times 0.8660 - 1.8 = 3.27 \text{ amp}$$

$$X_C = \frac{E_L}{I_2} = \frac{99.45}{3.27} = \mathbf{30.4\ ohms}$$

Q15-27. You are assigned to make a load survey of a factory having an aggregate load of 10,000 kva in a-c machinery. The equipment involved ranges between 50 and 2,500 kva. What instruments and testing apparatus would you require and what data would you consider essential? How would you make use of graphic representation to convey the results of your study to the management? Make any reasonable assumption necessary to qualify your answer.

ANSWER. A load survey might start with a determination of the annual *load duration curve* from past records, if available, giving the number of hours during the year for any particular load. If necessary, this load curve may be drawn up on a monthly or daily basis. From this curve, the yearly, monthly, or daily *load factors* may be established. This is done by calculating the ratio of average to peak load.

The *demand factor* would be found by dividing the maximum demand on a machine over, say, 24 hr by the rated connected load producing that demand. The *diversity factor* could be determined by dividing the maximum simultaneous demand load of the factory by the sum of the individual maximum demand loads during a 24-hr cycle. The *plant capacity factor* may be found by dividing the load by the installed capacity over, say, a year. The *use factor* can be found by dividing the load by the installed capacity during the period of actual operation. High load factor and low diversity factor are desirable, while the difference between load and capacity factor indicates the reserve capacity carried above the maximum load demand. The use factor would be a much better indication than the capacity factor when the factory is not in continuous operation throughout the year. The effect of variable load throughout the day should be considered, since power rates are based on both maximum demand and off-hour peaks.

For a study of the aggregate load it is necessary to know the load characteristics of each piece of machinery. Accordingly, the power input, the power output, the efficiency, the power factor, and the maximum demand must be ascertained for each individual piece of machinery. Knowing the current characteristics of the local public utility, the input may be measured with a wattmeter, the maximum demand with a demand meter, and the output by means of a Prony brake (for rotative machinery), another wattmeter (for transformers), or the rise in temperature of fluids (for heaters). The efficiency will be the ratio of output/input for each piece of equipment. The power factor may be obtained by use of a power-factor meter or by connecting an ammeter and voltmeter to the input line. Assuming the current to be three-phase and letting W equal the wattmeter reading, E equal the line voltage, and I equal the line current, the power factor would be equal to

$$\cos \theta = \frac{W}{EI\sqrt{3}}$$

The rated capacity of each machine would be obtained from the name plate, while temperatures in the case of heaters may be measured with mercury thermometers or thermocouples.

A load analysis should include a study of the cycle of production or the sequence of factory operations, both continuous and intermittent in order to concentrate the maximum demand at times when the cost of electricity is the least. The method of transmitting power, whether central or local together with the life of each machine, the possible use of synchronous motors or static condensers to improve the power factor, the removal of high inductive loads or defective machinery, etc., are all factors which should be considered.

Q15-28. A 30,000-kva 13.8-kv three-phase generator has a subtransient reactance of 15 per cent. The generator supplies two motors over a transmission line having transformers at both ends, as shown on the one-line diagram of Fig. 15-34. The motors have

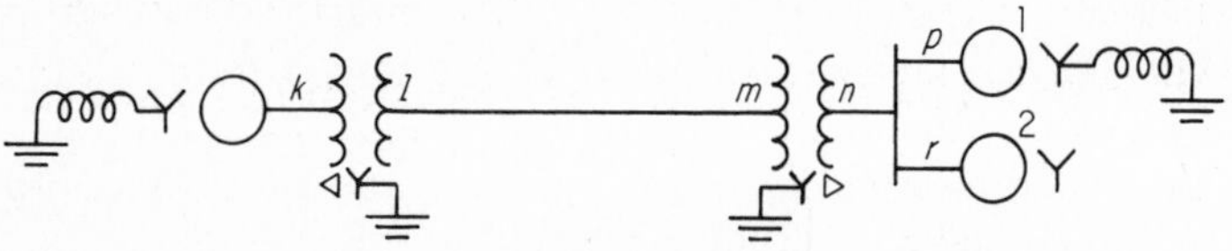

Fig. 15-34

rated inputs of 20,000 and 10,000 kva, both 12.5 kv with 20 per cent subtransient reactance. The three-phase transformers are both rated 35,000 kva, 13.2Δ-115Y kv with leakage reactance of 10 per cent. Series reactance of the transmission line is 80 ohms. Draw the reactance diagram with all reactances marked in per unit. Select the generator rating as base in the generator circuit.

ANSWER. A base of 30,000 kva, 13.8 kv in the generator circuit requires a 30,000-kva base in all other circuits and the following voltage bases:

In the transmission line: $13.8 \times 115/13.2 = 120$ kv
In the motor circuit: $120 \times 13.2/115 = 13.8$ kv

The reactances of the transformers must be converted from a base of 35,000 kva, 13.2 kv to a base of 30,000 kva, 13.8 kv, as follows:

$$\text{Transformer reactance} = 0.1 \times \frac{30{,}000}{35{,}000}(13.2/13.8)^2$$

$$= \mathbf{0.0784\ per\ unit}$$

The base impedance in the transmission line is

$$\frac{120^2 \times 1{,}000}{30{,}000} = 480 \text{ ohms}$$

and the reactance of the line is 80/480 = 0.167 per unit.

$$\text{Reactance of motor 1} = 0.2 \times \frac{30{,}000}{20{,}000}(12.5/13.8)^2 = \mathbf{0.246\ per\ unit}$$

$$\text{Reactance of motor 2} = 0.2 \times \frac{30{,}000}{10{,}000}(12.5/13.8)^2 = \mathbf{0.492\ per\ unit}$$

Fig. 15-35.

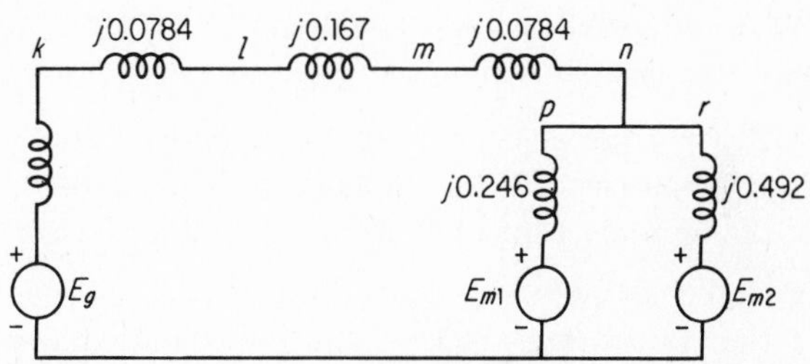

Fig. 15-35. Reactance diagram. Reactances are marked in per unit on the specified base.

Q15-29. Find the positive- and zero-sequence impedance per mile at 60 cycles for a single-circuit three-phase line consisting of No. 2/0 hard-drawn copper conductors with flat, horizontal spacing of 12 ft between centers. Assume $\rho = 100$.

ANSWER. For No. 2/0 seven-strand hard-drawn copper, $r' = 0.01252$ ft.

At 25°C, $R_a = 0.440$ ohm/mile

$$D_{eq} = \sqrt[3]{12 \times 12 \times 24} = 15.1 \text{ ft}$$
$$L = 0.7411 \log (15.1/0.0125) = 2.285 \text{ millihenrys/mile}$$
$$Z_1 = 0.440 + j2\pi 60 \times 2.285 \times 10^{-3} = 0.440 + j0.861 \textbf{ ohm/mile}$$

The self GMD of the equivalent composite conductor for zero sequence is

$$D_{aa} = \sqrt[9]{(0.01252)^3(12)^4(24)^2} = 1.42 \text{ ft}$$
$$Z_0 = 0.440 + 4.764 \times 60 \times 10^{-3} + j13.97 \times 60 \times 10^{-3} \log (2{,}790/1.42)$$
$$= 0.727 + j2.76 \textbf{ ohms/mile}$$

The ratio of zero-sequence inductive reactance to positive-sequence inductive reactance is 3.21.

Q15-30. A constant-voltage source (infinite bus) supplies a purely resistive 5,000-kw, 2.3-kv load and a 7,500-kva, 13.2-kv synchronous motor having a subtransient reactance of $X'' = 20$ per cent. The source is connected to the primary of the three-winding transformer described below:

Primary: Y-connected, 66 kv, 10,000 kva
Secondary: Y-connected, 13.2 kv, 7,500 kva
Tertiary: Δ-connected, 2.3 kv, 5,000 kva

Neglecting resistance, the leakage impedances are:

$$Z_{ps} = 7 \text{ per cent on 10,000-kva, 66-kv base}$$
$$Z_{pt} = 9 \text{ per cent on 10,000-kva, 66-kv base}$$
$$Z_{st} = 6 \text{ per cent on 7,500-kva, 13.2-kv base}$$

The motor and resistive load are connected to the secondary and tertiary of the transformer. Draw the impedance diagram of the

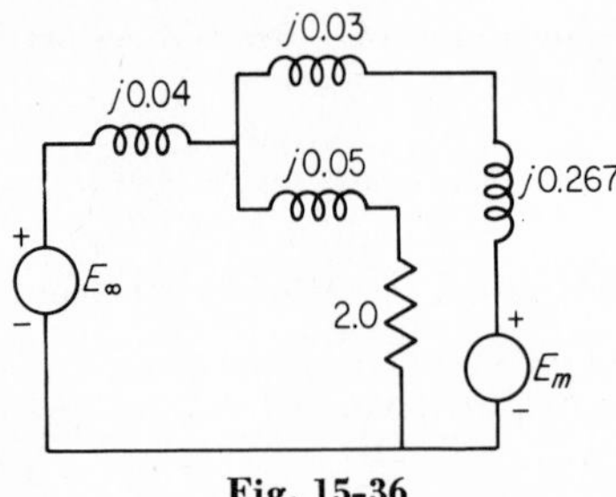

Fig. 15-36

system and mark the per-unit impedances for the base of 66 kv, 10,000 kva in the primary.

ANSWER. The constant-voltage source can be represented by a generator having no internal impedance. The resistance of the load is 1.0 per unit on a base of 5,000 kva, 2.3 kv in the tertiary. Exposed on a 10,000-kva, 2.3-kv base the load resistance is

$$R_L = 1.0 \times (10{,}000/5{,}000) = 2.0 \text{ per unit}$$

Changing the reactance of the motor to a base of 10,000 kva, 13.2 kv yields

$$X'' = 0.20 \times (10{,}000/7{,}500) = j0.267 \text{ per unit}$$

Fig. 15-36 is the required diagram.

varying strength whch will deflect the ray to give the form of the wave on the screen.

Q15-31. Determine the relative costs of copper in distributing direct current with a 120-volt two-wire system and with a 240/120-volt three-wire system, assuming the same amount of power to be transmitted over the same distance and with the same loss in both systems.

ANSWER. Let P_1 and P_2 be power loss over 120- and 240-volt systems, respectively

I_1 and I_2 be current flowing over 120- and 240-volt systems, respectively

E_1 and E_2 be voltage drop in 120- and 240-volt systems, respectively

R_1 and R_2 be resistance of 120- and 240-volt systems, respectively

W_1 and W_2 be weight of 120- and 240-volt systems, respectively

C_1 and C_2 be cost of 120- and 240-volt wire systems, respectively

C_3 be cost of 240-volt three-wire system

Then, since we assume the same power loss over both systems

$$P_1 = P_2$$

or

$$I_1^2R_1 = I_2^2R_2$$

Therefore

$$\frac{R_1}{R_2} = \frac{I_2^2}{I_1^2} = \frac{(P_2/E_2)^2}{(P_1/E_1)^2} = \frac{E_1^2}{E_2^2}$$

Now the resistance of a wire is given by $R = K(l/A)$, where K is the resistivity of the material, l is length, and A is the cross section of the wire. It is seen that for a given wire material, the distance l remaining constant, R varies inversely with A. Or $R_1/R_2 = A_2/A_1$. However, the weight W of a conductor varies directly with its cross section A and its cost C directly with its weight. Therefore, we may write

$$\frac{R_1}{R_2} = \frac{A_2}{A_1} = \frac{W_2}{W_1} = \frac{C_2}{C_1} = \frac{E_1^2}{E_2^2} = \frac{120^2}{240^2} = \frac{1}{4}$$

This shows that the cost of the 120-volt system C_1 would be four times that of the 240-volt system C_2, both these systems having two wires. Now, since the 240-volt system has three wires (conductors)

$$C_3 = C_2 + \tfrac{1}{2}C_2 = \tfrac{3}{2} \times \tfrac{1}{4}C_1 = \tfrac{3}{8}C_1$$

Therefore, the **240-volt three-wire system costs $\frac{3}{8}$ times as much as the 120-volt system.**

Q15-32. A pair of single-phase buses are copper tubes, each of 5-in. OD. The buses are parallel and placed 3 ft apart. They are supported on insulators located every 4 ft along each bus. What force will each support be called upon to withstand in the event of a short circuit involving a peak current of 160,000 amp at 60 cycles? If the current is in opposite directions in the two buses, what will be the direction of the forces on the support? See Fig. 15-37.

ANSWER. In very large generating stations detailed attention must be given the method and substance of bus-bar supports. This is necessary because of the tremendous values of short-circuit currents attendant to present-day generator station practice. In the literature may be found curves and formulas that are helpful in calculating the stresses set up under short-circuit conditions. In addition the designer must carefully consider the transmittal of these bus-bar stresses to transformers and other apparatus in the circuit involved under such conditions.

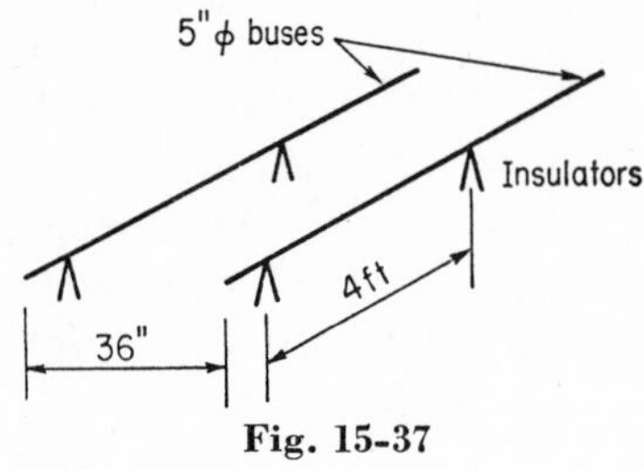

Fig. 15-37

The direction of the lateral short-circuit forces acting on two bus conductors will be one of repulsion when the currents in the two conductors are in opposite direction. When the currents are in the same direction in the two conductors, the force will be one of attraction. The proximity of two channels in a bus conductor arranged in box form requires the addition of spaced clamps at short intervals along the conductor to maintain a separation during short circuits.

Calculation of this lateral short-circuit force between conductors utilizes the basic short-circuit formula.

$$F = \frac{5.4 i_1 i_2 \times 10^{-7} \times k}{D} = \text{lb per ft of length}$$

where currents i_1 and i_2 = maximum direct currents or either (1) instantaneous peak values of alternating current or (2) maximum rms asymmetrical alternating current

D = separation between conductors, in.

k = shape factor of bus conductor

For determination of values of k refer to the "Kaiser Aluminum Bus Conductors Technical Manual," published by Kaiser Aluminum & Chemical Sales, Inc., Chicago, 1957. For the problem at hand k is equal to unity.

The NEMA standards for Power Switching Equipment, SG6-1954, has established conservative standards for the calculation of electromagnetic forces between two current-carrying conductors. The basis of this calculation is the fundamental formula for short-circuit forces previously indicated, using instantaneous peak-current values to establish the maximum force. This standard recognizes the fact that the forces thus calculated are, in most cases, higher than those that occur in practice. The NEMA standard tends to compensate for the possibility of increased force due to resonant vibration, since this factor is not taken into account. The corrected formula to be used in the solution to our problem is

$$F = M \times \frac{5.4 \times 1^2 \times 10^{-7}}{D} = \text{lb per ft of conductor}$$

The value of i is taken as maximum asymmetrical peak current and is used in conjunction with a multiplier M. When maximum peak current is substituted, $M = 1$ for a d-c or a single-phase fault on either a single-phase or a three-phase circuit. It is often desirable to use rms current values, either symmetrical or asymmetrical. Where these current values are available, they can be used in the formula with a corresponding value of M from Table 15-2. The condition for a symmetrical fault current is seldom attained in practice. An asymmetrical fault is the general rule, since the probability of the fault being initiated at the instant of peak voltage is very small.

$$F = 1 \times (5.4 \times 160{,}000^2 \times 10^{-7} \times 4)/36$$
$$= \textbf{1,535 lb} \text{ tending to separate buses}$$

TABLE 15-2. MULTIPLYING FACTORS[1]

Circuit	Current used	Relation to peak current	Multiplying factor M
Direct current	Maximum peak	1.00	1.00
1-phase alternating current or 1-phase of 3-phase	Maximum peak	1.00	1.00
1-phase alternating current or 1-phase of 3-phase	Rms asymmetrical	1.63	$1.63^2 = 2.66$
1-phase alternating current or 1-phase of 3-phase	Rms symmetrical	2.82	$2.82^2 = 8.0$
3-phase alternating current[2]	Maximum peak	1.00	$0.866 \times 1^2 = 0.866$
3-phase alternating current[2]	Rms asymmetrical	1.63	$0.866 \times 1.63^2 = 2.3$
3-phase alternating current[2]	Rms symmetrical	2.82	$0.866 \times 2.82^2 = 6.9$

[1] NEMA: "Standard for Power Switching Equipment," SG6-1954.

[2] For the center bus of a flat symmetrically spaced bus.

Q15-33. A bus-bar system for three-phase currents consists of three 1-in. diameter copper rods spaced 12 in. apart, center to center, in an equilateral arrangement. (*a*) What is the inductance per 100 ft? (*b*) What is the capacitance per 100 ft? (*c*) When one conductor carries 2,000 amp and the other two carry 1,000 amp, what are the forces between them per 100 ft? See Fig. 15-38.

ANSWER.

(*a*) From standard text the inductance is $L = 0.7411 \log_{10} D/D_s$ mh per mile,

where $D = 12$ in.

$D_s = \text{radius } \frac{6}{12} \times 0.7788$

Thus, $L = 0.7411 \log_{10} (12/0.3894) = 1.102$ mh per mile

Inductance per 100 ft $= 1.102/52.8 =$ **0.0209 mh** per 100 ft

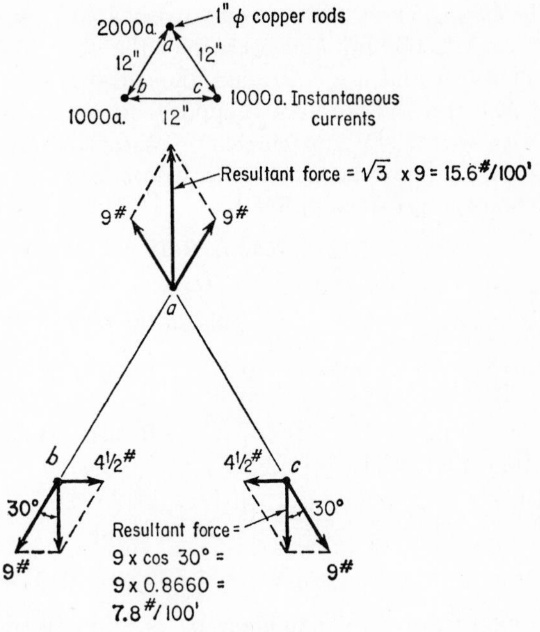

Fig. 15-38

(*b*) Capacitance per mile is given by $C = 0.03883/\log_{10} D/r$ μf per mile to neutral.

$$C = 0.03883/\log_{10} 12/0.5 = 0.03883/1.38 = 0.0281 \ \mu\text{f}$$

$$\text{Capacitance per 100 ft} = 0.0281/52.8 = \mathbf{0.000532\ \mu f} \text{ per 100 ft}$$

(*c*) Actually, the intent of this problem is not too clear. There are several possibilities.

1. The rms current in one phase is 2,000 amp and 1,000 amp in each of the others, and the problem, then, is to find the maximum force.

2. The currents given may be maximum values; hence, maximum force.

3. The currents given may be instantaneous values; hence we find the instantaneous force.

Since in a three-phase system $I_a + I_b + I_c = 0$, possibility 1 seems very improbable, since the 1,000 amp phases would have to be in phase and 180° out of phase with the 2,000 amp current.

The only arrangement which seems reasonable is that the I_{max} in each phase is 2,000 amp and because of the phase displacement when one phase current is a maximum, the current in the other two phases is 1,000 amp and both are in opposite directions (or 180° out of phase with the 2,000 amp current). With this interpretation the forces we find are instantaneous values and not maximum values. From the previous problem

$$F = M \times \frac{5.4 I_1 I_2 \times 10^{-7}}{D}$$

Again M is equal to unity. For repulsion the force is

$$F = 1 \times \frac{5.4 \times 2{,}000 \times 1{,}000 \times 10^{-7}}{12} \times 100$$

$$= \mathbf{9\ lb} \text{ per 100 ft } \mathbf{repulsion}$$

For attraction the force is

$$F = 1 \times \frac{5.4 \times 1{,}000 \times 1{,}000 \times 10^{-7}}{12} \times 100$$

$$= \mathbf{4.5\ lb} \text{ per 100 ft } \mathbf{attraction}$$

For the resultant forces refer to diagrams of forces herewith.

Q15-34. An ammeter A, a voltmeter V, and wattmeter W are connected as shown. The wattmeter is uncompensated, and the impedances of its coils may be considered as purely resistive with values of 0.4 and 8,000 ohms for the current and voltage coils, respectively. The impedances of the ammeter and voltmeter may also be considered as resistive, and their values are 0.1 and 10,000 ohms, respectively. Calculate the true power delivered to the load when the ammeter, voltmeter, and wattmeter indicate 2.0 amp, 220 volts, and 150 watts, respectively.

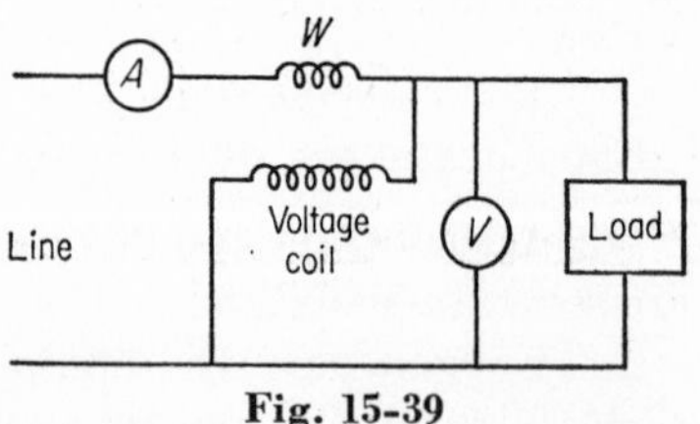

Fig. 15-39

ANSWER. Refer to Fig. 15-39. Note that the wattmeter potential circuit is connected directly across the load, but the wattmeter current coil carries the potential-coil current in addition to the load current. In fact, the wattmeter potential circuit may be considered as a small load connected in parallel with the actual load, whose power is to be measured. The power consumed by this potential circuit must be deducted, therefore, from the wattmeter reading. The true power taken by the load is, including the voltmeter,

$$P = P' - \frac{E^2}{R_p} - \frac{E^2}{R_v}$$

where P' = wattmeter reading
E = load voltage
R_p = resistance of wattmeter potential-coil circuit
R_v = voltmeter coil resistance

$$P = 150 - 220^2/8{,}000 - 220^2/10{,}000 = \mathbf{139.1\ watts}$$

It will be observed that a considerable percentage error would result in this case if the losses were neglected. There are wattmeters on the market which compensate for the loss in the wattmeter itself. A small auxiliary coil, connected in series with the moving-coil system, is interwound with the fixed coils so that a small countertorque is exerted, this countertorque being proportional to the power consumed by the potential circuit.

Q15-35. A 208-volt three-phase line supplies a system of impedances connected in Y. $Z_{ob} = 10 + j0$, $Z_{oa} = 10 + j10$, and $Z_{oc} = 10 + j10$. Calculate the line currents (see Fig. 15-40 for Y arrangement).

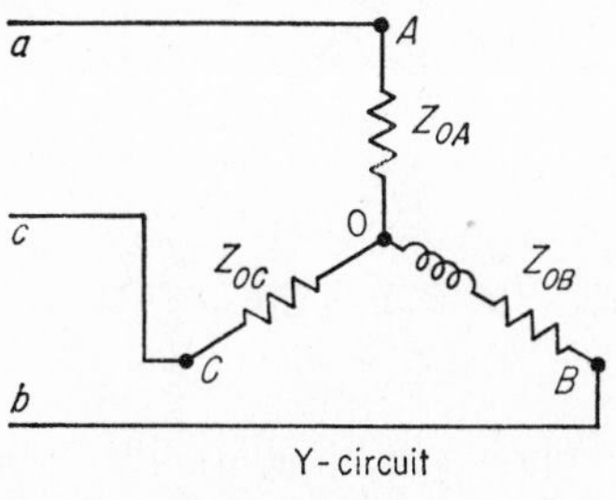

Y-circuit

Fig. 15-40

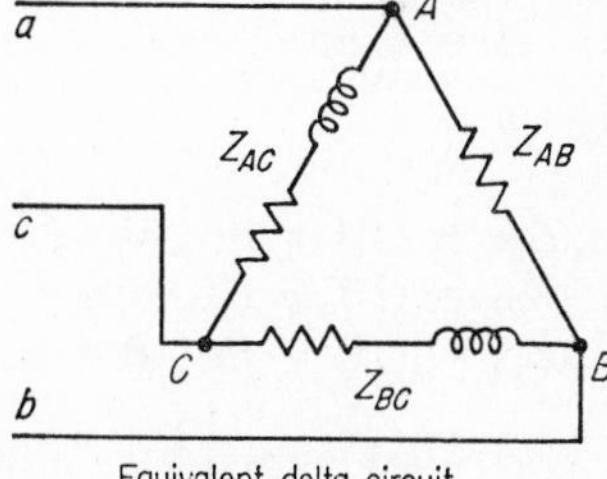

Equivalent delta circuit

Fig. 15-41

ANSWER. Often Δ-connected circuits are solved in terms of the equivalent Y circuit and vice versa. In this equivalent Δ connection (Fig. 15-41) the line currents and line-to-line voltages are the same as in the Y circuit. The Y-connected set of load impedances is replaced by a Δ-connected set, which, if viewed from its three terminals at a given frequency, is indistinguishable from the Y-connected set. All connections in one three-phase system need not be either Y or Δ. In practice both are used, the choice in any given situation being based on practical exigencies. In a given practical situation, one may be more desirable than the other.

The following equations apply when converting Y to Δ

$$Z_{ab} = \frac{Z_{oa}Z_{ob} + Z_{ob}Z_{oc} + Z_{oc}Z_{oa}}{Z_{oc}}$$

$$Z_{bc} = \frac{Z_{oa}Z_{ob} + Z_{ob}Z_{oc} + Z_{oc}Z_{oa}}{Z_{oa}}$$

$$Z_{ca} = \frac{Z_{oa}Z_{ob} + Z_{ob}Z_{oc} + Z_{oc}Z_{oa}}{Z_{ob}}$$

These may be rearranged and cleared to give their simplified versions.

$$Z_{ab} = Z_{oa}Z_{ob}f$$
$$Z_{bc} = Z_{ob}Z_{oc}f$$
$$Z_{ca} = Z_{oc}Z_{oa}f$$

The factor f is equal to $f = \frac{1}{Z_{oa}} + \frac{1}{Z_{ob}} + \frac{1}{Z_{oc}}$

Now substitute the values for impedances given in the problem in the equation for factor f.

$$f = \frac{1}{10 + j0} + \frac{1}{10 + j10} + \frac{1}{10 + j10} = 0.25 - j0.05$$

Now

$$Z_{ab} = (10 + j0)(10 + j10)(0.25 - j0.05) = 30 + j20$$
$$Z_{bc} = (10 + j10)(10 + j0)(0.25 - j0.05) = 30 + j20$$
$$Z_{ca} = (10 + j0)(10 + j0)(0.25 - j0.05) = 25 - j5$$

The corrected equivalent circuit is shown in Fig. 15-42. Since voltages are separated by 120° in a three-phase system, we can set one of the voltages across one pair of leads as reference and have it lead by 120°. Then by complex notation

$$E_{ab} = 208(1 - j0) = 208 - j0$$
$$E_{bc} = 208(-0.5 - j0.866) = -104 - j180$$
$$E_{ca} = 208(-0.5 + j0.866) = -104 + j180$$

Now we can determine the phase currents by Ohm's law.

$$I_{ab} = \frac{E_{ab}}{Z_{ab}} = \frac{208 - j0}{30 + j20} = \frac{6{,}240 - j4{,}160}{1{,}300} = 4.9 - j3.3$$

$$I_{bc} = \frac{E_{bc}}{Z_{bc}} = \frac{-104 - j180}{30 + j20} = \frac{-6{,}720 - j3{,}320}{1{,}300} = -5.17 - j2.55$$

$$I_{ca} = \frac{E_{ca}}{Z_{ca}} = \frac{-104 + j180}{25 - j5} = \frac{-3{,}500 + j3{,}980}{650} = -5.37 + j6.13$$

Finally, the line currents again by application of Ohm's law

$$I_a = I_{ab} - I_{ca} = (4.9 - j3.3) - (-5.37 + j6.13) = 10.27 - j9.43$$

$$I_b = I_{bc} - I_{ab} = (-5.17 - j2.55) - (4.9 - j3.3) = -10.7 + j0.75$$

$$I_c = I_{ca} - I_{bc} = (-5.37 + j6.13) - (-5.17 - j2.55) = -0.2 + j8.68$$

Eliminating j in each case

$$I_a = \sqrt{10.27^2 + 9.43^2} = \mathbf{13.9\ amp}$$

$$I_b = \sqrt{10.7^2 + 0.75^2} = \mathbf{10.7\ amp}$$

$$I_c = \sqrt{0.2^2 + 8.68^2} = \mathbf{8.68\ amp}$$

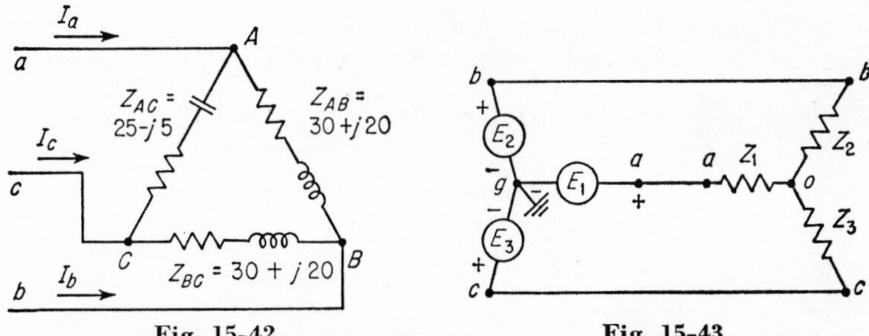

Fig. 15-42 **Fig. 15-43**

Q15-36. In the circuit shown in Fig. 15-43, the unbalanced three-phase generator supplies power to an unbalanced three-phase load. The circuit data are

$E_{ga} = E_1 = 120\underline{/0°}$ volts	$Z_1 = 10 + j0$ ohms
$E_{gb} = E_2 = 115\underline{/120°}$ volts	$Z_2 = 3 + j4$ ohms
$E_{gc} = E_3 = 125\underline{/-120°}$ volts	$Z_3 = 5 - j5$ ohms

Calculate the voltages V_{oa}, V_{ob}, V_{oc}.

ANSWER. Let us first determine the phase voltages.

$$E_{ga} = 120(1 + j0) = 120 - j0$$
$$E_{gb} = 115(-0.5 + j0.866) = -57.5 + j100$$
$$E_{gc} = 125(-0.5 - j0.866) = -62.5 - j108$$

Then the line voltages

$$E_{ab} = -E_{ga} + E_{gb} = -(120 - j0) + (-57.5 + j100) = -177.5 + j100$$
$$E_{bc} = -E_{gb} + E_{gc} = -(-57.5 + j100) + (-62.5 - j108) = -5 - j208$$
$$E_{ca} = -E_{gc} + E_{ga} = -(-62.5 - j108) + (120 - j0) = 182.5 - j108$$

The line currents may be calculated by application of well-known equations.

$$I_a = \frac{E_{ab}Z_3 - E_{ca}Z_2}{Z_1Z_2 + Z_2Z_3 + Z_3Z_1}$$
$$I_b = \frac{E_{bc}Z_1 - E_{ab}Z_3}{Z_1Z_2 + Z_2Z_3 + Z_3Z_1}$$
$$I_c = \frac{E_{ca}Z_2 - E_{bc}Z_1}{Z_1Z_2 + Z_2Z_3 + Z_3Z_1}$$

Substituting in the complex notation, we conclude the line currents to be

$$I_a = -4.5 + j2.75$$
$$I_b = 4.25 - j30.0$$
$$I_c = 0.25 + j27.26$$

The summation of $I_a + I_b + I_c$ checks out when equal to zero. Finally, the voltage drops are determined by Ohm's law application, $E = IR$.

$$V_{oa} = I_aZ_1 = (-4.5 + j2.75)(10 + j0) = \mathbf{52.8\ volts}$$
$$V_{ob} = I_bZ_2 = (+4.25 - j30.0)(3 + j4) = \mathbf{151.5\ volts}$$
$$V_{oc} = I_cZ_3 = (+0.25 + j27.26)(5 - j5) = \mathbf{193.1\ volts}$$

Q15-37. A single-phase relay is designed for service in a 110-volt 60-cycle circuit from which it draws an operating current of 0.1

amp at a power factor of 60 per cent. Calculate the resistance and power rating of the series resistor which will permit the use of this relay in 120-volt 25-cycle circuit. Neglect the iron losses.

ANSWER. This circuit may be considered as one of pure resistance and pure inductance in series. Under 60-cycle condition

$$Z = \frac{E}{I} = \frac{110}{0.1} = 1{,}100 \text{ ohms impedance}$$
$$R = Z \cos \theta = 1{,}100 \times 0.6 = 660 \text{ ohms}$$
$$X_L = Z \sin \theta = 1{,}100 \times 0.8 = 880 \text{ ohms}$$

Under 25-cycle condition

$$Z = 120/0.1 = 1{,}200 \text{ ohms impedance}$$
$$X_L = 880 \times {}^{25}\!/_{60} = 367 \text{ ohms}$$

Required resistance

$$\sqrt{Z^2 - X_L^2} = \sqrt{1{,}200^2 - 367^2} = 1{,}140 \text{ ohms}$$

Thus, add a resistor of 1,140 − 660 = **480 ohms,** with a rating equal to

$$0.1^2 \times 480 = \mathbf{4.8\ watts}$$

Q15-38. A d-c relay is to be energized. No direct current is available but a single-phase power source is available together with a full-wave contact rectifier. How will the operation of the relay on rectified alternating current compare with its operation on raw alternating current? Give reasons for your answer. Neglect the losses in the rectifier and in the relay for the purposes of this question.

ANSWER. Connecting a d-c relay to an a-c source will materially reduce the current flowing through the relay coils. In a hinged-arm relay there will probably be an annoying chatter due to the alternating current. The reduction in current already mentioned can be traced to the inductive reactance of the coil when the relay is operated on alternating current. When the relay is hooked up to a full-wave rectifier, its operation will be essentially normal. Once again, the resulting current flowing through the relay coils will be slightly less than that which will flow if the relay is connected to a battery circuit. In most practical cases, the rectified source of alternating current will be quite suitable for this relay.

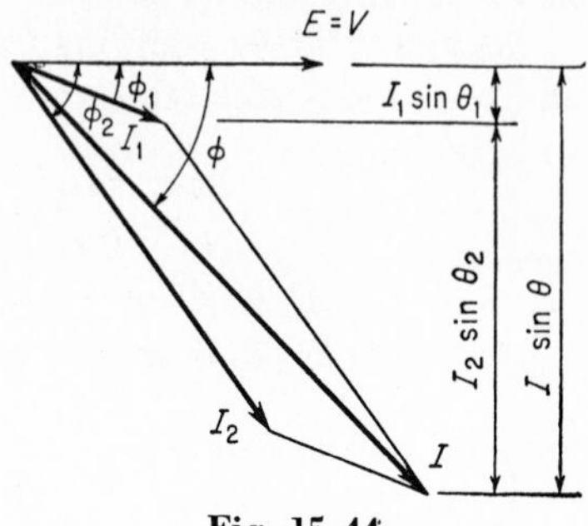

Fig. 15-44

Q15-39. Two alternators, operating in parallel, supply 2,500 kw to a load at a lagging power factor of 80 per cent. If one machine delivers 1,000 kw to a load at a lagging power factor of 95 per cent, (*a*) what is the power supplied by the second machine and (*b*) at what power factor does it operate?

ANSWER. Refer to Fig. 15-44. For part (*a*) the power supplied by the second machine is

$$P_2 = P_t - P_1 = 2{,}500 - 1{,}000 = \mathbf{1{,}500\ kw}$$

(*b*) Determination of the power factor

$$I_t = \frac{P_t}{E} \times \cos \theta_t \qquad \text{also } I_1 = \frac{P_1}{E} \times \cos \theta_1$$

$$I_2 = \frac{P_2}{E} \times \cos \theta_2$$

$$I_2 \sin \theta_2 = I_t \sin \theta_t - I_1 \sin \theta_1$$

Then by substitution in the above

$$\left[\frac{P_2}{E} \times \cos \theta_2\right] \times \sin \theta_2$$
$$= \left[\left(\frac{P_t}{E} \times \cos \theta_t\right) \times \sin \theta_t\right] - \left[\left(\frac{P_1}{E} \times \cos \theta_1\right) \times \sin \theta_1\right]$$
$$1{,}500 \tan \theta_2 = 2{,}500 \tan \theta_t - 1{,}000 \tan \theta_1$$
$$3 \tan \theta_2 = (5 \times 0.75) - (2 \times 0.329)$$
$$\tan \theta_2 = 3.092/3 = 1.031$$

from which we can find the value of θ_2 to be 45°51′, and

$$\cos \theta_2 = 0.696$$

Thus, the power factor of the second machine will be **69.6 per cent** lagging.

Q15-40. What is the purpose of a reverse-current cutout in an automobile electric system? Give a wiring diagram and explain how the device works.

ANSWER. Refer to Fig. 15-45. To prevent the battery from discharging through the generator when the battery voltage exceeds that of the generator, it is necessary to provide a reverse-current cutout, designed to open the circuit when the speed of the generator falls below that at which the battery charges. The relay consists of an iron core with a shunt and series winding. The shunt winding is connected permanently across the main generator terminals, one of which is usually grounded. It serves to magnetize the core and

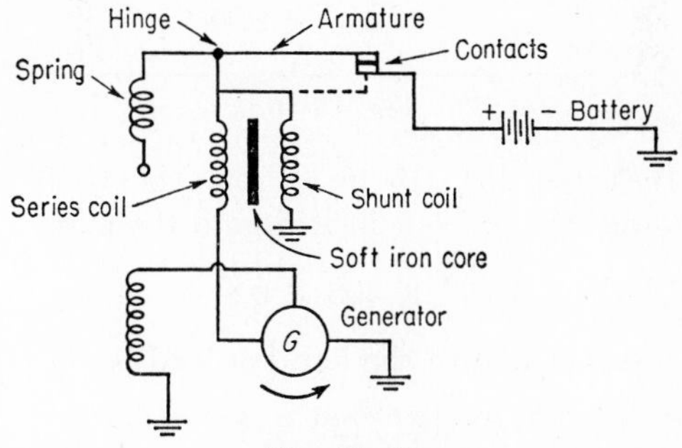

Fig. 15-45

a hinged armature as soon as the voltage is high enough and the magnetizing force great enough to overcome the tension of a spring which tends to hold the armature away from the core.

When the armature is attracted, it closes a pair of contact points, thereby closing the battery circuit and also the series winding circuit of the cutout. The additional flux generated by this winding presses the contacts more firmly together. Both the springs and stops that limit the travel of the armature of the relay are usually so set that the contacts close at 7 to 7.5 volts, equivalent to a charging rate of about 3 amp, with an ordinary type automobile relay for a 6- to 8-volt system. Contacts are open again at a discharge current not greater than 2.5 amp, when the series coil bucks the shunt coil and decreases the pull on the armature enough for the spring to pull it away from the core. In general, this actuation of the relay takes place at an engine speed of about 600 rpm, equivalent to a car speed, in high gear, of about 10 mph.

Q15-41. A three-phase four-wire 208/120-volt system supplies 100 kw at a lagging power factor of 80 per cent to a balanced motor load. In addition, 20 kw of power are supplied to each of two incandescent lamp loads, one connected between line A and the neutral, and the other between line B and the neutral. Calculate the current in each of the conductors A, B, C, and N of this system.

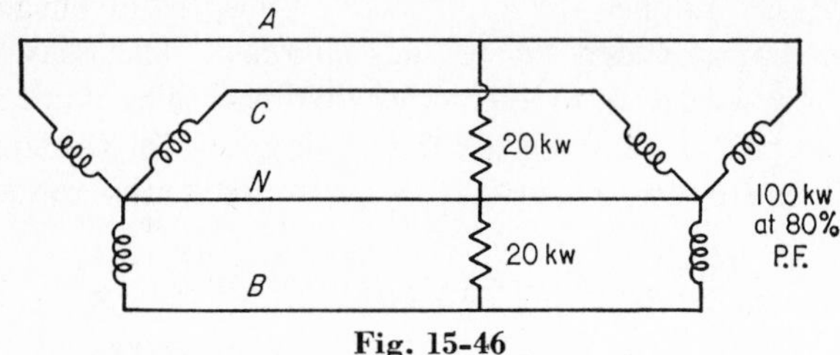

Fig. 15-46

ANSWER. Refer to Fig. 15-46. From the basic equation for three-phase power, the current in A due to the motor load is

$$100{,}000/(1.732 \times 208 \times 0.8) = 347 \text{ amp}$$

The current in line A due to the lighting load is

$$I'_a = 20{,}000/120 = 167 \text{ amp}$$

The total current in line A is found by the complex notation

$$347 + 0.8 \times 167 + j0.6 \times 167$$
$$347 + 134 + j100$$
$$481 + j100$$

from which $I = \sqrt{481^2 + 100^2} = \mathbf{491.3\ amp}$

This is the same as in line B, or 491.3 amp. The current in line C is only the motor load, or 347 amp. The current in N is the sum vectorially of $167\underline{/0} + 167\underline{/120}$.

$$2 \times {}^{167}\!/_{2} \times 1.732 = \mathbf{289\ amp}$$

Q15-42. For the circuit shown in Fig. 15-47, $E_1 = 120\underline{/0}$ volts. What will the ammeter read?

ANSWER. By the reciprocity theorem in any system composed of linear bilateral impedances, if an electromotive force E is applied between any two terminals and the current I is measured in any branch, their ratio (called the transfer impedance) will be equal to the ratio obtained if the positions of E and I are interchanged. The transfer impedance is

$$Z_t = \frac{120/\underline{0}}{0.028} = \frac{E}{I} = 4{,}280 \text{ ohms}$$

Then in the second case

$$I = \frac{120/\underline{30^\circ}}{4{,}280} = \frac{E}{Z_t} = \mathbf{0.028\ amp}$$

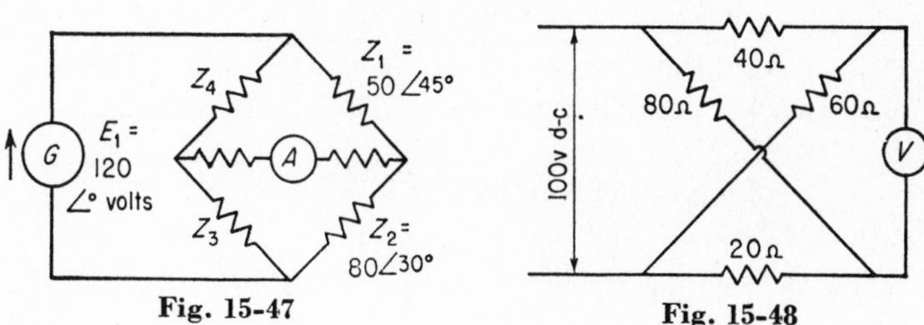

Fig. 15-47 **Fig. 15-48**

Q15-43. In the circuit shown in Fig. 15-48, what would the voltmeter read?

ANSWER. Rearrange to the new equivalent circuit shown in Fig. 15-49.

$$E_1 = {}^{40}\!/_{100} \times 100 = 40 \text{ volts}$$
$$E_2 = {}^{80}\!/_{100} \times 100 = 80 \text{ volts}$$
$$V = E_2 - E_1 = \mathbf{40\ volts}$$

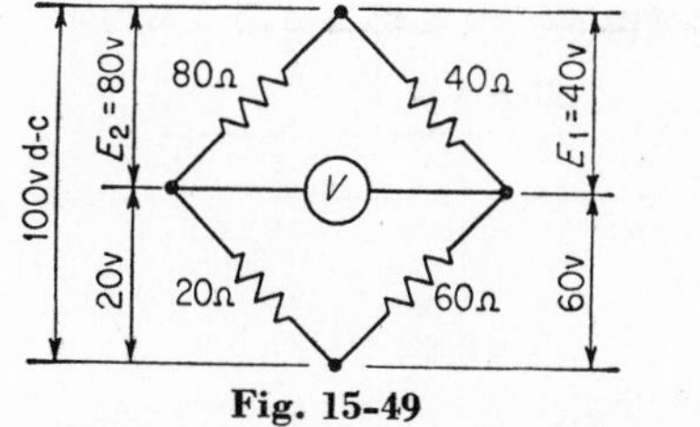

Fig. 15-49

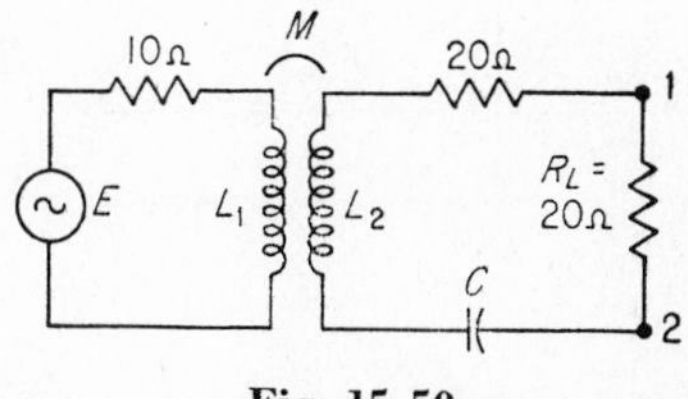

Fig. 15-50

Q15-44. In order to secure maximum power transfer to a load connected across terminals 1 and 2 of the device shown in Fig. 15-50, the load should have certain characteristics. What are these characteristics?

ANSWER. For maximum power transfer the impedances looking each way at the generator must be conjugates of each other, i.e., $R_g = R_L$ and $X_g = X_L$. Therefore, the impedances must be matched. In this case, the load must be **60 ohms pure resistance.**

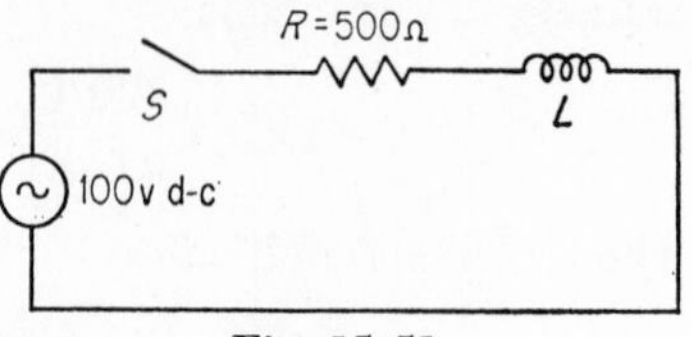

Fig. 15-51

Q15-45. In the circuit shown in Fig. 15-51, the time constant is 0.001 sec. What is the current in the circuit 0.001 sec after closing the switch?

ANSWER. The time constant $= L/R = 0.001$ sec. This is the length of time to reach 63.2 per cent of its final E/R value.

$$I = \frac{E}{R} \times (1 - \epsilon^{-(R/L \times t)})$$

$$= \frac{100}{500} \times (1 - \epsilon^{-1})$$

$$= \frac{100}{500} \times (1 - 0.367) = \mathbf{0.126\ amp}$$

Q15-46. A three-phase 2,300-volt cable supplies power to a bank of Δ-primary and Y-secondary transformers. The secondaries supply a 200-hp induction motor that operates at full load with 91 per cent efficiency and a power factor of 80 per cent. Also, connected to the 208/120-volt secondaries is a balanced lighting load of 90 kva at unity power factor. Neglecting the transformer losses, what current will flow in the cable? See Fig. 15-52.

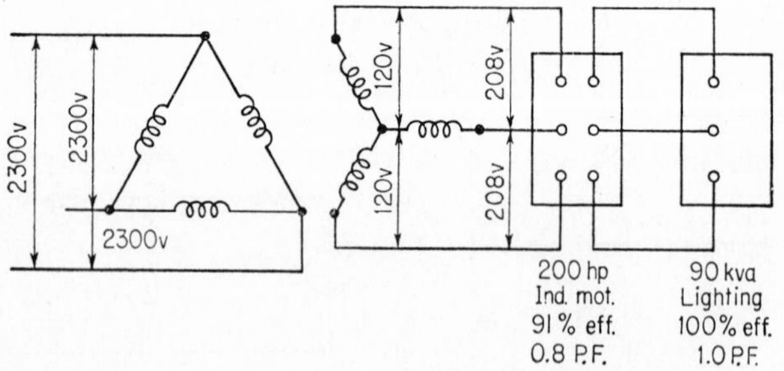

Fig. 15-52

ANSWER. $\cos^{-1} 0.8 = 36.9°$ and $\sin 36.9°$ is 0.6. Now considering the secondary currents

$$I_{\text{motor}} = \frac{\text{hp } 0.746}{\sqrt{3} \times E_L \times \text{eff} \times \text{pf}} = \frac{200 \times 0.746}{1.732 \times 208 \times 0.91 \times 0.80}$$
$$= 568 \text{ amp @ } \underline{/36.9°}$$
$$I_{\text{motor}} = 568(\cos 36.9° - \sin 36.9°) = 568(0.8 - j0.6)$$
$$= 455 - j341 \text{ amp}$$
$$I_{\text{lights}} = \frac{\text{kva} \times 1{,}000}{\sqrt{3}\, E \times \text{pf}} = \frac{90 \times 1{,}000}{\sqrt{3} \times 208 \times 1} = 250\underline{/0°}$$
$$I_{\text{lights}} = 250(\cos 0° + \sin 0°) = 250 \times 1 + 250 \times 0 = 250 + j0$$
$$I_{\text{total}} = I_m + I_{\text{lights}} = 455 - j341 + 250 + j0 = 705 - j341$$
$$= 782 \text{ amp}$$

This is at an angle $\underline{/-25.83}$.

$$\text{Transformer turns ratio} = 2{,}300/120 = 19.15$$
$$I_p = \text{primary phase current } \frac{I_t}{\text{trans. ratio}} = \frac{782}{19.15} = 40.8 \text{ amp}$$
$$I_L = \text{primary line current} = I_p\sqrt{3} = 40.8 \times 1.732 = \mathbf{70.8\ amp}$$

Q15-47. Determine the proper rating in amperes of the circuit protection devices on each of the following circuits as to (*a*) carrying capacity, (*b*) interrupting capacity.

1. A household 15-amp branch circuit.
2. Main circuit breaker or fuses for an isolated service station to be supplied alone from a 5-kva transformer having a 4 per cent impedance using a three-wire 115/230-volt system with solidly grounded neutral.
3. A circuit breaker is to be connected to a 100-kva 440-volt three-phase alternator, which supplies an isolated small system. The reactances of the alternator are as follows:

Positive sequence.....	125 per cent
Negative sequence....	12 per cent
Zero sequence........	6 per cent
Direct transient......	20 per cent
Direct subtransient...	12 per cent

ANSWER. 1. Use a 15-amp fuse or circuit breaker. The interrupting capacity of standard units (5,000 amp) is sufficient.

2. Rated current $= \text{kva} \times \dfrac{1{,}000}{E} = 5 \times \dfrac{1{,}000}{230} = 21.7$ amp

$$\text{Symmetrical current} = I_{sc} = \frac{100}{4}\left(\frac{\text{kva}}{\sqrt{3}\,\text{kv}}\right) = \frac{100}{4}\,21.7$$

$$= 542 \text{ amp}$$

$$\text{Asymmetrical current} = I_{asc} = 1.25 \times 542 = 677 \text{ amp}$$

Standard-current rated device would be 30 amp. The interrupting capacity of the standard breaker is 5,000 amp, which is well above the requirement.

3. The full-load current is $100{,}000/(\sqrt{3} \times 440) = 131$ amp. Refer to the AIEE Committee Report, *Elec. Eng.*, Nov. 1948. Low voltage breakers (600 volts and less) are often instantaneous in action and part contacts during the first half cycle. These circuit breakers, however, are rated on the basis of average current in the three phases, and circuits in which they are used rarely have X/R ratios exceeding 12. This corresponds to an average rms current of one-half cycle after fault which equals 1.25 times the symmetrical current. Such breakers may be applied on the basis of 1.25 times the three-phase initial symmetrical current, using subtransient reactance and including both synchronous and induction motors.

The symmetrical short-circuit current is found by

$$\frac{100 \times \text{kva base}}{\%X \times \sqrt{3} \times \text{kv}} = \frac{100}{\%X} \times I_{fl}$$

Using subtransient reactance

$$I_{sc} = \frac{100}{\%X} \times I_{fl} = {}^{100}\!/_{12} \times 131 = 1{,}090 \text{ amp}$$

The current contributed by synchronous and induction motors will not pass through the breaker; thus it equals zero. Using the 1.25 multiplier

$$I_{sc} \text{ asymmetrical} = 1.25 \times 1{,}090 = 1{,}360 \text{ amp}$$

The normal current rating may vary from 125 to 150 per cent of the generator rating depending on the type and character of the load. In this case the current might be 175 amp. The interrupting capacity of this size of breaker is 15,000 amp, which is well above the interrupting capacity required.

Q15-48. For the circuit shown determine the current in each resistance 0.02 sec after closing the switch. Initially the circuit was in a steady-state condition with the applied voltage indicated (see Fig. 15-53*a*).

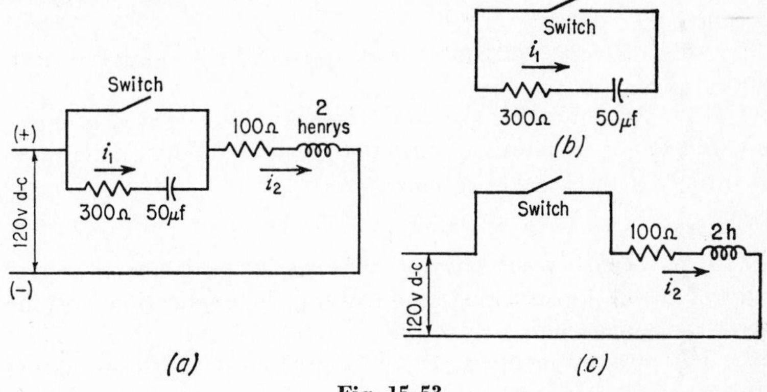

Fig. 15-53

ANSWER. At time t equal to zero (before switch is closed) $i_1 = i_2 = 0$. Also $e_c = 120$ volts.

The time constant $= t = RC = 300 \times 50 \times 10^{-6} = 0.015$ sec

For the circuit shown in Fig. 15-53*b*

$$\begin{aligned} i_1 &= \frac{E}{R} \times \epsilon^{-t/RC} \\ &= \frac{120}{300} \times \epsilon^{-0.02/(300 \times 50 \times 10^{-6})} \\ &= 0.4 \times \epsilon^{-1.333} \\ &= 0.4 \times 0.263 = \mathbf{0.1052\ amp} \end{aligned}$$

After the switch is closed 0.02 sec $L/R = 2/100 = 0.02$. Then refer to Fig. 15-53*c*.

$$\begin{aligned} i_2 &= \frac{E}{R} - \frac{E}{R} \times \epsilon^{-RT/L} \\ &= \frac{E}{R}(1 - \epsilon^{-RT/L}) \\ &= \frac{120}{100}(1 - \epsilon^{-(100 \times 0.02)/2}) \\ &= 1.2 \times (1 - \epsilon^{-1}) \\ &= 1.2(1 - 0.368) = \mathbf{0.76\ amp} \end{aligned}$$

Q15-49. (*a*) What is the NEC? (*b*) What is the legal status of NEC?

ANSWER.

(*a*) The NEC contains basic minimum provisions considered necessary for safety.

(*b*) NEC has no legal status, unless upheld by municipal ordinance.

Q15-50. An automatic circuit breaker is to be installed at a certain location in a power distribution system. What determines the ratings of the breaker specified?

ANSWER.

(*a*) The rated current-carrying capacity: the amount of copper in the breaker, the operating temperature, the application, and the size of the copper wire.

(*b*) The rated interrupting capacity: 600 volts and below rating is in amperes, above 600 volts rating is in kva.

(*c*) The circuit characteristics: short-circuit current, impedance of load, and generator capacity ahead of load.

Q15-51. Explain the meanings of or define the following terms as applied to wiring and wiring requirements. (*a*) Askarel, (*b*) cutout box, (*c*) demand factor, (*d*) dustproof, (*e*) dust-tight, (*f*) feeder, (*g*) outlet, (*h*) type SBW insulation, (*i*) type a-c cable, (*j*) hazardous location (class I).

ANSWER. Refer to latest edition of the NEC.

Q15-52. A 600-ohm resistor is to be matched to a 6,000-ohm generator at a frequency of 2,000 cps. A T arrangement of a tapped coil and a capacitor are to be used. Consider the elements to have negligible loss and determine the values of the inductance and capacitance to use (see Fig. 15-54).

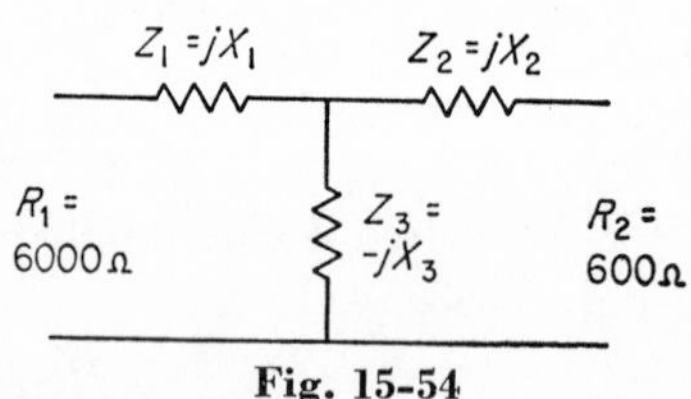

Fig. 15-54

ANSWER. This is a problem in impedance matching. The arrangement also represents a network consisting of a combination of generator and impedance elements, forming an active network.

Impedance matching is a condition to bring about maximum power transfer. There are a number of conditions necessary for maximum power transfer. At any frequency, a network containing generators and impedances can be replaced (by Thévenin's theorem) by a simple series circuit containing generators and impedances. If a generator and a load impedance are connected together and the generator impedance is fixed, the condition for maximum power transfer is that the load impedance be the conjugate of the internal impedance of the generator.

Refer to the figure. For simplification let $X_p = X_1 + X_3$ and let $X_s = X_2 + X_3$. Also assume $Z_3 = -jX_3 = -j3{,}000$ by complex notation.

$$X_s = \pm \sqrt{(R_2/R_1)(X_3^2 - R_1R_2)}$$
$$= \pm \sqrt{(1/10)[(9 \times 10^6) - (3.6 \times 10^6)]} = \pm 736$$
$$X_p = \pm \sqrt{(R_1/R_2)(X_3^2 - R_1R_2)} = \pm 7{,}360$$

by substitution of values.

Now
$$X_1 = X_p - X_3 = \pm 7{,}360 + 3{,}000 = -4{,}360, \text{ or } +10{,}360$$
$$X_2 = X_s - X_3 = -736 + 3{,}000 = 2{,}264, \text{ or } 3{,}736$$

The following relation from a standard text gives

$$\frac{R_1}{R_2} = \frac{X_1 + X_3}{X_2 + X_3}$$
$$10 = \frac{10{,}360 + (-3{,}000)}{X_2 + (-3{,}000)}$$

Solving for $X_2 = 3{,}736$. This checks. Now, let us assume that Z_3 is a pure X_c, and Z_1 and Z_2 are pure X_L's. Then

$$Z_1 = jX_1$$
$$Z_2 = jX_2$$
$$Z_3 = -jX_3$$

Then by substitution for X_1, X_2, and X_3

$$Z_1 = j10{,}360$$
$$Z_2 = j3{,}736$$
$$Z_3 = -j3{,}000$$

from which simply we obtain the following as results.

$$L_1 = \frac{X_2}{2\pi f} = \frac{10{,}360}{12{,}560} = \mathbf{0.825\ henry}$$

$$L_2 = \frac{3{,}736}{12{,}560} = \mathbf{0.297\ henry}$$

$$C = \frac{1}{2\pi f X_3} = \frac{1}{6.28 \times 2{,}000 \times 3{,}000} = \mathbf{0.0266\ \mu f\ equivalent}$$

Q15-53. The following news item, modified as to name and place, appeared in a large metropolitan newspaper recently.

Overloaded X-ray Line Sidetracks a Diagnosis

"A doctor declared today he had erroneously diagnosed stomach gas as ulcers because a neighbor plugged in an air conditioner connection into the 220-volt line serving his (the doctor's) X-ray machine. As a result, a warrant was issued for the arrest of Mr. A on the complaint of Dr. B. The doctor said that after making the faulty diagnosis because of a blurred X-ray print he discovered that Mr. A, who lives in the next apartment, had plugged in his air conditioner on the line. The deputy county attorney issued the warrant alleging grand larceny of electricity."

Assume you are a registered professional engineer and that you have been employed as an electrical expert by the lawyer who is to defend Mr. A:

(*a*) Outline the nature of the investigation you would conduct by indicating specifically what facts you would attempt to establish regarding the case..

(*b*) What bearing, if any, does the NEC have upon the case? Consider both direct and indirect possibilities.

(*c*) Is it physically possible that the operation of the air conditioner could have affected the X-ray prints? Explain.

ANSWER.

(*a*) Check the system electrically to find exactly how the air conditioner was connected. If the apartments were metered separately, Mr. A would be stealing electricity from Dr. B.

(*b*) There is nothing in the NEC about X-ray equipment being fed separately or in any special way. The feeder adequacy for the air conditioner should be investigated. See Art. 430, Motors and Controllers, and Sec. 4315 on Combination Load.

(*c*) X-ray print results are a function of milliamperes and time (tube current time in seconds). The control of an X-ray machine includes control for setting both. It is conceivable that if the air conditioner should start up during the interval between the setting of the machine and the taking of the picture, a poor picture would result. This can be likened to an underexposed photograph. Doctor B should realize that the X-ray picture is not properly exposed.

Q15-54. A pulsating direct current has a constant magnitude of 9 amp for 0.02 sec and is zero for 0.01 sec. The cycle is then repeated. What is its effective value?

ANSWER. A-c meters read effective current, rms. D-c meters read average current, d'Arsonval (refer to Fig. 15-55).

$$i(\text{rms}) = \sqrt{1/T \int_0^T i^2\,dt} = \sqrt{1/0.03 \int_0^T 9^2 \times 0.02} = \mathbf{7.36\ amp}$$

This would be read on the a-c meter.

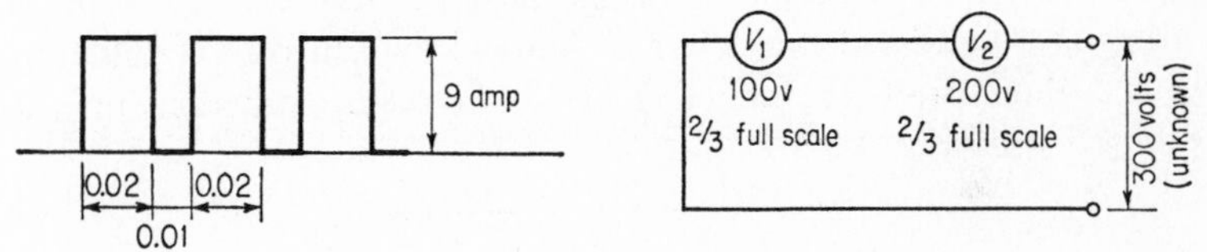

Fig. 15-55 **Fig. 15-56**

Q15-55. Two 1,000 ohms per volt voltmeters are connected in series across an unknown voltage source. The first voltmeter is connected to its 150-volt scale and reads 100 volts. The second voltmeter is connected to the 300-volt scale. What is the magnitude of the unknown voltage (see Fig. 15-56)?

ANSWER. Since both voltmeters have the same characteristics and since one reads two-thirds full scale, the other will also read two-thirds full scale, or 200 volts. Hence the total or unknown voltage is equal to 200 + 100 = **300 volts.**

Q15-56. In a bridge circuit with 10 volts applied, the sending instrument indicates 8 ma current. If the connections of the sensing instrument and the applied voltage are interchanged, what value of applied voltage will cause the sensing instrument to read 2 ma? See Fig. 15-57.

ANSWER. Using the reciprocity theorem: "In any system composed of linear bilateral impedances, if an electromotive force E is applied between any two terminals and the current is measured in

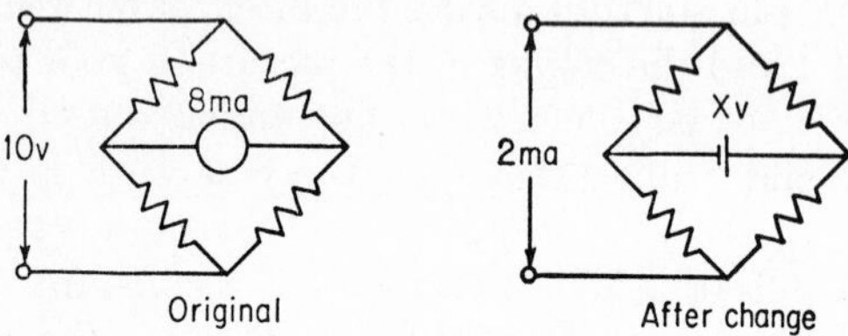

Fig. 15-57

any branch, their ratio (called the transfer impedance) will be equal to the ratio obtained if the positions of E and I are interchanged."

$$E = IZ \qquad \text{from which} \qquad Z = \frac{E}{I} = \frac{10}{8} \div 10^{-3} = 1{,}250 \text{ ohms}$$

For 2 ma, the applied voltage must be $\frac{2}{8} \times 10 =$ **2.5 volts**

Q15-57. An automobile storage battery has an open-circuit voltage of 6 volts and can deliver a short-circuit current of 400 amp. What is the current source or Norton's theorem equivalent for the battery? See Fig. 15-58.

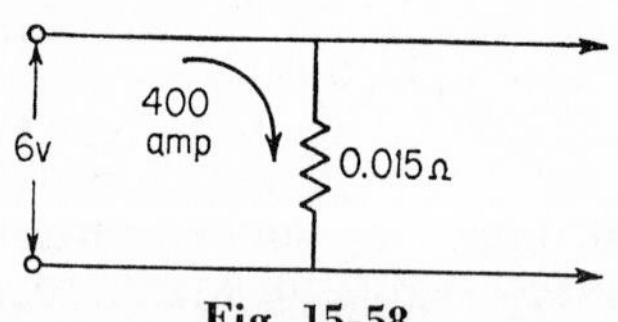

Fig. 15-58

ANSWER. Norton's theorem says: "The current in any impedance Z_r, connected to two terminals of a network, is the same as if Z_r were connected to a constant current generator whose generated current is equal to the current which flows through the two terminals when these terminals are short circuited, the constant current being in shunt with an impedance equal to the impedance of the network looking back from the terminals in question."

Thus $$R = \frac{E}{I} = \frac{6}{400} = \mathbf{0.015\ ohm}$$

Q15-58. What is meant by the term "complex frequency" as applied to an electric circuit?

ANSWER. This is a function of frequency versus time.

Q15-59. (*a*) Draw a riser diagram and show the number and size of the conductors and the size of the conduit for the main service, the feeders, and the subfeeders of a four-story office building

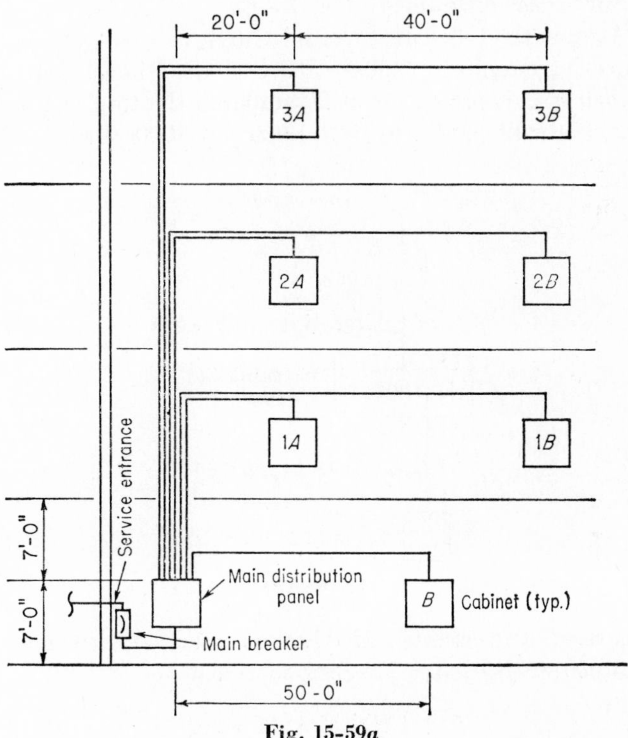

Fig. 15-59*a*

50 ft by 100 ft. The vertical rise from floor to floor is 14 ft. The three-phase four-wire 120/208-volt underground service enters the building 4 ft above the basement floor. The main breaker and the main distribution panel are to be located adjacent to the service entrance in the basement. From the main distribution panel to the floor branch circuit panel the horizontal run to the first cabinet is 20 ft and from the first to the second cabinet it is 40 ft. The

first and second cabinets on each floor and the basement branch circuit are each assumed to be of the same size, each supplying 15 lighting circuits, six receptacle circuits, and three spares. (*b*) What size main breaker should be used? (*c*) Determine the sizes of the feeder breakers in the distribution panel.

ANSWER.

(*a*) For a riser diagram see Fig. 15-59*a*.

(*b*) Assume the following factors to apply:

1. Lighting circuits are 1,600 watts each at ultimate load. Then the branch circuit protection at 20 amp and the total lighting load per branch circuit panel are 15 × 1,600 = 24,000 watts.

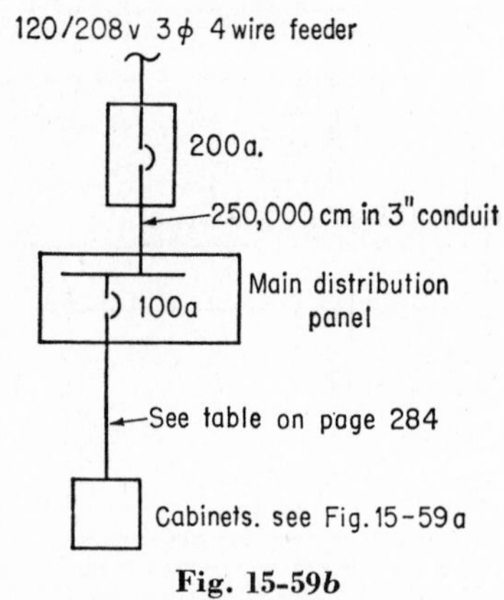

Fig. 15-59*b*

2. Six receptacle circuits with 20-amp protection with an ultimate load of 1,600 watts each. Then the total load is

$$6 \times 1{,}600 = 9{,}600 \text{ watts}$$

3. Three square circuits with 20-amp protection with an ultimate load per circuit are 16,000 watts. Then the total load is

$$3 \times 1{,}600 = 4{,}800 \text{ watts}$$

The total load on each branch circuit panel would be 38,400 watts, or 38.4 kw.

(*c*) See Fig. 15-59*b*.

Panel	Branch feeder, wire size	Conduit size	Length, ft
B	4-1/*c* No. 4	1½	50
1*A*	4-1/*c* No. 6	1¼	34
1*B*	4-1/*c* No. 2	2	74
2*A*	4-1/*c* No. 4	1½	48
2*B*	4-1/*c* No. 1	2	88
3*A*	4-1/*c* No. 3	1½	62
3*B*	4-1/*c* No. 1	2	102

The load per panel = 38.4 kw. NEC, article 220, par. 2203, allows demand factor of 70 per cent for loads in office buildings over 30 kw. Therefore, the panel loads are 38.4 × 0.70 = 26.88, say 27 kw. Allowing 1 per cent voltage drop in the branch feeders, the cable and conduit sizes are as shown in the table above.

$$\text{Wire area in cir mils} = \frac{2kLI}{\text{voltage drop}}$$

where k = resistivity of conductor material, ohms per cir mil–foot for copper, 11 approximately
L = length of circuit *one way*, ft
I = current flow, amp

Using the load demand calculated above as 27 kw, the amperes per panel on 208 volts (assuming load balanced over phases) will be

$$I = \frac{P}{E\sqrt{3}} = \frac{27{,}000}{208 \times 1.73} = 75 \text{ amp}$$

Then the area of wire will be as follows:

For basement B: $(22 \times 50 \times 75)/(0.01 \times 208)$
$= 793 \times 50 = 39{,}650$ cm ≡ No. 4 AWG
1A: $793 \times 34 = 26{,}762$ ≡ No. 6 AWG
1B: $793 \times 74 = 58{,}682$ ≡ No. 2 AWG
2A: $793 \times 48 = 38{,}064$ ≡ No. 4 AWG
2B: $793 \times 88 = 69{,}784$ ≡ No. 1 AWG
3A: $793 \times 62 = 49{,}166$ ≡ No. 3 AWG
3B: $793 \times 102 = 80{,}886$ ≡ No. 1 AWG

Q15-60. A telephone circuit makes power available at a pair of terminals. The open-circuit potential across the terminals is 1 volt and the impedance looking into the terminals is $500 - j500$ ohms. What is the maximum power that may be drawn from the circuit?

ANSWER. The greatest power transfer occurs when the receiver impedance equals the conjugate of the network impedance. Thus

$$Z_r = 500 + j500$$

Maximum power transfer

$$P_{\max} = \frac{E^2}{4R_L} = \frac{1^2}{4 \times 500} = \frac{1}{2{,}000} = \mathbf{0.0005\ watt}$$

Often, especially in communications circuits, the condition of maximum efficiency is not so much desired as the condition of maximum power transfer or output from a given source.

Thus, we see that the maximum power obtainable from a certain source is inversely proportional to its resistance. See "Electric Circuits" MIT-EE Staff, John Wiley & Sons, Inc., New York, 1943, p. 136.

Q15-61. Corresponding values of voltage and current for an electrical device are given by the oscillograms: Fig. 15-60. Find

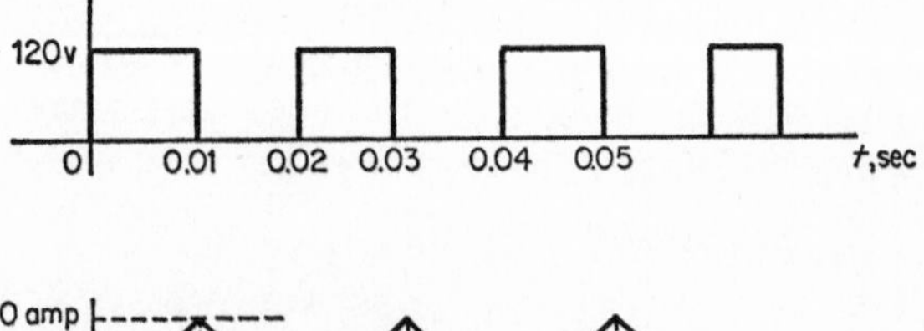

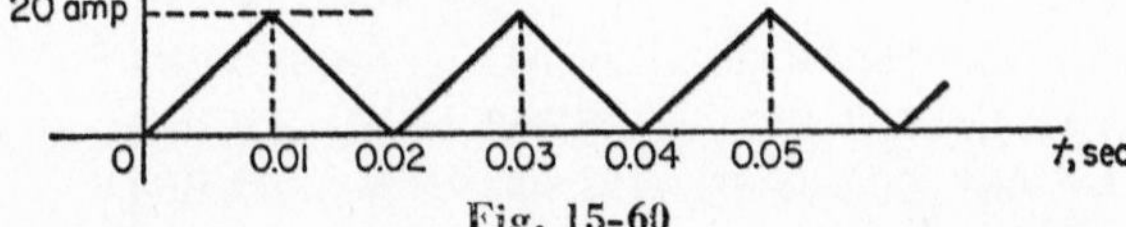

Fig. 15-60

the root-mean-square values of the voltage and current, the average value of the power taken, and the power factor.

ANSWER.

(a) $V_{\text{rms}} = \left(\frac{1}{T}\int_0^T V_t^2\,dt\right)^{1/2} = \left(\frac{120^2 \times 0.01}{0.02}\right)^{1/2} = \mathbf{84.88\ volts}$

$$I_{\text{rms}} = \left[\frac{1}{0.02}\int_0^{0.01} (2{,}000t)^2\,dt + \frac{1}{0.02}\int_{0.01}^{0.02} (40 - 2{,}000t)^2\,dt\right]^{1/2}$$

$I_{\text{rms}} = \mathbf{11.55\ amp}$

(b) Average power taken:

$$\frac{1}{T} = \int_0^T i(t)V_t\,dt = \frac{1}{0.02}(2)(\tfrac{1}{2})(0.01)(20) + \frac{120(0.01)}{0.02}$$

$P_{\text{avg}} = \mathbf{600\ watts}$

(c) Power factor:

$$\frac{P_{\text{avg}}}{P_{\text{rms}}} = \frac{600}{84.88 \times 11.55} = \mathbf{0.612}$$

Q15-62. The input to a circuit is 10 amp at 120 volts. A wattmeter reads 1.039 kw. What is the kvar?

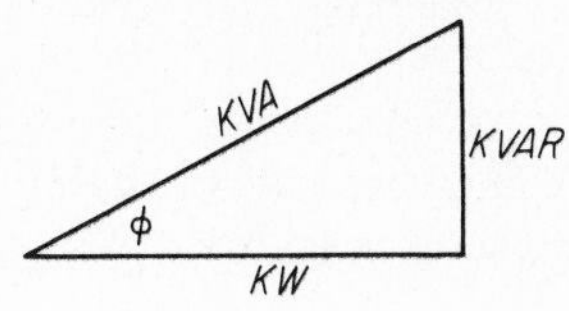

Fig. 15-61

ANSWER. Va = 120 × 10 = 1,200; kva = 1.200; kw = 1.039. Refer to Fig. 15-61. Power factor = $\cos\phi$ = 1.039/1.200 = 0.866. Kvar = kva $\sin\phi$ = 1.200 × 0.5 = **0.600.**

Chapter 16

TRANSFORMERS

16-1. Transformers. Transformers are widely used for changing the relationship between current and voltage in a-c circuits. The transformer is a static device consisting of two or more coils of wire formed around a common iron core. Transformation of energy under these conditions is not accompanied with a change in frequency.

A transformer may receive electric energy at one voltage and deliver it at a higher voltage; this is known as a *step-up* transformer. A transformer may receive energy at one voltage and deliver it at a lower voltage, a *step-down* transformer. A *one-to-one* transformer is one in which there is no change in voltage.

Transformers require less care and attention than almost any other kind of electric power apparatus. This is no reason, however, for neglecting them. The cost per kilowatt of transformers is low when compared with other apparatus. Their efficiencies are much higher.

A static transformer has no moving parts, and the windings may be immersed in oil for insulation against very high voltages. The oil should be tested for dielectric strength and the presence of sludge.

In the transmission of power the principal energy losses in a transformer are the I^2R losses which can be reduced by a decrease in current. Insulation losses vary with the square of the voltage, but are small by comparison. Thus, power may be transmitted at high voltage and low current. However, for reasons of safety relatively low voltages are desirable in power apparatus. The solutions to these apparently diametric problems can best be answered by transmitting power at high voltages with accompanying low losses, then stepping down the voltage and stepping up the current at the load with a transformer. This can best be shown by a practical example. In a large power plant energy may be generated at 13,000 volts, stepped up to 220,000 volts for transmission purposes, later stepped down to 33,000 volts for large-area distribution and eventually stepped down again to 2,300 volts for industrial loads and to 230 and 115 volts for small power and domestic uses.

In communications work the transformer is used to match impedances, i.e., as a means of changing the apparent impedance of a load to fit the particular requirements of the circuit. In this area it is not used to change voltages efficiently. With respect to the former, a typical problem would involve a matching transformer involving the delivery of audio-frequency power from a vaccum-tube source with, say, 50,000 ohms resistance to a speaker coil (voice coil) having, say, 50 ohms effective resistance.

16-2. Transformer Operation. The winding which receives electric power from the source is called the *primary* winding, and the one which delivers power to the load is called the *secondary* winding. Figure 16-1 shows a simple single-phase static transformer. The core is made up of laminations, usually rectangular stampings of silicon-sheet steel, clamped or bolted together. A continuous winding is placed on one leg of the iron core. Another winding is placed on the opposite leg. The first is known as the primary P, and the other as the secondary S. If it is a step-up transformer, the secondary has more turns than the primary. If it is a step-down transformer, the secondary has fewer turns, while a 1:1 transformer has the same number of turns in both windings. By means of the magnetic flux, an emf is induced in the secondary which is then capable of delivering current and energy. In a transformer either winding may be the primary, the other being the secondary, depending on which is connected to the source and which to the load.

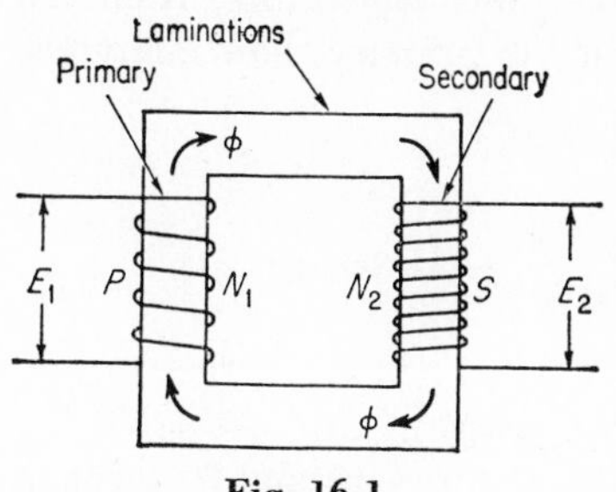

Fig. 16-1

The total induced emf in each winding of the transformer is proportional to the number of turns in that winding. Thus

$$\frac{E_1}{E_2} = \frac{N_1}{N_2} \tag{16-1}$$

where E_1 and E_2 are the primary and secondary induced voltages, respectively, and N_1 and N_2 are the number of turns in the primary and the secondary, respectively.

The emf induced is proportional to the flux, the frequency, and the number of turns.

$$E = 4.44fN\phi_{max} \times 10^{-8} \text{ volt} \tag{16-2}$$

where f = frequency, cps
N = number of turns
ϕ_{max} = maximum value of flux in core

The maximum flux density is also equal to $B_{max}A$, where B_{max} is maximum flux density and A is the core cross section. This equation is the more convenient to use, since transformer cores are designed on the basis of permissible flux density.

Q16-1. Design a transformer to step down the voltage from 120 to 10 volts. If 180 turns are used on the primary, how many are required on the secondary?

ANSWER. Using Eq. (16-1) we see

$$N_2 = N_1 \times \frac{E_2}{E_1} = 180 \times \frac{10}{120} = \mathbf{15\ turns}$$

Q16-2. The core of a 60-cycle transformer has a net cross section of 20 sq in., and the maximum flux density in the core is 60,000 lines per sq in. There are 700 turns in the primary and 70 turns in the secondary. What is the rated voltage of the primary and the secondary?

ANSWER.

$$E_1 = 4.44 \times 60 \times 700 \times 60{,}000 \times 20 \times 10^{-8} = \mathbf{2{,}230\ volts}$$
$$E_2 = 4.44 \times 60 \times 70 \times 60{,}000 \times 20 \times 10^{-8} = \mathbf{223\ volts}$$

16-3. Ampere Turns. We have previously learned that in a pure inductance the voltage leads the current by 90°. Thus, the impressed and induced voltages are out of phase by 180°, as shown in Fig. 16-2.

When a load is connected across the secondary of the transformer, a current I_2 will flow because of the induced voltage E_2. Depending on the characteristics of the load (its resistance and reactance), the amount that angle I_2 will lead or lag E_2 is affected. All in all it has been determined that, when a transformer is fully loaded, the current at no load I_0 is so small in comparison with I_1

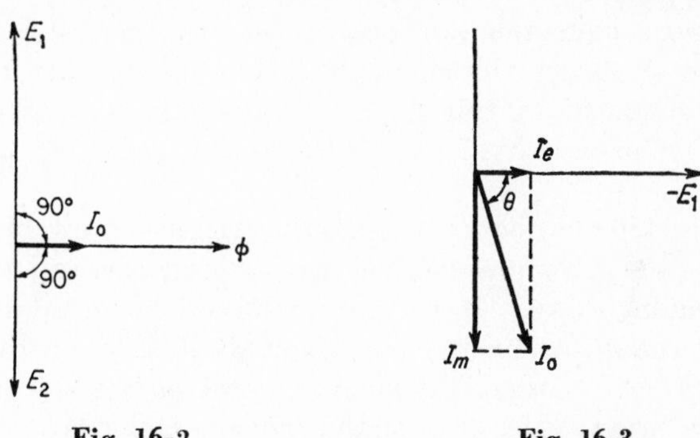

Fig. 16-2

Fig. 16-3

when loaded that it may be neglected. Under these fully loaded conditions the ampere turns of the primary are equal to the ampere turns of the secondary. See Fig. 16-3.

$$N_1I_1 = N_2I_2 \tag{16-3}$$

or

$$\frac{I_1}{I_2} = \frac{N_2}{N_1} \tag{16-4}$$

Q16-3. Explain how the input to a transformer accommodates itself to changes of the load.

ANSWER. An increase in load on a transformer means that more current is flowing through its secondary coils, whose demagnetizing effect on the core is equal to $0.4\pi N_2I_2$. For a given impressed voltage on the primary the primary current must be sufficient to produce the magnetizing flux and also to overcome the demagnetizing action of the secondary. Therefore, sufficient excess current is drawn automatically by the primary to offset the increased demagnetizing flux. Thus, the transformer input adjusts itself to a change in load on the secondary winding.

Q16-4. Under what conditions can two single-phase transformers of different ratings be connected in parallel on both primary and secondary sides so that they may share a load between them in proportion to their ratings?

ANSWER. When so connected the following conditions must be met:

The ratio of reactance to resistance must be the same.

The transformers must have the same ratios of transformation E_1/E_2, i.e., primary to secondary voltage.

They must have the same regulation, i.e., the same relative decrease in secondary voltage when the transformers are loaded.

Q16-5. Explain why transformers are rated in volt-amperes (va) or kilovolt-amperes, (kva) and justify your explanation with suitable proof.

ANSWER. The rating of an electric machine is established by considering the amount of heating that its insulation can withstand before breaking down. In turn, the heat equivalent of an electric current I, caused by an impressed voltage E, is given by Joule's law, $Q = 0.24EI \times \text{time}$. It is more useful to rate a transformer in va than in watts because of the fundamental relation that its primary and secondary va are equal, if we neglect minor core and copper losses which amount to 1 to 3 per cent. Since the power factor of the transformer itself at or near rating is 100 per cent, the rating in va enables the user to estimate output and input quickly. Moreover, if transformers were rated in kilowatts, the rating would not be representative of their performance. For consider the equation for power transmitted by a three-phase system

$$\text{Power} = EI \cos \theta \sqrt{3}$$

16-4. Leakage Reactance. It would be nice if transformers would act exactly in accordance with theory. For instance take the case of flux. Up until now we assumed that *all* the flux which links the primary also links the secondary. This is impossible in practice. Some of the flux bypasses the iron core and passes through the air around the core. The effect of this bypassing is measured by actual test.

Q16-6. A 500-kva 13,200/2,400-volt 60-cycle single-phase transformer has a 4 per cent reactance and a 1.0 per cent resistance. The leakage reactance and the resistance of the low-voltage winding are 0.250 and 0.055 ohm, respectively. The core loss under rated conditions is 1,800 watts.

(*a*) Calculate the leakage reactance and the resistance of the high-voltage winding (ohms).

(*b*) Calculate the efficiency of this transformer at full load and 85 per cent power factor.

ANSWER.

(*a*) If transformer losses are neglected and unity power factor is assumed

$$\frac{E_2}{E_1} = \frac{N_2}{N_1} = \frac{I_1}{I_2}$$

Then the ratio of transformation is $E_2/E_1 = 13{,}200/2{,}400 = 5.5$. Now let the following notations hold:

R_{02} = equivalent resistance referred to secondary
X_{02} = equivalent reactance referred to secondary
Z_{02} = equivalent impedance referred to secondary
R_2 = high side resistance
R_1 = low side resistance
X_2 = high side impedance
X_1 = low side impedance
E_2 = rated secondary voltage
E_1 = rated primary voltage
I_2 = rated secondary current
I_1 = rated primary current

The following equations hold for the transformer:

$$R_{02} = R_2 + R_1 n^2 \qquad Z_{02} = \sqrt{R_{02}^2 + X_{02}^2}$$

$$X_{02} = X_2 + X_1 n^2 \qquad \frac{Z_{01}}{Z_{02}} = n^2 = \left(\frac{N_1}{N_2}\right)^2$$

$$R_{02} = \text{resistance as decimal} \times \frac{E_2}{I_1}$$

$$X_{02} = \text{reactance as decimal} \times \frac{E_2}{I_1}$$

Since for a transformer "volts × amperes in" equals "volts × amperes out"

$$I_2 = \frac{\text{kva rated} \times 1{,}000}{E_2} = \frac{500 \times 1{,}000}{13{,}200} = 37.9 \text{ amp}$$

$$R_{02} = 0.01 \times \frac{13{,}200}{37.9} = 3.48 \text{ ohms}$$

$$X_{02} = 0.04 \times \frac{13{,}200}{37.9} = 13.92 \text{ ohms}$$

Then by rearrangement of the previous equations for equivalent resistance and reactance

$$X_2 = X_{02} - X_1n^2 = 13.92 - (0.25 \times 5.5^2)$$
$$= \mathbf{6.35\ ohms} \text{ leakage reactance}$$

$$R_2 = R_{02} - R_1n^2 = 3.48 - (0.055 \times 5.5)^2 = \mathbf{1.81\ ohms} \text{ resistance}$$

(b) $$\frac{\text{Core loss (copper loss)}}{\text{Rated volt} \times \text{ampere}} \times 100 = \text{copper loss in per cent } P_c$$

$$\frac{1{,}800}{500 \times 1{,}000} \times 100 = 0.36 \text{ per cent}$$

Full-load efficiency at 85 per cent power factor:

$$\frac{E_2I_2 \times \text{pf}}{E_2I_2 \times \text{pf} + \text{core loss} + I_2^2R_{02}}$$

$$\frac{13{,}200 \times 37.9 \times 0.85}{13{,}200 \times 37.9 \times 0.85 + 1{,}800 + 37.9^2 \times 3.48}$$
$$= 0.986, \text{ or } \mathbf{98.6\ per\ cent}$$

Check: $$\frac{100 \times 0.85}{100 \times 0.85 + 0.36 + 1} = 0.986, \text{ or } \mathbf{98.6\ per\ cent}$$

16-5. Regulation. The regulation of a constant-potential transformer is the difference between the no-load and rated-load values of the secondary terminal voltage, expressed in per cent of the rated-load secondary voltage, with the primary impressed terminal voltage adjusted to such a value that the transformer delivers the rated kva output at a specified power factor and at a rated secondary voltage. Thus,

$$\text{Regulation} = \frac{E_1(N_2/N_1) - E_2}{E_2} \times 100 \qquad (16\text{-}5)$$

Equation (16-5) is applicable to the primary side if the subscripts are changed, the regulation being the same in either case.

Q16-7. A 500 kva 11,000/2,300-volt 60-cycle transformer has been tested and gives the following data:

Open circuit test: $E = 2{,}300$ volts; $I = 2.3$ amp; $P = 4{,}000$ watts

Short circuit test: $E = 600$ volts; $I = 45.4$ amp; $P = 6{,}000$ watts

Determine the following:

(*a*) Copper loss
(*b*) Equivalent resistance referred to the primary
(*c*) Equivalent resistance referred to the secondary
(*d*) Equivalent impedance referred to the primary
(*e*) Equivalent reactance referred to the secondary
(*f*) Regulation at 0.8 power factor lagging

ANSWER.

(*a*) **6,000 watts,** the reading of the wattmeter.

(*b*) $R_{01} = \dfrac{6{,}000}{45.4^2} = \mathbf{2.92\ ohms}$

(*c*) $R_{02} = 2.92 \times \left(\dfrac{2{,}300}{11{,}000}\right)^2 = \mathbf{0.129\ ohm}$

(*d*) $Z_{01} = \dfrac{E_1}{I_1} = \dfrac{600}{45.4} = \mathbf{13.2\ ohms}$

(*e*) $X_{01} = \sqrt{Z_{01}^2 - R_{01}^2} = \sqrt{13.2^2 - 2.92^2} = 12.8$ ohms

$X_{02} = 12.8 \times \left(\dfrac{2{,}300}{11{,}000}\right)^2 = \mathbf{0.563\ ohm}$

(*f*) Rated primary current $I_1 = \dfrac{500{,}000}{11{,}000} = 45.5$ amp

$$E_1 = \sqrt{(11{,}000 \times 0.8 + 45.5 \times 2.92)^2 + (11{,}000 \times 0.6 + 45.5 \times 12.8)^2}$$

$E_1 = 11{,}500$ volts

$$\text{Regulation} = \frac{11{,}500 - 11{,}000}{11{,}000} \times 100 = \mathbf{4.56\%}$$

Note: In using Eq. (16-5) the value of $E_1(N_2/N_1) = 11{,}500$ volts.

16-6. All-day Efficiency. Transformers are in operation all day; i.e., they give 24-hr service even though they are on part load for a short time or a considerable part of the time. Under these conditions the performance of a transformer must be judged by its all-day efficiency. All-day efficiency is the ratio of the energy output divided by the energy input over the same 24-hr period.

Q16-8. The resistances of a 500-kva 11,000/2,300-volt 60-cycle transformer are respectively 0.8 and 0.04 ohm for the high and low-voltage sides. (*a*) The iron loss is 3,000 watts; calculate the full-load copper loss.

(*b*) This transformer is loaded daily as follows: 2 hr at full load, 3 hr at three-quarters load, and 6 hr at one-quarter full load. Calculate the all-day efficiency.

ANSWER.

(*a*) First determine the high-voltage and then the low-voltage winding copper losses. The sum of the two losses is the full-load copper loss.

$$\text{Rated current (primary)} = \frac{\text{kva rated} \times 1{,}000}{E_1 \text{ rated}} = \frac{500 \times 1{,}000}{11{,}000} = 45.5 \text{ amp}$$

$$\text{Copper loss} = I_1{}^2R_1 = 45.5^2 \times 0.8 = 1{,}650 \text{ watts for primary}$$

$$\text{Rated current (secondary)} = \frac{500 \times 1{,}000}{2{,}300} = 217 \text{ amp}$$

$$\text{Copper loss } I_2{}^2R_2 = 217^2 \times 0.04 = 1{,}890 \text{ watts for secondary}$$

$$\text{Total full-load copper loss} = 1{,}650 + 1{,}890 = \mathbf{3{,}540 \text{ watts}}$$

(*b*) Since the input equals the output plus the losses, the all-day efficiency may be written

$$\frac{\text{output} \times 100}{\text{output} + \text{losses}}$$

Also the copper losses vary directly as the fractional load squared. The energy output is found, assuming unity power factor

$$W_1 = 500 \times 2 + 500 \times \tfrac{3}{4} \times 3 + 500 \times \tfrac{1}{4} \times 6 = 2{,}875 \text{ kwhr}$$

The energy input (output plus losses)

$$W_2 = 2{,}875 + [3.54(\text{full-load copper loss}) \times 2 \times 1^2] + (3.54 \times 3 \times 0.75^2) + (3.54 \times 6 \times 0.25^2) + [3{,}000(\text{iron losses}) \times {}^{24}\!/_{1000}] = 86.38 \text{ kwhr}$$

$$\text{All-day efficiency} = \frac{2{,}875 \times 100}{2{,}875 + 86.38} = \mathbf{97\ per\ cent}$$

Note that conventional efficiency would be

$$\frac{500{,}000 \times 100}{500{,}000 + 3{,}000 + 3{,}540} = 98.7 \text{ per cent}$$

16-7. Core-type Transformers. There are generally two types of transformers, the core type and the shell type. Their only difference lies in the manner in which the iron and copper are arranged with respect to each other.

The core-type transformer has its winding wrapped around the iron core. Figure 16-1 is typical of this class. However, this arrangement would result in poor regulation, because of the large leakage flux for both primary and secondary. But with both the primary and the secondary wrapped around each leg, the leakage flux is reduced considerably. Insulation costs are kept to a minimum by wrapping the low-voltage coil next to the core leg with the high-voltage coil wrapped around that with only one layer of high-voltage insulation between. In the core type the mean length of the turn is less and the mean length of the magnetic path is greater than in the shell type. The core type is well adapted to high voltages.

Q16-9. A 50-kva 22,500/440-volt core-type transformer has a core area and length of 35 sq in. and 75 in. respectively. The core loss in transformers of this size and class is 0.32 watt per cu in.; the required magnetizing force is 11 amp-turns per in. with an additional allowance of 80 amp-turns for the joints in the core.

(*a*) Calculate the number of turns on the high- and low-tension windings in order that the maximum flux density shall be 66,000 lines per sq in.

(*b*) Calculate the power factor of this transformer at no load.

ANSWER.

(*a*) From Eq. (16-2) the number of turns on the high side is

$$N_1 = \frac{E_1 \times 10^8}{4.44fB_{\max}A} = \frac{22{,}500 \times 10^8}{4.44 \times 60 \times 66{,}000 \times 35} = \textbf{3,660 turns}$$

Using Eq. (16-1)

$$N_2 = N_1\left(\frac{E_2}{E_1}\right) = 3{,}660 \times \left(\frac{440}{22{,}500}\right) = \textbf{72 turns}$$

(*b*) The power factor of this transformer is given by the relation

$$\text{Power factor} = \frac{I_e}{I_0}$$

where I_e is the energy component of the exciting current I_0 and supplies the core losses. I_0 is the no-load current and is allied with I_e as $I_e = I_0 \cos\theta$. The magnetizing component I_m lags the impressed voltage by 90° and is equal to $I_0 \sin\theta$ (see Fig. 16-3, p. 292). The number of ampere turns required is (75 in. × 11) + 80 for core joints, or 905 amp turns. In most commercial transformers $I_0 = I_m$ very nearly. Assuming that rms values of current are involved

$$I_m = \frac{905}{3{,}660} = 0.248 \text{ amp}$$

$$I_e = \frac{\text{total core loss}}{E_1} = \frac{35 \times 75 \times 0.32}{22{,}500} = 0.0373 \text{ amp}$$

From Fig. 16-3 it may be seen that

$$I_0 = \sqrt{I_m^2 + I_e^2} = \sqrt{0.248^2 + 0.0373^2} = 0.25 \text{ amp}$$

and $$\text{Power factor} = 0.0373/0.25 = \mathbf{0.149}$$

16-8. Three-phase Transformers. Three-phase transformers are more compact than three single-phase transformers, and are, therefore, more commonly used in practice. Figure 16-4 shows a three-phase core-type transformer having a primary on each leg, secondaries not shown for simplicity's sake. Cores are placed 120° apart. This is not a practical arrangement and we see in Fig. 16-5 a more useful setup.

16-9. Transformer Connections. As a matter of definition, a *distribution transformer* is one whose rating is 500 kva or less. It is used for distributing power from high-voltage lines to locations where a lower voltage is required. Transformers rated above 500 kva are classed as *power transformers.*

Polarity is of interest on account of its bearing when paralleling or banking two or more transformers. In order to simplify the process of keeping a check on connections, all leads brought out

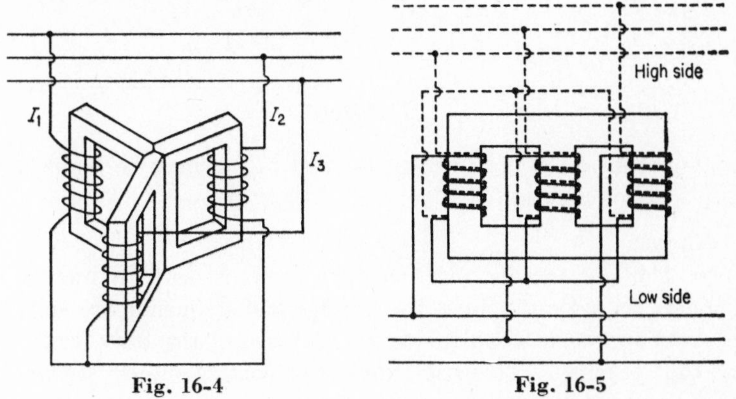

Fig. 16-4 **Fig. 16-5**

of the case or tank are marked by a system of letters and numbers. Refer to the name plate.

Standardization of polarity markings has been established by the AIEE. The letter H is used for the high-voltage leads and X for the low-voltage leads. The leads of either winding brought out of the case are to be numbered 1, 2, 3, etc. The highest and lowest numbers indicate full winding while the intermediate numbers mark fractions of windings or taps. In addition, all numbers are to be so applied that the drop from any lead having a lower number toward any lead having a higher number shall have the same sign at any instant. The chief three-phase transformer connections are as shown in Fig. 16-6.

Q16-10. A single-phase transformer is needed for a specific job, but only a three-phase transformer is available. If the frequency rating and voltage ratio are satisfactory, (*a*) how should the three-phase unit be connected for maximum single-phase output, (*b*)

	Three Phase Transformers	
Group 1 Angular displacement 0°	H2, H1, H3 X1, X2, X3	H2, H1, H3 X1, X2, X3
Group 2 Angular displacement 180°	H2, H1, H3 X3, X1, X2	H2, H1, H3 X3, X1, X2
Group 3 Angular displacement 30°	H2, H1, H3 X1, X2, X3	H2, H1, H3 X1, X2, X3
	H2, H3, H1 X2, X3, X1	H2, H3, H1 X2, X3, X1

Fig. 16-6

what percentage of its three-phase kva rating may be realized in single-phase operation?

ANSWER.

(*a*) The key to this question is the requirement for maximum single-phase output. Since the voltage and frequency are correct, the best approach would be to parallel two of the three windings on the primary. Likewise, the two corresponding secondary windings should be paralleled (see Fig. 16-7). It is most important that the polarity markings on the transformer leads be carefully checked when attempting this type of hookup.

(*b*) The power output using this hookup would be equivalent to two-thirds the normal three-phase kva rating.

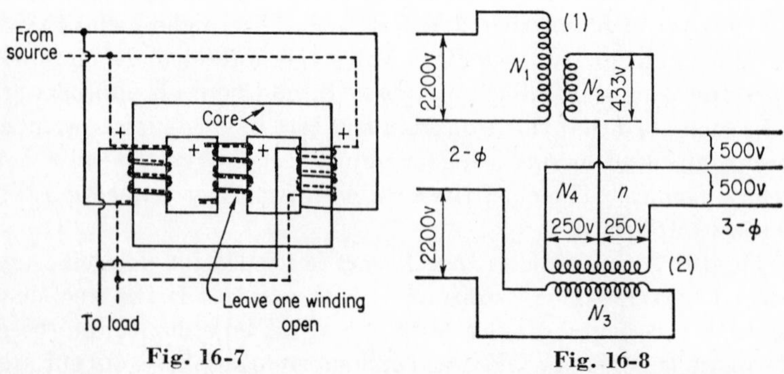

Fig. 16-7

Fig. 16-8

Q16-11. Show how to transform from 2,200 volts two-phase to 500 volts three-phase. Allowing 7 volts per turn of the windings, find the number of turns on each primary and on each secondary coil.

ANSWER. Refer to Fig. 16-8. The two transformers (1) and (2) are necessary to transform two-phase to three-phase current, with the secondary voltage, E_2 of the transformer (1), 86.6 per cent of the three-phase voltage, or

$$500 \times 0.866 = 433 \text{ volts}$$

The connections are as shown in Fig. 16-8, from which we obtain the relation $E_1/E_2 = N_1N_2$ and

$$\begin{aligned}
\text{Turns in primary of (1)} &= N_1 = 2{,}200/7 = \mathbf{314} \\
\text{Turns in secondary of (1)} &= N_2 = 433/7 \quad = \mathbf{62} \\
\text{Turns in primary of (2)} &= N_3 = 2{,}200/7 = \mathbf{314} \\
\text{Turns in secondary of (2)} &= N_4 = 500/7 \quad = \mathbf{72}
\end{aligned}$$

Provide with tap n so that there are 250 turns on either side of it.

16-10. Star-connected Three-phase Transformer. Refer to Fig. 16-9. The following relationships hold:

$$E = E_n \sqrt{3} \text{ volts} \qquad (16\text{-}6)$$

$$E_n = \frac{E}{\sqrt{3}} \qquad (16\text{-}7)$$

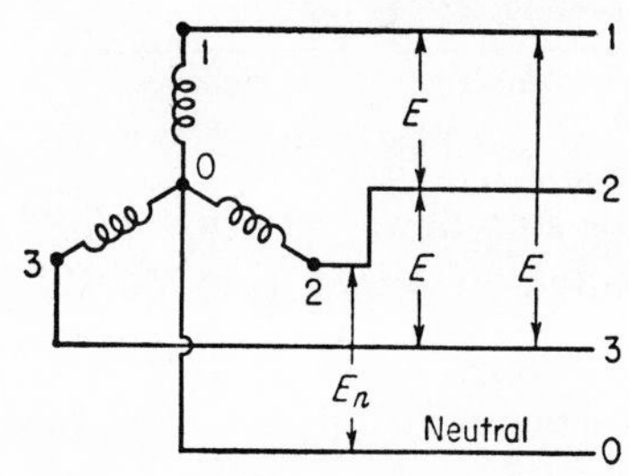

Fig. 16-9

Power conversion

$$P = 3E_nI \cos \theta \qquad (16\text{-}8)$$

$$P = \sqrt{3}\, EI \cos \theta \qquad (16\text{-}9)$$

16-11. Δ-connected Three-phase Transformer. Refer to Fig. 16-10. The following relationships hold:

$$I = i \sqrt{3} \text{ amp} \qquad (16\text{-}10)$$

$$i = \frac{I}{\sqrt{3}} \text{ amp} \qquad (16\text{-}11)$$

Power conversion

$$P = 3Ei \cos \theta \qquad (16\text{-}12)$$

$$P = \sqrt{3}\, EI \cos \theta \qquad (16\text{-}13)$$

16-12. Δ-Y Connection. The Δ-Y connection shown in Fig. 16-11 is a very useful connection for stepping up the voltage. It has a distinct advantage over the Δ-Δ connection (Fig. 16-12) in that for high voltages the insulation need not be so extensive. For a 100,000-volt system, Y-connected transformers need only be insulated for 58,000 (100,000 ÷ $\sqrt{3}$) volts, whereas the Δ-Δ connection would have insulation requirements for 100,000 volts.

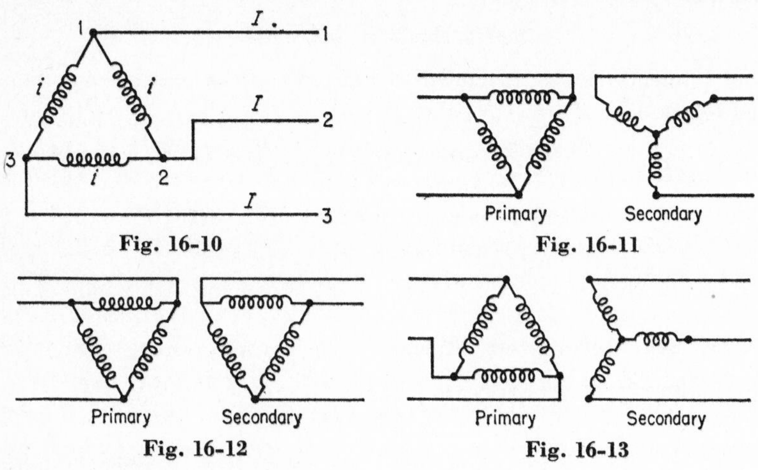

Fig. 16-10

Fig. 16-11

Fig. 16-12

Fig. 16-13

The Y-Δ arrangement is often used for step-down service. When two identical transformers are connected in Δ-star (Fig. 16-13), also identified as Δ-Y, then

$$\Delta \text{ volts} = \sqrt{3} \times \text{star volts} \tag{16-14}$$

$$\text{Star volts} = \frac{\Delta \text{ volts}}{\sqrt{3}} \tag{16-15}$$

Likewise

$$\text{Star amp} = \sqrt{3}\, \Delta \text{ amp} \tag{16-16}$$

$$\Delta \text{ amp} = \frac{\text{star amp}}{\sqrt{3}} \tag{16-17}$$

Q16-12. Two similar three-phase transformers are to be connected in parallel to supply a common load. The primary voltage is 1,200 volts and the ratio of transformation is 10:1. (*a*) State how the primary and secondary windings of these transformers should be connected in order that energy for lighting and power may be taken from the low-voltage side. (*b*) Describe briefly the protective measures and metering devices needed. (*c*) Outline the procedure in actually making connections recommended in (*a*) preceding.

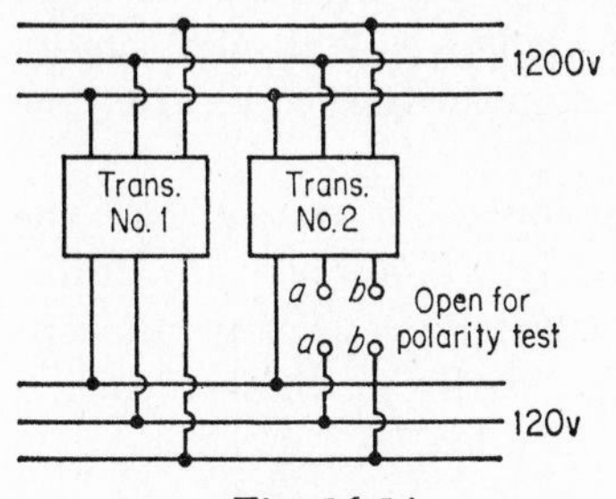

Fig. 16-14

ANSWER. See Fig. 16-14 for the actual hookup, answer (*a*).

(*b*) A rather expensive system for protecting the system of transformers is shown in Fig. 16-15. The circuit breakers might be of the overcurrent type designed to protect against 120 volts overload. Reactors might be inserted to protect against short circuits in the lines and voltage regulators to maintain constant voltage.

The transformer secondaries would be grounded. Current and potential instrument transformers would be needed to make it possible to read primary voltage and current. Instrument transformer casings and frames would also be grounded. A fuse would be placed in the potential transformer primary to protect the line from disturbance due to failure of the transformer. Lightning arresters

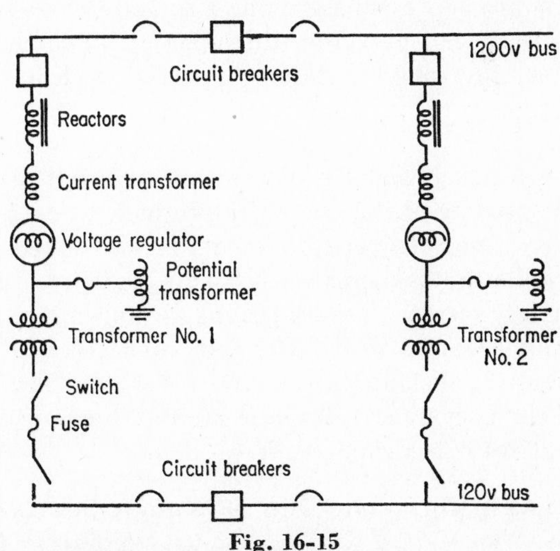

Fig. 16-15

would be placed in the primary to ground any induced charges in the line.

(*c*) After the transformers have been connected as indicated in Fig. 16-14, a voltmeter is connected to points *bb* first, and then to points *aa*. If a voltmeter reads zero, the polarity is correct and the leads may be spliced.

Q16-13. Show how a three-phase four-wire system may be obtained from a two-phase power supply. Show how the transformers are connected and indicate accurately at what points taps are to be taken or connections made. Explain whether such a three-phase system remains balanced when load is applied.

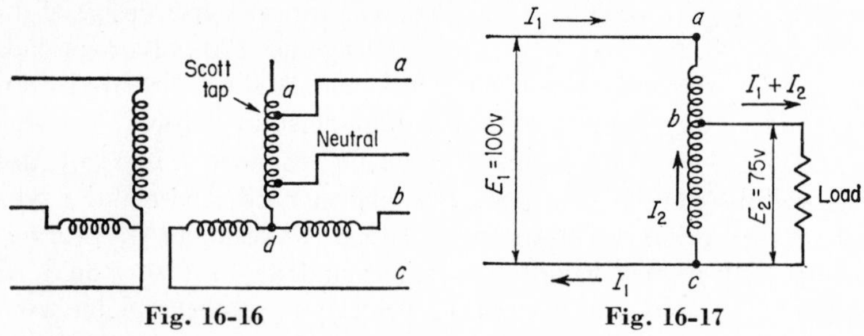

Fig. 16-16

Fig. 16-17

ANSWER. With the use of the Scott or T connection, it is possible to transform not only from three-phase to three-phase by means of two transformers but also from three-phase to two-phase or from two-phase to three-phase. Assume we start with a symmetrical two-phase three-wire system and transform to a three-phase four-wire system. The result will be as shown in Fig. 16-16, point *a* on the teaser being such that the distance *da* represents 87.6 per cent of the total winding of the teaser transformer. Coil to the left is the primary. The volts per turn are in the ratio of 100:86.6. The secondary will also be raised a corresponding amount. The neutral of the secondary system is two-thirds the way down the teaser transformer winding from *a* to *d*. In these connections, the voltages become slightly unbalanced even under balanced loads. This is because of the unsymmetrical phase relations among the voltages and currents in the individual coils.

16-13. Autotransformers. The autotransformer has but one winding. In Fig. 16-17 the load current is supplied by the secondary, and part of the current flows by direct conduction from the primary. The load current is the sum of I_1 and I_2. Such a transformer is also known as a *compensator*.

As for voltage and power relations in autotransformers the primary voltage is E_1 and the full winding *ac* receives the power. The secondary voltage is E_2. The ratio of transformation is E_1/E_2. Magnetizing current flows through winding *ac*. Thus, winding *ac* is known as the primary, while winding *bc* is the secondary. For sake of clarification assume for the following that

$$\begin{aligned} I_1 &= 15 \text{ amp} \\ I_2 &= 5 \text{ amp} \\ E_1 &= 100 \text{ volts} \\ E_2 &= 75 \text{ volts} \end{aligned}$$

Here the transformer is supplying a load at 75 per cent voltage. Now

$$\begin{aligned} \text{Power delivered to load} &= 75 \times 20 = 1{,}500 \text{ watts} \\ \text{Power in primary } ab &= 25 \times 15 = 375 \text{ watts} \\ \text{Power in secondary } bc &= 75 \times 5 = 375 \text{ watts} \end{aligned}$$

Power conducted must be $1{,}500 - 375 = 1{,}125$ watts. Thus, only one-fourth the total power involved is transformed. The relationship between voltages and turns is given by

$$\frac{E_1}{E_2} = \frac{N_1}{N_2} \tag{16-18}$$

The autotransformer is more economical than the regular transformer when the ratio of transformation is moderate. If m is the ratio of low voltage to high voltage, the ratio of the power transformed magnetically to the total power delivered is $(1 - m)$. The ratio of the power transformed to the total power is

$$\frac{N_1 - N_2}{N_1} = 1 - m \tag{16-19}$$

When the ratio of transformation is high, the high and low sides are directly connected together. This is a definite potential danger.

Q16-14. An autotransformer has a winding *ac* tapped at the center *b*. The resistance and leakage reactance of *bc* are 0.07 and 0.1 ohm respectively. If an emf of 10 volts is applied to *ac* when

bc is short-circuited, calculate the short-circuit current in the line. Show the vector diagram of the currents and voltages in the transformer (refer to Fig. 16-18).

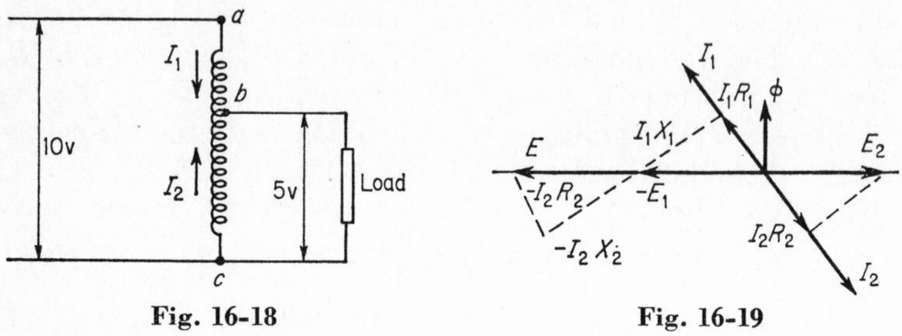

Fig. 16-18 **Fig. 16-19**

ANSWER. From problem $N_1 = N_2 = \frac{1}{2}$ total turns in primary ac. Induced voltage in secondary is

$$E_2 = E_1 \times \frac{N_2}{N} = 10 \times \tfrac{1}{2} = 5 \text{ volts}$$

where N = total turns in primary

$$E_2 = I_2 Z_2$$

From which we can solve for Z_2

$$Z_2 = \sqrt{0.07^2 + 0.1^2} = 0.122 \text{ ohm}$$

Therefore, $I_2 = 5/0.122 = 41$ amp. Also $I_1 = I_2 \times (N_2/N_1) = 41$ amp. Finally, the short-circuit current through supply is **41 amp.** The vector diagram is shown in Fig. 16-19.

Q16-15. A 120/180-volt autotransformer draws power at 120 volts and supplies power to a 2-kw load at 180 volts with a power factor of 0.80. An additional load of 1 kw is supplied at unity power factor from the 60-volt winding. Calculate the power factor and current drawn from the 120-volt line by the transformer, neglecting its losses.

ANSWER. Refer to transformer hookup (Fig. 16-20) and the vector diagram (Fig. 16-21). Load current

$$I_{LD1} = \frac{2{,}000}{180 \times 0.8} = 13.9 \text{ amp}$$

Supply current is given by $I_{LD2} = 1{,}000/(60 \times 1) = 16.6$ amp.

$$I_{LD3} \text{ (with respect to } I_{LD1}) = 3/2 \times 13.9 = 20.85 \text{ amp at 0.8 pf}$$
$$I_{LD3} \text{ (with respect to } I_{LD2}) = 1/2 \times 16.6 = 8.3 \text{ at unity pf}$$
$$I_{LD3} \text{ (resistive at 0.8 pf)} = 20.85 \times 0.8 = 16.7 \text{ amp}$$
$$I_{LD3} \text{ (reactive at 0.8 pf)} = 20.85 \times 0.6 = 12.51 \text{ amp}$$
$$I_{LD3} \text{ (resistive at unity pf)} = 8.3 \text{ amp}$$
$$I_{LD3} = 16.7 + 8.3 + j12.51$$
$$= \sqrt{25^2 + 12.51^2} = \mathbf{28\ amp} \qquad \text{pf} = 25/28 = \mathbf{0.893}$$

Q16-16. A 5-kva 60-cycle single-phase transformer has a full-load efficiency of 96 per cent and an iron loss of 40 watts under rated

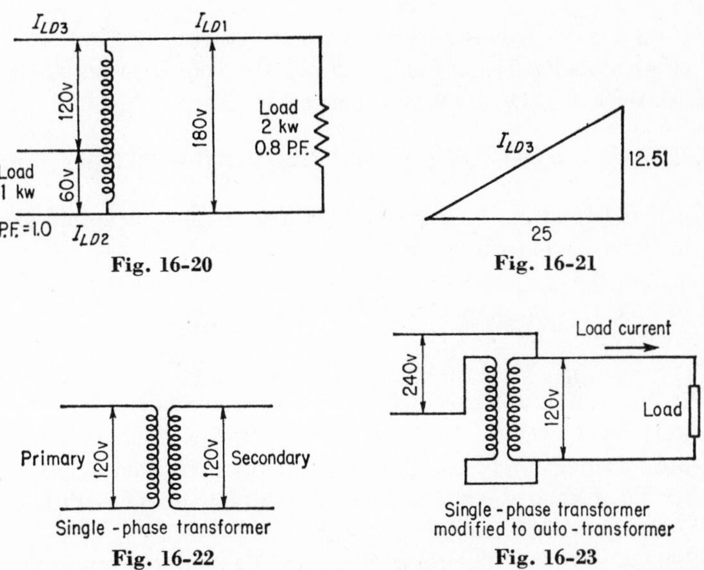

Fig. 16-20

Fig. 16-21

Fig. 16-22

Fig. 16-23

conditions. If the transformer is connected as an autotransformer operating from 240-volt mains and delivering 5 kw at unity power factor to a 120-volt circuit, calculate the efficiency of the operation and the current input to the high-voltage side.

ANSWER. Refer to Fig. 16-22. This shows a normal hookup as a single-phase transformer. Figure 16-23 shows the same unit as autotransformer. The windings are connected in series so that each winding has the same voltage (120 volts) as when connected up as a conventional transformer. Thus, the core loss would be the same under either condition. Because of the hookup the autotransformer windings will each carry but half the current (because of their combining effect at the mid-point) as the conventional arrangement. The copper losses will be one-fourth of those in the conventional transformer because they vary as the current squared. Neglect the losses and the magnetizing current.

For the standard transformer hookup: See also Q16-6.

$$\text{Efficiency} = \frac{\text{output}}{\text{output} + \text{losses}} = \frac{5{,}000}{5{,}000 + 40 + I^2R} \times 100 = 96$$

from which the I^2R losses work out to 168 watts.

For the autotransformer hookup: Since the copper losses are one-quarter those for the standard arrangement, then $^{168}\!/_4 = 42$ watts.

$$\text{Efficiency} = 5{,}000/(5{,}000 + 40 + 42) \times 100 = \mathbf{98.2\ per\ cent}$$

The current input to the high-voltage side is (now including losses but not the magnetizing current)

$$I_1 = \frac{\text{input}}{E_1} = \frac{5{,}082}{240} = \mathbf{21.1\ amp}$$

Q16-17. A 50-kva 2,200/220-volt transformer has an iron loss of 300 watts at 60 cycles. The resistances of its low and high potential windings are 0.005 and 0.5 ohm respectively. Calculate the input to this transformer when it is delivering full load to a circuit having a power factor of 80 per cent lagging.

ANSWER. In a transformer there are two principal losses: the core, or iron, loss and the copper loss. If we let E_1 and E_2 be primary and secondary voltages respectively, I_1 and I_2 their corresponding currents, R_1 and R_2 their corresponding resistances, m the ratio of transformation, $\cos \theta_2$ the power factor of the load, and P_1 and P_2 the input and output in watts, then for the problem at hand

$$m = \frac{E_1}{E_2} = \frac{2{,}200}{220} = 10$$

Since a transformer is rated by its output

$$I_2 = \frac{50 \times 1{,}000}{E_2} = \frac{50 \times 1{,}000}{220} = 227 \text{ amp}$$

Also $$I_1 = \frac{I_2}{m} = \frac{227}{10} = 22.7 \text{ amp}$$

The copper loss is

$$I_1^2(R_1 + R_2m^2) = 22.7^2 \times [0.5 + (0.005 \times 10^2)] = 516 \text{ watts}$$

For the power end

$$\begin{aligned} P_1 &= P_2 + \text{total losses} = E_2I_2 \cos \theta_2 + \text{copper loss} + \text{iron loss} \\ &= (50{,}000 \times 0.8) + 516 + 300 = \mathbf{40{,}816\ watts} \end{aligned}$$

16-14. Testing of Transformers. The most efficient method of testing is the *loss method.* As we have seen, the efficiency is found by the equation

$$\text{Efficiency} = \frac{\text{output}}{\text{output} + \text{losses}}$$

Transformer losses may be divided into two, core losses and copper losses. Core losses are for all intents and purposes constant for all loads and are the result of the eddy currents and hysteresis losses within the laminated iron core. Core losses are a direct function of frequency and flux density. These losses are determined by open-circuit test (see Fig. 16-24). In this test the input to the primary is measured while operated at rated voltage. The secondary is open.

Copper losses are determined by means of the short-circuit test. These losses are directly proportional to the square of the load current. The total copper losses are equal to the sum of those in the primary and in the secondary

$$P_c = I_1^2R_1 + I_2^2R_2 \tag{16-20}$$

Refer to Fig. 16-25 for the short-circuit test.

Q16-18. **(*a*) Explain how you would test a single-phase transformer. It is a 1,200-volt to 120-volt 10 kva 60-cycle unit and must be tested in order to determine: (1) its core loss at rated voltage and frequency, (2) its copper loss at rated output.**

(*b*) Explain how you would use these data to calculate the

efficiency of this transformer at full load and with a power factor of 80 per cent lagging.

ANSWER. (*a*) Refer to the previous discussion and transformer problems involving the determination of efficiency. It is known that the alternation of the core flux in a transformer produces eddy currents and hysteresis losses in the iron. These losses vary with the different exponents of flux density and frequency, and depend also on the wave shape of the impressed emf. In as much as the flux through the core of a constant-voltage transformer remains

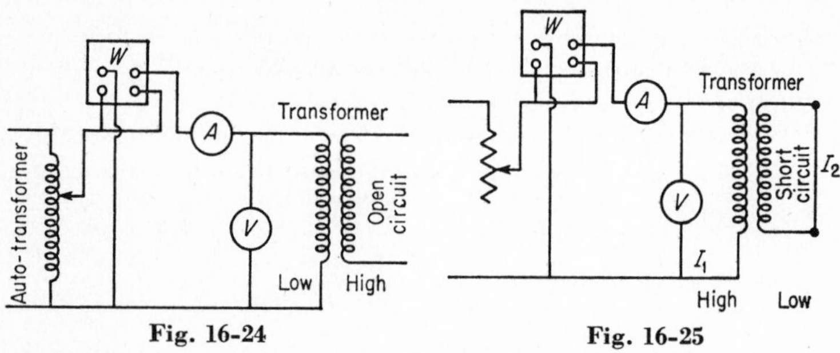

Fig. 16-24 **Fig. 16-25**

nearly constant for all practical purposes, the core loss determined at no load may be assumed as a constant loss over the entire range of loading. With increased load the main flux actually decreases, but the leakage flux increases. Therefore, the variation is very small and the effects neutralize. As previously indicated, the open-circuit test is used to make the determination (see Fig. 16-24). Either an autotransformer or a *drop wire* is shown as a means for regulating the input voltage. A voltmeter, an ammeter, and a wattmeter are connected in the primary circuit. The voltmeter reads the voltage across the primary, the ammeter reads the *no-load current*, and the wattmeter reads the power taken by the transformer under test conditions. The power supplies the primary I^2R loss and the transformer core loss. The exciting current is very small and can be neglected in its effect on the primary loss. Thus, the wattmeter measures the core loss.

If the primary voltage is varied and the core loss found for each variation, the results may be plotted in a curve. At the no-load condition the flux is practically proportional to the terminal voltage.

This is so because the impedance of the primary is negligible. It may be shown that the eddy-current loss varies as the square of the voltage, while the hysteresis loss varies as the 1.6 power of the voltage. As a result tests will show that the core loss will increase as the square of the voltage.

The *copper loss* at rated output is measured with the wattmeter likewise connected, but with the secondary short-circuited through an ammeter (see Fig. 16-25). Just enough voltage is supplied the primary to produce the rated current in the secondary. Note that the transformer is reversed to limit the line current so that a good voltage drop could be realized. In this way the results of the test would have high precision. In both tests instrument losses should be investigated; and, if necessary, corrections made.

(*b*) Efficiency at a certain power factor is given by

$$\frac{\text{rated output}}{\text{rated output} + \text{core loss} + \text{copper loss at rated output}}$$

or

$$\frac{E_1 I_1 \cos \theta_1}{E_1 I_1 \cos \theta_1 + \text{core loss} + \text{copper loss}}$$

Q16-19. A transformer is delivered to your plant. As the plant engineer you are required to check this transformer to see that the specifications in the purchase agreement are met. Outline the tests you would perform.

ANSWER. Modern transformers have a very high efficiency with accompanying small regulation. The losses, as a matter of fact, are so small that they are difficult to determine accurately by simultaneous measurement of output and input. For example, the efficiencies of moderate-size transformers are of the order of 98 per cent; i.e., the losses are about 2 per cent. This method of testing gives poor accuracy.

The method used exclusively in commercial testing of transformers is the loss method described above. It is not necessary to put transformers through all tests, but the breakdown test for insulation should be applied to all.

Should all tests be required these are as follows:

(*a*) Copper loss to determine efficiency.

(*b*) Iron core loss, hot and cold, to determine efficiency.

(*c*) Open circuit or exciting current.

(*d*) Regulation to determine magnetic leakage.

(*e*) Rise in temperature in its case and out for no load and full load; with and without oil. Polarity.

(*f*) Insulation test for ability to thoroughly and effectually insulate the secondary circuit from the primary.

Although the standard types of transformers of today are made on lines found by long experience to be the best for all purposes, and are subject to careful inspection and test at the factory in most cases, yet the various manufacturers have such different ideas as to the value of the different points that, in order to obtain fair bids when purchased, it is always best to prepare specifications. In addition, the purchaser should be prepared to conduct or check tests to determine whether the specifications have been met.

Large stations should have a complete outfit for testing. Smaller purchasers could hire an outside man to witness factory tests.

Q16-20. A farm whose electrical load can be represented by a resistance of 1 ohm in series with an inductive reactance of 1 ohm is supplied from an 11,000-volt single-phase line through a transformer with a turns ratio of 50:1. The resistance and leakage reactance of the transformer are 125 ohms and 250 ohms respectively when referred to its primary current, and its magnetizing current may be neglected. Determine the magnitude of the current taken from the secondary terminals of the transformer, the potential difference between those terminals, the magnitude and power factor of the primary current.

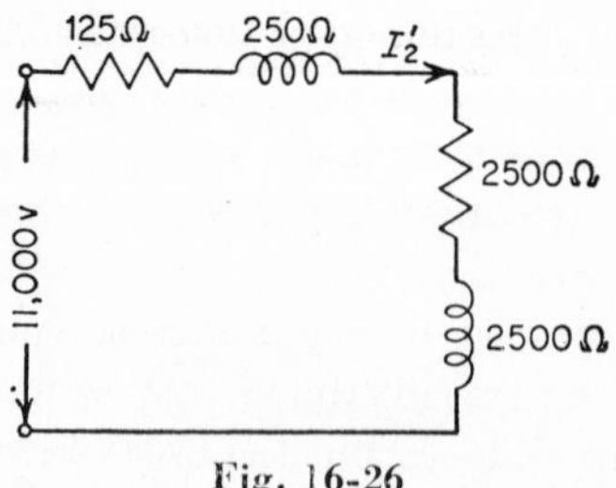

Fig. 16-26

ANSWER. If we refer all quantities to the primary side of the transformer, we can represent this by Fig. 16-26. Then

$$I_2' = \frac{11{,}000}{\sqrt{(2{,}500 + 125)^2 + (2{,}500 + 250)^2}} = \frac{11{,}000}{3{,}800} = 2.9 \text{ amp}$$

Therefore, the secondary current $I_2 = 2.9 \times 50 =$ **145 amp.** The potential difference between secondary terminals is E_2.

$$E_2 = \sqrt{1^2 + 1^2} \times 145 = \mathbf{205\ volts}$$

Primary current $I_1 = I_2' =$ **2.9 amp.**
Power factor is $\cos \phi = (2{,}500 + 125)/3{,}800 =$ **0.70.**

Q16-21. Calculate percentage voltage drop for a transformer with a percentage resistance of 2 per cent and a percentage reactance of 4 per cent of rating 550 kva, when it is delivering 450 kva of 0.6 power factor lagging.

ANSWER. Percentage voltage drop is given by the following:

$$\frac{(\text{per cent } R)\ I \cos \phi}{I_{\text{f.1.}}} + \frac{(\text{per cent } X)\ I \sin \phi}{I_{\text{f.1.}}}$$

where $I_{\text{f.1.}}$ is the full-load current and I is actual current. Therefore, percentage voltage drop is

$$(\text{per cent } R)\ \text{kw}/(\text{kva})_r + (\text{per cent } X)(\text{kva})_r/\text{kva rating}$$

Here kw $= 450 \times 0.6 = 270$. Also $(\text{kva})_r = 450 \times 0.8 = 360$. Then it follows percentage voltage drop is

$$(2 \times 270/360) + (4 \times 360/550) = \mathbf{3.60\ per\ cent}$$

Chapter 17

ALTERNATING-CURRENT TRANSMISSION LINES

17-1. Transmission Systems. For economic transmission of power, voltages must be high. As we have seen, high voltages may be realized with alternating current by use of transformers. Transmission voltages in the order of 20,000 volts and higher are too high for use in industrial and commercial installations. For purposes of distribution in these areas it may be stepped down to usable levels by means of transformers. Although experiments have indicated the use of thermionic tubes for high-voltage transmission of direct current, at present alternating current is almost always used for transmission purposes.

Single-phase power transmission lines may be used for such service, but for transmission of considerable power the polyphase systems are used. There are a number of attendant advantages to the use of polyphase systems. For instance, polyphase motors are a great deal cheaper and lighter than single-phase motors of equal rating. And as we have seen, the polyphase motor has better operating characteristics. Generator and converter ratings have a much greater rating when operating polyphase. In addition, the use of three-phase systems requires only 75 per cent as much copper as either the single-phase or two-phase systems.

Transmission voltages are usually determined by economic factors. The conductor cross section must be balanced against the cost of insulating the line and the size of transmission-line structures and of generating stations and substations. Figure 17-1 shows typical connections of a power system. A rough basis for determining transmission voltage is to use 1,000 volts per mile of line.

Referring to Fig. 17-1, there is imminent danger if extremely high voltages are transmitted through thickly populated areas to the substations. At substations located at the outskirts of the city the voltage is stepped down to values of 13,200 to 26,400 volts and carried underground into the city. Occasionally, these may be run overhead. Power is generally generated at 6,600 volts and then

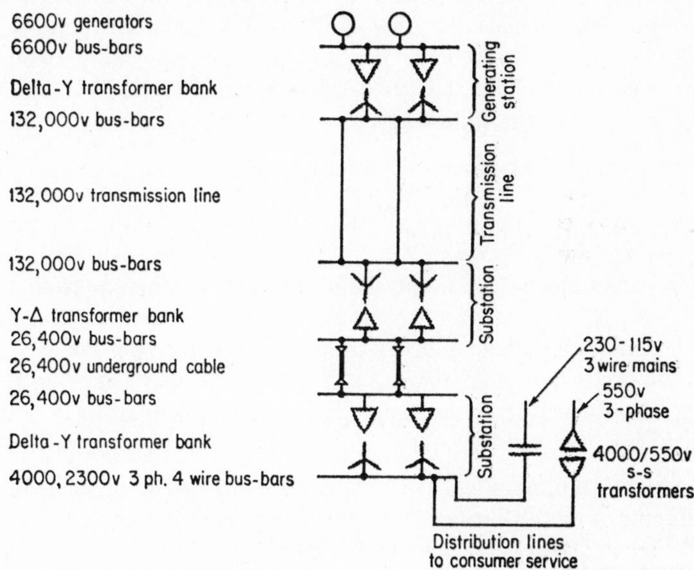

Fig. 17-1

stepped up to 132,000 volts. This is the transmission voltage as the result of Δ-Y transformer banks. Step-down transformers are then utilized to reduce voltage to 26,400 volts. By means of various substations throughout the city, the voltage is stepped down to 4,000 volts into a three-phase four-wire system where the voltage to the neutral is 2,310 volts. Here distribution is made to the consumers. At the point of consumption, step-down transformers are utilized for reduction to 230/115 volts.

17-2. Transmission-line Reactance for Single-phase System. It is a simple matter to make transmission-line calculations for d-c power, for only the resistance needs to be considered. However, for similar calculations using alternating current, we must consider the effect not only of the resistance, but, in addition, of the line reactance as well. When high voltages are used in cables and overhead lines, the capacitance between lines must be taken into consideration.

Q17-1. A single-phase transmission line is 50 miles long and consists of two 0000 solid conductors spaced 4 ft on centers. Find the inductance of the entire line and the reactance per conductor (*a*) at 25 cps and (*b*) at 60 cps. If a 250-amp line operating at 60 cps flows over this line, what is the total reactance drop?

ANSWER. Let the following nomenclature hold:

D = distance between conductors, in.
r = radius of conductor, in.
l = length of line, miles

The inductance per mile may be obtained from the equation

$$L' = 2\left(0.080 + 0.741 \times \log_{10}\frac{D}{r}\right) \times 10^{-3} \text{ henrys}$$

$$= 2\left(0.080 + 0.741 \times \log_{10}\frac{48}{0.230}\right) = 3.60 \text{ mh}$$

The total inductance is $L = 3.60 \times 50 =$ **180 mh,** or **90 mh** per conductor. Note that 0.230 in. is the radius of conductor.

The reactance per conductor at 25 cps is given by

$$X_1 = 2\pi \times 25\left(80 + 741 \log_{10}\frac{D}{r}\right) \times 10^{-6} \text{ ohms per mile}$$

$$= 2\pi \times 25\left(80 + 741 \log_{10}\frac{48}{0.230}\right) \times 10^{-6}$$

$$= 2\pi \times 25 \times 90 \times 10^{-3} = \textbf{14.1 ohms}$$

The reactance per conductor at 60 cps is

$$X_2 = 2\pi \times 60 \times 90 \times 10^{-3} = \textbf{33.8 ohms}$$

Total reactance drop with 250 amp 60 cps is

$$E = 33.8 \times 250 \times 2 \text{ conductors} = \textbf{16,900 volts}$$

Q17-2. A single-phase transmission line supplies 300 kw to a load at a voltage of 9,000 volts and a power factor of 80 per cent with current leading. What voltage is required back at the generator when the resistance and inductive reactance of the line are 12 and 13 ohms, respectively?

ANSWER. The following formula will be used.

$$E_g = \sqrt{(E_L \cos \theta_L + IR)^2 + (E_L \sin \theta_L - IX)^2}*$$
$$= \sqrt{(9{,}000 \times 0.8 + 41.7 \times 12)^2 + (9{,}000 \times 0.6 - 41.7 \times 13)^2}$$

from which E_g = **9,080 volts**

17-3. Transmission-line Reactance for Three-phase System. Here it is more convenient to consider the reactance of each conductor separately. If the spacing of the conductors is symmetrical, the flux produced by each conductor does not induce any emf in the circuit made up of the other two conductors.

Q17-3. A three-phase line consists of three 0000 solid conductors placed at the corners of an equilateral triangle, 4 ft on a side. Find the reactance drop per conductor per mile when a 25-cycle alternating current of 120 amp flows in the conductors.

ANSWER. From previous equations used in Q17-1 we see that

$$X = 2\pi \times 25(80 + 741 \log_{10} 48/0.23)10^{-6}$$
$$= 157(80 + 741 \times 2.32)10^{-6}$$
$$= 157 \times 1{,}800 \times 10^{-6} = \mathbf{0.282\ ohm}$$

17-4. Single-phase Line Calculations. When calculating the voltage drop in an a-c line, both resistance and reactance must be taken into consideration. Under these conditions the voltage

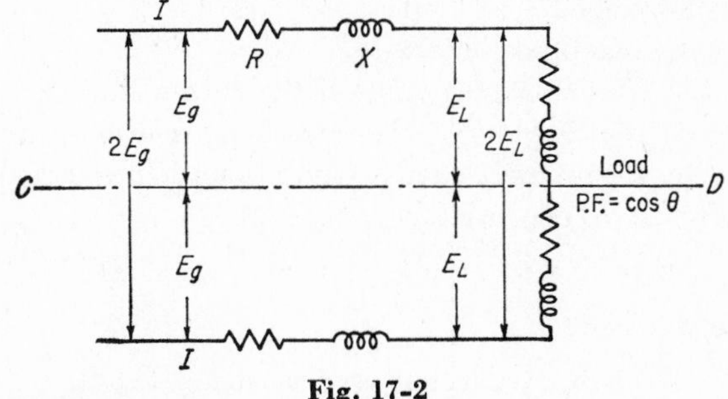

Fig. 17-2

* Use (+) when I lags E_L.

which supplies the resistance drop is in phase with current, while the voltage which supplies the reactance drop is in quadrature with the current and is leading. Figure 17-2 shows a single-phase transmission line. This system may be considered to have R ohms for resistance per wire and X ohms for reactance per wire. So as to simplify calculations, it is helpful to work to neutral in all cases. The load current is I amp at a power factor of $\cos \theta_L$. The total voltage at the load is $2E_L$. The voltage to neutral at the load is E_L. The sum voltage at the generator end is E_g.

Referring to Fig. 17-2, if this system is split longitudinally along line CD, we have two systems. One of these is shown in Fig. 17-3. Each of these transmits one-half total power. Also for this split system, the generator end and the load end voltages are one-half the voltage between the conductors. The voltage at each of the ends is then the voltage to neutral with the ground as the return conductor. If we can consider the load balanced, there will be no current flow in the return. Thus, we can consider the ground to have zero resistance and zero reactance. Now, the generator voltage E_g may be determined from the following

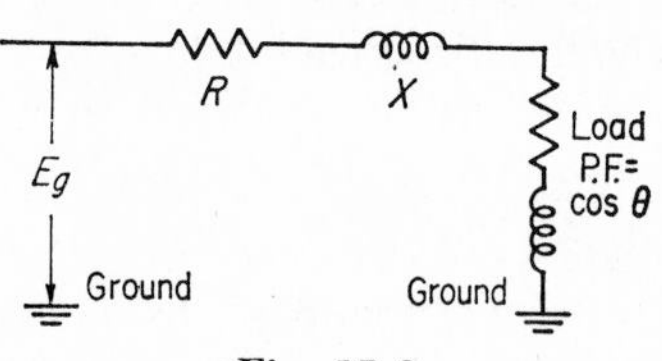

Fig. 17-3

$$E_g = \sqrt{(E_L \times \cos \theta_L + IR)^2 + (E_L \times \sin \theta_L \pm IX)^2} \quad (17\text{-}1)$$

where I = line current
R = line resistance, one wire
$\cos \theta_L$ = power factor at load
X = line reactance, one wire

Note that the IX has a *plus* value if the current lags E_L, a *minus* value if the current leads E_L. These voltage relations may also be determined by means of a complex notation E_L being taken along the axis of reals. Then, with current lagging

$$E_g = E_L + I(\cos \theta_L - j \sin \theta_L)(R + jX) \quad (17\text{-}2)$$

With current leading

$$E_g = E_L + I(\cos \theta_L + j \sin \theta_L)(R + jX) \quad (17\text{-}3)$$

The power factor at the sending or generator end of a single-phase transmission line is given by

$$\cos\theta_g = \frac{E_L \cos\theta_L + IR}{E_g} \tag{17-4}$$

The efficiency of the transmssion is

$$\eta = \frac{E_L I \cos\theta_L}{E_L I \cos\theta_L + I^2 R} \tag{17-5}$$

17-5. Three-phase Line Calculations. Assuming Y-connected load and sending (generator) point, we may set up Figs. 17-4 and

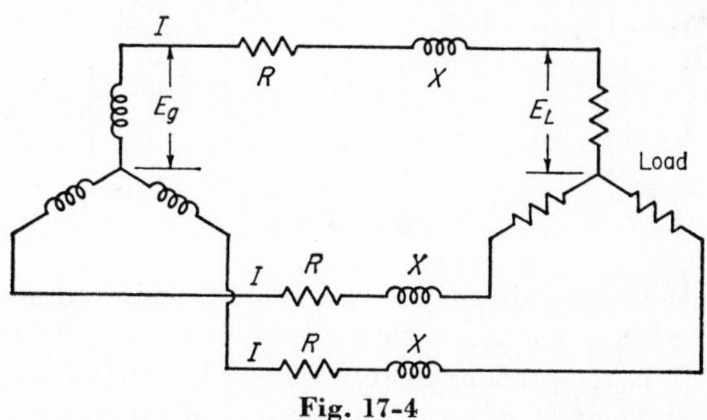

Fig. 17-4

17-5 as we did for single-phase systems. As we shall see, there is a more obvious advantage in working transmission-line problems to neutral with this system than with a single-phase system. In Fig. 17-4 each wire has a resistance of R ohms and a reactance of X ohms. The voltage to neutral at the generator end is E_g, that to neutral at load end is E_L.

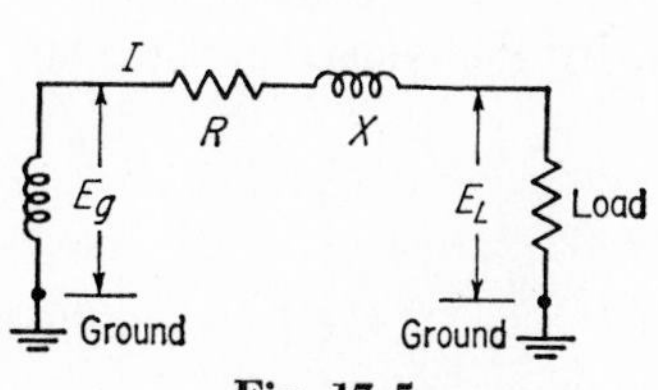

Fig. 17-5

To study the system and determine the line characteristics remove one phase (Fig. 17-5). Assuming balanced system load, relations in all three phases are the same, and so the results obtained for one phase hold true for all phases. Again there is no return current through the ground (neutral), and it displays no resistance or reactance under balanced load. The load need not be confined to a Y-connected load. This may be replaced with a Δ-connected load and no neutral. Procedure is to replace a Δ load with an equivalent Y load and to make computations for one phase alone.

17-6. Lines with Considerable Capacitance. Up to this point we have assumed negligible capacitance in transmission line study. Where lines are long and voltages high, the voltage rises from sending or generator end to load end because of the charging current. Such a system as described above may be shown as in Fig. 17-6.

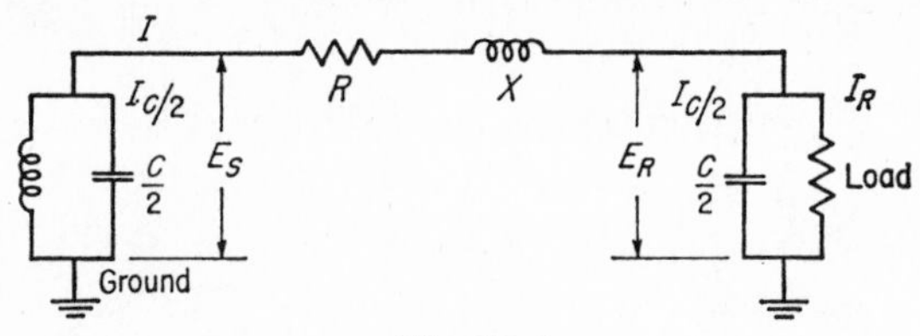

Fig. 17-6

Note that the total capacitance is divided in half and may be considered as being in parallel with the generator and with the load.

Q17-4. It is required to deliver 40,000 kw three-phase at 0.85 power factor, lagging current, at a distance of 140 miles, with a line loss not exceeding 10 per cent of the power delivered. The voltage at the load is 132,000 volts 60 cycles, and the lines are arranged at the apexes of an equilateral triangle, 13 ft on a side. Determine (*a*) the voltage between the conductors at the sending end; (*b*) the line regulation; (*c*) the total power supplied by the generating station; (*d*) the efficiency of the transmission.

ANSWER. The power per phase is

$$P = 40{,}000/3 = 13{,}330 \text{ kw}$$

The voltage at the load taken to neutral is

$$E_L = 132{,}000/\sqrt{3} = 76{,}200 \text{ volts}$$

The current per conductor at the load is

$$I_L = 13{,}330{,}000/(76{,}200 \times 0.85) = 206 \text{ amp}$$

The power loss per conductor is

$$13{,}330 \times 0.10 = 1{,}333 \text{ kw, or } 1{,}333{,}000 \text{ watts}$$

The conductor resistance per mile is

$$\frac{1{,}333{,}000/206^2}{140 \text{ miles}} = 0.224 \text{ ohm}$$

From the standard handbook we find that the conductor having the nearest resistance per mile is 250,000 cir mils. The corresponding resistance per mile for this conductor is 0.2278 ohm. Therefore, the total conductor resistance for the 140-mile run is

$$R = 140 \times 0.2278 = 31.9 \text{ ohms}$$

Again from a standard handbook we find that the reactance per conductor per mile for this size (250,000 cir mils) wire and 156-in. spacing is 0.804 ohm. Thus, the total reactance is

$$X = 140 \times 0.804 = 112.6 \text{ ohms}$$

The charging current at 60 cycles for this configuration and 100,000 volts to the neutral is from handbook data 0.534 amp per mile, and the total charging current of the line is

$$I_c = 0.534 \times (76{,}200/100{,}000) \times 140 = 57 \text{ amp}$$

But we have assumed that only one-half the charging current exists

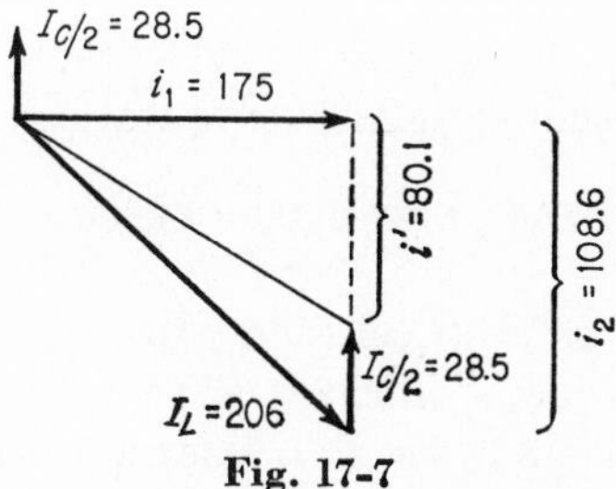

Fig. 17-7

at the load end so that the charging current flowing over the line is $57/2 = 28.5$.

In order to find the total line current, however, the 28.5 amp must be added vectorially to the 206 amp of the load current. The load current must, therefore, be resolved into an energy component $I_L \cos \theta_L = i_1 = 206 \times 0.85 = 175$ amp and a quadrature component $I_L \sin \theta_L = i_2 = 206 \times 0.527 = 108.6$ amp (see Fig. 17-7).

Since the quadrature component lags the load voltage by 90°, and the charging current leads the load voltage by 90°, the resultant quadrature component is $i' = 108.6 - 28.5 = 80.1$ amp. The total line current

$$I = \sqrt{175^2 + 80.1^2} = 192.5 \text{ amp}$$

If we let θ' be the angle between this current and the load-end voltage

$$\cos \theta' = 175/192.5 = 0.909 \qquad \theta' = 24.6°$$
$$\sin \theta' = 80.1/192.5 = 0.416$$

(*a*) The voltage to the neutral at the generator end

$$E_g = \sqrt{(76{,}200 \times 0.909 + 192.5 \times 31.9)^2 + (76{,}200 \times 0.416 + 192.5 \times 112.6)^2} = \mathbf{92{,}420 \text{ volts}}$$

The voltage between conductors $E = \sqrt{3} \times 92{,}420 =$ **160,000 volts**

(*b*) Line regulation $= (92{,}400 - 76{,}200)/76{,}200 = 0.213$, or **21.3 per cent**

(*c*) Line loss $P_c = 3 \times 192.5^2 \times 31.9 =$ **3,550,000 watts**

The total generator power required $P_g = 40{,}000 + 3{,}550 =$ **43,550 kilowatts**

(*d*) The line efficiency $\eta = 40{,}000/43{,}550 = 0.918$, or **91.8 per cent**

Q17-5. A source of electrical energy delivers 150 kw at 208 volts and a lagging power factor of 80 per cent to a three-wire three-phase line. This line delivers power to a balanced load at some distance. If each line wire has a resistance of 0.01 ohm, calculate the difference in potential between the line wires at the load.

ANSWER. Refer to Fig. 17-8. In order to solve this problem it may be assumed that the load is Y connected. From the general relation for a three-phase system, power $= EI \cos \theta$, from which we may write

$$I = \frac{150{,}000}{\sqrt{3}\,E\cos\theta} = \frac{150{,}000}{1.732 \times 208 \times 0.8} = 519 \text{ amp}$$

The angle θ may then be found to be 36°52′. The over-all circuit

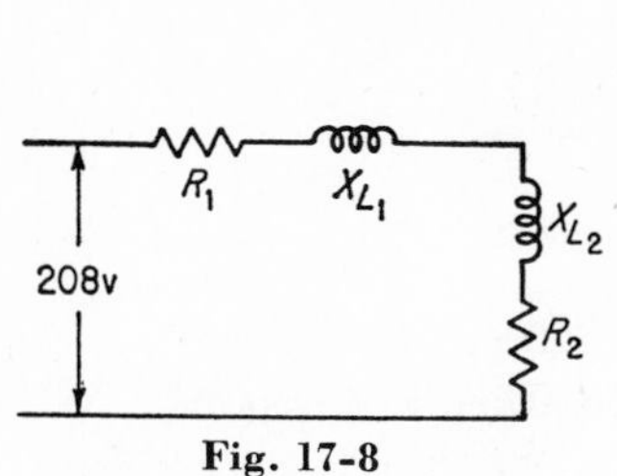

Fig. 17-8

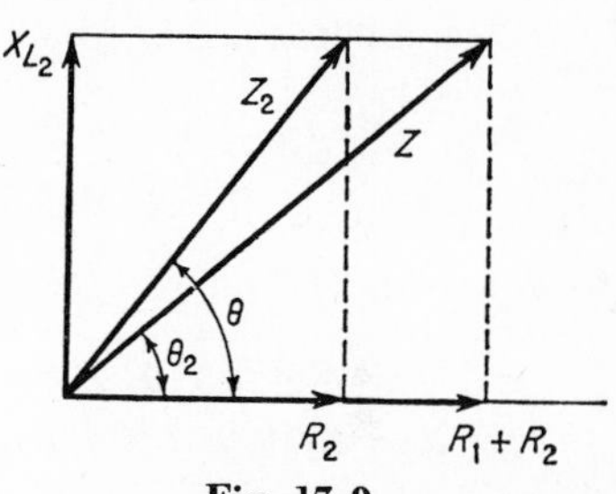

Fig. 17-9

impedance per phase is given by

$$Z = \frac{E}{I} = \frac{208/\sqrt{3}}{519} = 0.231 \text{ ohm}$$

Now referring to the circuit in Fig. 17-8 and the vector diagram in Fig. 17-9

$$R_1 + R_2 = Z\cos\theta$$

from which by rearrangement we obtain

$$R_2 = (0.231 \times 0.8) - 0.01 = 0.175 \text{ ohm}$$

Since the line inductive reactance X' is zero (balanced load), the load inductive reactance

$$X'' = Z\sin\theta = 0.231\sin\theta = 0.231 \times 0.6 = 0.139 \text{ ohm}$$

Load impedance Z'' is now found to be

$$Z'' = \sqrt{R_2^2 + X''^2} = \sqrt{0.175^2 + 0.139^2} = 0.223 \text{ ohm}$$

Phase voltage at load $= IZ'' = 519 \times 0.223 = 116$ volts

Potential difference at load $= 116 \times 1.732 =$ **201 volts**

Q17-6. A three-wire three-phase electric line delivers 500 kw at 6,000 volts and unity power factor. The line is two miles long and the cross section of each conductor is 0.023 sq in. What is the efficiency of operation of this line when the load is balanced, if the specific resistance of the conductors is 0.7 microhm per inch cube?

ANSWER. In using a three-wire system every effort is made to keep it balanced, since under such conditions the neutral carries no current. It is only when the system is unbalanced that the neutral is of some use and carries current. Then

$$R = K \times \frac{l}{A}$$

where $K = 0.71$

$$l = 5{,}280 \times 2 \times 12 = 126{,}720 \text{ in.}$$
$$R = 0.71 \times 126{,}720/(10^6 \times 0.023) = 3.86 \text{ ohms}$$
$$\text{Efficiency of operation} = (500{,}000/526{,}900) \times 100 = \mathbf{94.9\ per\ cent}$$

The efficiency was determined on the basis of output divided by output plus losses. The losses were determined as follows.

$$I = 500{,}000/(\sqrt{3} \times 6{,}000 \times 1) = 48 \text{ amp}$$
$$\text{Power lost in line per phase} = I^2R = 48^2 \times 3.86 = 8{,}900 \text{ watts}$$

For the three wires the total losses = $3 \times 8{,}900 = 26{,}900$ watts (losses).

Q17-7. A 460-volt three-phase circuit has a load of 75 kw. The length of the circuit is 400 ft and the three conductors are size 0 wire. The present power factor is 55 per cent which is to be raised to 0.90. What will be the saving in line loss?

ANSWER.

$$\text{Present kva} = \frac{\text{kw}}{\text{pf}} = \frac{75}{0.55} = 136$$
$$\text{Present amp} = (136 \times 1{,}000)/(1.732 \times 460) = 171 \text{ amp}$$
$$\text{New kva} = 75/0.9 = 83$$
$$\text{New amp} = (83 \times 1{,}000)/(1.732 \times 460) = 104 \text{ amp}$$

Resistance of copper wire 0 = 0.104 ohm per 1,000 ft. And since there are three conductors,

$$3 \times 0.104 = 0.312 \text{ ohm for three conductors @ 1,000 ft}$$

$$\text{Present kw loss in wire} = 171^2 \times 0.312 \times 400/1{,}000 \times 1/1{,}000 = 3.65 \text{ kw}$$

$$\text{New kw loss in wire} = 104^2 \times 0.312 \times 400/1{,}000 \times 1/1{,}000 = 1.35 \text{ kw}$$

$$\text{Saving in kw} = 3.65 - 1.35 = \mathbf{2.30\ kw}$$

Q17-8. A certain line, which has a resistance and reactance of 0.1 ohm and 0.3 ohm respectively, supplies a load at one end and is connected to a 440-volt generator at the other. The load consists of two parts, one drawing 100 amp at 100 per cent power factor

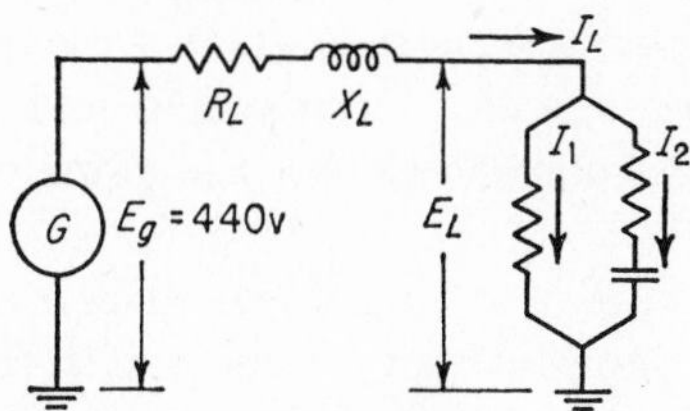

Fig. 17-10

and the other taking 75 amp at 80 per cent power factor, leading. Calculate the voltage at the load.

ANSWER. Refer to Fig. 17-10. From Kirchhoff's laws we see that

$$E_g = E_L + I_L Z_L \qquad I_L = I_1 + I_2$$

Using complex notation and E_L as reference axis, we have

$$\begin{aligned} I_1 &= 100 + j0 \\ I_2 &= 75(\cos\theta + j\sin\theta) = 75(0.8 + j0.6) \\ I_L &= (100 + j0) + (60 + j45) = 160 + j45 \\ E_g &= E_L + (160 + j45)(0.1 + j0.3) \\ &= (E_L + 2.5) + j52.5 \end{aligned}$$

Now by simple substitution and the usual approach as "sum of the squares"

$$440 = \sqrt{(E_L + 2.5)^2 + 52.5^2}$$

from which we obtain

$$E_L^2 + 5E_L - 190{,}900 = 0$$

or

$$E_L = \mathbf{435\ volts}$$

Chapter 18

LIGHTING

18-1. General. Today there is a consensus of opinion in industry as to what constitutes good lighting, the bible in this respect being the Illuminating Engineering Society (IES) recommendations. For best practice consideration is first given to good general lighting, with supplementary lighting for machine and work areas. Also included is a careful consideration of the paint colors for walls and equipment.

The trend is toward low brightness over larger lighting areas. Design with use of high-intensity sources is in the past. In offices the use of troffers and luminous ceilings is finding increasing favor.

Panel boards are often located in the webs of columns with the control wiring in the duct instead of in the conduit. According to the NEC, lighting-circuit voltages are limited to 150 volts to ground. However, in industrial plant work it is acceptable to use 300 volts with a service-approved lamp holder. Switches cannot be installed as a part of these fixtures because of the higher voltages. In addition, such fixtures are not permitted to be installed below 8 ft from the floor.

For supply the trend is to feed from the same feeds as the production machines. On the other hand, welding outlets are no longer being made a part of the machine circuit, but are being isolated from lighting.

This trend to use higher voltages for lighting appears to be centered around 277 volts. The fixtures are fed between the three-phase lines and the neutral of the 480-volt secondary distribution system. Dry-type transformers are also used by tapping into a 440-volt secondary distribution system for a secondary of 120/240 or 120/208 three-phase four-wire arrangement. The big advantage is a readily available source of 120 volts for small tools, etc.

Hazardous locations may be found in any plant, not only in oil refineries or chemical and powder plants. Thus, the lighting installation must be considered in this way, making use of properly applied fixtures for the job.

18-2. Lighting Design. There are many factors that enter the design considerations of any lighting installation. The one important item to be given considerable treatment is the proper quantity of illumination. This is the order in which such designs may be made.

(*a*) Analyze the seeing task
(*b*) Determine the particular illumination requirements
(*c*) Selection of lighting equipment
(*d*) Mathematically design the system

18-3. Lumen Method of Calculation. In this method there are six steps.

(*a*) *Determine the required level of illumination.* Minimum values may be obtained from handbooks and manuals. Higher levels may be needed, but experience is necessary.

(*b*) *Selection of system and luminaires.* There are several classifications for lighting systems. Westinghouse Electric Corporation lists these as direct, semidirect, general diffuse or direct-indirect, semidirect, and indirect.

(*c*) *Coefficient of utilization.* This is simply the ratio of lumens reaching the work divided by the total lumens generated by the lights. This factor takes into account the efficiency of illumination as to distribution, mounting heights of luminaire, room proportions, and coloring effect of walls and ceiling. The student is referred to the IES handbook and manuals published by such companies as Westinghouse and General Electric.

(*d*) *Maintenance factor.* This is estimated by a study of the conditions under which the system will operate, the atmospheric conditions, the frequency of the cleaning, and the depreciation of the lighting effect.

(*e*) *Calculation of the number of lamps and luminaires required.* The foot-candles (lumens per square foot reaching the working plane) are equal to the total lumens generated by the lamps, multiplied by the coefficient of utilization and the maintenance factor, all divided by the room floor area in square feet.

(*f*) *Determination of the location of the luminaires.* According to the Westinghouse manual on lighting for general lighting, luminaires are located symmetrically in each bay or pair of bays, although this practice is often modified by the position of the previous outlets, the general architecture of the area, the type of luminaire, or the nature and location of the seeing tasks to be performed.

18-4. Point-by-point Method. This method for determining lighting design is more accurate than the lumen method, but it is more complex. It will not be discussed further.

Q18-1. Specify the lighting installation and the wall and ceiling painting or treatment for a general office in which close work, computing, and designing are to be performed. The room is 30 ft long by 20 ft wide. The ceiling is 12 ft from the floor and the north wall (30 ft long) has windows that occupy 60 per cent of the wall area. There are no columns or beams in the office area.

ANSWER. Refer to the preceding discussion and with the use of the Westinghouse Handbook, P5-2, solve as follows.

(*a*) *Determine the required level of illumination.* According to this handbook use 50 ft-c

(*b*) *Select the lighting system and luminaires.* Use semiindirect four-lamp, 40-watt fluorescent white tubes No. 34 mounted approximately 24 in. below the ceiling and 10 ft above the floor.

(*c*) *Determine the coefficient of utilization.* Again refer to the Westinghouse Handbook. The room index is F, and with a white ceiling the average reflection factor is 0.88. Allowing for depreciation use the factor is 0.75. With green (light shade) or buff walls the average reflection factor is 0.70. Allowing for depreciation use the factor is 0.50. From a table using 0.75 for the ceiling and 0.50 for the walls the coefficient of utilization is found to be 0.43.

(*d*) *Maintenance factor.* In an office this should be good. Use 0.65.

(*e*). *Calculate the number of lamps and luminaires required.* The number of lamps is found by

$$\frac{\text{ft-c} \times \text{floor area (sq ft)}}{\text{lumens per lamp} \times \text{coef. of utilization} \times \text{maintenance factor}}$$

$$\frac{50 \times 30 \times 20}{2{,}480 \times 0.43 \times 0.65} = \frac{30{,}000}{692} = 43.3$$

Use 48 lamps

$$\text{No. of luminaires} = \frac{\text{number of lamps}}{\text{lamps per luminaire}} = \frac{48}{4} = 12 \text{ luminaires}$$

Specify the following: the ceiling should be white (reflective factor new, 0.88) and the walls, upper third or half, very light green or buff (reflective factor new, 0.75). The windows should be equipped with venetian blinds to let daylight in on clear days and reduce the transmittal loss on dark days, thus keeping in the reflected light.

(*f*) *Determine the location of the luminaires.* For general lighting luminaires are usually located symmetrically in each bay or pair of bays. These positions are often, however, modified by the position of previous outlets, etc. Where working positions are fixed, it is sometimes desirable to position the luminaires with reference to the particular areas where high light intensities are required. In schools, offices, or other areas where work is performed throughout the room, a high degree of uniformity is desirable. In such cases the spacing from the wall is normally one-half the distance between the luminaires. If desks or work benches are next to the walls, one-third the luminaire spacing is customary. In order to provide even distribution of illumination for an area, it is necessary to observe certain limitations of "spacing-to-mounting height" ratios. These values are indicated for the various types of luminaires in *coefficient of utilization tables.* The spacings shown are a maximum, and closer spacings may be necessary in order to obtain the required levels of illumination. Industrial high bay units and one-or-two lamp fluorescent luminaires, in particular, commonly require closer spacings than those indicated in the table. Where a large number of fluorescent luminaires are to be installed, continuous-row mounting has the advantage of simplifying the wiring and reducing the apparent number of sources.

Careful consideration should be given to the ceiling brightness which may result when high-wattage lamps are used in luminaires with short suspension hangers. This is especially true when indirect luminaires are installed in relatively low ceiling areas. For luminaire arrangement see Fig. 18-1.

Q18-2. A drafting room measuring 100 by 20 ft with an 11-ft ceiling is to be lighted by fluorescent tubes. Specify the average intensity of illumination on the working plane desirable for such a room and design the illumination to fit your specification. State the power and luminous rating, distribution, and hanging of the luminaires.

ANSWER. Refer to the previous problem and solution. Make use of catalogues and handbooks as necessary. An excellent brochure on lighting design may be obtained from the manufacturers

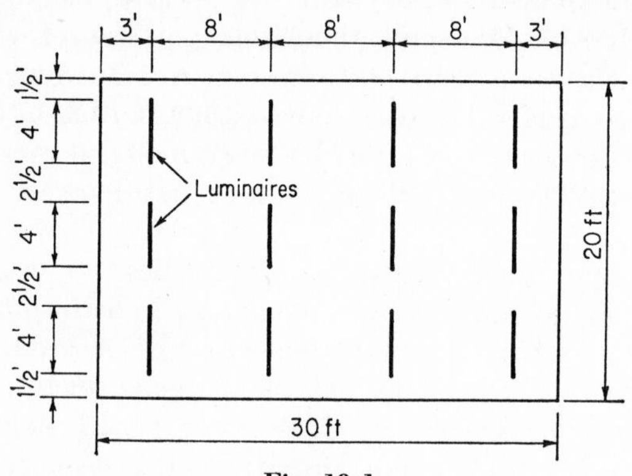

Fig. 18-1

of lighting fixtures and lamps. Below are listed several brochures available from General Electric, Nela Park, Cleveland 12, Ohio.

25 Industrial Lighting Solutions
Lighting for Stores
Lamp Information Bulletin
Essential Data for General Lighting Design
Foot-candles in Modern Lighting Practice
General Electric Lumen Counter

Reference should also be made to the General Electric Bulletin LS-145 entitled "22 Office and Drafting Room Lighting Solutions."

(*a*) *Recommended illumination* is 75 ft-c in service. Place the drafting boards parallel to the run of lights, so that they may be viewed crosswise to the axis of the luminaires. The fixtures must be suspended. Paint the walls light green and the ceiling white.

(*b*) *Select the lighting system and the luminaires.* Use direct-indirect luminaire with opaque side panels and louvers. These are to have two-lamp 40-watt fluorescent white tubes No. 34 mounted about 24 in. from the ceiling.

(c) *Determine the coefficient of utilization.* Use room index *E*. With a white ceiling the average reflection factor is 0.88. The depreciation use factor is 0.75. For light green walls the reflection average is 0.70. From tables the coefficient of utilization is found to be 0.43.

(d) *Maintenance factor.* Use recommended 0.75.

(e) *Number of lamps and luminaires.*

$$\text{No. of lamps} = (75 \times 20 \times 100)/(2{,}500 \times 0.43 \times 0.75) = 189 \text{ lamps}$$

$$\text{No. of luminaires} = {}^{189}\!/_{2} = 94.5 \qquad \text{Try 95.}$$

See Fig. 18-2 and use 96 luminaires. In addition install venetian blinds. Arrange the luminaires in a continuous trough.

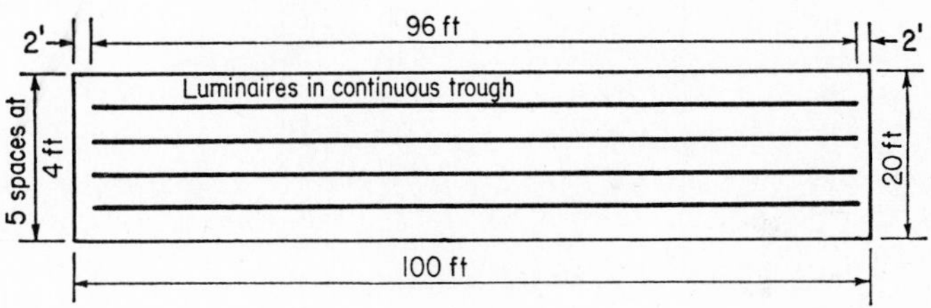

Fig. 18-2

Q18-3. Two sources of supply of 100-watt incandescent lamps are available to a large purchaser. The first offers lamps with an efficiency of 16 lumens per watt and a life of 750 hr, at a cost of 20 cents each installed. The second offers lamps with an efficiency of 13 lumens per watt and a life of 1,200 hr, at 19 cents each installed. The cost of electricity is 3 cents per kwhr. Calculate which supply of lamps is more economical in the use of electric energy, and estimate the saving.

ANSWER. The basis is 1,000 lumens. The question asks which supply is more economical in the use of electric energy. Therefore, ignoring life and installed cost

	A	*B*
Watts required per 1,000 lumens......	1,000/16 = 62.5	1,000/13 = 76.9
Cost per 1,000 lumens per hour..........	62.5 × 0.03/1,000 = \$0.001875	76.9 × 0.03/1,000 = \$0.00231

Therefore, type *A* is more economical and will save per lumen per hour

$$0.00231 - 0.001875 = \mathbf{\$0.435 \times 10^{-6}}$$

Q18-4. When tested in a distribution photometer, a certain luminaire gave the following results:

Lower hemisphere		Upper hemisphere	
Degrees	Candle power	Degrees	Candle power
0 (nadir)	100	90	59 (check point)
10	98	100	56
20	95	110	58
30	89	120	64
40	82	130	70
50	77	140	74
60	68	150	80
70	64	160	81
80	60	170	82
90	58	180	80

What is the lumen output of the luminaire?

ANSWER.

Mid-zone angle	Candle power	Zonal constant	Zonal lumens (candle power × zonal constant)
0	100		
5	99	0.0954	9.4
15	96.5	0.2835	27.3
25	92	0.4629	42.6
35	85.5	0.6282	53.3
45	79.5	0.7744	61.4
55	73	0.8972	65.5
65	65.5	0.9926	65.0
75	62	1.0579	65.5
85	59	1.0911	64.3
95	57	1.0911	62.2
105	56.5	1.0579	59.6
115	60	0.9926	59.6
125	67	0.8972	60.1
135	71.5	0.7744	55
145	77.5	0.6282	48.7
155	80.5	0.4629	37.3
165	82	0.2835	23.3
175	81	0.0954	7.72
Total..........	.....	......	868.12

Mean spherical candle power $868.12/4\pi = \mathbf{69}$

Q18-5. A parking lot in an industrial plant is 400 ft long and 200 ft wide. Choose the type and location of the lighting for this lot. Lighting may be provided from sundown to sunrise.

ANSWER. Three types of lamps are popular for outdoor lighting: mercury-vapor, fluorescent, and filament. The filament-type lamp is usually more economical for installations using less than 1,000 lighting hours per year, i.e., under 3 to 4 hr per night. Since this installation requires lighting from sundown to sunrise, filament-type lamps would not be economical because they would be used more than 4 hr per night. Thus, the lighting should be provided by either a mercury-vapor or fluorescent lamp. Select mercury-vapor floodlight-type lamps because they provide economical lighting with little maintenance. From manufacturer's recommendations and

Illuminating Engineering Society studies, the recommended lighting level for parking areas is 0.5 to 2.0 ft-c. Use a level of 2.0 ft-c for this area. And the watts per square foot is 0.10. For lighting-fixture locations use center poles with multiluminaire mountings. See Fig. 18-3.

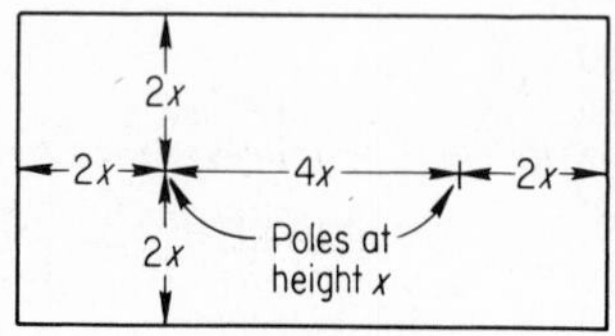

Fig. 18-3

Now compute the number of lights required. Use the relation: number of floodlights required = (area lighted, sq ft) (recommended ft-c)/(floodlight lumen rating) (MF), where MF = maintenance factor = 0.165 for open floodlights and 0.75 for enclosed floodlights. Typical mercury-vapor floodlight lumen ratings range from 4,200 lumens for a 500-watt lamp to 18,000 lumens for a 1,500-watt lamp. Assume that 1,000-watt, 9,500-lumen, enclosed, clear, general-

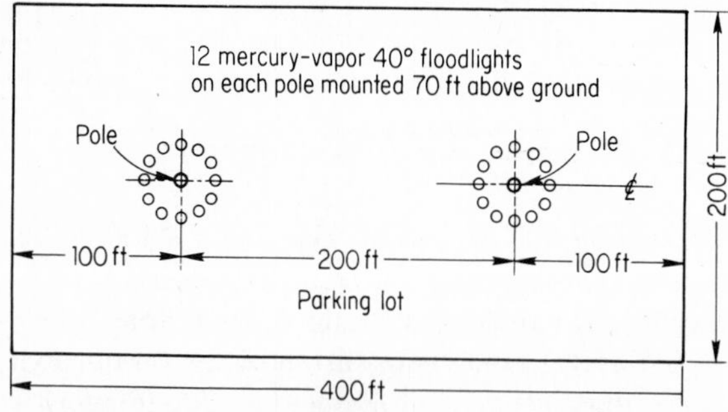

Fig. 18-4

service floodlights are used. Then, the number of floodlights required = $(400 \times 200)(2.0)/(9{,}500)(0.75) = 22.5$, say 24 lamps to obtain an even number of lamps to be divided between two supporting poles.

Lamp arrangement is recommended to be 24 on a 2-pole arrangement. This would lend itself to the mounting of 12 lamps on each pole. Now refer to a lamp manufacturer's lamp catalog. These data show that a 40° beam floodlight mounted 55 ft above the ground will illuminate an area 104 ft long and 65 ft wide, or an elliptical area of 2,650 sq ft. Hence, 24 lamps will illuminate an area of $24 \times 2{,}650 = 63{,}600$ sq ft. The area of the parking lot is $400 \times 200 = 80{,}000$ sq ft. Thus, the lamps will not illuminate the entire parking light.

Raising the lamps to 70 ft above the ground provides an illuminated area of 4,700 sq ft per lamp, or a total of $24 \times 4{,}700 = 112{,}800$ sq ft. Also, the lamps will be within twice their height of $2 \times 70 = 140$ ft from the area perimeter because they are mounted 100 ft from the perimeter, Fig. 18-4. Since the area covered by the floodlights exceeds the ground area, the beams will overlap. This is a desirable condition because it provides more uniform illumination.

Chapter 19

ELECTRONIC TUBES AND CIRCUITS

19-1. Short History. The electronic tube is an adaptation of the incandescent lamp which was invented by Thomas Edison in about 1883. It is based on the principle of utilizing the emission of electrons from a heated wire. Edison, although discovering this phenomenon, found no practical use for it, and it was Dr. J. A. Fleming of London who in 1904 discovered its rectifying characteristics and its adaptability to the reception of radiotelegraphic signals. For this reason, the first electronic (vacuum) tubes were called Fleming valves. These were primarily used as detectors. The Fleming valve consisted of a filament and a plate. Modern electronic tubes contain, in addition to a grid added by Dr. Lee De Forest in 1909, other elements. The more modern name given to the first electronic tube consisting of a filament and a plate is *diode*. Today we have the three-element tube with a cathode (filament), an anode (plate), and a grid bearing the name *triode*. In addition, we have the *tetrode* (four-element tube) and the five-element tube known as the *pentode*.

Early in the history of electronic tube development and application, tubes were designed for general service. At that time the triode was used as a radio-frequency amplifier, an intermediate-frequency amplifier, an audio-frequency amplifier, an oscillator, or a detector. It was soon apparent that such diversity could not be expected from one tube design with equal advantage and with the best possible results. Present trends of tube design are the development of specialty types. Today we have multielectrode tubes, some with seven electrodes, twin diodes, twin-diode triodes, pentagrid-converter types, etc.

19-2. Edison Effect. In the course of his experiments with electric lamps, Edison discovered that if a cold metal plate were placed in an electric lamp bulb, a current could be passed through the evacuated tube between the plate and the heated element, or electrode, when a voltage was applied between them. Edison discovered that the flow of current was unidirectional from the plate to the cathode only. This is known as rectification. No matter how high he applied the voltages, Edison found that current would not flow in the opposite direction. He also discovered that the current flow took place only when the plate was made positive. When the current through the filament was turned off and the filament cooled, the flow of current through the tube ceased. Also, when the current flow through the filament was varied, there resulted changes in the current flow between the filament and the plate (refer to Fig. 19-1).

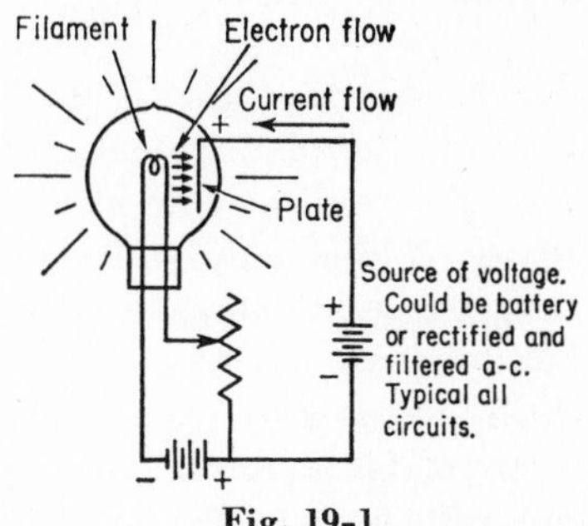

Fig. 19-1

19-3. Electron Theory of Vacuum Tubes. It is well known that when any piece of metal is heated to a white heat it will emit electrons. Electrons carry a negative charge while the heated body maintains a positive charge, caused by the loss of electrons. When so freed from the heated metal or wire (filament), the electrons leave the cathode surface and form an invisible cloud in the space around it. The negative electrons thrown off are immediately attracted back to the cathode by its positive charge and dance off into space again. This action takes place at terrific speeds. Any positive electric potential within the evacuated envelope of an electron tube offers a strong attraction to the electrons. Such a positive electric potential can be supplied by an anode (positive electrode) located within the tube in proximity to the cathode. In our discussion note the interchangeability of the terms filament and cathode, both meaning the same thing.

When the heated element is a filament enclosed in an airtight globe, the space surrounding the incandescent filament is charged with negative electrons as long as the filament remains heated; the greater the heat of the filament, the stronger will be the negative space charge within the bulb. Now, if a second element is inserted into this negatively charged space, and at the same time it is connected in series with the filament circuit and given a positive charge from a d-c source, a current will flow *from* the plate to the filament. However, this is not the manner in which the electrons flow as mentioned previously.

Because the plate is more positive than the filament, the electrons will flow from the filament to the plate. If the plate potential is reversed, from positive to negative, there will result no current flow through the tube. The negative charge on the plate will repel (unlike electric charges attract, like charges repel) the negative electrons, and no current can flow in either direction.

An increase in filament voltage (which increases the emission of negative electrons) will increase the flow of plate current until a *saturation* point is reached. At this point, the filament temperature is such that no more electrons can be thrown off per second and the space within the tube has reached a state of charge which neutralizes any increased attraction for the electrons. Beyond this saturation point an increase in the filament voltage has no effect, except to destroy the filament itself.

With a fixed charge of the filament, an increased plate voltage will also increase the flow of plate current up to a point of saturation. Beyond this point no increased plate current can be obtained with increased plate voltage.

Although there is an electron flow from the filament to the plate, there is a current flow from the plate to the filament as measured by an ammeter. This is due to the fact that an ammeter shows the direction in which the positive charge of electricity flows. And since electrons are negative, their direction of flow is not registered by an ammeter.

19-4. Thermionic Emission. The evaporation of electrons described above is known as *thermionic emission.* Any device which utilizes such an effect is a *thermionic tube.* There is no gaseous conduction, because there is no gas present, nor does it play any part in this device. In fact, the more completely a tube is evacuated, the more efficiently the device works. Thermionic tubes are thus freed as far as possible from gas; that is, they operate in a *hard* vacuum.

Such a tube is also known to be a *hard tube.* A tube that has leaked gas or contains a gas is known as a *soft tube.*

19-5. Two-electrode Tube As a Rectifier. When a two-electrode tube is connected in an a-c circuit, it functions as a *rectifier.* Alternations in the opposite direction are cut off. The well-known tungar rectifier, used for charging storage batteries from an a-c supply, utilized a tube of this type. The current resulting from such a hookup is pulsating but unidirectional. This is known as *half-wave rectification.* This rectifying action is identical in principle with that of the mercury-arc rectifier and the tungar rectifier. Fleming, in 1905, was the first to recognize this rectifying property of a two-electrode tube. Fleming obtained a patent on its use as a detector (rectifier) of high-frequency oscillations, which became one of the fundamental patents in electron-tube development.

By using two rectifiers *full-wave rectification* is obtained. Condensers connected across the load give good smoothing out of ripples.

19-6. Triodes. To control the space charge of the electron tube, a third electrode, called the *grid,* is placed between the cathode and the plate. The grid consists of a coiled or crimped wire which is placed in the tube in such a position that it is directly in the path of the flow of electrons from the filament to the plate. If this grid is energized with a third source of current, the polarity of this charge on the grid will have a marked effect on the space charge of the tube. The purpose of the grid is to control the flow of plate current (see Fig. 19-2).

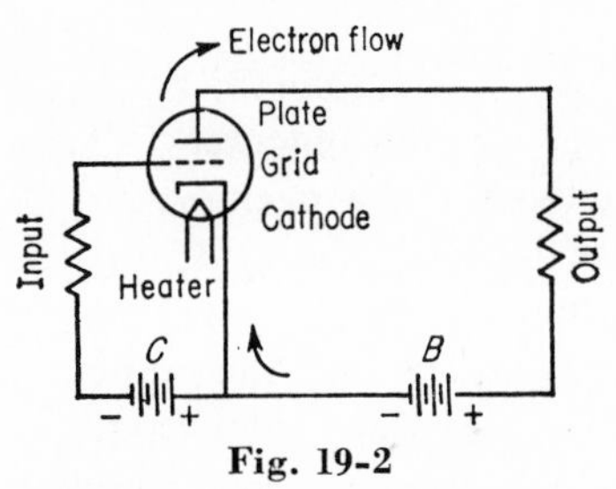

Fig. 19-2

When the grid has a positive charge, the positive attraction is increased and the electron flow away from the filament will be increased with a resulting increased current flow.

If given a negative charge, the grid will repel the negative electrons emitted from the filament, allowing only a few to reach the plate through the openings in the grid structure. This reduces the plate current flow to below normal. If given a negative charge to the saturation point, the grid will completely neutralize the plate charge and no current will flow, although the filament and plate voltages are the same as before. This is often called the "trigger action" of the grid. When a tube is used as an *amplifier*, this negative charge is used to reduce the plate current flow.

19-7. The Triode As an Amplifier. With a negative voltage (bias) on the grid, we saw that the plate current was less than would flow if the grid had no potential at all. If this negative potential is decreased slightly, the plate current increases because the opposition of the grid to the flow of electrons is decreased. Thus, we can see that we can change the plate current by merely changing the grid potential. This effect is more pronounced than by changing the plate current and with less potential requirements on the grid. Then a small change in the grid potential produces the same change in the current through the tube as a large change in plate current. Moreover, as long as the grid voltage is kept negative, no power is required in the grid circuit in order to control a large change of power in the plate circuit. Any device which requires a small amount of power to control a large amount of power is called an amplifier.

The grid, plate, and cathode of a triode form an electrostatic system, each electrode acting as one plate of a small condenser. These capacitances are given the name of interelectrode capacitances. Of utmost importance is the capacitance between the grid and the plate. In high-gain radio-frequency amplifier circuits, this acts in many cases to produce undesirable coupling between the input circuit, the circuit between the grid and the plate, and the output circuit, the circuit between the plate and the cathode. Such coupling is undesirable in an amplifier because it may cause instability and unsatisfactory performance. Here is the area where the tetrode finds application.

19-8. Tetrode. By the addition of a screen grid the triode becomes a tetrode, or four-element tube. The screen grid acts as an electrode shield between grid 1 and the plate, thus reducing the grid-to-plate capacitance. The added screen grid has another feature which makes it highly desirable. Over a certain range it makes the plate current practically independent of the plate voltage. Because the plate current in a screen-grid tube is largely independent of the plate voltage, much higher amplification can be realized with a tetrode than with a triode. Feedback and a tendency toward instability are markedly reduced.

Q19-1. (*a*) What are the three principal emitter materials used on cathodes of vacuum tubes? (*b*) What emitter material is used with the indirectly heated type of cathode? (*c*) What two factors limit the electron emission in a vacuum tube?

ANSWER.

(*a*) Pure metals: tungsten, molybdenum, tantalum, thorium.
Atomic film: thoriated tungsten.
Oxide coated: barium, strontium-oxide coating.

(*b*) The oxide-coated type of cathode is indirectly heated by a pure tungsten filament running through the inside of a hollow cylinder. The cylinder is the actual emitter, which in this case is indirectly heated to approximately 1560°F by the filament.

(*c*) The temperature of the emitter and the electric field. The vacuum tube depends for its action on a stream of electrons that act as current carriers. To produce this stream of electrons the special metals listed above must be present in every tube. But at room temperatures the "free" electrons in the metallic emitter cannot leave its surface because of certain restraining forces that act as a barrier. To leave the surface of the emitter the electrons must have sufficient energy imparted to them from some external source of energy. By heating the cathode the electrons will attain sufficient speed and energy to escape from the surface of the emitter. Alternating or direct current may be used for either method of heating. By applying a positive charge to the plate of a diode, say, an electric field is established between the cathode and the plate that attracts electrons emitted from the cathode.

19-9. Pentode. When a *suppressor grid* is added to the tetrode, there is an effect that provides for greater improvement of operation as a receiving tube. Anode characteristics of a typical pentode show the anode current to be practically independent of anode voltage. The undesirable characteristics of a tetrode are thus eliminated. In receivers there is much saving on power requirements.

For a more complete discussion on tube types the student is

referred to the RCA Receiving Tube Manual, Technical Series RC-18.

19-10. Electron-tube Characteristics. These may be shown in curve form or as tabulated data. When given in curve form the information may be used to determine tube performance and further tube characteristics.

Tube characteristics are either static or dynamic. Dynamic characteristics are of greater value for they are indicative of the performance capabilities of a tube under actual working conditions.

Static characteristics may be explained by plate-characteristic curves and transfer- (mutual) characteristic curves. *Plate-characteristic* curves are obtained by varying the plate voltage and measuring

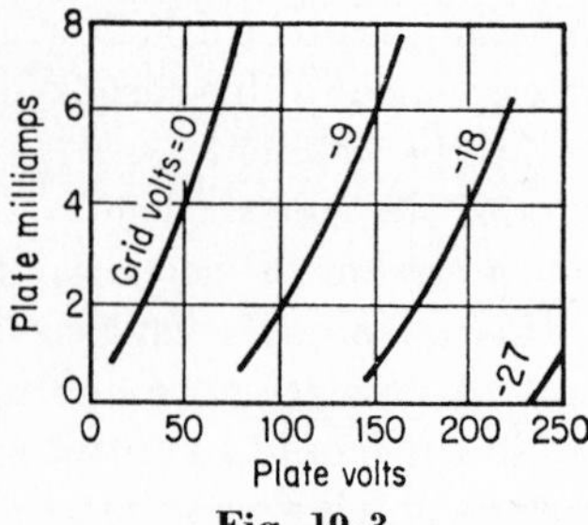

Fig. 19-3

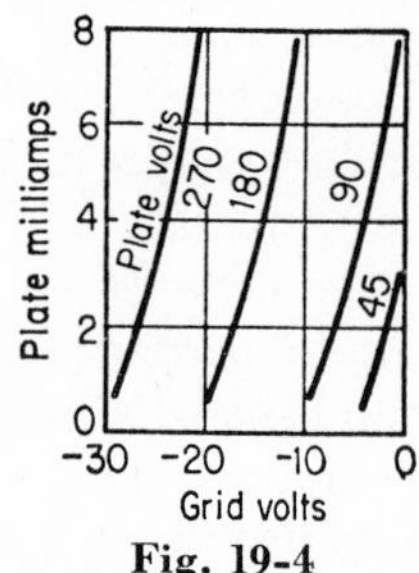

Fig. 19-4

the plate current for different grid bias voltages. The *transfer-characteristic* curve is obtained by varying the grid bias voltage and measuring the plate current for different plate voltages. Figure 19-3 shows a plate-characteristic curve and Fig. 19-4 shows the transfer-characteristic family of curves for the same tube.

What are the dynamic characteristics of a tube? These include: amplification factor, plate resistance, control-grid–plate transconductance, and certain detector characteristics. All may be shown by curves.

19-11. Amplification Factor. This is known as μ and is the ratio of the change in plate voltage to a change in control-electrode voltage in the opposite direction. It is a measure of the relative effectiveness of the control grid in overcoming the electrostatic field produced by the plate. By comparing the plate-voltage change to the grid-voltage change for the same change in plate current, we can determine their effectiveness relatively. This is amplification factor.

As we mentioned before, the operation of a triode may best be understood from a study of its characteristics curves. These are curves which show how the plate current of the tube varies as the plate voltage and grid voltage are varied. Figure 19-5 shows the circuit connections which may be used to obtain these characteristics.

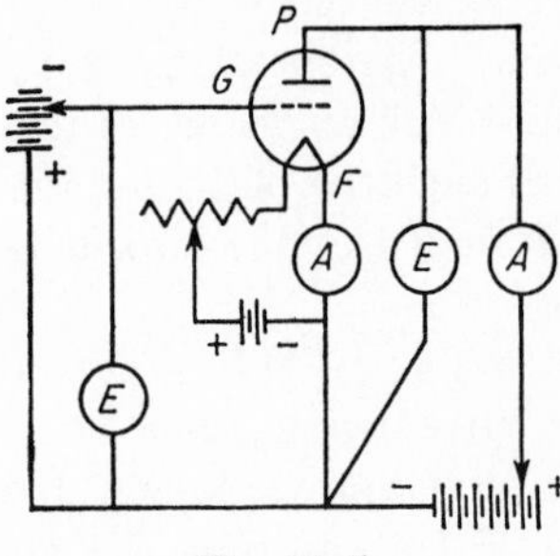

Fig. 19-5

Since there are three quantities that will vary, we can best show tube characteristics from more than one curve. If we keep the grid voltage constant while we vary the others over their entire range, and repeat the procedure for a number of different values of the fixed quantity, we obtain a family of curves as shown in Fig. 19-6 (see also Fig. 19-3).

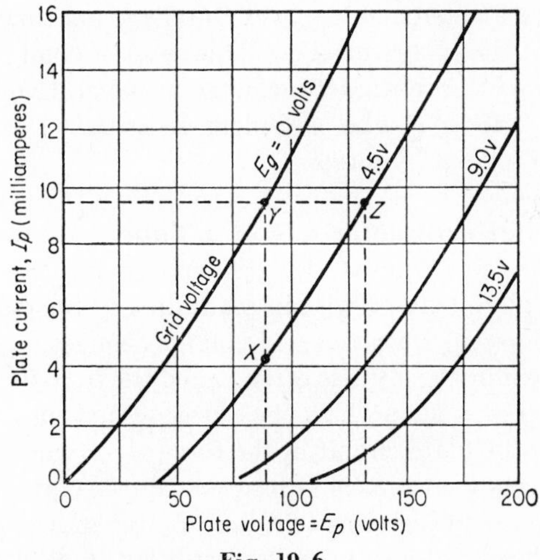

Fig. 19-6

In Fig. 19-6 the curve at the left is taken with no voltage on the grid and is, therefore, equivalent to operating the tube as a diode. Now let us suppose we have 90 volts impressed on the plate and zero voltage on the grid, or zero grid bias. We see that this will produce a current of 9.4 ma in the plate circuit, operating at point Y.

If we now impress a voltage on the grid of minus (−)4.5 volts, keeping plate voltage constant, we must go to the curve corresponding to minus (−)4.5 volts on the grid. The operating point, therefore, moves from Y to X, and the plate current decreases to 4 ma. Moving along the line YX does not change the plate voltage, so that the entire change of plate current from 9.4 to 4 ma is due to the change in grid voltage or grid bias.

If we hold the grid voltage constant at minus (−)4.5 volts and increase the plate voltage until the plate current is again 9.4 ma, we see from the curve that at point Z the plate voltage is now 135 volts. With no grid bias it required only 90 volts to force 9.4 ma through the tube, but with minus (−)4.5 volts grid bias, it required 135 volts to pass the same current through the tube. In other words, a change of 4.5 volts in the grid circuit is equivalent to a change of 135 − 90, or 45, volts in the plate circuit. Since the plate voltage must change ten times as much as the grid voltage, it is evident that a change in the grid voltage is ten times as effective as a change in the plate voltage. This ratio of change in the plate voltage required to counteract a change in the grid voltage is called *amplification factor* of a tube. Modern tubes have μ values running from three to several hundred.

19-12. Transfer Characteristic of a Tube. By plotting curves of the plate current against the grid voltage for various constant values of the plate voltage we can read directly the change in plate current which occurs when the grid voltage changes. Such a family of curves shown in Fig. 19-7 is often called the *transfer characteristic* of the tube. For it shows how the effect of varying grid voltage is transferred to the plate circuit in the form of varying plate current.

If no alternating current is impressed, the tube operates at point X on the curve, corresponding to E_g and I_p with E_p constant. Now, if we apply an a-c voltage e in series with the grid battery voltage, these two voltages merely add at each instant with the result that while the grid voltage always remains negative, it oscillates about the bias voltage with an amplitude equal to the amplitude of the alternating voltage. The plate current always remains positive but oscillates about I_p in sync with the alternating voltage on the grid. Figure 19-7 shows how an oscillating grid voltage produces an oscillating plate current.

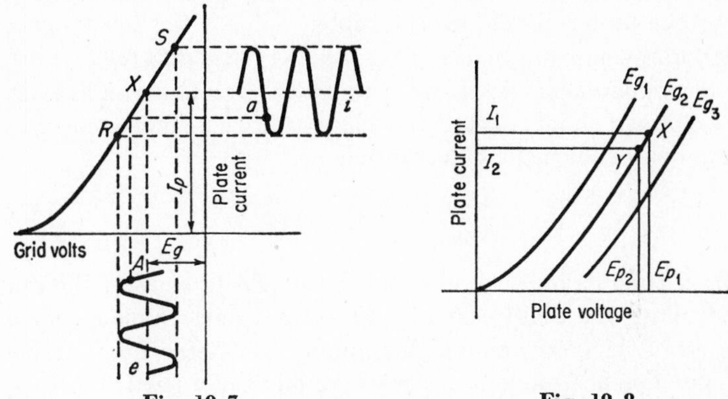

Fig. 19-7 **Fig. 19-8**

19-13. Plate Resistance of a Tube. The amount of amplification which a tube has can be determined from its plate characteristics. In order to find the voltage gain of an amplifier tube, we must know the internal resistance, or *plate resistance*, of the tube. Because the plate current is always varying during operation we cannot apply Ohm's law to find the tube resistance. We must use a value of resistance which correctly reflects the relation between the change in plate voltage and the change in plate current. The plate resistance of a tube is a change in plate voltage which produces a given change in the plate current, divided by a change in the plate current resulting. Thus (Fig. 19-8) points X and Y represent slightly different values of the plate voltage and the plate current on a curve of a constant grid voltage. From the definition above we see that the plate resistance R_p is

$$R_p = \frac{E_{p1} - E_{p2}}{I_{p1} - I_{p2}} \tag{19-1}$$

If the plate characteristics of a vacuum tube are known, the plate resistance of the tube can be computed by the above formula. Usually, the plate resistance is known from measurements reported by the manufacturer. Thus, if a change of 0 1 ma (0.0001 amp) is produced by a plate-voltage variation of 1 volt, the plate resistance is 1/0.0001 = 10,000 ohms.

19-14. Mutual Conductance. There is a third constant used to describe the properties of electron tubes. This is the control grid-to-plate transconductance, many times called the *mutual conductance*. Mutual conductance is described as the ratio of plate-current change to grid-voltage change, when the plate voltage remains constant. The symbol for mutual conductance is g_m.

$$g_m = \frac{\Delta I_p}{\Delta E_g} \text{ (with } E_p \text{ constant)} \qquad (19\text{-}2)$$

Mutual conductance is a measure of the effectiveness of the grid in controlling the plate current. Since in many problems a low resistance of the plate and a high amplification factor are desirable, the value of g_m is a rough measure of the merit of a particular tube. Since the mutual conductance has the dimensions of the reciprocal of resistance, it is measured in micromhos, typical values being 500 to 5,000 micromhos.

Q19-2. What is the relationship between the amplification factor μ, the plate resistance R_p, and the mutual conductance g_m?

ANSWER.
$$\mu = R_p g_m$$

Q19-3. Explain how the plate resistance, the amplification factor, and the mutual conductance may be calculated for a three-element electron tube from a family of plate-characteristic curves for the tube.

ANSWER. Refer to Fig. 19-6. Such curves may be found in tube manuals. For a grid bias of minus (−)9 volts, a plate current of 5.5 ma, and a plate voltage of 140 volts let us determine the requirements of the problem.

Plate resistance. Find the change in E_p for a small change in I_p, with E_g held constant at −9 volts. From the curve

$$R_p = \frac{150 \text{ at } I_p \text{ of } 6 - 135 \text{ at } I_p \text{ of } 4}{(6 - 4) \times 10^{-3}} = \mathbf{7{,}500\ ohms}$$

Amplification factor. Find the change in E_p caused by a small change in E_g with I_p held constant at 5.5 ma. You will find that at I_p of 5.5 ma and E_g of 8 (interpolate), E_p is 137.5 volts. Then at the same I_p, but an E_g of 10, E_p is 160. Now

$$\mu = \frac{160 - 137.5}{-8 - (-10)} = \mathbf{11.25}$$

Mutual conductance. At 140 volts find the change in I_p occasioned by a small change in E_g.

$$g_m = \frac{(6.5 \text{ at } E_g \text{ of } 8 - 5 \text{ at } E_g \text{ of } 10) \times 10^{-3}}{-8 - (-10)}$$
$$= 0.75 \times 10^{-3}, \text{ or } \mathbf{75\ micromhos}$$

Note that it is not possible to check these figures with the use of the relationship $\mu = R_p g_m$ because of interpolation.

Q19-4. A vacuum tube draws a plate current of 6 ma when operating at 200 volts at the plate and minus (−)6 volts at the grid bias. If the grid voltage is reduced to −4 volts with the same plate voltage, the plate current is doubled. In order to bring the plate current back to its original value, the plate voltage had to be reduced to 150 volts. Find: (*a*) the amplification factor, (*b*) the mutual conductance, and (*c*) the plate resistance.

ANSWER. Refer to Fig. 19-6, but interpolation is not required because the differences are already given in the problem.

Amplification factor

$$\mu = \frac{200 - 150}{(-)4 - (-)6} = \mathbf{25}$$

Mutual conductance

$$g_m = \frac{(12 - 6) \times 10^{-3}}{(-)4 - (-)6} = 3 \times 10^{-3} \text{ ohms, or } \mathbf{3{,}000\ micromhos}$$

Plate resistance

$$R_p = \frac{\mu}{g_m} = \frac{25}{(3 \times 10^{-3})} = \mathbf{8{,}333\ ohms}$$

Q19-5. Draw a set of typical plate-current versus grid-voltage curves for a three-element radio tube and show by means of the curves how the tubes may be used for rectification or for amplification.

ANSWER. Refer to Figs. 19-9 and 19-10.

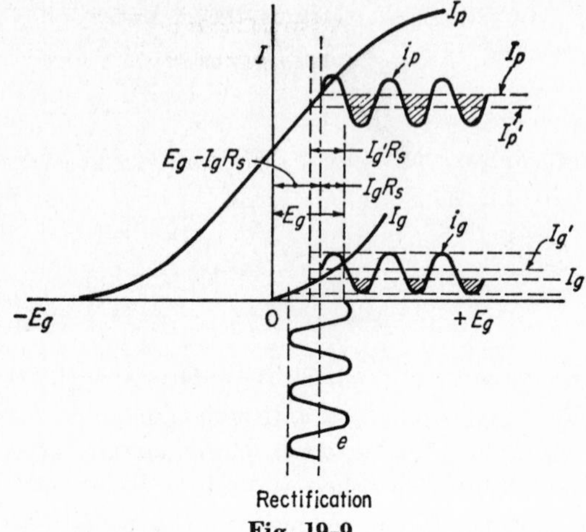

Rectification

Fig. 19-9

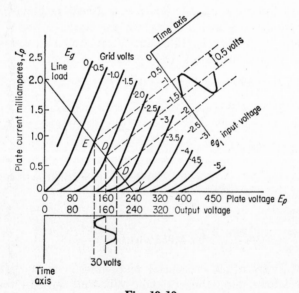

Fig. 19-10

Rectification (detection). When electric waves of radio frequency are emitted from a transmitting antenna, corresponding oscillations are excited in the antenna circuit of any receiving station within range that can be adjusted to resonance with the frequency of the transmitter. With suitable tuning devices, and a rectifier, known as a detector, these electric waves can be rectified and amplified so that they can be heard through earphones or speakers.

The three-electrode tube may also detect in a manner indicated. The connections are as shown in Fig. 19-11. A high resistance R_s of from 1 to 5 megohms is connected in series with and adjacent to the grid. This resistance is shunted by a small condenser C_s whose capacitance is between 50 and 200 μf. This hookup is known as grid-leak detection. The grid is polarized positively by the voltage E_g, so that a current I_g flows in the grid circuit, as shown in Fig. 19-9. This current flowing through the high resistance R_s produces in it a voltage drop I_gR_s, so that the effective polarization of the grid is $E_g - I_gR_s$. The corresponding plate current is I_p. An alternating voltage e in the grid circuit will produce an alternating current i_g in the grid circuit, whose negative portions, shown shaded, are of lesser magnitude than its positive portions. Hence, the average grid current is increased from I_g to I'_g. This decreases the polarization of the grid from $E_g - I_gR_s$ to $E_g - I'_gR_s$. The alternating component i_g is by-passed through the condenser C_s. The average plate current is decreased from I_p to I'_p with a superimposed alternating current i_p. With reference to I_p as an axis, the positive portions of i_p are less in magnitude than the negative portions, shown shaded. When the impressed voltage e is modulated, this change in the plate current $I_p - I'_p$ will follow the variations of the modulating current. The radio-frequency plate current i_p is bypassed through the condenser C.

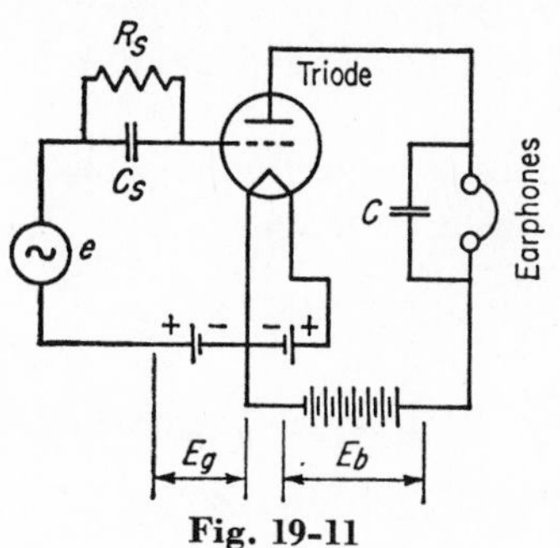

Fig. 19-11

The large curvature of the grid-current characteristic, the large slope of the plate-current characteristic, and the fact that the high grid-leak resistance may be made very large, all combine to make this type of detection the most sensitive of all.

Amplification. The three-element tube is most frequently used in industry and in the laboratory as an amplifier of low a-c voltages. Figure 19-10 shows how curves may be used to show this. Curves for a type 6FS triode are shown. Here we have produced a voltage amplification of 60 times the input voltage without destroying the shape of the sine curve. We can, of course, produce a voltage amplification with a transformer, but there the output power must always be somewhat less than the power input. However, with the vacuum tube, we are able to add power and produce a much greater output than input.

Q19-6. A power triode with fixed values of excitation and plate voltage is working into a resistance load as a power converter. If the maximum power output occurs with a load resistance of 2,000 ohms, what is the approximate plate resistance of the tube?

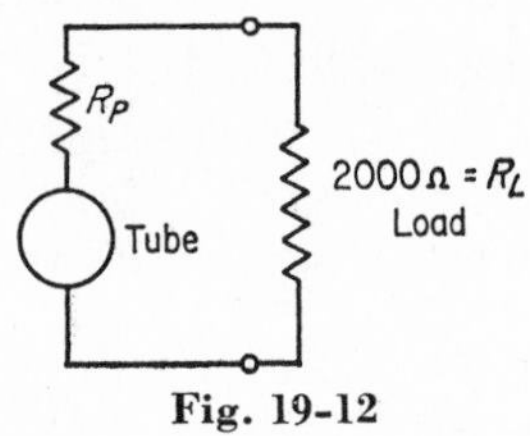

Fig. 19-12

ANSWER. Let us assume a linear power amplifier because of lack of a characteristics curve. For a linear circuit, the maximum power-transfer theorem shows that for a maximum power output, the load resistance must equal the generator resistance. Thévenin's theorem states that the tube in the problem may be considered as a constant-voltage source in series with its plate resistance, as shown in Fig. 19-12.

Since maximum power takes place when the load is adjusted to equal 2,000 ohms resistance, the resistance of the plate must also equal 2,000 ohms. For maximum undistorted power a load resistance equal to two times the plate resistance is recommended.

19-15. Summary. The combined effect of the plate and the control grid determines the plate current. The control grid has a much greater effect on the plate current than on the plate voltage. As a result an amplification of the input signal is realized. The cutoff bias is the lowest value of grid voltage that has a telling effect on the plate current.

The grid current will flow for positive grid voltages, but this results in an undesirable power consumption in the grid circuit.

Triode characteristic curves are based on the functional relationships between the plate voltage, the grid voltage, and the plate current. When placed together in graphic form they represent a family of curves which are very useful in practice. In the linear portion of these curves changes in the grid or plate voltage cause equal changes in the plate current, in order to effect undistorted amplification.

19-16. Transistors. Recently the transistor has come into its own. It is nowhere near so bulky as the electron tube. It has no filament and therefore needs no heating power. Like the electron tube the transistor is able to rectify and amplify. While the electron tube utilizes electrons flowing in a vacuum, the transistor utilizes electrons moving in a *solid*, a semiconductor. Publicity aimed at glamorizing the solid-state transistor has fostered a popular misconception—that transistors will eventually make electron tubes obsolete. The fact is that the two kinds of devices complement each other. Tubes can operate at much higher temperatures; transistors have longer theoretical lifetimes. Transistors are very small and sturdy, and so are some tubes. In short, for the foreseeable future, there is every indication that *both* will become more important. Each has unique features; neither can completely replace the other.

Transistors are only one family of semiconductors; new ones are constantly being discovered.

Q19-7. Compare and contrast the spacistor, transistor, and the vacuum tube. Show diagrams or sketches for each. How does each work?

ANSWER. *Spacistor.* Refer to Fig. 19-13. A voltage is applied between the base and the collector in a direction to produce a high electric field and almost no current. A voltage applied to the injector causes electrons to enter the region of a high field. The electrons flow rapidly to the collector contact. This current (flow of electrons) is modulated by a signal applied to the modulator as shown. Since the modulator uses only a small current to cause the current between the injector and the collector to fluctuate, amplification results.

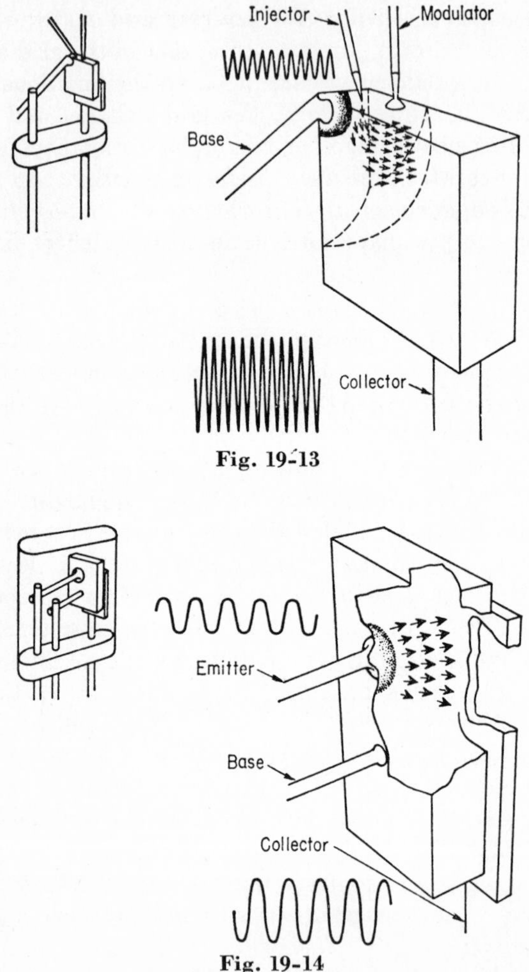

Fig. 19-13

Fig. 19-14

Transistor. Refer to Fig. 19-14. In a typical transistor, a negative voltage is applied between the emitter and the positive collector. The small fluctuating signal to be amplified is applied between the emitter and the base as shown (under the base). This causes the negative emitter voltage to fluctuate accordingly. The more negative the emitter, the greater the current between the emitter and the collector. The transistor amplifies because the input resistance is lower than the output resistance, while the input current is substantially equal to the output current. The result is a net signal-power gain.

Vacuum tube. The cathode is heated by a filament heater. A large number of electrons boil off the negative and are attracted to the positive plate (see Fig. 19-15). The small fluctuating signal to be amplified is applied to the negative grid as shown. This causes the negative grid voltage to fluctuate accordingly. The more negative the grid, the smaller the current flow between the cathode and the plate. Thus the grid acts as a valve, or shutter, a very small signal controlling a relatively large current. The large output signal (shown under the plate) is the amplified counterpart of the small input signal.

Note: The spacistor's chief features are: it will amplify up to 10,000 Mc, 50 times higher than transistors; it will operate up to 500°C, twice the heat for transistors. Table 19-1 gives some comparative characteristics of the three devices.

TABLE 19-1. COMPARATIVE CHARACTERISTICS

	Vacuum tube	Spacistor	Transistor
Frequency limit	High (1,000 Mc)	High (10,000 Mc)	Medium (250 Mc)
Heater power	Required	None	None
High-temperature materials	Available	Available	Not available
Theoretical life	Limited	Unlimited	Unlimited
Vacuum envelope	Required	None	None
Circuit weight and space	High	Low	Low
Strategic materials	Required	None	None
Complexity of multiple-stage circuitry	Low	Low	High
Input and output impedances	High	Very high	Low

This high-temperature performance is extremely important in rocket and guided-missile instrumentation. It will also result, undoubtedly, in many as yet undreamed of industrial instrument applications. Operating on a fraction of the power required for vacuum tubes and having no filament to heat up or burn out, the spacistor can be tightly packaged into subminiature instrument assemblies.

The operating principles of spacistors can be compared to those of transistors by dropping a single ink drop into two glasses—one empty, the other filled with water. The drop rapidly reaches the bottom of the empty glass, similar to the rapid dispersion of charged particles in spacistors. But in the water-filled glass, the drop will

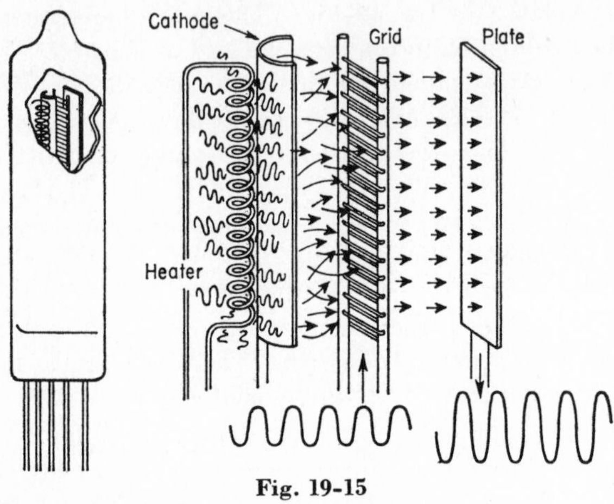

Fig. 19-15

slowly diffuse until it finally reaches the bottom—similar to the relatively slow dispersion of charged particles in a transistor.

19-17. Amplification. Electronic devices have very little to work with. A person speaking naturally delivers to his immediate surroundings only fifteen millionths of a watt, and only a very small portion of this energy is picked up by a sound-responsive instrument such as a microphone. In a great many industrial applications of electronics circuits normal currents or voltages are of the order of thousandths or millionths of an ampere or volt.

Before such feeble signals can be put to work, they must be amplified many, many times and at the same time the original wave shape must be preserved. The basic principle of amplification has been reviewed earlier so that we can immediately interest ourselves in amplifier operation.

19-18. Amplifier Operation. All amplifiers are broadly classified into four classifications. An amplifier circuit is an electron-tube circuit or transistor circuit which builds up an a-c signal applied to its input. If the output voltage of an amplifier is much greater than the input voltage, the circuit is a voltage amplifier. The ratio of the output voltage to the input voltage is called the amplification, or gain, of the amplifier. In addition there are also power amplifiers. In these circuits there is a power build-up for use in an output circuit. When a number of amplifiers are hooked up in series, i.e., hooked up in *cascade* so that the output of one serves as the input to the next stage, the early stages are used to build up the voltage, while the last builds up the power to operate a set of earphones or speakers. Each stage is said to be coupled to the next.

19-19. Classification of Amplifiers. Four classes of amplifiers are recognized by engineers, and as covered by definitions standardized by the Institute of Radio Engineers. This classification depends primarily on the fraction of the input cycle during which the plate current is expected to flow under rated full-load conditions. The classification is rated according to service, all as listed under *American Standard Definitions of Electrical Terms*—ASA No. C42, p. 234, *AIEE*, New York, 1941.

Class A Amplifier—A class A amplifier is an amplifier in which the grid-bias and alternating grid voltages are such that plate current in a specific tube flows at all times. *Note*—To denote that grid current does not flow during any part of the input cycle, the suffix 1 may be added to the letter or letters of the class identification. The suffix 2 may be used to denote that grid current flows during some part of the cycle.

Class AB Amplifier—A class AB amplifier is an amplifier in which the grid-bias and alternating grid voltages are such that plate current in a specific tube flows for appreciably more than half but less than the entire cycle. *Note*—See note under class A amplifier.

Class B Amplifier—A class B amplifier is an amplifier in which the grid-bias is approximately equal to the cut-off value so that the plate current is approximately zero when no exciting grid voltage is applied, and so that plate current in a specific tube flows for approximately one-half of each cycle when an alternating grid voltage is applied. *Note*—See note under class A amplifier.

Class C Amplifier—A class C amplifier is an amplifier in which the grid-bias is appreciably greater than cut-off value so that the plate current in each tube is zero when no alternating grid voltage is applied, and so that plate current flows in a specific tube for appreciably less than one-half of each cycle when an alternating grid voltage is applied. *Note*—See note under class A amplifier.

The term "cutoff bias" used in these definitions is the value of the grid bias at which the plate current is some very small value.

For radio-frequency (rf) amplifiers which operate into a selective tuned circuit, as in radio-transmitter applications, or under requirements where distortion is not an important factor, any of the above classifications of amplifiers may be used. For audio-frequency (af) work amplifiers in which distortion is not to be acceptable, only class A amplifiers permit single-tube operation. The distortion can be reduced with single-tube operation by the use of *inverse feed-back* circuits. With class A amplifiers, reduced distortion with improved power performance can be realized by the use of a push-pull stage for audio service. With class AB and class B amplifiers, a balanced amplifier stage using two tubes is required for audio service.

Q19-8. A power amplifier uses a 2A3 tube to supply a 300-ohm load from a 2:1 transformer. Assuming a minus (−) 50 volts grid bias, select an operating point on the plate characteristic and compute the maximum power that can be delivered to the load with class A1 operation about this point. What a-c voltage is required on the grid to produce this power?

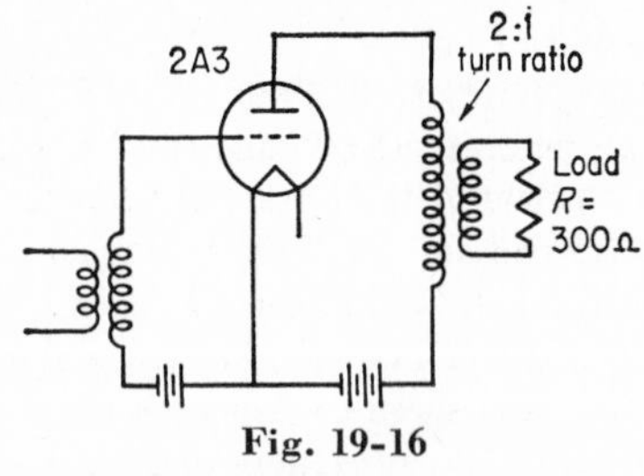

Fig. 19-16

ANSWER. Refer to Fig. 19-16. Using the nomenclature from the RCA Receiving Tube Manual let the grid bias E_c be minus (−) 50 volts, the plate voltage e_b be 300 volts, and the plate current I_b be 100 ma. On the plate characteristic in the tube handbook a load line with slope $1/R_L$ is shown drawn through the $e_b - I_b$ point with 40 volts maximum signal. Then I_b maximum = 190 ma and I_b minimum = 20 ma.

$$\% \text{ 2d harmonic} = \frac{\frac{1}{2}(I_{b\,\text{max}} + I_{b\,\text{min}}) - I_{b0}}{I_{b\,\text{max}} - I_{b\,\text{min}}} \times 100$$

$$\frac{\frac{1}{2}(190 + 20) - 100}{190 - 20} \times 100 = 3 \text{ per cent, too small}$$

Try with 45 volts maximum signal. Then $I_{b\,max} = 205$ ma and $I_{b\,min} = 15$ ma. From which the per cent second harmonic is found similarly to be 5.25, too small.

Try with 50 volts maximum signal. Then $I_{b\,max} = 220$ ma and $I_{b\,min} = 10$ ma. From which the per cent second harmonic is found as above to be 7.15 per cent. This is within the 10 per cent limit and constitutes maximum signal.

The maximum a-c plate current is given by the following.

$$I_p = \frac{I_{b\,max} - I_{b\,min}}{2} = \frac{220 - 10}{2} = 105 \text{ ma}$$

$$I_p = \frac{105}{\sqrt{2}} = 74.5 \text{ ma (rms)}$$

The a-c power delivered to the load

$$I_p{}^2R_L = 0.0745^2 \times 1{,}200 = 6.65 \text{ watts}$$

The d-c power delivered to the load

$$I_{b0}{}^2R_L = 0.1^2 \times 1{,}200 = 12 \text{ watts}$$

The amount of harmonic generation that may be acceptable in amplifiers for sound reproduction depends on the sensitivity of the human ear to waveform distortion. A value of 10 per cent total harmonic generation is commonly used as the permissible upper limit by engineers in the industry. The solution to the above problem assumes that no harmonics higher than the second exist in the plate current. The student is referred to the following sources for further reading: F. Massa, Permissible Amplitude Distortion of Speech in an Audio Reproducing System, *Proc. IRE*, 21, pp. 682–689, 1933.

Q19-9. A relay is to be controlled by a 2A3 tube. The relay coil has a resistance of 150 ohms. The relay is actuated when the coil current is 20 per cent above the normal plate current for the tube. The relay drops out when the current drops to 5 per cent above the normal tube current.

(*a*) Sketch a suitable circuit and determine all magnitudes of voltage and resistances to be used.

(*b*) What change in grid voltage is necessary to close the relay?

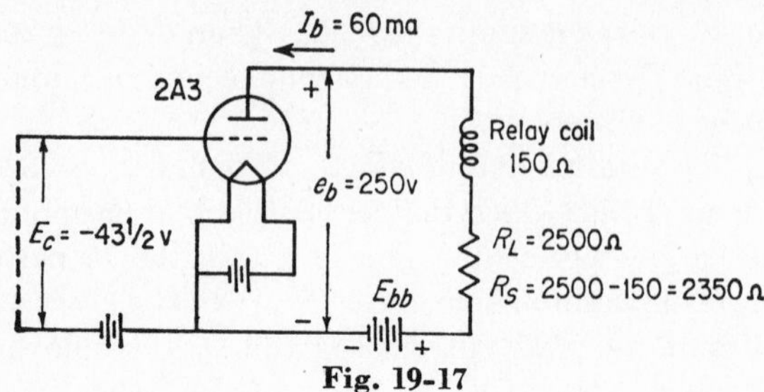

Fig. 19-17

ANSWER. Refer to Fig. 19-17. In addition to the nomenclature used in the previous problem, let us assume g_m is 5,250 micromhos. Then let

$$e_b = 250 \text{ volts}$$
$$R_L = 2500 \text{ ohms}$$
$$E_c = -43.5 \text{ volts or } -45 \text{ volts to center tap of a-c filament}$$
$$I_b = 60 \text{ ma}$$
$$g_m = 5{,}250 \text{ micromhos}$$

(a) See Fig. 19-17.

(b) The transconductance g_m is the rate of change of the plate current with the cathode voltage with the plate voltage constant. Thus

$$g_m = \left.\frac{dI_b}{dE_c}\right|_{e_b \text{ constant}}$$
$$dE_c = \frac{dI_b}{g_m}$$

For 20 per cent increase in the current, $dI_b = 0.2 \times 0.060 = 0.012$ amp.

$$dE_c = \frac{0.012}{5{,}250 \times 10^{-6}} = \frac{12}{5.25} = 2.29 \text{ volts}$$

It is assumed that the change is 20 per cent and not the difference between the pickup and the dropout. Now, for 5 per cent increase in the current

$$dI_b = 1.05I_b = 1.05 \times 0.060 = 0.003, \text{ or } 3 \text{ ma}$$
$$dE_c = \frac{dI_b}{g_m} = \frac{3 \times 10^{-3}}{5.25 \times 10^{-3}} = \frac{3}{5.25} = 0.572 \text{ volt}$$

Finally,

Relay picks up at 43.5 − 2.3 = **41.2 volts**

Relay drops out at 43.5 − 0.57 = **42.9 volts**

Q19-10. The ratio of the output voltage to the input voltage of an amplifier at a certain frequency is 60:1. What is the decibel voltage gain of the amplifier?

ANSWER. It is often advantageous to express the voltage amplification in terms of decibels. The gain G in decibels is defined as ten times the common logarithm of the ratio of the power output to the power input, or

$$G = 10 \log_{10} \frac{|E_o|^2/R_o}{|E_s|^2/R_i} = 10 \log_{10} \left(\frac{|E_o|}{|E_s|}\right)^2 \frac{R_i}{R_o}$$

$$20 \log_{10} \frac{|E_o|}{|E_s|} + 10 \log_{10} \frac{R_i}{R_o}$$

If the input resistance R_i is equal to the output resistance R_o, the gain of the amplifier is

$$G = 20 \log_{10} |A|$$

Note that the output voltage is E_o while the input signal is E_s. Now for the solution to the problem at hand we assume $R_i = R_o$. And

$$G = \text{db} = 20 \log_{10} {}^{60}\!/\!_{1} = 20 \times 1.7782 = \mathbf{35.564}$$

Q19-11. A pair of triodes operating in a class A balanced amplifier deliver power by means of an output transformer to a 10-ohm load. Explain how you would calculate the turn ratio of this transformer for maximum power transfer to the 10-ohm load.

ANSWER. Refer to Fig. 19-18 showing push-pull amplifier hookup. A push-pull amplifier is used to obtain a power output about equal to the sum of each tube rating. It is used for the final stage and has the advantage of delivering maximum power with least distortion. Assuming the plate resistances of each tube are equal, for maximum power output the effective load resistance R_L for a tube must be approximately equal to twice the plate resistance of the tube R_p.

Turn ratio n is equal to N_1/N_2. From a consideration of a standard text on applied electronics, the ideal output transformer reflects the load resistance R_L into the plate circuit of the tubes as

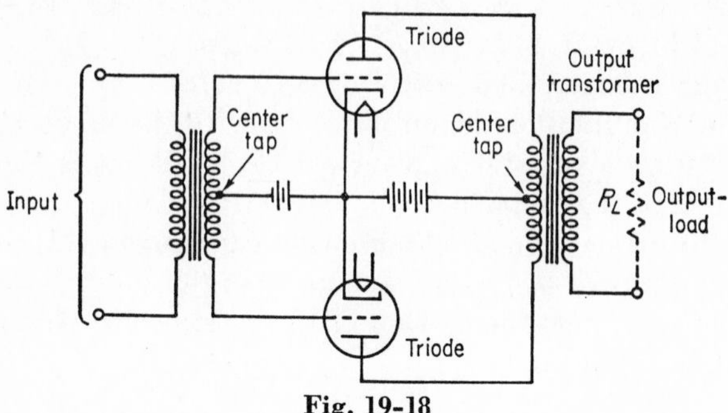

Fig. 19-18

the value $(N_1/N_2)^2R_L$, from which we can develop the final expression to be

$$2R_p = \left(\frac{N_1}{N_2}\right)^2 R_L$$

Thus

$$n = \left(\frac{2R_p}{R_L}\right)^{1/2}$$

Q19-12. (*a*) Explain how you would calculate the power output of two triodes employed in a class AB_1 balanced amplifier.

(*b*) If the slope of the composite characteristics of a pair of triodes connected as a balanced amplifier corresponds to a resistance of 1,000 ohms and the load is a 10-ohm resistance, what should be the turns ratio of the output transformer?

ANSWER. (*a*) From the RCA Receiving Tube Manual, Technical Series RC-18, p. 22: "In class AB_1 push-pull amplifier service using triodes, the operating conditions may be determined graphically by means of the plate family $E_p - I_p$ curve if the operating plate voltage E_o is given." In this service, the dynamic load line does not pass through the operating point P as in the case of the single-tube amplifier, but through the point D in Fig. 19-19. Its position is not affected by the operating grid bias provided the plate-to-plate load resistance remains constant.

Under these conditions the grid bias has no appreciable effect on the power output. The grid bias cannot be neglected, however, since it is used to find the zero-signal plate current and, from it, the zero-signal plate dissipation. Because the grid bias is higher in class AB_1 than in class A service for the same plate voltage, a higher signal voltage may be used without grid current being drawn and, therefore, a higher power output is obtained than in class A service.

In general, for any load line through point D (Fig. 19-19) the plate-to-plate load resistance in ohms of a push-pull amplifier is R_{pp} equals $4E_o/I'$, where I' is the plate current value in amperes at which the load line as projected intersects the plate current axis and E_o is in volts. This formula is another form of the one given

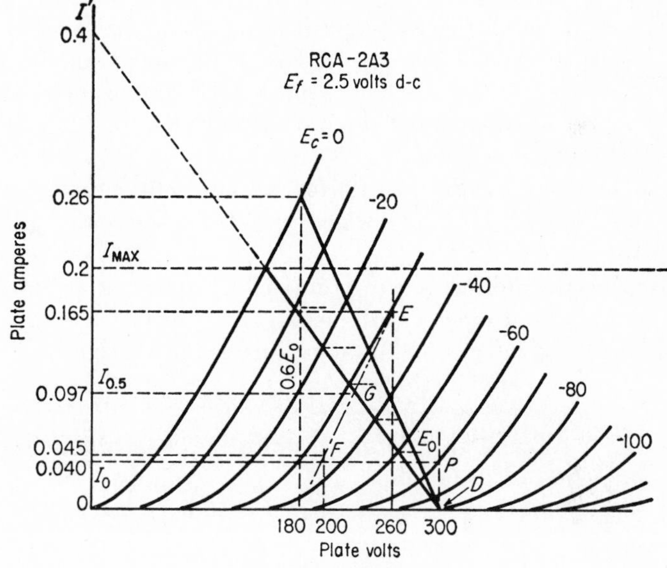

Fig. 19-19

under the push-pull class A amplifiers, $R_{pp} = 4(E_o - 0.6E_o)/I_{max}$, but it is more general. The power output is given as $(I_{max}/\sqrt{2})^2 \times R_{pp}/4$, where I_{max} is the peak plate current at zero grid volts for the load chosen. The maximum-signal average plate current is $2I_{max}/\pi$, or $0.636I_{max}$; the maximum-signal average power input is $0.636I_{max} \times E_o$.

It is desirable to simplify these formulas for a first approximation. This simplification can be made if it is assumed that the peak plate current I_{max} occurs at the point of the zero-bias curve corresponding approximately to $0.6E_o$, the condition for maximum power output.

The simplified formulas are

$$P_o \text{ (for two tubes)} = \frac{I_{\max} \times E_o}{5}$$

$$R_{pp} = \frac{1.6E_o}{I_{\max}}$$

where E_o is in volts, $I_{\max}$ is in amp, R_{pp} is in ohms, and P_o (power output) is in watts.

It may be found during subsequent calculations that the distortion or the plate dissipation is excessive for this approximation; in that case, a different load resistance must be selected using the first approximation as a guide and the process repeated to obtain satisfactory operating conditions.

(*b*) Refer back to Fig. 19-18 for a push-pull circuit. Also $N_1/N_2 = (2{,}000/10)^{1/2}$. From which we obtain the turns ratio n to be **14.1.**

The output transformer serves a three-fold purpose; (*a*) it provides for maximum power output with almost any tube or loading, since its turns ratio can be selected, (*b*) it eliminates the d-c component of power in the load, and (*c*) the device serves to isolate the tube from the load circuit.

Q19-13. A triode is transformer coupled to a load which has an equivalent value of 2,000 ohms in the plate voltage. The plate voltage is 250 volts and the plate current is 50 ma when there is no signal applied to the grid and the bias is minus (−)40 volts. When operated at maximum output, the instantaneous plate current varies from 100 to 10 ma. What are

(*a*) the power output

(*b*) the plate efficiency

(*c*) the second harmonic distortion

ANSWER. (*a*) Refer to Fig. 19-20. Let the following hold for solution:

$$R_L = 2{,}000 \text{ ohms}$$
$$E_{bo} = 250 \text{ volts}$$
$$e_c = 40 \text{ volts}$$
$$I_{b\,\max} = 100 \text{ ma}$$
$$I_{b\,\min} = 10 \text{ ma}$$

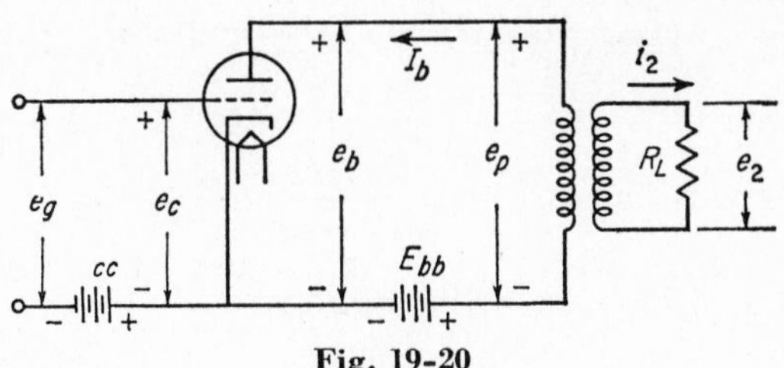

Fig. 19-20

Since the load is transformer coupled, no d-c component appears in the load current.

$$I_{b\,\max} = I_{bo} + I_{po} + \sqrt{2}\,I_{p1} + \sqrt{2}\,I_{p2}$$

$$\sqrt{2}\,I_{p1} = \frac{I_{b\,\max} - I_{b\,\min}}{2} = \frac{100 - 10}{2} = 45 \text{ ma}$$

$$\sqrt{2}\,I_{p2} = I_{po} = \frac{1}{2}\left(\frac{I_{b\,\max} + I_{b\,\min}}{2} - I_{bo}\right) = \frac{1}{2}\left(\frac{100 + 10}{2}\right) - 50$$

$$= 2.5 \text{ ma}$$

Since it is conventional to refer to power from the fundamental frequency component only

$$\sqrt{2}\,I_{p1} = 45 \text{ ma} \qquad \text{thus} \qquad I_{p1} = \frac{45}{\sqrt{2}} = 31.8 \text{ ma}$$

$$\text{Power output} = P_o = \frac{1}{2\pi}\int_0^{2\pi} I_p{}^2 R_L \, d\omega t = I_{p1}{}^2 R = 0.0318^2 \times 2{,}000$$

$$= \mathbf{2.02 \text{ watts}}$$

(*b*) Plate efficiency $= \eta_p = \dfrac{P_o}{E_b I_b} = \dfrac{2.02}{250 \times 0.0525} = \dfrac{2.02}{13.13}$ $= 0.154$, or **15.4 per cent.** Note $I_b = I_{bo} + I_{po} = 50 + 2.5 =$ 52.5 ma.

(*c*) Per cent second harmonic

$$\text{Per cent 2d harmonic} = \frac{\sqrt{2}\,I_{p2}}{\sqrt{2}\,I_{p1}} \times 100 = \frac{2.5}{45} \times 100$$

$$= \mathbf{5.5 \text{ per cent}}$$

From K. Henney, "Radio Engineering Handbook," 4th ed., McGraw-Hill Book Company, Inc., New York, 1950.

$$\text{Power output} = \frac{(I_{max} - I_{min})^2 R}{8} = \frac{(0.100 - 0.010)^2 \times 2{,}000}{8}$$

$$= 0.090^2 \times 250 = \mathbf{2.02\ watts}$$

$$\text{Per cent 2d harmonic} = \frac{(I_{max} + I_{min})/2 - I_o}{I_{max} - I_{min}} \times 100$$

$$= \frac{(100 + 10)/2 - 50}{100 - 10} \times 100 = \frac{55 - 50}{90} \times 100 = \frac{5}{90} \times 100$$

$$= \mathbf{5.55\ per\ cent}$$

19-20. Rectification. This is the process by which alternating current is converted to direct current. Alternating current is more easily generated and transmitted over long distances. However, electron tubes need to work on d-c supply for all their electrodes, except their filaments, which can be heated by alternating current as well. It is most convenient to change alternating and direct current by means of rectifiers.

Rectifier units are capable of changing alternating current into a pulsating form of direct current. Then, in order to smooth out the pulsations, additional filter circuits are required. A complete power pack also contains a voltage divider for providing a source of direct current for plate voltages or even a voltage regulator.

By elimination of the negative half-cycles or alternations of the alternating current, voltage rectifiers can change alternating current into pulsating direct current. All that needs to be done is to smooth out the pulsations with a filter system.

19-21. Diode As a Rectifier. The rectifying action of a diode finds important application in supplying a receiver with d-c power from an a-c line and supplying high-voltage direct current from a high-voltage pulse. The function of the filter is to smooth out the ripple of the tube output.

As previously discussed, we went into the matter of the diode as a rectifier. This is true of all diodes regardless of type. Diodes come in many forms. They may come as vacuum tubes, crystals or semiconductors, or metallic types, such as copper oxide and selenium rectifiers. Although the rectifier circuits that follow illustrate primarily electron-tube diodes, any of these other types may be substituted equally well. This latter type needs no filament supply, since they have no filaments. This is the reason why selenium rectifiers have become increasingly popular in television and radio receivers.

19-22. Half-wave Rectifier. When used singly, a diode is known as a *half-wave rectifier* because it will permit current to flow only during the positive half-cycle of the applied alternating current. A half-wave rectifier circuit with input and output waveforms is shown in Fig. 19-21. This is quite a simple circuit. Although

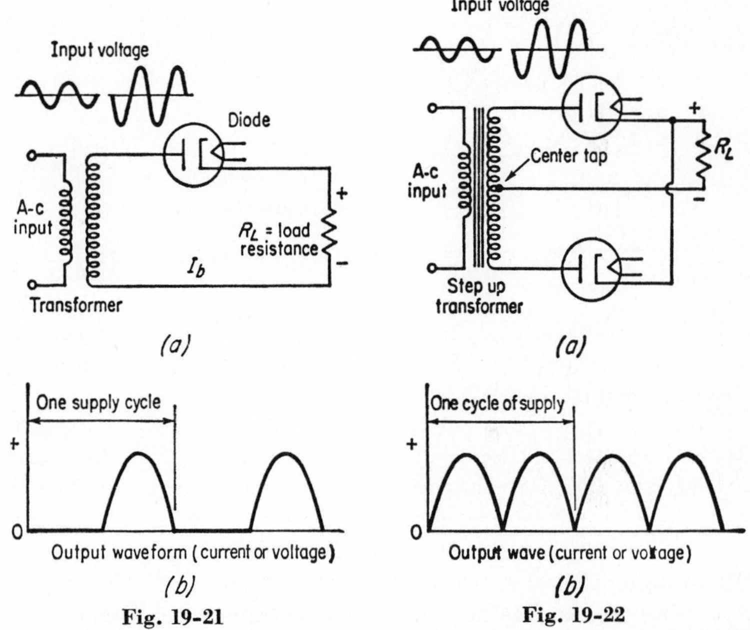

Fig. 19-21

Fig. 19-22

unidirectional, the current or voltage is not direct current because of the pulsations. The efficiency of the half-wave rectifier is low and is used for applications where the current drain is small. In addition, elaborate filter systems are required to smooth out the a-c ripple.

19-23. Full-wave Rectifier. If two diodes are employed, we have a *full-wave rectifier*. A hookup with waveforms is shown in Fig. 19-22. Figure 19-23 shows how filters work on smoothing out the a-c ripple in a full-wave rectifier. The efficiency of a full-wave rectifier is much greater than the half-wave arrangement. Because no direct current passes through the transformer, this prevents a d-c magnetization of the core of the transformer. As a result the transformer for a full-wave rectifier is much smaller than the similar unit

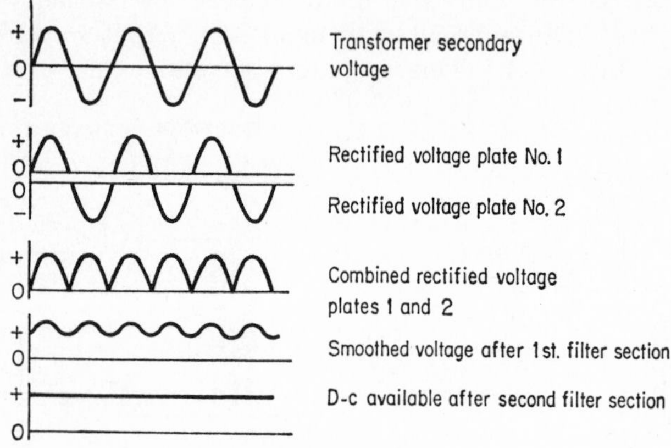

Fig. 19-23

in a half-wave circuit. The full-wave circuit is the standard for low-power applications.

19-24. Tungar Rectifier. It frequently happens that the source of power available for charging storage batteries is alternating current, which cannot be used directly on the battery, as it would not charge it. Many types of rectifiers are manufactured for adapting alternating current for charging storage batteries. They may be given three general classifications: tube, vibrator, and chemical cell. One of the best known tube devices is the tungar rectifier. Half-wave and full-wave tungar rectifiers are commercially available.

Q19-14. A full-wave rectifier without a filter has three ammeters in series in its output circuit. These ammeters are: a hot-wire meter, a permanent-magnet meter, and an iron-vane meter. How will the readings of the three instruments compare? State how you would determine the efficiency of such a rectifier by measurement.

ANSWER. Refer to Fig. 19-24. The hot-wire meter will give a reading dependent on the heating of its wire which will be proportional to the mean value of the current squared, the rms value.

The iron-vane meter will also register the rms, or effective, value, provided the shape of the rectified wave is uniform.

The permanent magnet meter, on the other hand, will give an average value of the current, since its reading is proportional to the torque on the coil, which, in turn, varies with the current through the coil. It follows that the coil reading of this meter will be smaller than that of either the hot-wire or the iron-vane meter.

In order to obtain the efficiency of the rectifying unit, a permanent magnet ammeter is used, since its current reading will be that of the equipment ordinarily charged, such as batteries. This

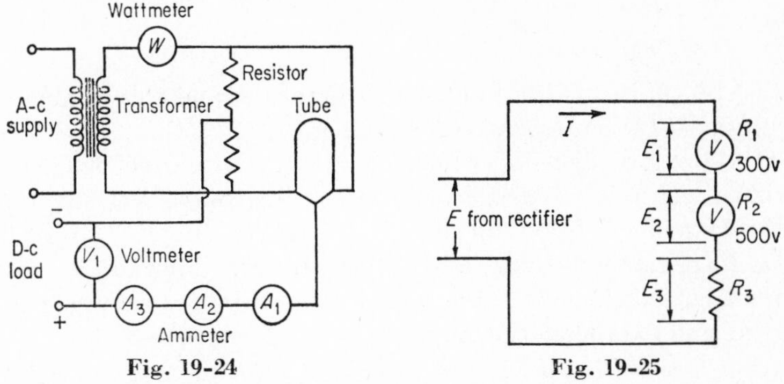

Fig. 19-24 Fig. 19-25

ammeter A together with a voltmeter V and an electrodynamometer-type of wattmeter W are hooked up as shown in Fig. 19-24. Then if P denotes the power input, E the d-c voltage, and I the ammeter reading,

$$\text{Rectifier efficiency} = \frac{EI \times 100}{W}$$

Q19-15. The voltage at the output side of a rectifier is of the order of 700 volts. There are on hand a 300-volt voltmeter of 50,000 ohms resistance and a 500-volt voltmeter of 500,000 ohms resistance. In addition, a 200,000-ohm resistor is available. If these units are arranged in series and the unknown voltage is impressed upon the combination, what is the value of the rectifier output if the 500-volt voltmeter indicates 450 volts?

ANSWER. Refer to Fig. 19-25.

$$I = \frac{E_1}{R_1} = \frac{E_2}{R_2} = \frac{E_3}{R_3} = \frac{E}{R}$$

from which we obtain

$$E = \frac{E_2}{R_2} \times R \qquad \text{also} \qquad R = R_1 + R_2 + R_3 = 750{,}000 \text{ ohms}$$

Then

$$E = \frac{450}{500{,}000} \times 750{,}000 = \textbf{675 volts}, \text{ the rectifier voltage}$$

Also

$$E = E_1 + E_2 + E_3 = 45 + 450 + 180 = 675 \text{ volts}$$

$$E_1 = \frac{450}{500{,}000} \times 50{,}000 = 45 \text{ volts}$$

$$E_3 = \frac{450}{500{,}000} \times 200{,}000 = 180 \text{ volts}$$

Note: The current is the same for all branches, and the voltage is proportional to the resistance it straddles.

Q19-16. A diode rectifier tube has a plate resistance of 600 ohms (assumed constant) and is connected as a half-wave rectifier to a load resistance of 1,400 ohms. The energizing supply voltage is 440 volts effective 60-cycle, alternating current. Find (*a*) the rms rectified current, (*b*) the average load current, (*c*) the average load voltage, and (*d*) the rectification efficiency.

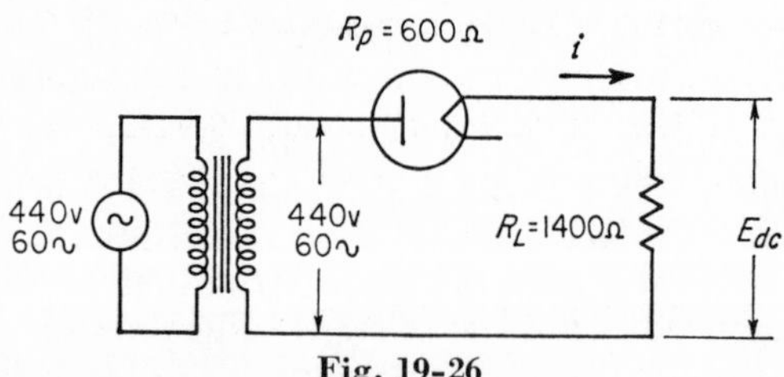

Fig. 19-26

ANSWER. Refer to Fig. 19-26. This shows the circuit as wired. Figure 19-27*a* shows an equivalent circuit and Fig. 19-27*b* shows a waveform.

(*a*)

$$I_m = \frac{E_{sm}}{R + R_o} = \frac{\sqrt{2} \times 440}{1{,}400 + 600} = 0.311 \text{ amp}$$

$$I_{\text{rms}} = \sqrt{\frac{1}{2}\pi \int_0^{2\pi} i^2\omega t\, d\omega t} = \sqrt{\frac{I_m{}^2}{2\pi} \int_0^{2\pi} \sin^2 \omega t\, d\omega t} = \frac{1}{2} I_m = \frac{0.311}{2}$$

$$= \textbf{0.155 amp}$$

(*b*) Average load current

$$I_{dc} = \frac{I_m}{\pi} = \frac{0.311}{3.14} = \mathbf{0.099\ amp}$$

(*c*) Average load voltage $= E_{dc} = I_{dc}R = 0.099 \times 1{,}400 =$ **138.5 volts.**

Check

$$E_{dc} = \frac{E_{sm}}{\pi(1 + R_o/R)} = \frac{\sqrt{2} \times 440}{\pi(1 + 600/1{,}400)} = \frac{622}{4.84} = \mathbf{138.5\ volts}$$

(*d*) Rectification efficiency

$$P_{dc} = \text{power input} = I_{dc}{}^2R = 0.099^2 \times 1{,}400 = 13.7 \text{ watts}$$
$$P_i = \text{power input} = I_{rms}{}^2(R + R_o) = 0.155^2 \times (1{,}400 + 600) = 48 \text{ watts}$$

$$\text{Rectifier efficiency} = \frac{P_{dc}}{P_i} = \frac{13.7}{48} \times 100 = \mathbf{28.7\ per\ cent}$$

19-25. Mercury-arc Rectifier. An early mercury-arc rectifier was a weird sight, indeed. When in service, the mercury vaporized and then condensed, running down the sides of the tube back into the pool of mercury. The principal use for this type of rectifier is for charging storage batteries of large capacity. The rating of these rectifiers is limited to 50 kw.

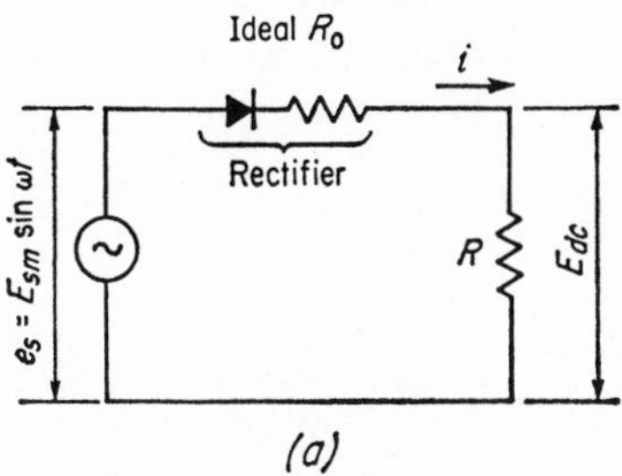

(a)

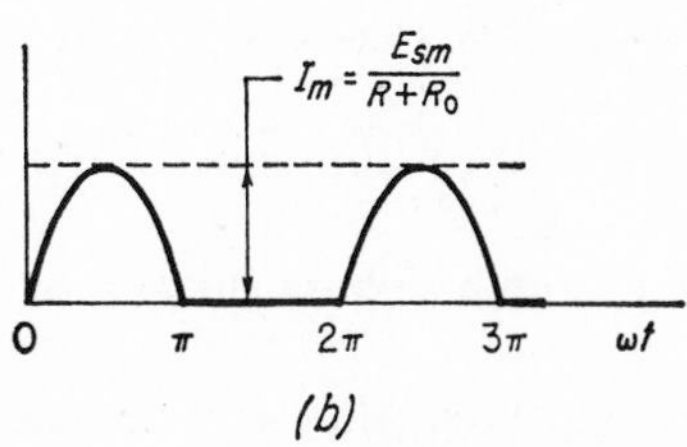

(b)

Fig. 19-27

In tungar rectifiers, the tungsten constitutes the cathode and the ionized atoms of an inert gas constitute the current carriers, producing the space charge as well. Since the cathode operates at high temperatures, it slowly volatilizes and its life is limited. With mercury-arc rectifiers, the mercury performs two functions: it is the cathode and produces the mercury vapor from which the necessary positive and negative ions are produced. Since the mercury returns to the cathode pool, there is no deterioration of the cathode with use.

The voltage drop in the arc depends on the operating conditions. Ordinarily the arc drop is of the order of 18 to 22 volts, but it may be as high as 30 volts in rectifiers of large power rating, where the cross section of the arc is restricted. This arc drop remains almost constant over a considerable range of the instantaneous values of current. In well-designed units of large capacity the arc drop is almost independent of the load current.

Mercury-arc rectifiers may be single-phase or polyphase. Due to the fact that a large current can be rectified by this means, it is being used widely to replace synchronous converters and motor-generator sets as a means for converting alternating to direct current.

Q19-17. A three-phase mercury rectifier supplies a resistance load with a negligible inductance. For a load current of 50 amp average

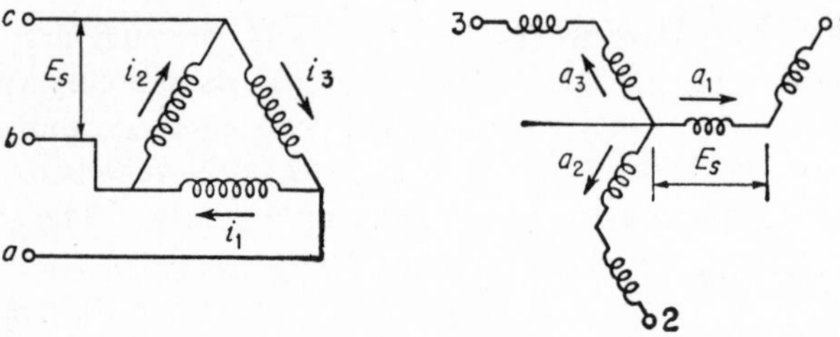

Fig. 19-28

and 120 volts average, the tube drop is 15 volts. The system is supplied from a 460-volt three-phase 60-cycle system by three transformers connected in Δ on the primary and zigzag star on the secondary.

(*a*) Sketch the current waveform of a tube approximately to scale. What are its average and rms values?

(*b*) Sketch the current waveform of a primary phase current and compute its rms value.

(*c*) Sketch the current waveform of a primary line current and compute its rms value.

(*d*) Specify the ratings of each primary and secondary winding as to voltage and current.

ANSWER. Refer to Fig. 19-28 for circuit. $I_{dc} = 50$ amp and $E_L = 120$ volts. The tube drop is 15 volts.

(a) Refer to Fig. 19-29. Anode average value

$$I_{\text{avg}} = \frac{I_{dc}}{3} = \frac{50}{3} = 16.6$$

$$\text{Effective value} = \frac{I_{dc}}{\sqrt{3}} = \frac{50}{\sqrt{3}} = \mathbf{28.8\ amp}$$

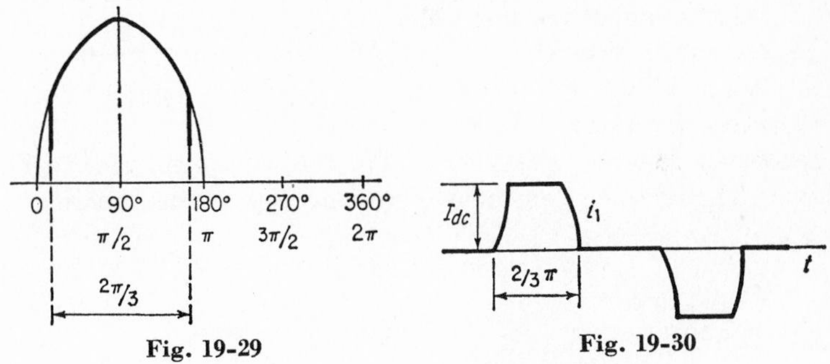

Fig. 19-29

Fig. 19-30

(b) Refer to Fig. 19-30. Primary phase current

$$\text{Effective (rms) value} = \sqrt{1/2\pi \int_0^{2\pi} a^2\, d\omega t}$$

$$= \sqrt{2/2\pi (I_{dc}^2\, 2\pi/p)} \qquad \text{where } p \text{ is three-phase}$$

$$= \sqrt{2/3 I_{dc}^2} = \sqrt{2/3 \times 50^2} = 40.8 \text{ amp}$$

This value is based on a 1:1 ratio. Now from a standard text

$$\frac{E_d}{E_s} = \frac{\text{average rectified voltage}}{\text{effective value of voltage}} = 1.17$$

$$E_s = \frac{120 + 15}{1.17} = 115 \text{ volts}$$

For a transformer ratio of 4:1, the primary current $40.8/4\sqrt{3} =$ **5.88 amp.**

(c) Refer to Fig. 19-31. Primary line current (rms) i'_a

$$\sqrt{3}\, i'_a = \sqrt{1/2\pi (4\pi/3 \times I_{dc}^2 + 2\pi/3 \times 4I_{dc}^2)}$$

$$= \sqrt{2/3\, I_{dc}^2 + 4/3\, I_{dc}^2}$$

$$= \sqrt{2\, I_{dc}^2} = \sqrt{2} \times 50 = 70.7 \text{ amp on a 1:1 ratio}$$

On a 4:1 ratio $= i_a = 70.7/4\sqrt{3} =$ **10.2 amp**

(d) Primary winding: $i_1 =$ **5.88 amp, 460 volts**
Secondary winding: $i_{s1} =$ **28.8 amp, 115 volts**

Q19-18. A mercury-vapor rectifier rated at 5 amp peak, 1.5 amp average current, is used in a single-phase half-wave circuit to charge a 6-volt battery having 0.02-ohm internal resistance. The a-c supply is a 20-volt rms source. The tube-breakdown and tube-drop voltages are 10 volts.

(*a*) What value of resistor will be needed in the circuit?

(*b*) What is the efficiency of the battery charging?

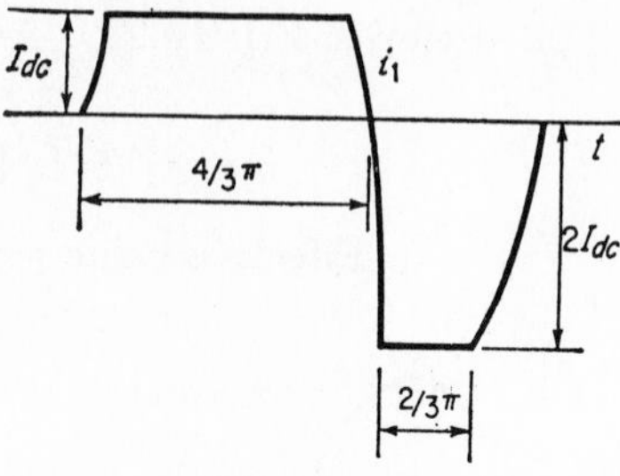

Fig. 19-31

ANSWER. Refer to Fig. 19-32. For the waveform refer to Fig. 19-33. The rectifier element is represented by an ideal rectifier in

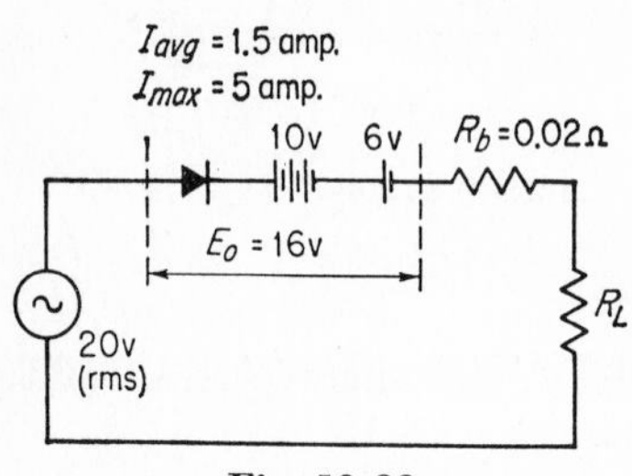

Fig. 19-32

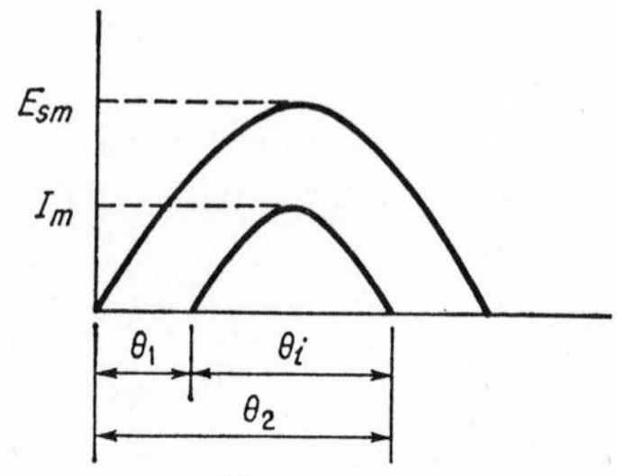

Fig. 19-33

series with a battery of voltage E_o. The waveform of the supply voltage is $e_s = E_{sm} \sin \omega t$.

$$E_{sm} = \sqrt{2}\, E_{\text{rms}} = \sqrt{2} \times 20 = 28.3 \text{ volts}$$

The angle in the supply-voltage cycle during which the current flows changes with the ratio $\alpha = E_o/E_{sm}$, or $\alpha = 16/28.3 = 0.565$.

$$\theta_1 = \sin^{-1} \alpha = \sin^{-1} 0.565 = 34.4°$$
$$\theta_2 = 180 - \theta_1 = 180 - 34.4 = 145.6°$$
$$\theta_i = \theta_2 - \theta_1 = 145.6 - 34.4 = 111.2°$$

(*a*) To limit the maximum current to 5 amp, the total resistance R is

$$R = R_b + R_L = \frac{E_{sm} - E_o}{I} = \frac{28.3 - 16}{5} = \frac{12.3}{5} = 2.46 \text{ ohms}$$

$$R_L = R - R_b = 2.46 - 0.02 = \mathbf{2.44\ ohms}$$

(*b*) I_{dc}, the rectified current, is found to be

$$I_{dc} = \frac{1}{2\pi}\int_{\theta_1}^{\pi-\theta_1} i d\omega t = \frac{I'_m}{2\pi}\int_{\theta_1}^{\pi-\theta_1} (\sin \omega t - \alpha)\, d\omega t$$

$$= \frac{I'_m}{2\pi}[\cos \omega t + \alpha\omega t]_{\sin^{-1}\alpha}^{\pi-\sin^{-1}\alpha}$$

$$= \frac{I'_m}{2\pi}(\sqrt{1-\alpha^2} - \alpha \cos^{-1}\alpha)$$

α is in radians, and where

$$I'_m = \frac{E_{sm}}{R} = \frac{28.3}{2.46} = 11.55 \text{ amp}$$

$$I_{dc} = \frac{11.55}{2\pi}(0.827 - 0.565 \times 0.97) = \mathbf{1.03\ amp}$$

For the power input the effective current must be determined. The formula is

$$\text{Power input} = P_i = I^2R + I_{dc}E_o$$

First determine I

$$I = I'_m \sqrt{\tfrac{1}{2}\pi \int_{\theta_1}^{\pi-\theta_1} (\sin \omega t - \alpha)^2\, d\omega t}$$

$$= I'_m \sqrt{\tfrac{1}{2}\pi[(1 + 2\alpha^2)\cos^{-1}\alpha - 3\alpha\sqrt{1-\alpha^2}]}$$

$$= 11.55\sqrt{\tfrac{1}{2}\pi[(0.364 \times 0.97) - 3(0.565 \times 0.827)]} = 2.01 \text{ amp}$$

$$\text{Power input} = 2.01^2 \times 2.46 + 1.03 \times 16 = 26.3 \text{ watts}$$

$$\text{Power used for battery charging } P_p = I_{dc}E_c = 1.03 \times 6 = 6.18 \text{ watts}$$

$$\text{Efficiency of battery charging} = \frac{P_p}{P_{in}} \times 100 = \frac{6.18}{26.3} \times 100 = \mathbf{23.5\ per\ cent}$$

Q19-19. In a given ideal *p-n* junction, a forward bias of 1 volt produces a certain current at a temperature of 300°K. What forward bias is required if this current is to be doubled?

ANSWER. The current when a forward bias v is applied is given by

$$I = I_s(E^{ev/kt} - 1) \tag{1}$$

The current when a forward bias s is applied is given by

$$2I = I_s(E^{es/kt} - 1) \qquad (2)$$

Dividing (2) by (1), we obtain

$$2(E^{ev/kt} - 1) = E^{es/kt} - 1$$

Since the following values are also given:

$$\begin{aligned} e &= 1.6 \times 10^{19} \text{ coulomb} \\ k &= 1.38 \times 10^{-23} \text{ joule/°R} \\ T &= 300°K \\ v &= 1 \end{aligned}$$

then,

$$\begin{aligned} E^{ev/kt} &= E^{38.7} \gg 1 \\ 2E^{38.7} &= E^{es/kt} \end{aligned}$$

Taking ln's, 38.7 ln 2 = es/kt = 38.73. Finally,

$$s = 39.493/38.8 = \textbf{1.02 volts forward bias}$$

Q19-20. When the grid and plate voltages of a vacuum tube triode are −1 and 100 volts respectively, the plate current is 2.5 ma, and when these voltages are −4 and 200 volts respectively, the plate current is found to be 6 ma. Assuming that the dependence of the plate current on the grid and plate voltages obeys the three-halves-power law, determine the amplification factor μ.

ANSWER. Plate current $i_b = k(e_b + \mu e_c)^{3/2}$. When the term $(e_b + \mu e_c)$ is negative, no current will flow. When

$$e_b \text{ (plate voltage)} = 100 \text{ volts}$$

and e_c (grid voltage) = −1 volt, i_b = 2.5 ma.

Therefore

$$\begin{aligned} 2.5 \times 10^{-3} &= k(100 - \mu)^{3/2} \\ 6.0 \times 10^{-3} &= k(200 - 4\mu)^{3/2} \end{aligned}$$

Thus

$$\left(\frac{6}{2.5}\right)^{2/3} = 1.793 = \frac{200 - 4\mu}{100 - 1\mu}$$

$$179.3 - 1.793\mu = 200 - 4\mu$$

$$2.207\mu = 20.7 \qquad \text{and consequently} \qquad \mu = \mathbf{9.33}$$

Q19-21. A series LCR circuit with R = 4 ohms, L = 100 μh and C = 200 $\mu\mu$f is connected to a constant-voltage generator of variable frequency. Calculate the resonant frequency, the value of Q, and the frequencies at which half the maximum power is delivered.

ANSWER. Resonant frequency

$$\tfrac{1}{2}\pi\sqrt{LC} = \tfrac{1}{2}\pi\sqrt{100 \times 200} = \mathbf{1{,}125\ kc\ per\ sec}$$

Value of Q is given by and is equal to

$$Q = 2\pi f_r \frac{L}{R} = 6.28 \times 1{,}125 \times \frac{0.1}{4} = \mathbf{177}$$

where $2\pi f_r = \omega_r$.

$$\text{Proceeding,}\quad Q = \omega_o \frac{L}{R} = \frac{\omega_o}{\omega_2 - \omega_1} = \frac{f_o}{f_2 - f_1} = \frac{f_o}{\text{Bandwidth}}$$

$$\text{Therefore,}\quad f_2 - f_1 = \frac{f_o}{Q} = \frac{1{,}125}{177} \times 10^3 = 6.4 \text{ kc per sec}$$

$$\frac{f_2 - f_1}{2} = \frac{6.4}{2} = 3.2 \text{ kc per sec}$$

Finally,

upper half-power frequency $= 1{,}125 + 3.2 =$ **1,128.2 kc per sec.**

And

lower half-power frequency $= 1{,}125 - 3.2 =$ **1,121.8 kc per sec.**

Q19-22. A parallel resonant circuit includes a 50-$\mu\mu$f capacitor and has a bandwidth of 250 kc per sec. Calculate the maximum impedance of the circuit. C equals 50×10^{-12} farad.

ANSWER. At a certain frequency the circuit will have a maximum value of impedance R. Bandwidth is the difference in cycles between the two frequencies where the impedance is given by $R/\sqrt{2}$ and can be shown to be $1/(2\pi CR)$. Thus, the bandwidth is 250×10^3 cps $= 1/(2\pi CR)$, from which the impedance is found to be $R =$ **12,740 ohms.**

Q19-23. Calculate the input impedance of the circuit, Fig. 19-34, at a frequency of 1 megacycle per sec. Coefficient of coupling may be taken as 0.1.

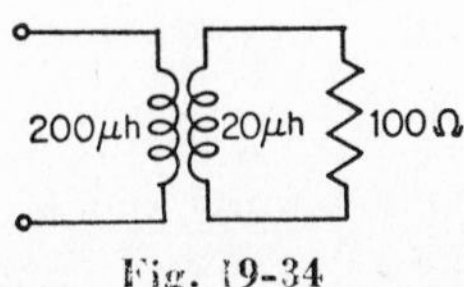

Fig. 19-34

ANSWER. Input impedance $= Z_p - \omega^2 M^2/Z_s$. Z_p = primary impedance and Z_s = secondary impedance. Then

$$\begin{aligned} Z_p &= jWL_p = j2\pi \times 10^6 \times 200 \times 10^{-6} = j400L \\ Z_s &= 100 + jWL_s = 100 + j2L \times 10^6 \times 20 \times 10^{-6} \\ Z_s &= 100 + j40L \\ (WM)^2 &= (2L \times 10^6)^2 \times k^2 L_1 L_2 \\ &= 4L^2 \times 10^{12} \times (1/10)^2 \times 4 \times 10^3 \times 10^{-12} = 160L^2 \end{aligned}$$

$$\text{Input impedance} = 100j40L + \frac{160L^2}{100 + j40 + j40L}$$

$$= \mathbf{6.1 + j1{,}250\ ohms}$$

Q19-24. An amplifier has a gain of 20 without feedback. If 10 per cent of the output voltage is fed back by means of a resistive negative-feedback circuit, find the actual amplification.

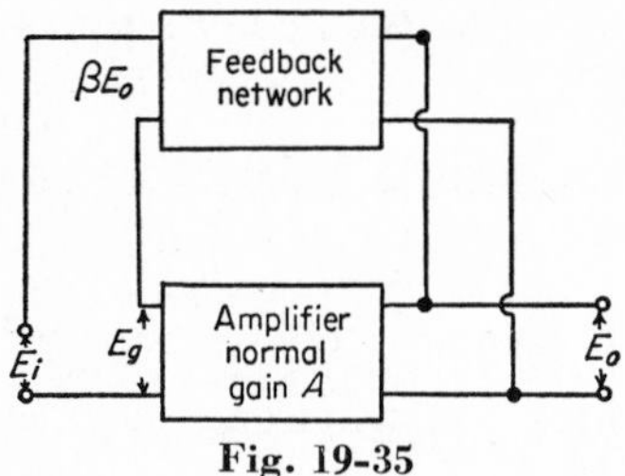

Fig. 19-35

ANSWER. Refer to Fig. 19-35. Then $E_o = AE_g$, $E_g = E_i + \beta E_o$, $E_o/E_i = A_f = A/(1 - \beta A)$. And the gain with feedback is

$$\frac{20}{1 + (1/10) \times 20} = \mathbf{6.7}$$

Q19-25. An amplifier employing a pentode with an amplification factor of 1,000 and a mutual conductance of 5 ma per volt has a 200,000-ohm load resistor. Calculate the voltage amplification (*a*) without feedback, (*b*) with 5 per cent negative voltage feedback.

ANSWER.

(*a*) Without feedback

$$A = \frac{\mu R_1}{r_a + R_1} = \frac{1{,}000 \times 200 \times 10^3}{(200 + 200)10^3} = \mathbf{500}$$

(*b*) With 5 per cent negative voltage feedback

$$A_f = \frac{A}{1 + 25} = \frac{500}{26} = \mathbf{19.2}$$

Q19-26. An audio-frequency amplifier has a nominal gain of 120 and gives an output of 60 volts to its output transformer, with 10 per cent second harmonic distortion. How much feedback must be used to reduce distortion to 1 per cent? Also find the additional gain required ahead of the feedback amplifier so as to give the same output voltage.

ANSWER. From given data, the original input voltage is $60/120 = 0.5$ volt. Also the

$$\text{original distortion voltage} = 10 \times 60/100 = 6 \text{ volts}$$

To meet conditions, the distortion voltage has to be reduced 10 times. Then, from the relationship $1 - (\beta \times 120) = 10$, we calculate feedback factor β to equal $-\mathbf{0.075}$. Likewise, the gain also will be reduced by a factor of 10. Thus, $A_f = 120/10 = 12$. The additional distortionless gain required upstream of the feedback amplifier is $120/12 = \mathbf{10}$. This amplifier must supply a signal voltage of $60/12 = \mathbf{5}$ **volts**. β is feedback factor.

Q19-27. A series circuit of R, L, and C is resonant to an angular velocity ω of 10^6 radians per sec. The fractional bandwidth, $(\omega_2 - \omega_1)/\omega_r$, is 0.5 between quarter-power points. The quality factor Q of the coil used is 5 at resonance and its inductance is 200×10^{-6} henry. Calculate the necessary values of R and C for this circuit.

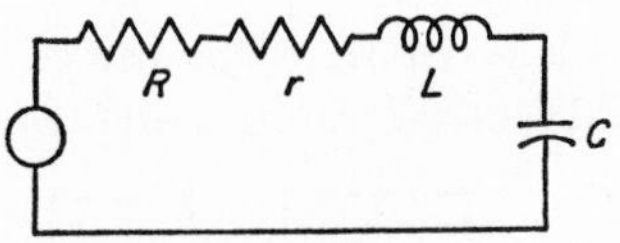

Fig. 19-36

ANSWER. Refer to Fig. 19-36. Also $R_T = R + r$, $Q = \omega L/r$, and

$$r = \frac{\omega L}{Q} = \frac{10^6 \times 200 \times 10^{-6}}{0.5} = 40 \text{ ohms}$$

Since $\omega C = 1/\omega^2 L$,

$$C = \frac{1}{(10^6)^2 \times 200 \times 10^{-6}} = \mathbf{0.005\ \mu f}$$

At $1/n$ power point the fractional bandwidth is given by

$$\frac{\omega_2 - \omega_1}{\omega_r} = \frac{R_T\sqrt{n-1}}{\omega_r L_T}$$

Solving for R_T,

$$R_T = \frac{\omega_2 - \omega_1}{\omega_r} \times \frac{\omega_r L_T}{\sqrt{n-1}}$$

where $\omega_r = 10^6$

$$L_T = 200 \times 10^6$$
$$n = 4$$

$$\frac{\omega_2 - \omega_1}{\omega_r} = 0.5$$

$$R_T = 0.5\,\frac{10^6 \times 200 \times 10^{-6}}{\sqrt{4-1}} = \frac{100}{\sqrt{3}}$$

But $R_T = R + r$ and $R = R_T - r = (100/\sqrt{3}) - 40 =$ **17.8 ohms**

Q19-28. A push-pull amplifier employing triodes is to operate into a 500-ohm resistive load. The individual plate-circuit characteristics of the tubes are such that the composite characteristics for the amplifier are straight lines with a slope corresponding to 2,000 ohms. The maximum composite current with zero bias on one tube is 0.1 amp. Neglecting the effects of distortion and assuming a sinusoidal input voltage to the grids, calculate: (*a*) the turns ratio (total primary to secondary) required to the ideal output transformer for maximum power output without grid current in the tubes; (*b*) the value of this maximum power.

ANSWER. Refer to Fig. 19-37. The use of two tubes in parallel provides the possibility of obtaining twice the output power of a

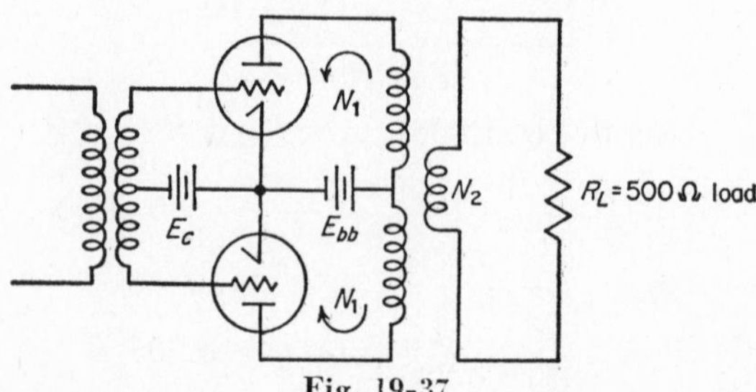

Fig. 19-37

single tube with the same distortion. To realize maximum power, the load resistance that is transferred to the primary coil should equate twice the plate resistance R_p of one tube. Assumption is that both tubes are identical. In this problem R_L must equal 2,000 ohms.

(*a*) Turns ratio $= \sqrt{2{,}000/500} = N_1/N_2 = \mathbf{2}$. Note from Fig. 19-37 that $N_1 = \frac{1}{2}$ turns of primary and $N_2 =$ turns of secondary. Turns ratio of entire push-pull transformer $= 2 \times 2 = \mathbf{4}$.

(*b*) Power output maximum $= (0.1^2 \times 4^2 \times 500)/8 =$ **10 watts.**

Q19-29. The circuit shown in Fig. 19-38 is to resonate at $10^6/2\pi$

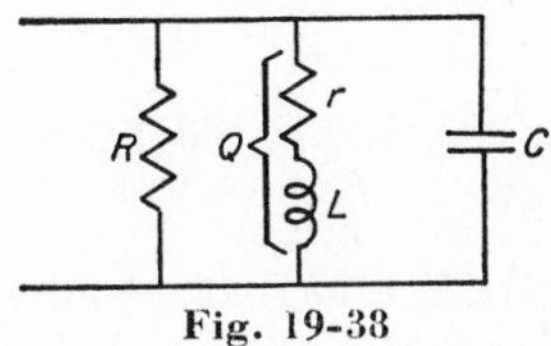

Fig. 19-38

cps and have a bandwidth between half-power points of $20{,}000/2\pi$ cps. The coil has an inductance of 0.001 henry and a Q factor of 100 which may be taken as constant over the band between the half-power points. Calculate the necessary values of R and C.

ANSWER. This is a condition of parallel resonance and the susceptance of the capacitive branch must equal the susceptance of the inductive branch in magnitude and size.

$$Q = \frac{\omega_r L}{r} = \frac{2\pi(10^6/2\pi)10^{-3}}{r}$$

from which $r = 10^3/100 = 10$ ohms. Continuing,

$$\omega_r C = \frac{X_L}{r^2 + X_L^2} = \frac{\omega_r L}{r^2 + (\omega_r L)^2}$$

Therefore,
$$C = \frac{L}{r^2 + (\omega_r L)^2} = \frac{10^{-3}}{100 + (10^6 \times 10^{-3})^2}$$
$$= \frac{10^{-3}}{10^6} = 10^{-9}\ \textbf{farad}$$

For parallel resonance bandwidth $= \Delta f = f_2 - f_1 = 1/(2\pi RC)$. Also, for half-power point $n = 2$,

$$G_T = \frac{2\pi C(f_2 - f_1)}{\sqrt{n-1}}$$

$$G_T = \frac{2\pi \times 10^{-9}(20{,}000/2\pi)}{\sqrt{2-1}} = 20 \times 10^{-6}$$

$$G_R = G_T - G_L = 20 \times 10^{-6} - \frac{r}{r^2 + X_L{}^2} = 20 \times 10^{-6} - \frac{10}{100 + 10^6}$$

from which G_R is found to be equal to 10^{-5}. Therefore,

$$R = \frac{1}{G_R} = 10^5 = \mathbf{100{,}000\ ohms}$$

Q19-30. Discuss the variation that would occur in the quiescent plate current of the vacuum-tube circuit of that shown if the resistor R_g became open-circuited. If R_g became short-circuited, would the quiescent plate current be affected? Could the device function as an amplifier under either of these circumstances?

ANSWER. Refer to Fig. 19-39. The figure shows biasing connections for a vacuum-tube-triode amplifier showing a grid-leak

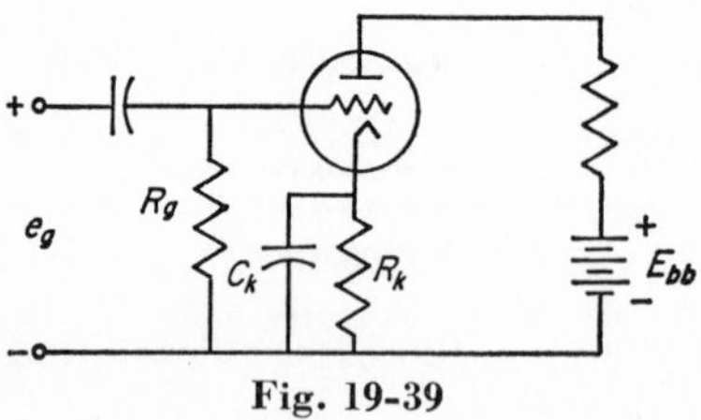

Fig. 19-39

resistance and a d-c blocking capacitor. If R_g became *open-circuited*, electrons would collect on the grid. Since the grid is not a collector, a large negative charge would be built up. This would then make the grid very negative with respect to the cathode and hence significantly change the quiescent point and much less plate current would flow. If R_g became *short-circuited*, the quiescent point would not be affected because the cathode-to-grid bias is achieved using the cathode resistor R_k.

The amplifying capabilities of the circuit would be severely curtailed in either case. For R_g open-circuited, it is possible that the tube would be sent into cutoff and hence no amplification would occur. For R_g shorted, a negligible signal would appear across the grid-cathode input. No amplification would take place.

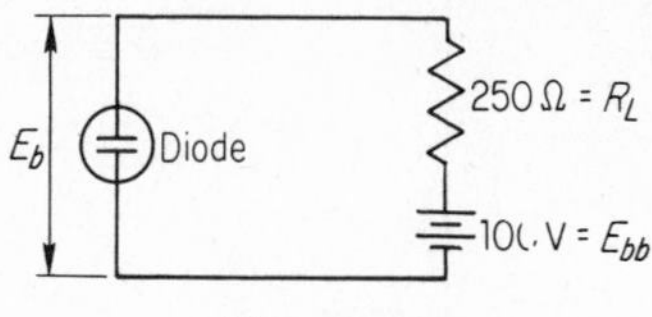

Fig. 19-40

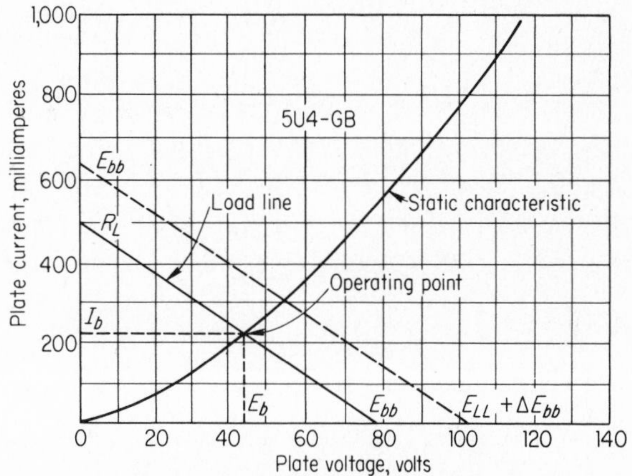

Fig. 19-41. Plate characteristic.

Q19-31. A 250-ohm resistor is placed in series with a 5U4-GB diode and a 100-volt battery, with rated filament voltage applied. Calculate the plate current, the plate voltage, and the plate dissipation.

ANSWER. Refer to Figs. 19-40 and 19-41. Then $E_b = E_{bb} - I_bR_L$ and $I_b = -E_b/R_L + E_{bb}/R_L$. Draw load line on static characteristic to determine the operating point. Actually, solving two equations, one in graphical form, $I_b = f(E_b)$. The other is load line. Load line intersects

$$I_b = 0 \qquad E_b = E_{bb} = 100 \text{ volts}$$
$$E_b = 0 \qquad I_b = E_{bb}/R_L = 100/250 = 0.4 \text{ amp or } 400 \text{ ma}$$

From plate-characteristic diagram, Fig. 19-41,

$$E_b = \mathbf{45\ volts} \qquad I_b = \mathbf{210\ ma} \qquad P_b = I_bE_b = 0.210 \times 45 = \mathbf{10.1\ watts}$$

Supplement **F**

INDUCTION MOTORS

Questions and Answers

F-1. Explain how the operating characteristics of a three-phase motor of the induction type are affected (*a*) when the applied voltage is reduced; (*b*) when the frequency of the supply system is reduced.

ANSWER.

(*a*) An illustration of the effect of reducing the voltage on an induction motor is found by starting it with autotransformers. Upon reducing the impressed voltage both the rotor current and torque are reduced. The action may be analyzed as follows: A reduction in the voltage applied to the stator causes a reduction in the value of the revolving field. Since the rotor current is proportional to the revolving field and the starting torque varies with the product of the field and rotor current, the torque will diminish as the square of the applied voltage. Thus, both current and torque decrease with the speed which, being independent of voltage, remains constant.

(*b*) For a given induction machine, having a certain number of poles, the speed of the revolving field is directly proportional to the current frequency, so that a reduction in frequency causes a decrease in speed. The starting torque is inversely proportional to the synchronous speed and, therefore, increases with a reduction in frequency. However, since the flux density in the magnetic circuit increases rapidly with a drop in frequency, the hysteresis loss becomes excessive, decreasing the motor efficiency. For this reason the induction motor should not be operated on a frequency less than the name plate or design rating.

F-2. A 5-hp four-pole 60-cycle three-phase induction motor takes 4,500 watts from the power supply. The rotor and stator or copper losses are 92 and 292 watts, respectively. The friction and windage losses are 72 watts, and the core loss is 225 watts. Calculate the torque, efficiency, and speed.

ANSWER. In a four-pole induction motor there are four poles per phase, or two pairs of poles per phase, so that it takes two cycles of current for the field to travel all the way around. The speed of the field, then, is

$$N_s = 60/2 \times 60 = 1{,}800 \text{ rpm synchronous speed}$$

This would be the speed if we disregard the slip. Now the motor output is equal to the input minus the losses, or

$$4{,}500 - (92 + 292 + 72 + 225) = 3{,}819 \text{ watts}$$

This wattage is equal to 3,819/746 = 5.11 hp equivalent. Now to determine the torque from Eq. (14-5)

$$(5.11 \times 33{,}000)/(2\pi \times 1{,}800) = \mathbf{14.9\ lb\text{-}ft}$$

$$\text{Motor efficiency is } 3{,}819/4{,}500 \times 100 = \mathbf{84.8\ per\ cent}$$

F-3. A three-phase Δ-connected induction motor operating on a 440-volt line has a lagging power factor of 0.75 and an efficiency of 0.85. It is delivering 20 hp. What is the line current and the current in the motor phase winding?

ANSWER.

$$\text{Motor input} = \frac{20 \times 746}{0.85} = 17{,}500 \text{ watts} = EI\sqrt{3}\cos\theta$$

$$\text{Line current } I = \frac{17{,}500}{E\sqrt{3}\cos\theta} = \frac{17{,}500}{440 \times 1.732 \times 0.75} = \mathbf{30.8\ amp}$$

Phase current i for Δ-connected motor = 30.8/1.732 = **17.8 amp.**

F-4. A three-phase Y-connected induction motor operating on 220 volts has a power factor of 70 per cent and an efficiency of 82 per cent when delivering 8 hp. What is the current in the phase windings and what is the voltage across each?

ANSWER.

$$\text{Motor output} = 8 \times 746 = 5{,}970 \text{ watts}$$
$$\text{Motor input} = 5{,}970/0.82 = 7{,}280 \text{ watts}$$

For a Y-connected motor, the motor input is equal to

$$EI\sqrt{3}\cos\theta = 7{,}280 \text{ watts}$$

Now $i = I = 7{,}280/(220 \times 1.732 \times 0.70) =$ **27.3 amp,** phase current. The voltage across each phase winding is

$$\frac{220}{1.732} = \mathbf{127\ volts}$$

F-5. A 20-hp three-phase induction motor operates at full load with 440 volts across its terminals. It takes 25 amp per line, lagging 30° behind the phase voltage. Determine the efficiency of this motor.

ANSWER.

$$\text{Efficiency} = \frac{\text{output}}{\text{input}} = \frac{20}{(440 \times 25 \times 1.732 \times \cos 30)/746} = 0.905$$

Thus, the efficiency is **90.5 per cent.**

F-6. Two 30-hp three-phase 60-cycle 208-volt induction motors are offered for a specific duty. Explain how you would test these machines to determine the better of the two, keeping in mind that the cost of the test must be a minimum. Name the tests to be conducted, give the directions for each test, state the significance of each, and list the equipment required.

ANSWER. The performance of an induction motor is an index of its suitability for a given task. The motor performance consists of its operating and starting characteristics. By comparing these characteristics, as determined by tests, it is possible to choose the better of the two motors.

The operating characteristics include efficiency, power factor, torque, and speed measured and plotted over a series of loads on the motor. A simple way of determining them is indicated by Fig. F-1. With a Prony brake as load, readings of the power input, input current, impressed voltage, and speed are taken for eight different torque values varying from zero to 150 per cent of the rated load. The two wattmeters are used to measure the power input. At low loads one wattmeter will read negative, whereupon its terminals are reversed and its reading subtracted (instead of being added) to that of the other wattmeter. The input current is measured by three ammeters and the impressed voltage by a voltmeter which is transferred from one line to the other as required. The speed is obtained by means of a tachometer, and the torque measured by a Prony brake.

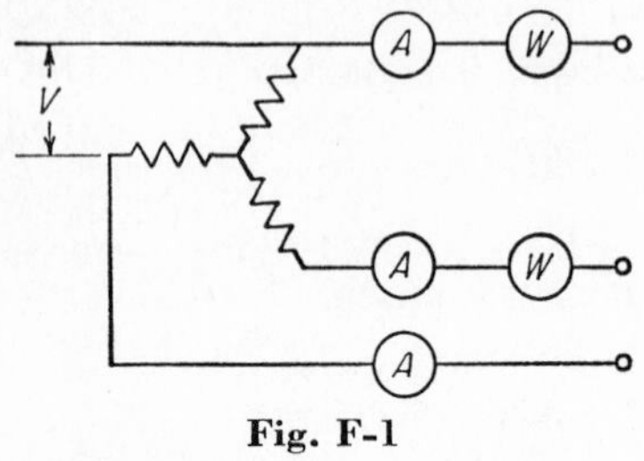

Fig. F-1

Per cent efficiency is calculated by the following.

$$\text{Efficiency} = \frac{\text{output}}{\text{input}} = \frac{2\pi NT \times 746 \times 100}{33{,}000 \times \text{power input}}$$

where T = torque, lb-ft
N = rpm of motor

The power factor is given as

$$\cos\theta = \frac{\text{power}}{EI\sqrt{3}}$$

where power is in watts

I = input current
E = impressed voltage

Of the operating characteristics the efficiency is the most important because it is an index of the operating cost. The power factor shows the character of the current in the line feeding the motor, and has a direct effect on the size of wire for the line. Torque and speed determine whether the motor can be used to drive a certain piece of equipment.

Starting characteristics consist of torque, rotor current, and stator input taken for several different values of the starting resistance. Voltmeter, ammeters, and wattmeters are connected to the stator and rotor circuits, as shown in Fig. 14-17 for the rotor circuit (Fig.

14-16 is for stator circuit). The rotor is locked with the Prony brake, and the voltage applied to the stator is reduced to one-half the motor-rated voltage to reduce or prevent overheating. The instrument readings of torque, rotor, and stator currents and the calculated value of the stator power factor are plotted for different positions of the external resistance. There is a value of the starting resistance which gives the maximum starting torque.

A comparison between the maximum starting torques decides which of the two motors is better suited to the given task, while the power factor and stator current are an index of the economy of the motor in starting up. Where the load is expected to vary widely, the pull-out torque, beyond which the motor ceases to run, is also determined. This is done by adding a load on the motor with the Prony brake, noting the speed at which the pull-out torque takes place at a rated impressed voltage.

F-7. What are the general induction-motor torque relationships?

ANSWER. With 40 per cent normal volts impressed on the stator terminals, a polyphase induction motor will take approximately 100 per cent normal full-load current and develop about 30 per cent full-load torque. At 60 per cent normal voltage it will take 250 per cent full-load current and develop 52 per cent full-load torque. At 80 per cent normal voltage the current will be about 400 per cent full-load value and the torque about 125 per cent that at full load. For the 100 per cent normal voltage the current will be 600 per cent full load and the torque 200 per cent full load. These will vary with the manufacturer and the design.

F-8. A dual-voltage three-phase induction motor has its leads burned off in the connection box. The leads cannot be identified. How would you go about identifying the leads in the position they come out through the motor frame into the connection box?

ANSWER. If your motor is a Y-connected motor it has nine leads. Figure F-2*a* shows how these may be connected and numbered for the purpose of discussion. You can identify the three coils that are common on the inside by a lamp, ringer, continuity tester, or other means. Then tag these three leads as 7, 8, and 9, respectively (see Fig. F-2*a*). Next, pick out the other three coils and keep their ends in pairs. Run the motor on the lower voltage with lines on 7, 8, and 9. Next connect one end of the coil to 7 and measure the voltage from the outer end to 8 and 9. If the voltages are balanced, you have the correct coil to go with 7, i.e., 1–4. Now reverse the coil and check the voltages again; it will still be balanced. If the voltage is higher than the first connection, tag the end that is on 7 with 4. The other end will be 1. Do likewise with the other phases, and all your numbers will be standard.

Since most manufacturers bring out the nine leads of dual-voltage motors through a single opening, the individual leads are tagged to correspond with the accompanying Fig. F-2*b*. If the leads are improperly tagged or burned off as stated in the problem, they may also be retagged by measuring the approximate resistance between the leads. For example, the resistance between 7 and 8, 8 and 9, or 7 and 9 would be approximately twice that of 1 and 4, 2 and 5, or 3 and 6. Referring to Fig. F-2*c* it is apparent that the leads 7, 8, and 9 are the opposite ends of the coils making the star. After the leads 1 and 4, 2 and 5, 3 and 6 have been tagged for trial and the leads 4 and 7, 5 and 8, 6 and 9 are connected together, a check should be made for reversal of windings 1 and 4, 2 and 5, 3 and 6. This will be indicated by noise, low torque, and heating when loaded. Once the single-circuit star connection is made and properly tagged, the connection diagram can be used for the two-circuit or low-voltage connection or vice versa.

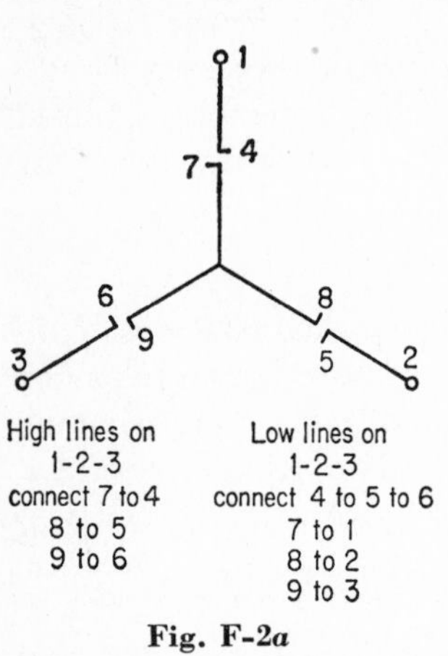

Fig. F-2*a*

F-9. A three-phase 208-volt 60-cycle squirrel-cage motor has 12 leads. These leads are not marked. Outline a procedure by means of which you would determine the proper arrangement of these leads so that the machine could be connected in Δ.

ANSWER. We must make assumptions as we go along, because the question does not give complete information.

A 208-volt motor with 12 leads means the pole groups are disconnected. The motor is rated at 208 volts, so that for maximum rating of 208 volts, the pole groups must be connected in series. It is a three-phase motor, which means three leads per phase or two groups per phase. On the surface this makes it look like a two-pole motor but that cannot be ascertained without a test. The number of poles does not matter at this stage of the testing procedure. The first operation is to trace the leads, using a low-voltage source with

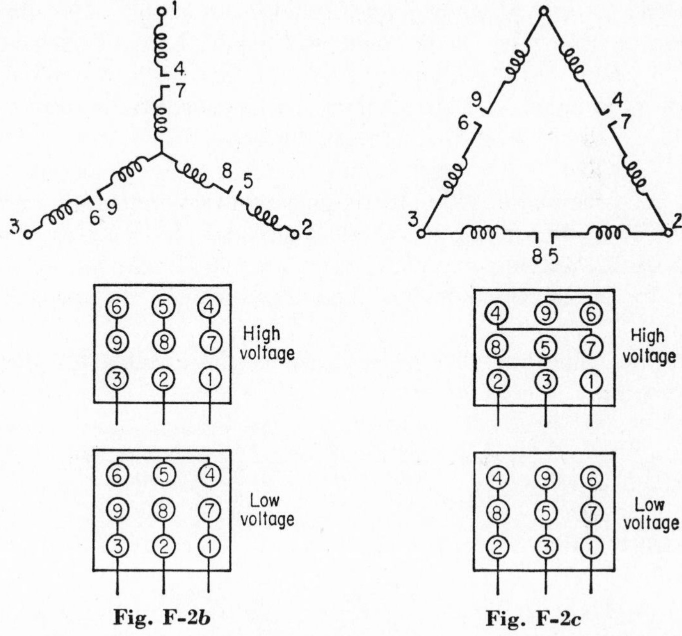

Fig. F-2*b* Fig. F-2*c*

a lamp in series (continuity test), and mark the lead pairs. Then remove one end bell and the rotor.

The next operation is to connect each pair of leads to a battery and with a compass inside the stator locate the pole and determine whether it is a south or north pole. Do this with each set of leads and mark the leads with the polarity of the battery. Mark the stator with the pole as indicated by the compass. Here is where you must determine the number of apparent poles connected to each pair of leads, because their position determines which two groups are to be connected in series. When this connection is made, a north pole must be diametrically opposite a south pole in the same phase group, and you then end up with six free leads.

The six free leads should be identified with battery polarity marks, so that the three phases can be connected properly. Remember that when the Δ connection is made, the phases should be in such relation to each other that they produce a rotating field.

Test the Δ connection by holding a small squirrel-cage rotor inside the stator and connecting the motor on about half voltage. The rotor will turn in one direction as it is moved around the stator circle by hand. Any stalling or reverse rotation indicates an incorrect phase relation and the Δ must be reconnected by reversing the set of leads leading to the reversed group.

F-10. A three-phase 440-volt six-pole 60-cycle induction motor delivers 50 hp to a pump. The efficiency and power factor of the motor are 92 and 90 per cent, respectively. Calculate the current drawn by the motor under this load and estimate the speed and torque.

ANSWER. The line current is given by the formula for three-phase motors.

$$I = \frac{\text{hp} \times 746}{\sqrt{3}\,E \times \text{eff} \times \text{pf}} = \frac{50 \times 746}{1.732 \times 440 \times 0.92 \times 0.90} = \mathbf{59.1\ amp}$$

From the equation

$$\text{Frequency} = \frac{\text{poles} \times \text{sync speed}}{120}$$

we can rearrange and solve for the synchronous speed.

$$N_s = \frac{120 \times 60}{6} = 1{,}200 \text{ rpm}$$

Now we must take into account the slip. This may be assumed to be 2 per cent. Then the rotor speed is $1 - 0.02 \times N_s$, or

$$1 - 0.02 \times 1{,}200 = 1{,}176 \text{ rpm}$$

This is the motor speed under actual operating conditions. Now for estimating the torque we use the familiar formula

$$\text{hp} = \frac{2\pi NT}{33{,}000}$$

By rearrangement

$$T = \frac{50 \times 33{,}000}{6.28 \times 1{,}176} = \textbf{225 lb-ft}$$

F-11. A 5-hp four-pole 60-cycle three-phase induction motor with a wound rotor draws 4,590 watts from the line. What is the speed of this motor when delivering 4.95 hp, if the core loss is 250 watts, the stator copper loss is 390 watts, and the friction and windage loss is 70 watts?

ANSWER.

Stator input (from problem) = 4,590 watts
Stator copper loss = 390 watts
Core loss = 250 watts
Rotor input (4,590 − 390 − 250) = 3,950 watts
Friction and windage = 70 watts
Load (4.95 × 746) = 3,690 watts
Rotor copper loss (3,950 − 3,690 − 70) = 190 watts

$$\text{Slip} = \frac{\text{rotor copper loss}}{\text{rotor input}} = \frac{190}{3{,}950} = 0.048, \text{ or } 4.8 \text{ per cent}$$

$$N_s = \frac{120 \times 60}{4} = 1{,}800 \text{ rpm}$$

$$\text{Rotor speed} = (1 - 0.048) \times 1800 = \textbf{1,714 rpm}$$

F-12. A four-pole 60-cycle three-phase induction motor draws 8,200 watts from the line. The losses in this machine are: stator copper loss = 300 watts, rotor copper loss = 160 watts,

core loss = 400 watts

friction and windage loss = 130 watts. Calculate the torque and speed.

ANSWER.

$$\begin{aligned} \text{Stator input (from problem)} &= 8{,}200 \text{ watts} \\ \text{Stator copper loss} &= 300 \text{ watts} \\ \text{Core loss} &= 400 \text{ watts} \\ \text{Rotor input } (8{,}200 - 300 - 400) &= 7{,}500 \text{ watts} \\ \text{Friction and windage} &= 130 \text{ watts} \\ \text{Rotor copper loss} &= 160 \text{ watts} \\ \text{Load } (7{,}500 - 160 - 130) &= 7{,}210 \text{ watts} \end{aligned}$$

The horsepower load is easily found from 7,210/746 = 9.67 hp. And the slip is 160/7,500 = 0.021, or 2.1 per cent.

$$N_s = \frac{120 \times 60}{4} = \mathbf{1{,}800\ rpm} \text{ sync speed}$$

$$\text{Rotor speed under operating conditions} = (1 - 0.021) \times 1{,}800 = \mathbf{1{,}762\ rpm}$$

$$\text{Torque} = \frac{9.67 \times 33{,}000}{6.28 \times 1{,}762} = \mathbf{28.5\ lb\text{-}ft}$$

F-13. A 5-hp six-pole three-phase 60-cycle induction motor operates with slip of 2 per cent and requires 11.0 amp and 3,500 watts when driving its usual load. When operating without load, this machine requires 4.3 amp and 290 watts. When the rotor of this machine is blocked, 440 watts at 52 volts are required to circulate a current of 14 amp. Calculate the horsepower output, the torque exerted, and the efficiency of this motor when driving its usual load.

ANSWER. Under *no-load conditions* the entire power input is equal to the stray losses, i.e., core losses, friction and windage losses, copper loss. For the *blocked-rotor condition* the power input equals the copper losses plus the iron losses. At the *usual load*

$$\text{hp output} = \frac{\text{input} - \text{losses}}{746}$$

Then, since the copper losses vary as the current squared, the losses are found.

$$\text{Copper loss at no load} = (4.3/14)^2 \times 440 = 41.5 \text{ watts}$$
$$\text{Copper loss at usual load} = [^{11}\!/_{14}]^2 \times 440 = 272 \text{ watts}$$
$$\text{Stray-power losses} = 290 - 41.5 = 248.5 \text{ watts}$$
$$\text{Total losses at usual conditions} = 272 + 248.5 = 520.5 \text{ watts}$$
$$\text{hp output} = (3{,}500 - 520.5)/746 = \mathbf{4\ hp}$$
$$N_s = (120 \times 60)/6 = 1{,}200 \text{ rpm}$$
$$\text{Rotor speed at operating conditions} = 1{,}200 - (0.02 \times 1{,}200) = \mathbf{1{,}176\ rpm}$$
$$\text{Torque} = \frac{4 \times 33{,}000}{6.28 \times 1{,}176} = \mathbf{18\ lb\text{-}ft}$$
$$\text{Efficiency} = \frac{\text{output}}{\text{input}} = \frac{4 \times 746}{3{,}500} = 0.85, \text{ or } \mathbf{85\ per\ cent}$$

F-14. The losses at full load in a 10-hp 220-volt three-phase induction motor are as follows: stator copper loss = 4.1 per cent, friction and windage = 4.0 per cent, core loss = 3.0 per cent,

$$\text{rotor copper loss} = 4.8 \text{ per cent}$$

Calculate the slip and efficiency of this machine at half and at full load.

ANSWER. The core, friction, and windage losses are constant over the entire load range. Assume 1 per cent stator copper loss and 1 per cent rotor copper loss at no load. Then at *half load*

$$\text{Efficiency} = \frac{5 \times 746}{(5 \times 746) + \underset{\substack{\text{(Friction and}\\ \text{windage plus}\\ \text{core loss)}}}{(0.07 \times 5 \times 746)} + \underset{\substack{\text{(Total copper}\\ \text{loss at}\\ \text{one-half load)}}}{(0.055 \times 5 \times 746)}}$$

$$= \frac{1}{1 + 0.07 + 0.055} = 0.89, \text{ or } \mathbf{89\ per\ cent}$$

$$\text{Slip} = \frac{\text{rotor } I^2R \text{ loss}}{\text{secondary input}}$$

Also the secondary input is equal to output − motor stator copper loss. Thus, in accordance with the above we see

$$\text{Slip} = \frac{\{[1 + (4.8 - 1)/2] \times 0.01\} \times (5 \times 746)}{[(5 \times 746)/0.89] - \{[1 + (4.1 - 1)/2] \times 0.01\} \times (5 \times 746)} = 0.0265$$

Thus, in percentage the slip is equal to **2.65 per cent.** The above follows, since

$$1 + (4.8 - 1)/2 \times 0.01 = 0.029 \text{ by interpolation}$$
$$1 + (4.1 - 1)/2 \times 0.01 = 0.026 \text{ by interpolation}$$
$$\text{Total} = 0.055$$

F-15. A 30-hp three-phase 440-volt squirrel-cage motor operates at full load with an efficiency of 0.92, a power factor of 0.90, and a slip of 0.02. Calculate the torque, the speed, and the current drawn by this machine at full load. The number of poles is six. The motor runs on 60-cycle current.

ANSWER.

$$N_s = (120 \times 60)/6 = 1{,}200 \text{ rpm}$$
$$\text{Slip } s = 0.02 \times 1{,}200 = 24 \text{ rpm}$$

Rotor speed $= 1{,}200 - 24 =$ **1,176 rpm** operational speed under load

The input is the output/the efficiency. Also we know that the input is given by

$$\sqrt{3}\, EI \cos \theta$$

Then by rearrangement and calculating for I

$$I = (30 \times 746)/(0.92 \times 1.732 \times 440 \times 0.90) = \mathbf{35.5\ amp}$$
$$T = (30 \times 33{,}000)/(6.28 \times 1{,}176) = \mathbf{134\ lb\text{-}ft}$$

F-16. A 10-hp 550-volt 60-cycle three-phase induction motor has a starting torque of 160 per cent of full-load torque and a starting current of 425 per cent of full-load current. (*a*) What voltage is required to limit the starting current to the full-load current value? (*b*) If the motor is used on 440-volt lines and a 60-cycle system, what are the starting torque and starting current expressed in per cent of the full-load values?

ANSWER.

(*a*) The starting inrush current varies directly as the applied voltage. Thus, in this problem: 550/4.25 = **130 volts** will limit the starting current.

(*b*) We already know that the starting torque varies as the square of the voltage. Thus, at 440 volts

$$[440/550]^2 \times 1.60 = 1.02, \text{ or } \mathbf{102\ per\ cent} \text{ of the rated torque}$$
$$440/550 \times 4.25 = 3.4, \text{ or } \mathbf{340\ per\ cent} \text{ of the rated current}$$

F-17. A 500-hp 4,160-volt three-phase induction motor is required for service in an atmosphere likely to contain appreciable quantities of inflammable gases. (*a*) Specify the type of frame for this motor and the types of starting and control equipment to ensure safe operation. (*b*) Make a circuit diagram to show how all the equipment specified is to be connected. (*c*) Point out any special features that are required in the equipment specified because of the presence of these gases.

ANSWER. This installation must be made in accordance with the requirements of the NEC for atmospheres containing explosive or inflammable gases or vapors. See the latest edition of the NEC for applications in hazardous locations. The locations are further broken down into division 1 and division 2 areas.

(*a*) The motor frame must be totally enclosed, either fan-cooled or nonventilated, and suitable for installation in class I, division 1 areas. As for the starting and control equipment for this area, complete explosion-proof equipment is of three general types:

Cast-iron enclosures, having machined flanges and bolts around the periphery on at least 6-in. centers (600 volts only).

The threaded-pipe arrangement, frequently referred to as the top-hat design (600 volts only).

Oil-immersed, in which all arcing contacts are under at least 6 in. of oil for all voltages. Oil-immersed equipment finds considerable application in corrosive areas, since, in effect, it is a type of hermetic sealing.

(*b*) To make up the connections, rigid metal conduit with threaded explosion-proof joints and explosion-proof boxes and fittings should be employed. For each threaded joint at least five threads should be made up. When it is necessary to use flexible connections, as at motor terminals, approved connections should be used. The conduit must be sealed in an approved manner to prevent the passage of the vapors from one part of the electric installation to another. Seals should be placed at least but no more than 18 in. from enclosures such as motors and switchgear (see Fig. F-3).

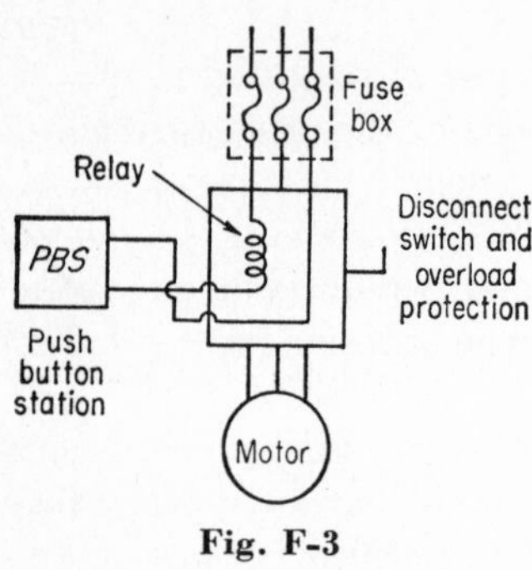

Fig. F-3

(*c*) The one special feature is the requirement that flame must not be able to pass from the enclosure (motor or switchgear) to the outside. Such equipment is flameproof.

It is worth while to point out here that what constitutes division 1 and division 2 areas is generally not too well known. *For class I, division 1 areas any one of the following conditions exists:*

A hazardous gas or vapor concentration in the surrounding air during normal operation of the process equipment.

A flammable atmospheric concentration occurs frequently because of maintenance, repairs, or leakage.

Failure of process, storage, or other equipment is likely to cause an electric system failure simultaneously with the release of flammable gas or liquid.

The area is not freely ventilated and is one in which flammable liquids or vapors are handled or processed in other than a suitable, well-maintained piping system without valves, screwed or flanged fittings, and meters; or they are stored in other than suitable closed containers.

The area below the surrounding elevation or grade is such that flammable liquids or vapors may accumulate therein.

For division 2 areas the following holds:

The location is freely ventilated and is one in which flammable liquids or vapors are processed or handled in closed systems other than suitable, well-maintained piping systems consisting only of pipes, valves, fittings, and meters from which they can escape only during abnormal conditions, such as accidental blowing of a gasket or rupture of a pipe; or they are stored in other than suitable closed containers.

The location is adjacent to a division 1 area, or vapor can be communicated to the location through trenches, pipes, conduits, or ducts.

When positive mechanical ventilation is used, failure or abnormal operation of the ventilating equipment permits atmospheric vapor mixtures to build up to hazardous concentrations.

In general, it may be said that for division 1 areas the criterion for these locations is that they are likely to be hazardous under normal conditions. The criterion for division 2 areas is that they are likely to be hazardous only under abnormal conditions, such as failure or rupture of the equipment.

F-18. A 440-volt 60-cycle four-pole three-phase wound-rotor induction motor is directly connected to a pump which delivers 1,000 cfm of water against an effective head of 8.7 ft. Under this load the motor draws 15.62 kw at a power factor of 0.92. When operated without load, the motor draws 803 watts. The stator and rotor resistances per phase are 0.202 and 0.022 ohm, respectively. The effective turns ratio between the stator and rotor is 4:1. Calculate the efficiency of the pump.

ANSWER. The pump efficiency is equal to the output divided by the input. Thus

$$\text{Output hp} = \frac{\text{cfm} \times \text{density of water} \times \text{total effective head}}{33{,}000}$$

$$= \frac{1{,}000 \times 62.4 \times 8.7}{33{,}000} = 16.45 \text{ hp}$$

The input to the pump is equal to the motor output. This in turn is equal to the motor input minus the losses. The motor losses will now be determined. The primary or secondary copper losses are equal to the product of the number of phases, the effective primary or secondary copper resistance per phase, and the square of the current per phase. The total copper loss is the sum of the primary and secondary copper losses. Thus,

$$\text{Stator copper loss} = 3I_s^2R_s$$
$$\text{Rotor copper loss} = 3I_r^2R_r$$

The missing current values must first be determined. In order to do this, we may assume that the motor is one with its stator and rotor windings Y-connected. The stator current is

$$I_s = \frac{\text{watts input}}{\sqrt{3}\,E \cos\theta} = \frac{15{,}620}{\sqrt{3} \times 440 \times 0.92} = 22.2 \text{ amp}$$

$$I_r = \text{turns ratio} \times I_s = 4 \times 22.2 = 88.8 \text{ amp}$$

Now referring back to the equations for stator and rotor copper losses

$$\text{Stator copper loss} = 3 \times 22.2^2 \times 0.202 = 298 \text{ watts}$$

$$\text{Rotor copper loss} = 3 \times 88.8^2 \times 0.022 = 521 \text{ watts}$$

Since the friction, windage, and iron losses are constant throughout the motor operation

$$\text{Friction, windage, and iron losses} = 803 \text{ watts}$$

$$\text{Total motor losses} = 298 + 521 + 803 = 1{,}622 \text{ watts}$$

The motor output is now

$$\text{motor output} = 15{,}620 - 1{,}622 = 13{,}998 \text{ watts}$$

$$\text{Finally, the pump efficiency} = 16.45 \times 746/13{,}998 = 0.879, \text{ or } \mathbf{87.9 \text{ per cent}}$$

F-19. A 50-hp 440-volt three-phase induction motor delivers during the starting period 150 per cent of the normal torque and draws 550 per cent of the rated current with the rated voltage. The full-load power factor and full-load efficiency of this motor are 80 and 90 per cent, respectively. If the starting torque of the load is only 50 per cent of the rated torque of the motor, and an autotransformer is used as a starting unit, what should be the starting voltage to the motor? What is the starting current in the line?

ANSWER. Since the starting current is proportional to the impressed voltage, and since the starting torque is proportional to the square of the impressed voltage, then if we let E_o be the reduced impressed voltage, we can find 50 per cent normal torque compared to 150 per cent at the rated voltage as follows: First determine the rated input current from the standard relationship.

$$\frac{50 \times 746}{1.732 \times 440 \times 0.8 \times 0.9} = 68 \text{ amp}$$

$$\text{Starting current at rated voltage} = 5.50 \times 68 = 375 \text{ amp}$$

Now let us go on to find the reduced impressed voltage E_o. The following holds.

$$\frac{\text{Starting torque } (0.5)}{\text{Torque at rated voltage } (1.5)} = \frac{E_o^2}{E^2}$$

where E = impressed voltage

$$\frac{0.5}{1.5} = \frac{E_o^2}{440^2}$$

from which we find

$$E_o = \frac{440}{1.732} = \textbf{255 volts} \text{ starting}$$

$$\text{Starting current at this voltage} = I_o = \frac{375 \times 255^2}{440^2} = \textbf{126 amp}$$

$$\text{Starting current in line} = (255 \times 217)/440 = \textbf{126 amp}$$

F-20. A motor-driven pump is required to deliver 400 gpm of water against a head of 200 ft for 2,000 hr each year. Bids are offered by two concerns for pumps having an expected life of 15 years with a salvage value equal to the cost of removal. The cost of interest, taxes, and insurance may be taken as 8 per cent of the purchase price, and the cost of power is 1.7 cents per kwhr. Bid *A* guarantees an over-all efficiency of 79 per cent at full load, whereas bid *B* guarantees an over-all efficiency of 75 per cent. How much more is the client justified in paying for the pump and motor with the higher efficiency?

ANSWER. The hydraulic horsepower required for pumping against the head given is given by the formula: gpm × lb per gal × ft head/33,000

$$(400 \times 8.33 \times 200)/33{,}000 = 20.2 \text{ hydraulic hp}$$

$$\text{kw equivalent} = (20.2 \times 746)/1{,}000 = 15.1 \text{ kw}$$

For bid *A* (at 79 per cent efficiency): 15.1/0.79 = 19.1 kw input
For bid *B* (at 75 per cent efficiency): 15.1/0.75 = 20.1 kw input

The annual savings in cost of electric energy by bid A is found to be

$$2{,}000 \times 0.017 = \$34.00 \text{ saved per year per kw}$$

If we let X equal the additional capital investment made by bid A, then

$$0.08\,X = \text{additional annual fixed charges}$$

And $X/15$ = recovery of X. Then the balance or "break-even" point is additional annual fixed charges + recovery = 34

$$0.08\,X + X/15 = 34. \quad \text{Solving for } X \text{ it is found to be } \mathbf{\$232}$$

F-21. Water is required at 50 cfm at a point 120 ft above the surface of a lake. The pipe friction is estimated as 10 ft. The pump efficiency is 75 per cent; the efficiency of the three-phase 440-volt induction motor driving the pump is 90 per cent at a power factor of 0.95. Calculate: (*a*) the annual power cost of driving this pump 10 hr a day for 300 days a year, assuming the cost of power to be 0.5 cent per kwhr; (*b*) the current drawn by the motor.

ANSWER.

(*a*) $$\text{kw input} = \frac{(120 + 10) \times 50 \times 62.4}{33{,}000 \times 0.75 \times 0.90} \times 0.746 = 13.56 \text{ kw}$$

$$\text{Power costs} = 10 \times 300 \times 13.56 \times 0.5 = \mathbf{\$203}$$

(*b*) $$\text{Current drawn} = \frac{13.56 \times 1{,}000}{1.732 \times 440 \times 0.95} = \mathbf{18.7\ amp}$$

F-22. A 60-cycle 16-pole three-phase motor has a slip of 6 per cent at full load. At what speed does the armature turn at this loading?

ANSWER. $$\text{Slip } s = 450 \times 0.06 = 27 \text{ rpm}$$

$$\text{Armature speed at full load} = 450 - 27 = \mathbf{423\ rpm}$$

F-23. A certain industrial process requires 60 gal of water per min at 90°F. Well water is available at 50°F and must be raised 120 ft. A 110-volt supply is available and a direct connected motor pump is to be used with an electric heating coil inserted in the pump discharge. For this load the pump efficiency is 67 per cent and the motor efficiency is 80 per cent. Electric energy sells for 2 cents a kwhr. Neglecting the friction and heat losses in the pipeline and wiring, determine the following: (*a*) the power to pump the water; (*b*) the power to heat the water; (*c*) the total cost of energy per gallon water processed.

ANSWER.

(*a*) Power required to raise water

$$\frac{60 \times 8.33 \times 120}{0.80 \times 0.67 \times 33{,}000} \times 0.746 = \mathbf{2.53\ kw}$$

(*b*) Heat to raise the water temperature

$$60 \times 8.33 \times 1 \times (90 - 50) = 20{,}000 \text{ Btu per min}$$

$$\text{Power required} = \frac{20{,}000 \times 778 \times 0.746}{33{,}000} = \mathbf{352\ kw}$$

(*c*) Total energy cost per gallon

$$\frac{2.53 + 352 \times 2¢}{60 \text{ gpm} \times 60} = \mathbf{0.197\ cents\ per\ gal}$$

F-24. Explain how you would reverse the direction of rotation of each of the following types of motors: (*a*) a compound d-c motor; (*b*) a single-phase repulsion-induction motor; (*c*) a capacitor motor; (*d*) a three-phase three-wire induction motor. What precautions must be taken to ensure proper operation?

ANSWER.

(*a*) *Compound d-c motor.* Reverse the polarity of the armature and series field as a unit. This will retain the compounding action. Watch out for the fact that in many machines the armature connection also includes the series field; i.e., the series field is not always brought out separately, so that sometimes, when reversing the armature polarity, you will change the cumulative effect.

(*b*) *Single-phase repulsion-induction motor.* Reverse by shifting the brushes from one side of the electric neutral to the other by shifting the line of the short-circuited brushes 90°. This should not be done while the machine is running, as that would not reverse the direction of rotation because the motor will be running primarily as an induction motor.

(*c*) *Capacitor motor.* Reverse by connecting the capacitor across the other winding. No trouble will result if this is done before starting, as long as the two windings are alike. This is usually the case for capacitor motors, meant to operate with the capacitor in the circuit. If the motor is a capacitor-start induction-run motor, the starting winding may be built for intermittent duty only, and it will burn up if it is made the running winding. The starting winding in this type of motor is switched out of the circuit when the motor gets up to speed. This is accomplished by means of a centrifugal switch. Another way (for both cases) would be to reverse the connection of the capacitor and its series winding as a unit.

(*d*) *Induction motor.* This is the simplest and easiest of all to reverse. Merely interchange any two of the three leads. No precautions are necessary, assuming, of course, that the load can safely be reversed, and that the shaft-locking nut cannot unwind.

For proper operation check the lubrication and ventilation. Do not place the motor in hot areas when not running.

F-25. Indicate how the no-load and locked-rotor data for a 50-hp 440-volt three-phase squirrel-cage induction motor may be obtained, and show how these data may be used to determine the circle diagram for the machine. Show clearly how the locked-rotor point is determined on this diagram and how the torque, slip, and efficiency may be calculated.

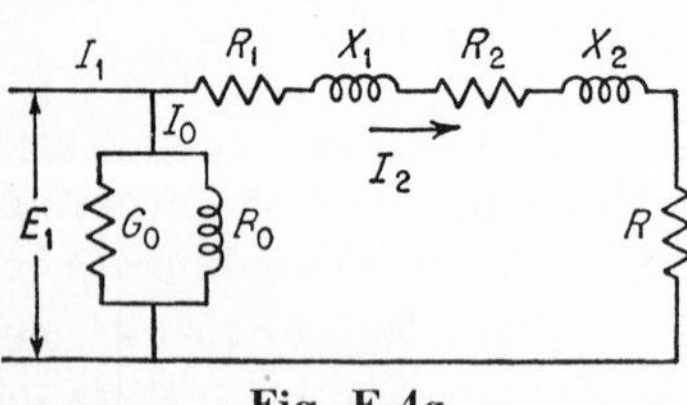

Fig. F-4*a*

ANSWER. Although it is possible to determine the operating characteristics of an induction motor mathematically by means of an equivalent circuit (see Fig. F-4*a*) it is simpler and more convenient to use a circle diagram. In Fig. F-4*b* the shunt circuit is connected outside the stator impedance, and the shunt current I_o does not flow in the stator impedance. Except in the smaller motors this introduces no practical error.

From the equivalent circuit the *circle diagram* may be constructed. In a series circuit it may be remembered that, if the reactance remains constant and the resistance varies, the locus of the current vector is a circle. Hence, if the energy current and the quadrature current vectors are plotted, one as a function of the other, their vector sum always being the total current, the locus of their resultant is a circle. In the circuit of Fig. F-4*a* the current I_o to the shunt circuit is constant. This current I_o which supplies the exciting current, the hysteresis and eddy current, and the friction and windage losses does not flow into the stator winding. In the circuit the reactances X_1 and X_2 and the resistance R_2 are all substantially

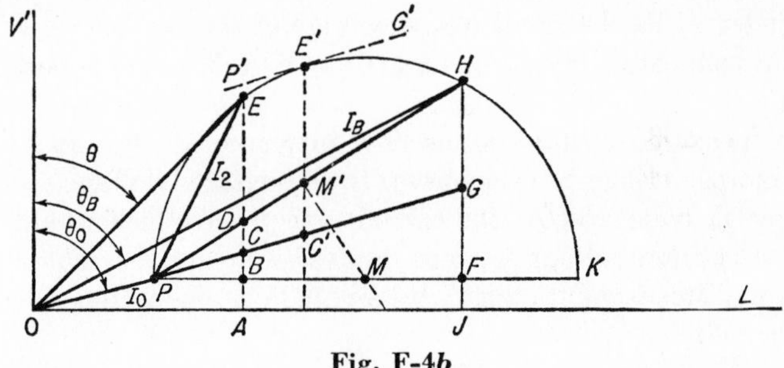

Fig. F-4*b*

constant, but R varies with the loading. Hence, as previously stated, the locus of the current I_2 is a circle.

Since the total current I to the motor from the line is the sum of this variable current I_2 and the constant current I_o, the locus of I is also a circle. Thus, in Fig. F-4*b* with changes of loading, the locus of the motor current I (point E) is the circular arc $PEHK$. This diagram gives approximate results in that the impedance drop and the copper loss in the stator due to magnetizing and core-loss currents are neglected.

The position of the voltage vector V' is taken along the Y axis. Data for the construction of this diagram are obtained from an open-circuit and a short-circuit (or blocked) test. Using the data obtained from these two tests, the operation of the motor may be determined with a fair degree of accuracy by the use of such a circle diagram.

Now for the *procedure for determining the circle diagram* continue as follows. First run the motor at a rated voltage without load, and the line voltage V, the line current I_o, and the total wattage P_o are measured. The no-load power-factor angle θ_o can then be determined ($\cos \theta_o = P_o/\sqrt{3}\, VI_o$ for a three-phase motor). The voltage per phase is laid off vertically in Fig. F-4b, and the no-load current I_o (per phase) is laid off at an angle θ_o from V' and lagging. Now block the rotor.

So that the current may be kept within reasonable limits, the supply voltage per phase is reduced to voltage e', which should be such value as to give a short-circuit current approximately equal to the rated current. The phase current $I_{B'}$, the total power P', and the phase voltage e' are then measured under these conditions.

Now let V' be the rated phase voltage of the machine. $V' = V$ for a Δ-connected machine, and $V' = V/\sqrt{3}$ for a Y-connected machine.

The measured current $I_{B'}$ is now increased in the ratio of the rated motor voltage V' (per phase) to the reduced voltage e'. This will give I_B equal to OH, the current per phase which would exist, if the rated line voltage V were impressed across the motor when blocked. This current lags V' by an angle θ_B, for which

$$\cos \theta_B = \frac{P'}{nI_{B'}e'}$$

$$I_B = I_{B'} \times \frac{V'}{e'}$$

where n = number of phases involved

Now draw OL making an angle of 90° with OV' in a clockwise direction. I_B equal to OH is laid off, making an angle θ_B with OE'. Points P and H on the circle are determined in the same way.

Draw line PH, and then draw PK parallel to OL. It is not necessary to know point K in order to construct the diagram.

With PK as diameter, draw a semicircle through points P and H. The center M of this semicircle is found by erecting a perpendicular MM' at the center of PH. The intersection of MM' with PK provides us with the center M of the circle. With MP as a radius and M as center, the semicircle $PEHK$ may be drawn. PK is the diameter of the semicircle, and its length in amperes is $PK = V'/(x_1 + x_2)$, where E' is the phase voltage and x_1 and x_2 are the stator and rotor reactances per phase, referred to the stator.

Drop perpendicular HJ from H to OL. HF is divided by G into two segments, such that $HG/GF = I_2{}^2R_2/I_1{}^2R_1$; that is, in proportion to the secondary and primary resistances as a 1:1 ratio of the rotor to the stator turns is assumed. Now draw line PG.

At any load current I, I_2 ($= PE$) is the secondary current, being equal to $I - I_o$ vectorially. EA is the energy component of the current I, and, therefore, the total power input per phase

$$P_1 = EA \times V'$$

The core and friction losses are

$$\begin{aligned} P_c &= BA \times V' \text{ per phase} \\ \text{Primary copper loss } I_1^2R_1 &= BC \times V' \text{ per phase} \\ \text{Secondary copper loss } I_2^2R_2 &= CD \times V' \text{ per phase} \\ \text{Output } P &= DE \times V' \text{ per phase} \\ \text{Efficiency} &= \frac{DE}{AE} \\ \text{Torque } T &= CE \text{ (to scale)} \\ \text{Slip } s &= \frac{CD}{CE} \\ \text{Power factor} = \cos\theta &= \frac{EA}{I} \end{aligned}$$

Draw $P'G'$ parallel to PG and as a tangent to the circle at E'.

$$\text{Breakdown torque } T_B = C'E' \text{ (to scale)}$$

The circle diagram (Fig. F-4*b*) is drawn for but one phase of the motor. The values of power, losses, and torque must be multiplied by n if the motor has n phases.

The torque scale may be found as follows: The torque is equal to a constant times the power, divided by the speed, the value of the constant depending on the units adopted. The power output per phase is $P = V' \times DE$. The rotor speed $N_2 = N(1 - s)$, where N is the synchronous speed in rpm.

$$N_2 = N\left(1 - \frac{CD}{CE}\right) = \frac{N(CE - CD)}{CE} = \frac{N \times DE}{CE} \tag{1}$$

The torque developed per phase

$$T' = K \times \frac{P}{N_2} = K \times \frac{E' \times DE}{N \times DE/CE} = \frac{K \times V' \times CE}{N} \tag{2}$$

where K is a constant

$V' \times CE$ is *the total power per phase delivered to the rotor*

The total power delivered to the rotor by n phases

$$P_2 = nV' \times CE \qquad \text{watts}$$

The horsepower output

$$\text{hp} = \frac{nDE \times V'}{746} = \frac{2\pi N_2 T}{33{,}000} \tag{3}$$

where T is the total torque

But $$N_2 = \frac{N \times DE}{CE} \qquad \text{from Eq. (1)}$$

Substituting in Eq. (3)

$$\frac{nDE \times V'}{746} = \frac{2\pi(N \times DE)T}{CE \times 33{,}000}$$

$$T = 7.04 \times \frac{nV' \times CE}{N} \qquad \text{lb-ft}$$

$$K = 7.04n \tag{4}$$

Since the number of phases n, the voltage E', and the synchronous speed N are usually fixed, the torque $T = K' \times CE$, where

$$K' = \frac{7.04nV'}{N}$$

F-26. A 20-hp 208-volt 1,750-rpm three-phase 60-cycle induction motor draws 5.5 times the full-load current and develops 1.6 times the full-load torque when starting under the rated voltage. If the full-load current is 52 amp, calculate (*a*) the starting voltage which will produce a full-load torque, (*b*) the starting current produced by this starting voltage, (*c*) the primary current drawn by an auto-transformer starter.

ANSWER. You will recall that for an induction motor its starting current will vary directly as the impressed voltage, and as for the torque developed in starting, this will vary as the square of the voltage impressed.

(*a*) $$\frac{1.6 \times \text{full-load torque}}{\text{full-load torque}} = \frac{208^2}{\text{starting voltage}^2}$$

By rearrangement and solving for starting voltage $= 208\sqrt{1/1.6} =$ **165 volts,**

(b) Starting current $= {}^{165}\!/_{208} \times 5.5 \times$ full-load current
$= {}^{165}\!/_{208} \times 5.5 \times 52 =$ **226 amp**

(c) Primary current $= {}^{165}\!/_{208} \times 226 =$ **179 amp**

F-27. An induction regulator consisting of two 120-volt and two 48-volt coils is to be connected to deliver 160 volts when the input is 120 volts. Show by means of a diagram how the regulator should be connected to provide the maximum power output.

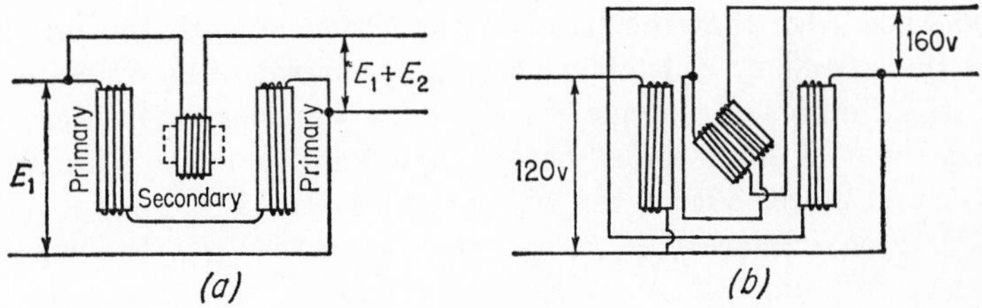

Fig. F-5

ANSWER. *Foreword.* Without auxiliary apparatus it is practically impossible to maintain a proper voltage at all the distribution points of a system. This is so because with a fixed voltage at the station bus bars, the voltage at the ends of the short feeders will ordinarily be greater than the voltage at the ends of the longer feeders. Owing to the ohmic and reactive drops in the lines, the voltage at the end of the feeder may vary considerably with the load on the feeder. In order to maintain a more constant voltage at the distribution point, without using an excessive amount of copper, an induction regulator is often connected to each feeder. This machine maintains the voltage at the distribution point practically constant.

An induction regulator is actually a transformer with its primary and secondary coils wound as in an induction motor. The primary is connected across the source of the voltage, while the secondary is left open to form one side of the output line, as shown in Fig. F-5*a*.

While the primary coils remain stationary, the secondary coils may be rotated to the position indicated by the dotted line. The action is as follows: With a voltage E_1 impressed on the primary, a magnetizing flux is set up which flows through the secondary. This flux, then, induces a current in the secondary, proportional to its number of turns, and generates a voltage E_2 which, by virtue of the connections shown, is added to E_1, giving the sum $E_1 + E_2$ as the output voltage. If the secondary is turned 90°, the flux set up by the primary will not thread it and E_2 becomes zero, so that the output voltage becomes E_1, the impressed voltage.

Reverting to the problem at hand, there are evidently two ways in which both the primary and the secondary coils may be hooked up in their respective circuits, i.e., in series or in parallel. It is common knowledge that the parallel connection of both the primary and the secondary coils allows a greater current to be drawn for a constant impressed voltage E_1, since the impedance Z_1 is greater when the coils are in series than when in parallel. Therefore, for a maximum power output the coils must be placed in parallel (refer to Fig. F-5*b*). Since $E_1 + E_2 = 120 + 48 = 168$ volts for a position of the maximum voltage, it will be necessary to turn the secondary to obtain 160 volts.

F-28.

(*a*) The efficiency of a 550-volt three-phase induction motor is 90 per cent when the line current is 100 amp per wire at a power factor of 0.92. Calculate the output of the motor.

(*b*) Explain why an induction motor having a very low rotor resistance with respect to its reactance gives a poor starting torque.

ANSWER.

(*a*)

$$\text{Let } E = \text{impressed line voltage}$$
$$I = \text{line current}$$
$$\cos\theta = \text{power factor}$$

Then, the motor input $= EI\sqrt{3}\cos\theta$ for a three-phase system.

$$550 \times 100 \times 1.732 \times 0.92 = 87{,}700 \text{ watts}$$
$$\text{Motor output} = \text{input} \times \text{efficiency (decimal)}$$
$$= 87{,}700 \times 0.90 = 79{,}000, \text{ or } \mathbf{79\ kw}$$

(*b*) With a very low motor resistance with respect to the motor reactance, the current in the bars will lag considerably behind the impressed voltage, so that the maximum current at any given time will not coincide with the maximum voltage. Now, the torque developed is proportional to the product of current and flux, according to the expression: Torque $= k\phi I$. Since the flux density is greatest where the emf is greatest, its maximum will not coincide with the current maximum, and the torque will be proportionately less.

A lagging current also means that the motor conductors carrying current in a positive direction will be situated in a negative magnetic field with further lessening of the developed torque.

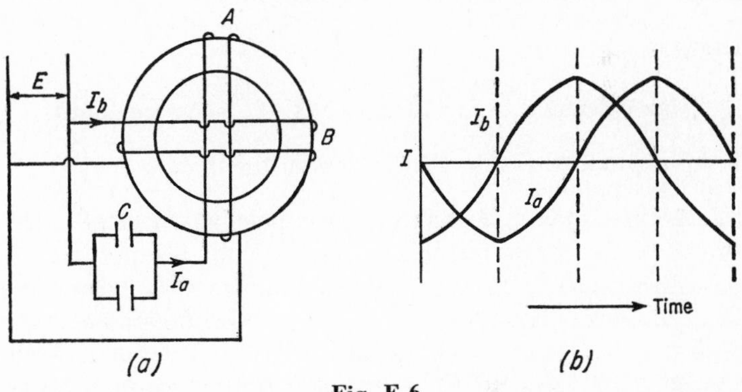

Fig. F-6

F-29. Explain how you would reverse the direction of rotation of a capacitor-start capacitor-run induction motor.

ANSWER. One way of starting a single-phase induction motor consists in "splitting" the single-phase power supply E into two phases, A and B. On the capacitor-start capacitor-run induction motor this is accomplished by connecting the windings of phase A with two capacitors C in parallel, as indicated by the diagram (Fig. F-6*a*). Both capacitors are used in starting the motor, but one of them is disconnected by a centrifugal device when the motor reaches 75 per cent of its synchronous speed.

In order to reverse the direction of rotation, the electric connections of either phase A or B are reversed.

In any induction motor the torque developed by the currents in the rotor revolves in the same direction as the magnetic field set up by the stator. Figure F-6b shows I_b, the current from phase B, preceding the current I_a of phase A in a particular stator conductor. Reversing the connections of phase A, say, will place I_a ahead of I_b in that conductor after a while. This is true of any other conductor in the stator, causing the rotating field set up by I_a and I_b to reverse. A reversed field produces a torque in the opposite direction and the rotation changes direction.

F-30. A four-pole induction motor operating from 60-cycle mains is running at its name-plate full-load speed of 1,620 rpm. What is the highest possible efficiency of this machine?

ANSWER.

$$\text{Max efficiency} = (1{,}620/1{,}800) \times 100 = \mathbf{90\ per\ cent}$$

The slip is proportional to the I^2R loss in the motor.

F-31. A three-phase, Y-connected, six-pole motor draws 27,700 watts at 0.885 power factor from a 240-volt, 60-cps line. The exciting component of phase current is 10 amp, lagging the phase terminal voltage by 70°. Friction and windage is 400 watts. Stator resistance is 0.08 ohm; stator reactance is 0.4 ohm; rotor resistance is 0.01 ohm; rotor reactance is 0.04 ohm; transformation ratio a is 3.0. Determine the rotor speed and the efficiency.

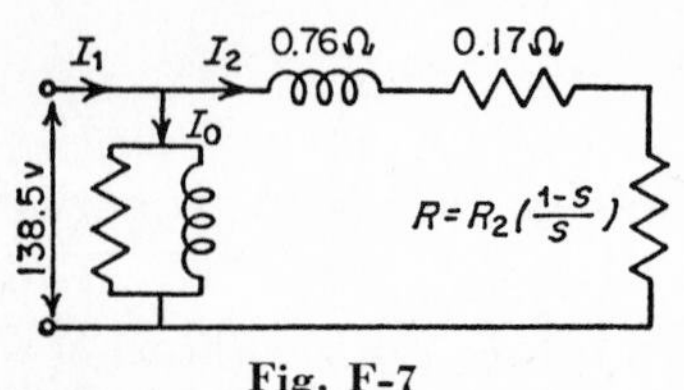

Fig. F-7

ANSWER. First determine the constants of the equivalent circuit. Refer to Fig. F-7. Also

$$R_1 = 0.08,\ x_1 = 0.4,\ \text{each given}$$
$$R_r = 0.01 \qquad \text{whence } R_2 = 9(0.01) = 0.09\ \text{ohm}$$
$$x_r = 0.04 \qquad \text{whence } x_2 = 9(0.04) = 0.36\ \text{ohm}$$
$$E_1 = 240(3^{1/2})\underline{/0°} = 138.5\underline{/0°}$$
$$I_1 = 27{,}700/(3 \cdot 138.5 \cdot 0.855) = 75.3\ \text{amp}$$
$$\dot{I}_1 = 75.3\underline{/-\cos^{-1} \cdot 0.855} = 75.3\underline{/-27.7°}$$
$$\dot{I}_1 = 75.3\underline{/-27.7°} = 66.6 - j35.0$$
$$\dot{I}_\phi = 10\underline{/-70°} = \underline{/3.42 - j9.40}$$
$$\dot{I}_2 = \dot{I}_1 - \dot{I}_\phi = 63.18 - j25.6 = 68.0\underline{/-22.1°}$$

Total impedance of load branch is

$$R_1 + \frac{R_2}{S} + j(x_1 + x_2) = \frac{138.5\underline{/0°}}{68.0\underline{/-22.1°}}$$

$$2.04\underline{/22.1°} = 190 + j0.780$$

Hence $$R_1 + \frac{R_2}{S} = 1.90 = 0.08 + \frac{0.09}{S} = 1.90$$

$$S = \frac{0.09}{1.82} = 0.0495$$

Synchronous speed $N_1 = (120 \times 60)/6 = 1{,}200$ rpm
Operating speed $N = (1 - S)N_1 = 0.950 \times 1{,}200 =$ **1,140 rpm**

The power output may be computed either as $P_{\text{input}} - \Sigma$ losses or as $P_{\text{converted}} - p_{f+w}$.

$$P_{\text{core}} = 3 \times 138.5 \times 10 \cos 70 = 1{,}420 \text{ watts}$$

$$p_{\text{copper}} = 3 \times 0.17 \times 68.0^2 = 2{,}360$$

$$p_{f+w} = 400$$

$$\Sigma \text{ losses} = 4{,}180 \text{ watts}$$

Efficiency $= [(27{,}700 - 4{,}180)/27{,}700] \times 100 =$ **84.8 per cent**

F-32. A three-phase, 100-kva, 2,300-volt, 60-cps, four-pole, Y-connected alternator having a synchronous impedance of $1.2 + j20 = 20\underline{/86.6°}$ ohms per phase delivers 34.5 kw at 0.866 power factor, current lagging, to a 2,300-volt line. Its d-c field current is 20 amp. (*a*) Compute the voltage regulation of the alternator when operating as above, and (*b*) find what value of d-c field current is required for the alternator to deliver an output of 34.5 kw at unity power factor. Assume linear saturation curve. (*c*) Determine the output power when the d-c field current is 20 amp and the driving torque is increased so as to cause the excitation voltage E_f to lead the phase terminal voltage V_t by 15°. (*d*) Find the field current to give unity power factor when the machine is operating as a motor, drawing 69 kw power input from the 2,300-volt line.

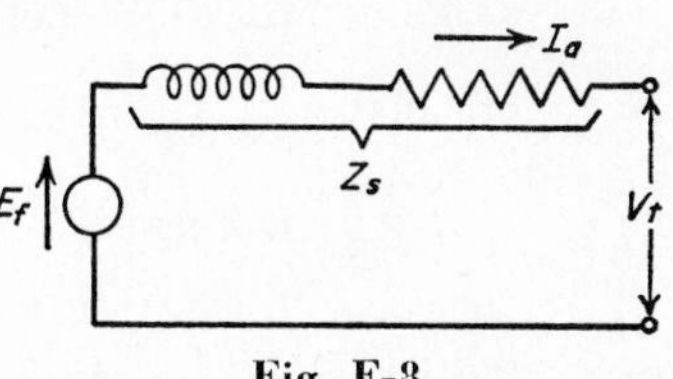

Fig. F-8

ANSWER. Refer to Fig. F-8. Let

$$V_t = \frac{2{,}300}{3^{1/2}}\underline{/0°} = 1{,}328\underline{/0°} \text{ volts}$$

(a) $$\dot{I}_a = \frac{34{,}500}{3^{1/2} \times 2{,}300 \times 0.866} \underline{/-\cos^{-1} 0.866} = 10.0\underline{/-30^\circ}$$

$$\dot{E}_f = 1{,}328\underline{/0^\circ} + (20\underline{/86.6^\circ})(10.0\underline{/-30^\circ}) = 1{,}447\underline{/6.6^\circ}$$

$$\text{V.R.} = \frac{V_{t\,\text{no load}} - V_{t\,\text{load}}}{V_{t\,\text{load}}} \times 100 \qquad \text{and} \qquad V_{t\,\text{no load}} = E_f$$

$$\text{V.R.} = \frac{1{,}447 - 1{,}328}{1{,}328} \times 100 = \mathbf{8.97\ per\ cent}$$

(b) $$\dot{I}_a = \frac{34{,}500}{3^{1/2} \times 2{,}300} \times \underline{/0^\circ} = 8.66\underline{/0^\circ} \qquad \text{since } \dot{V}_t = 1{,}328\underline{/0^\circ}$$

$$\dot{E}_f = 1{,}328\underline{/0^\circ} + (20\underline{/86.6^\circ})(8.66\underline{/0^\circ}) = 1{,}349\underline{/7.4^\circ}$$

An excitation voltage of 1,447 volts is produced by a field current of 20 amp [from part (a)]. Hence, for an excitation voltage of 1,349 volts,

$$I_f = 20(1{,}349/1{,}447) = \mathbf{18.64\ amp}$$

(c) E_f = **1,447** volts [same as in part (a)], because the field current is the same.

$$\dot{E}_f = 1{,}447\underline{/15^\circ} \text{ (considering } \dot{V}_t = 1{,}328\underline{/0^\circ}) = 1{,}398 + j374$$

$$\dot{I}_a = \frac{\dot{E}_f - \dot{V}_t}{\dot{Z}_s} \times \frac{70 + j374}{20\underline{/86.6^\circ}} = 19.05\underline{/-7.2^\circ}$$

$$P_{\text{out}} = (3)(1{,}328)(19.05 \cos 7.2^\circ) = \mathbf{75{,}100\ watts}$$

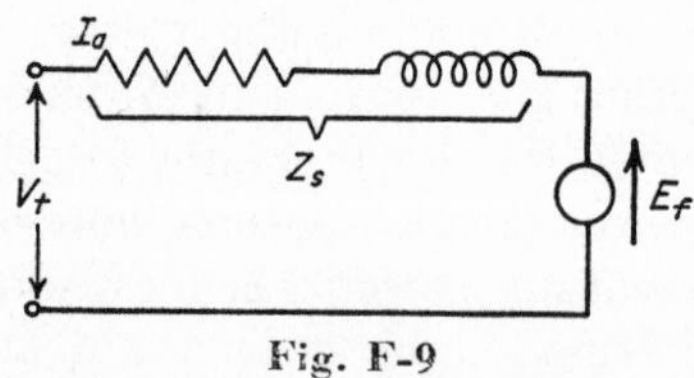

Fig. F-9

(d) Refer to Fig. F-9. Let $\dot{V}_t = 1{,}328\underline{/0^\circ}$, and then

$$\dot{I}_a = \frac{69{,}000}{3^{1/2} \times 2{,}300} \times \underline{/0^\circ} = 17.32\underline{/0^\circ}$$

$$\dot{E}_f = \dot{V}_t - \dot{Z}_s\dot{I}_a = 1{,}328 + j0 - (1.2 + j20)(17.32 + j0)$$
$$= 1{,}350\underline{/-14.88^\circ}$$

$\dot{I}_f = 20(1{,}350/1{,}447) =$ **18.7 amp** assuming a linear saturation curve

Supplement **G**

TRANSFORMERS

Questions and Answers

G-1. A 2,200/110-volt 60-cycle transformer has two 110-volt coils connected in series. Across the first of these two coils is connected a resistive load drawing 15 amp. Across the second coil is connected a pure capacitor drawing 10 amp. Across the two 110-volt coils in series is connected an 80 per cent lagging power-factor circuit drawing 20 amp. Neglecting the exciting current and the regulation of the transformer, calculate the current to the 2,200-volt coil and the power factor of the system.

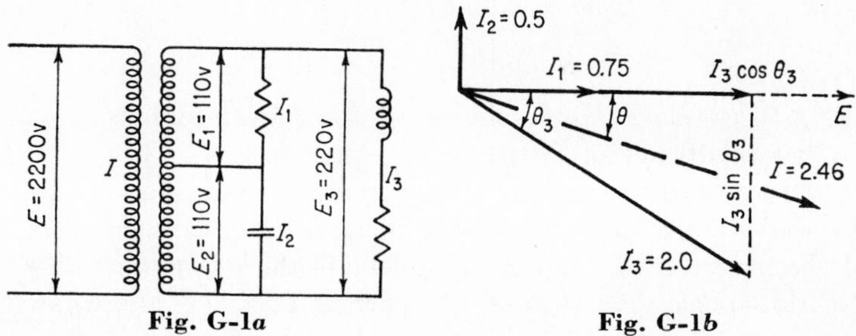

Fig. G-1*a* **Fig. G-1*b***

ANSWER. Refer to Fig. G-1*a*. The vector diagram is as shown in Fig. G-1*b*.

$I_1 = 15$ amp at 100 per cent power factor
$I_2 = 10$ amp at zero power factor
$I_3 = 20$ amp at 80 per cent power factor lagging
$I = I_3 \dfrac{E_3}{E}$ corresponding primary current relation

With reference to the vector diagram

$I_1 = 15 \times (110/2{,}200) = 0.75$ amp at 100 per cent power factor
$I_2 = 10 \times (110/2{,}200) = 0.50$ amp at 0° leading power factor
$I_3 = 20 \times (220/2{,}2000)$
$= 2.0$ amp at 80 per cent lagging power factor

Then, since the exciting current is to be neglected,

$$I = I_1 + I_2 + I_3$$
$$I_1 \sin \theta_1 = 0.75 \times 0 = 0$$
$$I_2 \sin \theta_2 = 0.5 \times 1 = 0.5 \text{ amp}$$
$$I_3 \sin \theta_3 = -2 \times 0.6 = -1.2 \text{ amp}$$
$$\text{Total } I \sin \theta = -0.7, \text{ since } 0.5 - 1.2 = -0.7$$

Totaling up all the $I \cos \theta$'s, we get

$$I_1 \cos \theta_1 = 0.75 \times 1 = 0.75$$
$$I_2 \cos \theta_2 = 0.5 \times 0 = 0$$
$$I_3 \cos \theta_3 = 2 \times 0.8 = 1.6$$
$$\text{Total } I \cos \theta = 2.35$$

Tan $\theta = 0.7/2.35 = 0.298$, θ is equal to 16°36′, $\cos \theta = 0.958$. Finally, $I = 2.35/0.958 =$ **2.46 amp.** The power factor ($\cos \theta$) is **95.8 per cent.**

G-2. A three-phase transformer substation delivers 100 kw of power to a balanced load of 80 per cent power factor. The power is received at 13,800 volts and is delivered at 2,300 volts. A single bank of V-connected transformers is used. Neglecting the losses and the magnetizing current, (*a*) calculate the minimum rating of each transformer; (*b*) calculate the current ratings of the windings of each transformer; (*c*) calculate the power factor at which each transformer operates.

ANSWER. At no load with only two transformers, three equal three-phase voltages coexist around the secondaries, and a three-phase transformation is possible. Therefore, with only two transformers this condition holds. Even under balanced loads the voltages may become slightly unbalanced. This is not serious in commercial transformers, as their regulation is seldom greater than 2 or 3 per cent. Now refer to Fig. G-2*a*.

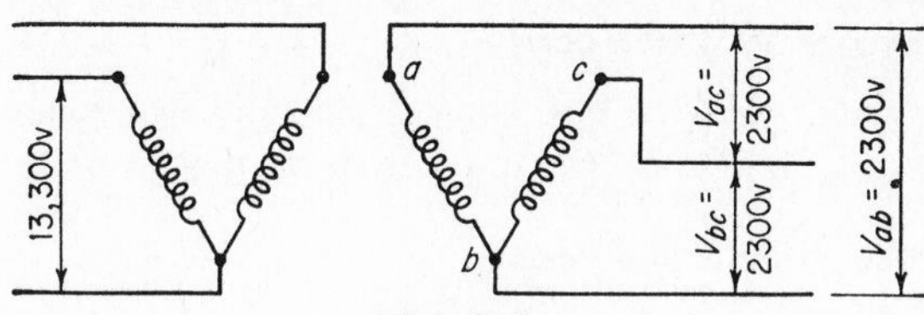

Fig. G-2a

(a) The minimum rating of each transformer is $\sqrt{3} \times$ output if connected in closed Δ.

$$\text{kva output} = \frac{P}{\cos\theta} = \frac{100}{0.8} = 125 \text{ kva}$$

The output for each transformer (closed Δ) $= 125/3 = 41.67$ kva
The rating for each transformer (V connection)

$$\sqrt{3} \times 125/3 = \mathbf{72.1\ kva}$$

(b) The line current for the balanced load is given by the following.

$$I = 100{,}000/(1.732 \times 2{,}300 \times 0.8) = \mathbf{31.4\ amp}\ \text{current rating}$$

(c) Refer to vector diagram (Fig. G-2b). The power factor is lagging and you can neglect the losses and the magnetizing-current

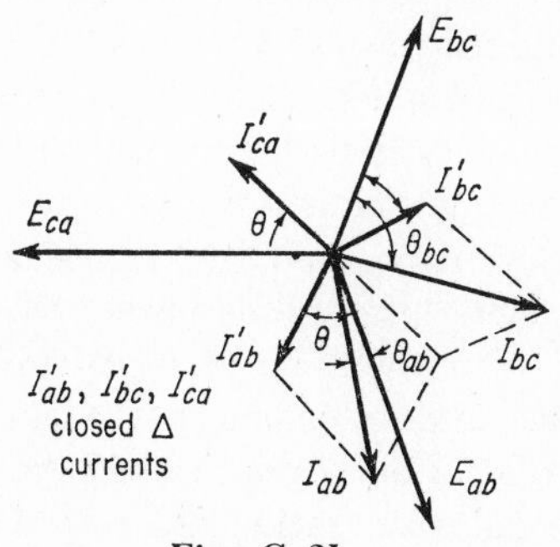

Fig. G-2b

effects. The power-factor angle is $\theta \pm 30° \cdot \cos\theta_{ab} = \cos(\theta - 30)$, or equal to

$$\cos\theta\cos 30 + \sin\theta\sin 30 = 0.8 \times (3/2)^{1/2} + 0.6 \times 1/2 = 0.993$$

Power factor_{ab} = **99.3 per cent**

$$\cos \theta_{bc} = \cos (\theta + 30) = \cos \theta \cos 30 - \sin \theta \sin 30$$
$$\cos \theta_{bc} = 0.8 \times (3/2)^{1/2} - 0.6 \times 1/2 = 0.393$$

Power factor_{bc} = **39.3 per cent**

G-3. Under what conditions can two single-phase transformers of different ratings be connected in parallel on both the primary and the secondary sides so that they may share a load between them in proportion to their ratings?

ANSWER. For two single-phase transformers to share the load in proportion to their ratings, when connected in parallel, the following conditions must be met:

(*a*) The ratio X/R must be the same, where X is the reactance and R is the resistance of each transformer. In general, R and X will be directly proportional to the rating of a transformer, so that they will both be twice as large for a 50 kva transformer as for a 25 kva unit.

(*b*) The transformers must have the same ratio of transformation E_1/E_2, where E_1 is the primary voltage and E_2 the secondary voltage.

(*c*) They must have the same regulation, i.e., the same relative increase in secondary voltage, when the transformers are loaded.

G-4. List and show seven three-phase transformer connections commonly used in industry.

ANSWER. Many three-phase connections are possible both for single-phase units connected in banks and for three-phase transformers. The more familiar connections are given below. A knowledge of polarity, ratio, and impedance is important before connecting transformers into three-phase banks. Transformers must be connected with the same polarity for each. The same transformation ratios and impedances are also necessary if the voltage and the current of phases are to be kept balanced. Simple tests determine the polarity, ratio, and impedance, although the latter requires variable voltage.

(*a*) *Y-Y with connected neutrals* (see Fig. G-3*a*). This arrangement has the advantage that the windings are subjected to the line voltage divided by $\sqrt{3}$. The Y-Y connection without connected neutrals is rarely used since high unequal voltages may result from an unbalanced load. In addition, there is high voltage and distorted wave from a suppressed third harmonic component. Note that the low side is called L, not X.

(*b*) Δ-Δ *connection* (see Fig. G-3*b*). This arrangement has full-line voltage on the windings, but the current in each phase is the

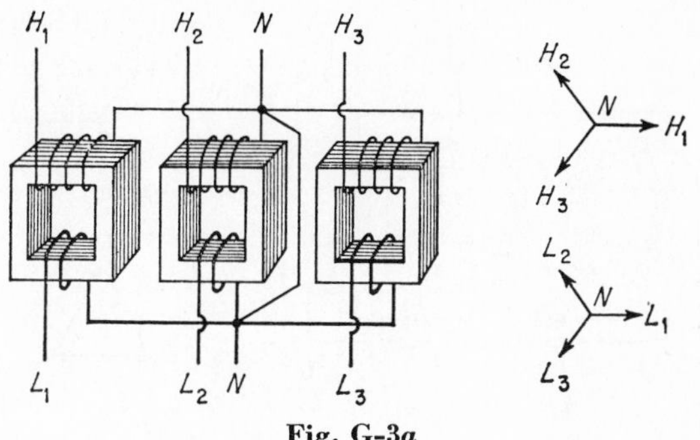

Fig. G-3*a*

line current divided by $\sqrt{3}$. A Δ-Δ connection is thus suited to moderate voltage, large current operation. The ratio of transformation and impedance of transformers must be the same to prevent circulating the current and keep the phase currents equal.

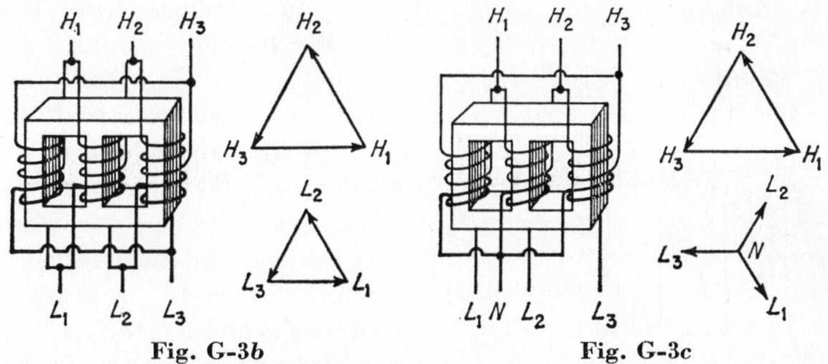

Fig. G-3*b* **Fig. G-3*c***

Trouble will not result from third harmonics since third harmonic current can flow in Δ of either winding.

(*c*) Δ-*Y connection*. This is used to step up the voltage for transmission since Δ-Y ratio is $1:\sqrt{3}$. This hookup is also commonly used to step down to utilization voltage with a four-wire system such as secondary 208 line-to-line, 120 line-to-neutral, or $^{480}/_{277}$. Third harmonic does not cause trouble. With secondary neutral connected to the load neutral, there will be very little unbalancing of the load. Δ-Y banks cannot be banked with Δ-Δ (see Fig. G-3*c*).

(*d*) *Y-Δ connection* (see Fig. G-3*d*). This is for all intents and purposes the reverse of Δ-Y. Such an arrangement is mainly used

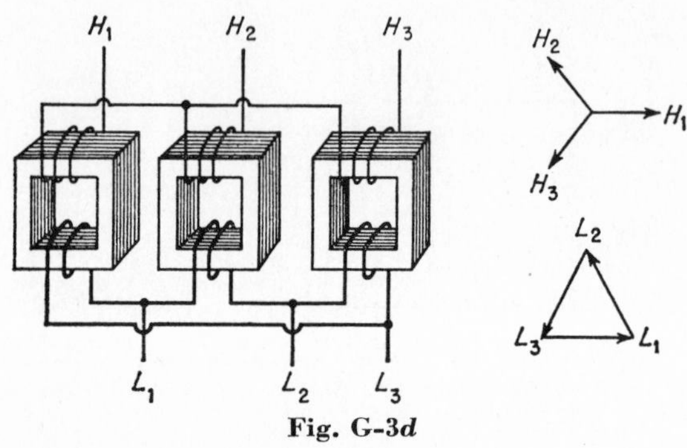

Fig. G-3*d*

for stepping down from the transmission voltage. Third harmonics are taken care of by secondary Δ, or by connecting the primary neutral to the power source neutral. Δ-connected secondaries can, in emergencies, be operated with two transformers.

(*e*) *Autotransformers* (see Fig. G-3*e*). As previously mentioned this transformer has but one winding. The secondary voltage is obtained from taps. The ratio of transformation is the total number of turns to the number of turns carrying difference between the secondary and primary currents. Autotransformers can also be used for voltage step-up. Where the ratio of transformation is low, as in reduced-voltage motor starting, autotransformers are lighter than others, but they can't be used where the secondary must be isolated.

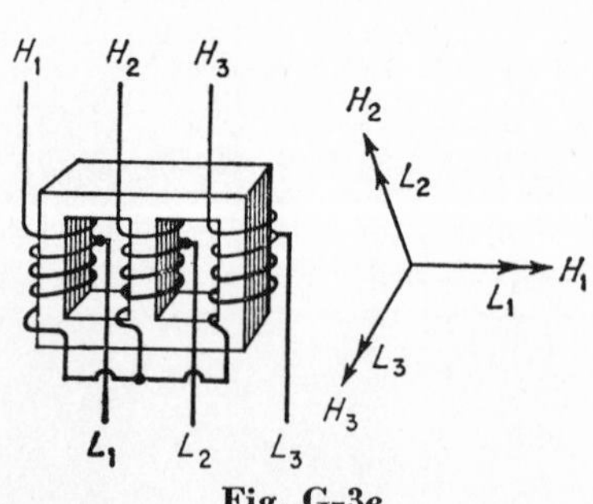

Fig. G-3*e*

(*f*) *Phase conversion* (see Fig. G-3*f*). The T (Scott) connection is used to convert two to three-phase. Phase *AB* of the two-phase side is induced by the three-phase winding *ab*. The voltage in *CD* is induced by the vector sum of the three-phase voltages *bd* and *da*. The same connection converts two-phase to three-phase. If need be, a 50 per cent tap on the teaser may be used for the neutral of the four-wire system. For a 1:1 ratio an autotransformer is generally used.

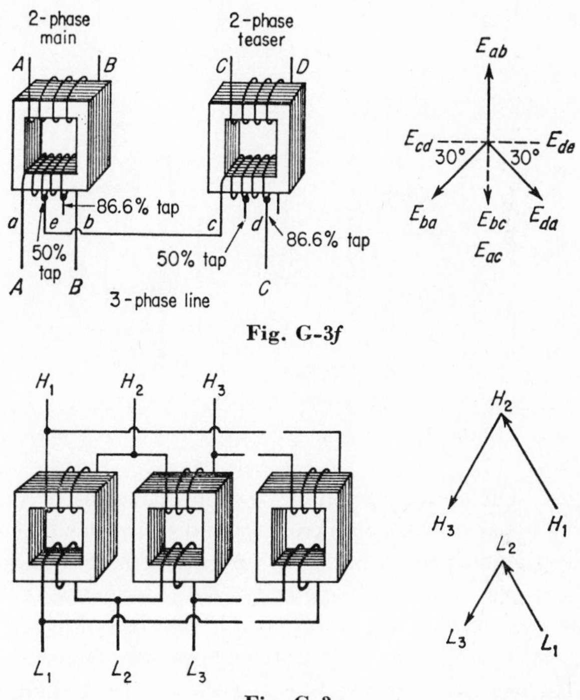

Fig. G-3*f*

Fig. G-3*g*

(*g*) *Open-Δ* (see Fig. G-3*g*). This is an emergency connection and gives approximately balanced secondary voltage. However, the bank capacity is only 58 per cent of three-transformer bank. Each winding must now carry a full-load current rather than a line current divided by $\sqrt{3}$. The power factor is also reduced. Two single-phase autotransformers may be connected up as open Δ.

G-5. Name three transformer tests and show their hookup by diagrams.

ANSWER. *Polarity test.* When V_t is less than V_a, the polarity is subtractive and the leads are marked as shown. When V_t is more than V_a, the polarity is additive and the secondary leads are reversed. If a standard transformer known to have a correct polarity and the same ratio as the test transformer is available, the simplest method for testing for polarity is to connect the primaries and secondaries of the transformers in parallel, placing a fuse in series with the

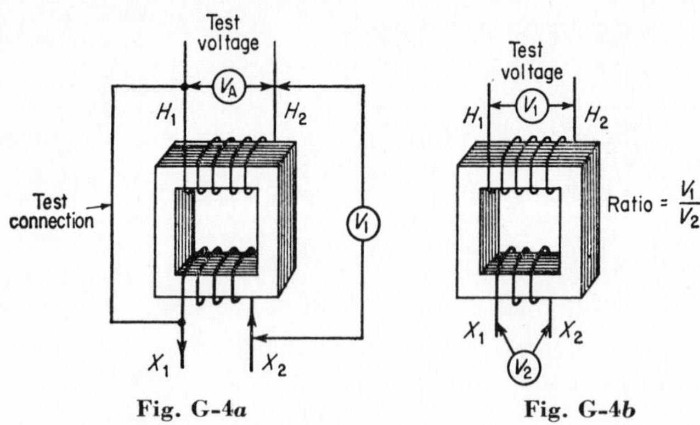

Fig. G-4*a* Fig. G-4*b*

secondaries. On applying voltage to the primaries of the transformers, if they are of the same polarity and ratio, no current should flow in the secondary circuit and the fuse will remain intact. If the transformers are of opposite polarity, the connection will short-circuit one transformer on the other, and the fuse selected should, therefore, be small enough to blow before the transformers are injured. In nearly all transformers there will be a slight current in the secondaries when connected as above (Fig. G-4*a*). This current is known as the exchange current and should be less than 1 per cent of the normal full-load current of the transformer.

Ratio test. Refer to Fig. G-4*b*. As we already know, the ratio of transformation is the primary voltage divided by the secondary voltage. The applied voltage and instrument range should permit accurate readings of V_1 and V_2. A ratio test is made as a check against possible mistakes in winding the coils and connecting up. The ratio test is made at a fractional part of the full voltage at no-load current, and should not be substituted for a regulation test.

An error of 1 or 2 per cent is permissible for this test because of partial loading.

Impedance test. Refer to Fig. G-4*c*. In this test the primary voltage is adjusted until the secondary carries the rated current. The transformer impedance is equal to the applied test voltage divided by the short-circuit current.

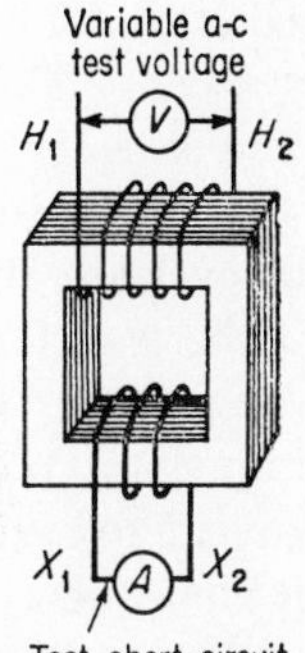

Fig. G-4c

G-6. In a chemical plant it is necessary to increase by 20 per cent the voltage to a series of electrolytic cells. These are supplied by a number of shunt-wound rotary converters. State how you would effect this change and explain why your method is preferable to others that are possible.

ANSWER. The rotary or synchronous converter is a machine designed to convert alternating to direct current. The reverse is also possible. Its construction is such that for a given machine the ratio of alternating to direct current emf is constant unless special features such as series windings or split poles are incorporated in the converter design. It is assumed in this problem that the converters have no provision for changing the alternating current to direct current, inherent in their design. Consequently, the d-c voltage to the cells may be increased in the following ways:

(*a*) By using an induction regulator.

(*b*) By coupling a synchronous booster to each converter shaft and connecting them in series on either the a-c or d-c side.

(*c*) By using transformers on the a-c side with different taps on the primaries, requiring the opening or the short-circuiting of the transformers as the case may be.

(*d*) By rewinding the converters to provide series windings or split poles.

Clearly methods (*b*) and (*d*) are not feasible since they are too expensive. Method (*a*) could be used if it were not for the slow response of the induction regulators. Therefore, method (*c*) appears to be the best choice.

One transformer might be made to serve all converters. This would result in a simple and inexpensive installation. It would be highly efficient in operation because transformers operate at efficiencies of about 97 or 98 per cent. In addition, the combination would be easy to operate because of its possible centralized control. This latter advantage would enable the operator to change the voltage on all cells by simply changing the transformer taps.

G-7. A 25-cycle 1,000-kva transformer is applied to a 60-cycle system. The full-load efficiency of the transformer at 25 cycles is 98 per cent, half of the loss being in the copper and half in the core. The hysteresis loss and the eddy-current loss equal each other at 25 cycles. What would be the rating of this transformer at 60 cycles if the transformer were operated at rated voltage?

ANSWER. Since the copper and iron-core losses play such an important part in the efficiency of a transformer, their effects are considered significant in the solution of this problem. The core loss is that energy which is absorbed by the transformer when the secondary circuit is open, and is the sum of the hysteresis and eddy-current loss in the core and a slight loss in the primary coil, which is generally neglected in the measurements. The hysteresis loss is caused by the reversals of the magnetism in the iron core. This loss varies as the 1.6 power of the voltage with constant frequency. Let us proceed to review the effects of frequency on hysteresis and eddy-current losses. For the sake of simplicity assume a unity power factor.

For operation at 25 cycles. Since the total losses are made up of the copper and core losses, we first determine the input from which we then find the total losses.

$$\text{Input} = \frac{\text{output}}{\text{eff.}} = \frac{1{,}000}{0.98} = 1{,}020.4 \text{ kw}$$

$$\text{Total losses} = \text{input} - \text{output} = 1{,}020.4 - 1{,}000 = 20.4 \text{ kw}$$

Since the core loss is given as one-half the total losses, we obtain

$$\text{Core loss} = P_c = 20.4/2 = 10.2 \text{ kw}$$

Now, since the copper loss is also one-half of the total losses, then the copper loss is also equal to 10.2 kw. Since the core loss is the sum of the hysteresis and eddy-current losses, this value must be broken down two ways. Steinmetz has shown that the hysteresis loss in an iron core may be determined from the empirical formula

$$P_h = k_h f V B_m^x \times 10^{-7} \text{ kw}$$

where V = volume of core

B_m^x = maximum value of flux density in core

This may be reduced to the following simple relationship.

$$P_h = k_h f \phi_m^{1.6} = 5.1 \text{ kw}$$

This is so since the hysteresis and eddy-current losses were given as equal. The eddy-current loss may be given as follows, since this loss varies as the square of the voltage.

$$P_e = k_e f^2 \phi_m^2 = 5.1 \text{ kw}$$

where k_h = hysteresis constant

f = frequency

k_e = eddy-current constant

ϕ_m = transformer-core flux maximum value

Now we shall determine the values of k_h and k_e by rearranging the proper equations above.

$$k_h = \frac{5.1 \text{ kw}}{f\phi_m^{1.6}} = \frac{5.1}{25\phi_{25}^{1.6}}$$

$$k_e = \frac{5.1 \text{ kw}}{f^2\phi_{25}^2} = \frac{5.1}{25^2 \times \phi_{25}^2}$$

where ϕ_{25} is maximum flux at 25 cycles.

Since both transformers are operated at the rated voltage, then $E_{25} = E_{60}$. Also since the emf induced is proportional to the flux, the frequency, and the number of turns, then the following relations hold.

$$E = 4.44 f \phi_m N \times 10^{-8}$$

$$E_{25} = 4.44 \times 25\phi_{m_{25}} N \times 10^{-8}$$

$$E_{60} = 4.44 \times 60\phi_{m_{60}} N \times 10^{-8}$$

$\phi_{m_{25}}$ in terms of $\phi_{m_{60}}$ may now be expressed as

$$\phi_{m_{60}} = {}^{25}\!/_{60} \times \phi_{m_{25}}$$

The hysteresis loss will change with a change in frequency, but the eddy-current loss will not change. This may be seen from the following substitutions.

$$P_h = \frac{5.1}{25\phi_{m_{25}}{}^{1.6}} \times 60\phi_{m_{60}}$$

$$= \frac{5.1}{25 \times \phi_{m_{25}}{}^{1.6}} \times 60 \times \left(\frac{25}{60} \times \phi_{m_{25}}\right)^{1.6} = 5.1 \times \left(\frac{25}{60}\right)^{0.6} = 3.01 \text{ kw}$$

$$P_e = \frac{5.1}{25^2 \times \phi_{m_{25}}{}^2} \times 60^2 \times \left(\frac{25}{60} \times \phi_{m_{25}}\right)^2 = 5.1 \text{ kw}$$

In determining the rating at 60-cycle operation, the copper loss may be increased at the expense of the reduction in hysteresis loss. Thus

$$\text{Copper loss at 60 cycles} = 10.2 + (5.1 - 3.01) = 12.29 \text{ kw}$$

We know that copper losses are proportional to the square of the current I^2R and that the transformer rating varies directly with the current, kva. Thus, we can write directly the simple relationship

$$\text{Rating at 60 cycles} = \sqrt{\frac{\text{copper loss at 60 cycles}}{\text{copper loss at 25 cycles}}} \times \text{rating at 25 cycles}$$

$$\text{Rating at 60 cycles} = \sqrt{\frac{12.29}{10.2}} \times 1{,}000 = 1.1 \times 1{,}000 = \mathbf{1{,}100 \text{ kva}}$$

G-8. Discuss the dry-type transformer as applied to standard industrial machinery and commercial installations.

ANSWER. The dry-type transformer, 600 volts and less, 2 to 25 kva, enables designers to adapt standard industrial machinery to practically any of the existing electric service voltages used in industrial plants and commercial installations. As a design unit the modern dry-type transformer is compact, easily installed, and economical to operate. Based on a temperature rise, there are two general classifications of dry-type transformers. Those with class A insulations of organic material, such as paper and cotton, are applied on basis of a 55°C temperature rise at full load in an ambient temperature of 40°C. Transformers with class B insulations of inorganic material, such as fiber glass, mica, and porcelain with some organic varnishes and binders, are used for an 80°C temperature rise at full load in an ambient temperature of 40°C.

Dry-type transformers are designed for a rated temperature rise at a rated load for continuous operation. For intermittent loads, such as motor starting, an excess rating can be obtained on a short-time basis. However, the transformer must have sufficient capacity to provide the motor with a minimum starting voltage at the instant of starting, when the maximum current is drawn. In general, dry-type transformers, designed for a given frequency, can be used on a higher but not on a lower frequency. A transformer designed for 60 cycles is expected to reach its rated temperature rise at full rated load with rated voltage applied. When operated on a circuit with lower frequency, the magnetization in the iron core, exciting current, and core loss all increase, and the transformer overheats. This condition is only alleviated by reducing the applied voltage in proportion to the reduction in frequency. A reduction of load has no effect here.

G-9. How many ratios does a transformer have? What is the polarity of a transformer?

ANSWER. A transformer has two ratios: (*a*) the voltage ratio is the ratio of the rms primary terminal voltage to the rms secondary terminal voltage at specified loading conditions; (*b*) the turn ratio is the ratio of the number of turns in the high-voltage winding to the number of turns in the low-voltage winding. The two ratios are equal when the transformer is unloaded. The ratios differ when the transformer is loaded, because of the voltage drops caused by the inductance and resistance of the windings. This difference in ratios is regulation, which is usually expressed as a percentage of the full-load voltage.

In most dry-type transformers used for general purpose, the windings terminate in cable leads. The leads extend into the wiring compartment, to which a conduit can be connected. The cable leads do not usually extend outside of the wiring compartment in a fixed position. Such transformers do not have polarity. Polarity refers to the voltage-vector relations between terminals external to the enclosure, when the terminals are taken in a definite order and viewed from a definite position.

G-10. Discuss the importance of keeping a check on transformer connections.

ANSWER. Each winding has a voltage induced in it by the magnetic flux in the core. Since the flux is mutual, the induced voltages will be in the same direction in the windings. If the windings are wound on the core in the same manner or direction, the induced voltages will appear at the terminals in the same direction. The start of the high-voltage winding is usually marked H_1 and the finish H_2. The start of the low-voltage winding is then marked X_1 and the finish X_2. If the highest lead, numbered H, is connected to

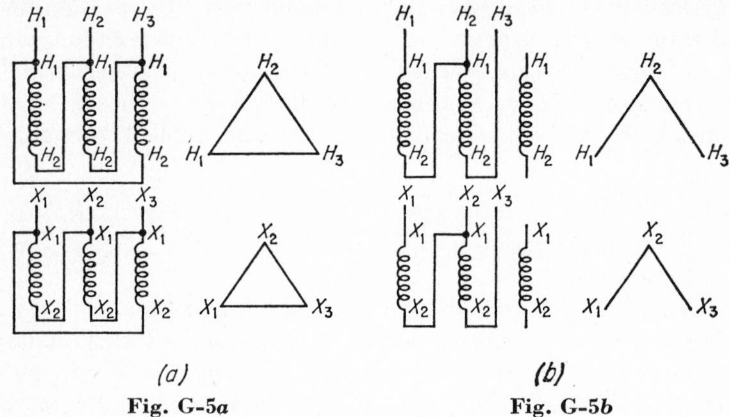

Fig. G-5*a* Fig. G-5*b*

the lowest lead, numbered X, the voltages will add throughout the windings.

Since the lead markings indicate the direction of the windings and the direction of the voltages, they are important when two identical transformers are connected in parallel or series, or when two or more are connected in polyphase bank. For transformers connected in parallel to operate satisfactorily and share the load equally, they should have the same voltage ratios and the same values of reactances and resistances.

G-11. With three single-phase transformers connected in Δ-Δ, show the connection before and after one winding becomes damaged. What is the resulting voltage?

ANSWER. In Fig. G-5*a* the transformers are shown connected in Δ-Δ. After one of the windings is burned out and the other two are connected up, the result is as shown in Fig. G-5*b*. In the Δ connection, the line voltage is the Δ voltage, or the voltage of each single-phase transformer. The line current is displaced 30° from the voltage and is equal to 1.732 times the current in the Δ. The three transformers in the Δ-Δ connection should have the same ratios, reactances, and resistances to share the load equally and to avoid a flow of circulating current. If one transformer becomes damaged, the two remaining can be operated in open Δ, as shown in Fig. G-5*b*. In open Δ they provide 58 per cent of the original rating.

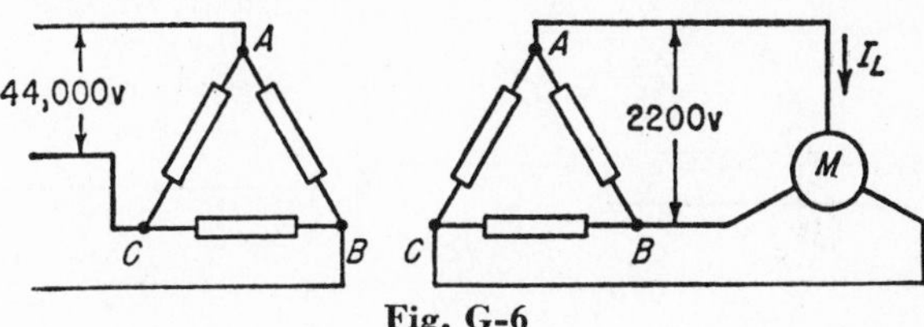

Fig. G-6

G-12. A 2,200-volt three-phase 500-hp induction motor operates at 94 per cent efficiency at full load and has a power factor of 90 per cent. The motor is supplied from a Δ-Δ–connected bank of single-phase transformers. If one of the transformers fails, what kva is carried by each of the remaining transformers when the motor delivers a full output? Assuming the primary of the transformer bank to be connected to a 44,000-volt system, determine the primary line current to the bank.

ANSWER. Refer to Fig. G-6. From previous considerations the line current for the system is given by

$$I_L = \frac{500 \times 746}{1.732 \times 2{,}200 \times 0.90 \times 0.94} = 116 \text{ amp}$$

If the maximum current output of one transformer is I amp, the line current that can be drawn from the closed Δ is $\sqrt{3}\, I$. However, with one transformer removed (or failing) as in open Δ, the maximum line current may not exceed I amp. Thus, the line currents from the open-Δ system will continue to have the same value of 116 amp. Under these conditions, the kva rating of each transformer would be given by

$$(116 \times 2{,}200)/1{,}000 = \mathbf{254.5\ kva}$$

Now, since the input is equal to the output, or "watts in" equal "watts out," the current flowing in the primary or high-voltage lines is

$$116 \times (2{,}200/44{,}000) = \mathbf{5.8\ amp}$$

G-13. A small transformer has an eddy-current loss of 18 watts and a hysteresis loss of 72 watts at no load when its rated voltage

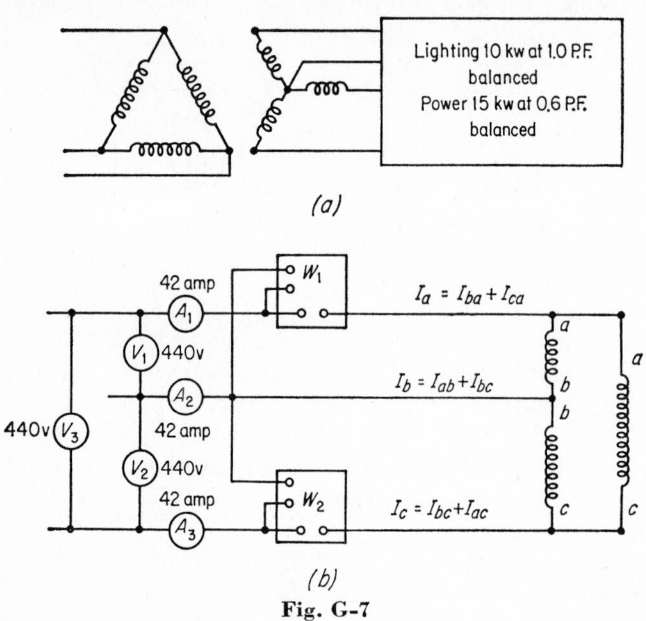

Fig. G-7

of 240 volts at a frequency of 60 cps is applied. What will be the total core loss if 240 volts at a frequency of 50 cps is applied?

ANSWER. The hysteresis loss is independent of the frequency and will remain 72 watts. The eddy-current loss varies as the square of the frequency, and at 50 cps this becomes

$$P_e = 18 \times [50/60]^2 = 18 \times 0.695 = 12.5 \text{ watts}$$

Therefore, the core loss at 50 cps is the sum of both the hysteresis and eddy-current losses.

$$\text{Core loss} = 72 + 12.5 = \mathbf{84.5\ watts}$$

G-14. A Δ-Y–connected bank of three-phase transformers is used to reduce the voltage from 440 volts to 208/120 volts for use for small motors and lighting. If the lighting load amounts to 10 kw, balanced at unity power factor, and the motor load amounts to 15 kw, balanced at 0.6 power factor lagging, what would be the readings of meters, i.e., voltmeters, ammeters, and wattmeters properly connected in the 440-volt circuit to read the input power to the transformers? Assume no losses for the transformer and show by sketch the meters and their connections (see Fig. G-7*a*).

ANSWER. For the motor load, $\cos \theta' = 0.6$ and $\sin \theta' = 0.8$. This is equivalent to an angle of θ' equal to 53°. Then, reactive kva is found.

$$\text{Reactive kva} = 15 \tan 53° = 15 \times 1.33 = 20$$
$$\text{Total kva} = \sqrt{25^2 + 20^2} = 32$$
$$\text{Current in primary } I_p = 32 \text{ kva}/(\sqrt{3} \times 440) = 42 \text{ amp}$$

The angle from the total kva determination may be found as $\theta = \tan^{-1}(20/25)$, or θ is 38.7°. From this $\sin \theta$ equals 0.625 and $\cos \theta$ equals 0.780. Now refer to Fig. G-7*b*.

$$\begin{aligned} W_1 &= E_L I_L \cos(30 + \theta) \\ &= 440 \times 42 \times \cos(30 + 38.7) \\ &= 440 \times 42 \times 0.363 = \mathbf{6{,}710\ watts} \\ W_2 &= 440 \times 42 \times \cos(30 - \theta) \\ &= 440 \times 42 \times \cos(30 - 38.7) \\ &= 440 \times 42 \times 0.988 = \mathbf{18{,}280\ watts} \\ W_1 + W_2 &= 6{,}710 + 18{,}280 = 24{,}990 \text{ watts} \end{aligned}$$

G-15. An ideal transformer has a primary coil P and two secondaries S_1 and S_2. The turns ratio between P and S_1 is 10:1, while that between P and S_2 is 5:4. Across S_1 is connected an impedance, $Z_1 = 10 + j20$ ohms. Across S_2 is connected a parallel combination of R and C. What must be the value of R and the reactance of C so that the impedance, seen when looking into P, is $500 + j0$ ohms?

ANSWER. Refer to Fig. G-8 given. Set up the equivalent circuit shown in Fig. G-9 and solve by complex notation.

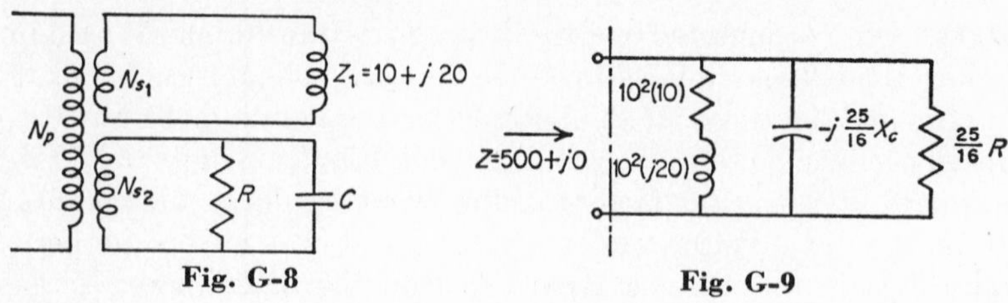

Fig. G-8 Fig. G-9

$$500 + j0 = \frac{1{,}000(1 + j2)[-j(25/16)\,X_c](25/16\,R)}{1{,}000(1 + j2)[(25/16)\,(R + jX_c)] + [-j(25/16) \times (25/16)\,RX_c]}$$

$$500 + j0 = 640(1 + j2)(R - jX_c) - jRX_c = 2(1 + j2)(-jX_cR)$$

$$500 + j0 = 640[(R + 2X_c) + j(2R - X_c)]$$

$$= jRX_c - j2RX_c(1 + j2) = 4RX_c - jRX_c$$

Equating j terms, $640(2R - X_c) = -RX_c$ (1)

Equating real terms, $640(R + 2X_c) = 4RX_c$ (2)

$$\frac{(1)}{(2)} = \frac{2R - X_c}{R + 2X_c} = -\frac{1}{4} \qquad \text{Therefore,} \qquad X_c = \frac{9}{2}R$$

Substituting in Eq. (1), $640(2R - 9/2\,R) = -R \times 9/2\,R$, from which $R = $ **355 ohms.**

$$X_c = \frac{9}{2}R = \frac{9}{2}355 = \textbf{1,600 ohms}$$

G-16. The transformer shown in Fig. G-10 may be considered ideal. The turns ratio or $N_2/N_1 = 1.5$ and $N_3/N_1 = 2$. The impedances

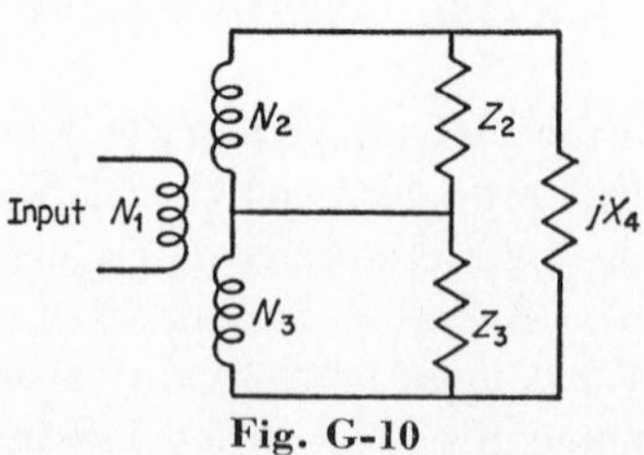

Fig. G-10

are $Z_2 = 100(1 + j)$ ohms and $Z_3 = 50(1 + j3)$ ohms. Calculate the magnitude and sign of X_4 which will make the power factor at the input equal to unity.

ANSWER. Referring impedances to primary

$$Z_2' = \left(\frac{N_1}{N_2}\right)^2 Z_2 = \left(\frac{1}{1.5}\right)^2 [100(1 + j1)] = \frac{100(1 + j1)}{2.25}$$

$$Z_3' = \left(\frac{N_1}{N_3}\right)^2 Z_3 = \left(\frac{1}{2}\right)^2 [50(1 - j3)] = \frac{50(1 - j3)}{4}$$

$$jX_4' = \left(\frac{N_1}{N_2 + N_3}\right)^2 jX_4 = \left(\frac{1}{N_2/N_1 + N_3/N_1}\right)^2 jX_4 = \left(\frac{1}{1.5 + 2}\right)^2 jX_4$$

$$jX_4' = \frac{jX_4}{3.5^2} = \frac{jX_4}{12.25}$$

$$\frac{1}{Z_o'} = \frac{1}{Z_2'} + \frac{1}{Z_3'} + \frac{1}{jX_4'} = \frac{2.25}{100(1 + j1)} + \frac{4}{50(1 - j3)} + \frac{12.25}{jX_4}$$

$$\frac{1}{Z_o'} = \frac{2.25(1 - j1)}{100(1 + j1)(1 - j1)} + \frac{4(1 + j3)}{50(1 - j3)(1 + j3)} - j\frac{12.25}{X_4}$$

$$\frac{1}{Z_o'} = \frac{2.25(1 - j1)}{100 \times 2} + \frac{4(1 + j3)}{50 \times 10} - j\frac{12.25}{X_4}$$

$$\frac{1}{Z_o'} = \frac{2.25(1 - j1)}{200} + \frac{4(1 + j3)}{500} - j\frac{12.25}{X_4}$$

At unity power factor reactive = 0 (j items = 0).

$$-j\frac{2.25}{200} + j\frac{12}{500} - j\frac{12.25}{X_4} = 0$$

$$-\frac{11.25}{1{,}000} + \frac{24}{1{,}000} = \frac{12.25}{X_4} \qquad \text{from which } X_4 = \mathbf{955\ ohms}$$

G-17. A load having an impedance of $1{,}000(1 + j5)$ ohms receives power from a generator having an internal impedance of $100(1 - j2)$ ohms and an emf of 100 volts, as shown in Fig. G-11.

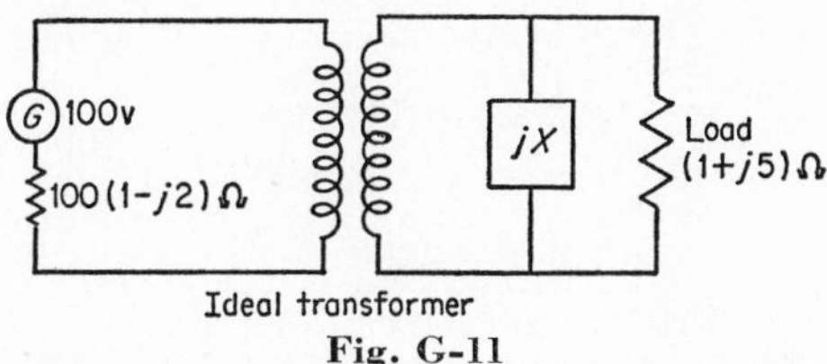

Fig. G-11

(*a*) Calculate the maximum power that may be transferred to the load when both the pure reactance X and the turns ratio N_1/N_2 of the ideal transformer are variable.

(*b*) Calculate the value of X and of the turns ratio N_2/N_1, necessary for this maximum power transfer.

ANSWER.

(*a*) Maximum power transferred

$$\frac{E_g^2}{4R_g} = \frac{100^2}{(4 \times 100)} = \mathbf{25\ watts}$$

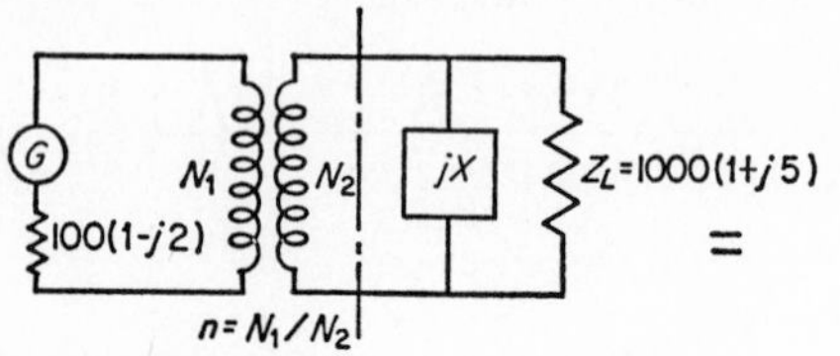

Fig. G-12

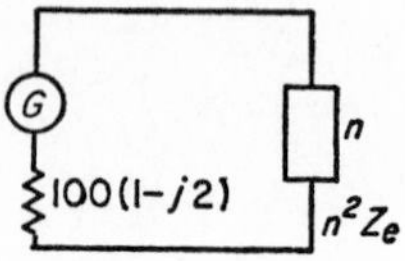

Fig. G-13

(*b*) Set up circuit in Fig. G-12 and its equivalent circuit, Fig. G-13. Then

$$Z_e = \frac{jX(1{,}000 + j5{,}000)}{1{,}000 + j(X + 5{,}000)}$$
$$= \frac{jX(1{,}000 + j5{,}000)[1{,}000 - j(X + 5{,}000)]}{1{,}000^2 + (X + 5{,}000)^2}$$
$$Z_e = \frac{1{,}000X^2 + j(26{,}000 + 5X)1{,}000}{1{,}000^2 + (X + 5{,}000)^2}$$
$$n^2Z_e = 100(1 + j2) = 100 + j200$$

Therefore,

$$\frac{1{,}000n^2X^2}{1{,}000^2 + (X + 5{,}000)^2} = 100$$
$$\frac{1{,}000n^2(26{,}000 + 5X)}{1{,}000^2 + (X + 5{,}000)^2} = 200$$

from which X is found to be equal to **−8,666 ohms.** Now by substitution,

$$\frac{1{,}000n^2(-8{,}666)^2}{1{,}000^2 + (-8{,}666 + 5{,}000)^2} = 100$$

Further

$$\frac{10n^2(-8{,}666)^2}{1{,}000^2 + (-8{,}666 + 5{,}000)^2} = 1$$
$$10n^2(-8{,}666)^2 = 1{,}000^2 + (-3{,}666)^2$$

$n^2 = 14.5/750$ and $n = \sqrt{14.5/750} = 3.8/27.4 = \mathbf{0.1387}$

Note $n = N_1/N_2$.

G-18. The characteristic curve of a nonlinear circuit element is shown in Fig. G-14. A voltage of $e = 10 + 5 \sin \omega t$ is applied to

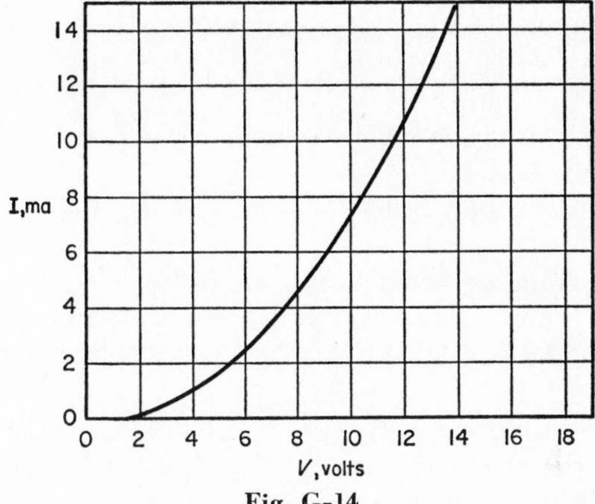

Fig. G-14

this element. You are required to:

(*a*) Determine the magnitude of the direct current that flows through the circuit element.

(*b*) Determine the peak value of the fundamental frequency of the current wave.

(*c*) Determine the peak value of the second harmonic of the current wave.

(*d*) Determine the peak value of the third harmonic of the current wave.

ANSWER.

(*a*) The basic task in this problem is the superimposition of the sinusoidal voltage upon the nonlinear (current) circuit element. For different values of ωt we obtain the corresponding values of e. Superimposing these values upon the nonlinear circuit element, we obtain corresponding values of i_b in milliamperes (ma). The obtained values can be tabulated as follows:

ωt	$\sin \omega t$	$5 \sin \omega t$	e	$i_b(I)$	ma
0	0	0	10.0	I_b	8.0
30	0.5	2.5	12.5	$I_{1/2}$	13.0
90	1.0	5.0	15.0	I_{max}	19.0
210	−0.5	−2.5	7.5	$I_{1/2}$	4.5
270	−1.0	−5.0	5.0	I_{min}	2.0
360	0	0	10.0	I_b	8.0

The equation for the alternating current wave is

$$i_b = B_o + B_1 \cos \omega t + B_2 \cos 2\omega t + B_3 \cos 3\omega t \qquad (1)$$

where the magnitude of the d-c component $= I_b + B_o$.

Also
$$B_o = \frac{1}{6}(I_{\max} + 2I_{\frac{1}{2}} + 2I_{-\frac{1}{2}} + I_{\min}) - I_b \qquad (2)$$

Substituting from the above tabulation in (2),

$$B_o = \frac{1}{6}(19.0 + 2 \times 13.0 + 2 \times 4.5 + 2.0) - 8.0 = 1.33$$

The total direct current is $I_b + B_o = 8.0 + 1.33 =$ **9.33 ma.**

(*b*) From the above equation for i_b the peak value of the fundamental frequency of the current wave is the factor B_1.

$$B_1 = \frac{1}{3}(I_{\max} + I_{\frac{1}{2}} - I_{-\frac{1}{2}} - I_{\min})$$

$$B_1 = \frac{1}{3}(19.0 + 13.0 - 4.5 - 2.0) = \mathbf{8.5\ ma}$$

(*c*) From the same equation above the peak value of the second harmonic is factor B_2.

$$B_2 = \frac{1}{4}(I_{\max} - 2I_b + I_{\min})$$

$$B_2 = \frac{1}{4}(19.0 - 16.0 + 2.0) = \mathbf{1.25\ ma}$$

(*d*) Again from the same equation above the peak value of the third harmonic is factor B_3.

$$B_3 = \frac{1}{6}(I_{\max} - 2I_{\frac{1}{2}} + 2I_{-\frac{1}{2}} - I_{\min})$$

$$B_3 = \frac{1}{6}(20.0 - 2 \times 13.0 + 2 \times 4.5 - 2.0) = \mathbf{0\ ma}$$

Therefore, the total current wave will be

$$i_b = 8.0 + 1.33 + 8.5 \cos \omega t + 1.25 \cos 2\omega t + 0$$

Supplement **H**

ALTERNATING-CURRENT TRANSMISSION LINES

Questions and Answers

H-1. A 5,000-kva base transmission line has a resistance and reactance of 0.02 and 0.12 per unit, respectively. If a 6,000-kva inductive load having a power factor of 0.6 lagging is connected to the receiving end, while the sending end voltage is maintained at 1.10 per unit, what kva of a synchronous machine running light would

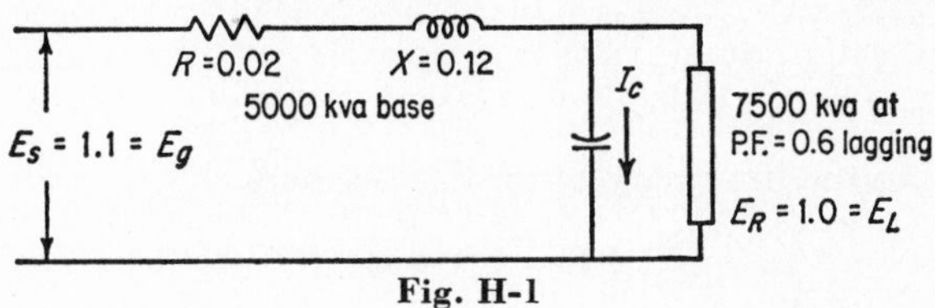

Fig. H-1

be necessary to maintain 1.00 per-unit voltage at the load end? Assume that the synchronous machine has no losses (See Fig. H-1).

ANSWER. If a 5,000-kva base is used

$$\begin{aligned} I_L &= 1.5(0.6 - j0.8) \\ &= 0.9 - j1.2 \end{aligned}$$

If I_{LX} is the reactive component of the line current and is assumed to be leading

$$E_s = E_R + Z_L I_L \qquad \text{where} \qquad Z_L = r + jx = 0.02 + j0.08$$

or

$$E_s = E_R + (0.02 + j0.08)(0.90 + jI_{LX})$$

If $E_s = 1.1$ and $E_R = 1.0$, then the substitution in the above equation gives

$$1.1 = 1.0 + (0.018 - 0.08I_{LX}) + j(0.072 + j0.02I_{LX})$$

The imaginary term is small compared to the real part and may be neglected.

$$1.1 = 1.018 - 0.08I_{LX}$$

And solving for I_{LX}

$$I_{LX} = (1.1 - 1.018)/-0.08 = 0.082/-0.08$$
$$= -1.025 \text{ per unit and lagging}$$

Since the original imaginary term was -1.2

$$I_c = -1.025 - (-1.2) = 0.175 \text{ per unit leading}$$

At no load

$$E_s = E_R + Z_L I_L = 1.1 = 1.0 + (0.02 + j0.08)jI_c$$

This equates to

$$1.1 = 1.0 - 0.08I_c + j0.02I_c$$

Neglecting the imaginary term, this becomes

$$1.1 = 1.0 - 0.08I_c$$

Solving for I_c

$$I_c = (1.1 - 1.0)/-0.08 = -1.25 \text{ lagging}$$

At no load the synchronous condenser will be underexcited (operating at lagging power factor) and will have a capacity of

$$\text{Motor kva} = 1.25 \times 5{,}000 = 6{,}250 \text{ kva}$$

At full load the condenser will be overexcited (operating at leading power factor) and will be operated at

$$(0.175/1.25) \times 100 = 14 \text{ per cent capacity}$$

This will give a resulting capacity of $0.14 \times 6{,}250 =$ **875 kva**

Using a 7,500-kva base

$$I_L = 0.6 - j0.8 \qquad \text{also} \qquad r = 0.02(7{,}500/5{,}000) = 0.03$$

and

$$X = 0.08(7{,}500/5{,}000) = 0.12$$

Then

$$E_s = E_R + (0.03 + j0.12)(0.60 + jI_{LX})$$
$$1.1 = 1.0 + (0.018 - 0.12I_{LX}) + j(0.072 + 0.03I_{LX})$$

Neglecting the imaginary term

$$1.1 = 1.018 - 0.12I_{LX}$$

Solving for I_{LX}

$$I_{LX} = (1.1 - 1.018)/-0.12 = -0.683 \text{ per unit lagging}$$
$$I_c = -0.683 - (-0.8) = +0.117 \text{ leading}$$

At no load

$$1.1 = 1.0 + (0.03 + j0.12)jI_c$$
$$1.1 = 1.0 - 0.12I_c + j0.03I_c$$

Neglecting the imaginary term

$$1.1 = 1.0 - 0.12I_c$$

from which we obtain I_c.

$$I_c = 0.10/-0.12 = -0.833 \text{ per unit lagging}$$

At no load the synchronous condenser will be underexcited (operating at lagging power factor) and will have a capacity of

$$\text{Condenser kva} = 0.833 \times 7{,}500 = 6{,}250$$

At full load the condenser will be overexcited (operating at leading power factor) and will operate at $(0.117/0.833) \times 100 = 14$ per cent capacity. This is equivalent to $0.14 \times 6{,}250 = 875$ kva. This condenser will therefore be able to maintain the desired voltage control under the condition of the removal of the 7,500-kva load.

H-2. A three-phase transmission line has a resistance of 10 ohms per wire and an inductive reactance of 80 ohms per wire. The load current is 90 amp and the power factor of the load is 80 per cent lagging. The sending (generator) end voltage of the line is 44,000 volts line to line. What is the receiving (load) end voltage? See Fig. H-2*a*.

ANSWER. Refer to the vector diagram in Fig. H-2*b*. The phase-to-neutral voltage is $E_s = 44{,}000/1.732 = 25{,}400$ volts. Since the configuration of wires, admittance, or charging current were not given, this line may be considered to be short. Let θ_R be at the receiving end and θ_s at the sending end. Then $\cos\theta_R = 36.87°$

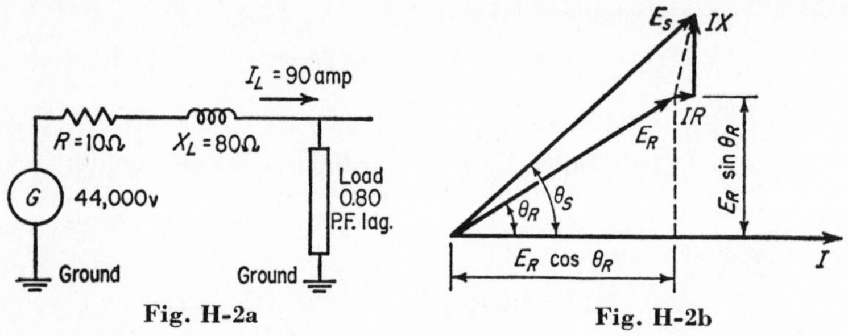

Fig. H-2a **Fig. H-2b**

and $\sin\theta_R = \sin$ of $36.87 = 0.6$. Now

$$E_s^2 = (E_R \cos\theta_R + IR)^2 + (E_R \sin\theta_R + IX)^2$$
$$25{,}400^2 = (E_R \times 0.8 + 90 \times 10)^2 + (E_R \times 0.6 + 90 \times 80)^2$$

Solving for E_R, this is found to equal **19,810 volts** (line-to-neutral voltage). The line voltage at the receiving end is equal to $E_R \times 1.732 =$ **34,300 volts.** The angle θ_s may be determined from the following calculation.

$$\theta_s = \frac{\cos^{-1}(E_R \cos\theta_R + IR)}{E_s} = \frac{19{,}810 \times 0.8 + 90 \times 10}{25{,}400} = 0.66$$

or

$$\theta_s = 48.7°$$

H-3. Each conductor of a three-phase three-wire transmission line has an impedance of $15 + j20$ ohms at 60 cycles. The impressed emf between the line conductors is 13,200 volts. The load connected to this line is balanced and takes 1,000 kw at a lagging power factor. The current per conductor is 70 amp. What is (*a*) the efficiency of transmission, (*b*) the difference of potential between the line wires at the load, (*c*) the power factor of the load?

ANSWER.

(*a*) Refer to Fig. H-3. The efficiency of transmission is

$$\frac{\text{Power received}}{\text{Power sent}} \times 100 = \frac{1{,}000 \times 1{,}000}{70^2 \times 15 \times 3 + \text{(power received)}} \times 100 = \textbf{82 per cent}$$

(*b*) From the three-phase calculations the value of E_R (also load end) may be determined from a complex notation as

$$\begin{aligned} E_R &= E_s - I_L Z_L = (13{,}200/1.732) - I_L(\cos\theta - j\sin\theta)(R_L + jX_L) \\ &= 7{,}625 - 70(0.763 - j0.645)(15 + j20) \\ &= 7{,}625 - 1{,}705 - j390.1 \\ &= \sqrt{(7{,}625 - 1{,}705)^2 + 390.1^2} = 5{,}930 \text{ volts} \end{aligned}$$

The difference in potential wanted is 1.732 × 5,930 = **10,266 volts**

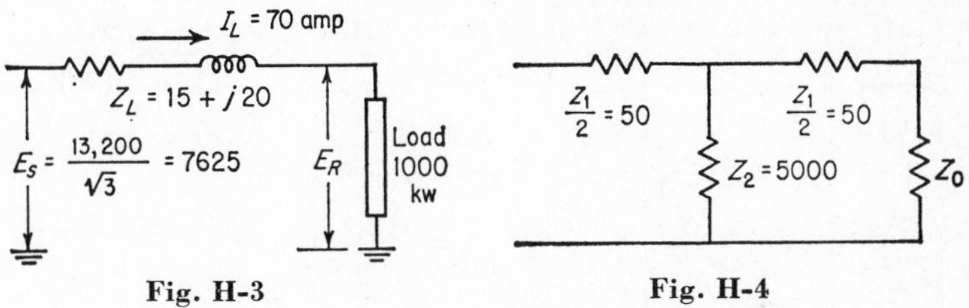

Fig. H-3

Fig. H-4

Note: The power factor at the sending end of the line is $\cos\theta_s$. This has been determined from the relation

$$\cos\theta_s = \frac{(1{,}000 \times 1{,}000) + (70^2 \times 15 \times 3)}{1.732 \times 13{,}200 \times 70} = 0.763$$

From which we can find $\sin\theta_s$ to be 0.645.

(*c*) $\cos\theta_L = 1{,}000{,}000/(1.732 \times 10{,}265 \times 70) =$ **0.805**

H-4. A T network has series arms of 50 ohms each, and the shunt arm is 5,000 ohms. How may this network be terminated in order to simulate an infinite line, and what would be the attenuation of a T network in decibels?

ANSWER. Refer to Fig. H-4. In order to simulate an infinite line terminal, the impedance must equal its characteristic impedance Z_o. The form below is also basic in the study of filters.

$$Z_o = \sqrt{Z_1 Z_2 \left(1 + \frac{Z_1}{4Z_2}\right)}$$

For the problem at hand

$$Z_o = \sqrt{100 \times 5{,}000[1 + 1/4 \times (100/5{,}000)]} = \mathbf{710\ ohms}$$

The attenuation in decibels may be determined from the following equation.

$$\text{attenuation} = 20 \times \log_{10} \frac{I_1}{I_2}$$

or $$\text{attenuation} = 20 \times \log_{10} \frac{50 + 5{,}000 + 710}{5{,}000} = \mathbf{1.2\ db}$$

H-5. A certain four-terminal transmission device has a characteristic impedance of 600 ohms and a propagation constant of $0.5 + j1.0$. When terminated with its characteristic impedance, the ratio of the input voltage to the output voltage is what?

ANSWER.

$$\frac{I_1}{I_2} = \frac{E_1}{E_2} = \epsilon^{\alpha + j\beta} = \epsilon^{0.5 + j1.0}$$

$$\frac{E_1}{E_2} = \epsilon^{0.5} = \mathbf{1.65} = \sqrt{2.72}$$

H-6. For the device in problem H-5 above, the voltage attenuation in decibels is of what value?

ANSWER.

$$\text{Attenuation db} = 10 \times \log_{10} \left(\frac{E_1}{E_2}\right)^2$$

$$\text{db} = 20 \times \log_{10} \frac{E_1}{E_2} = 20 \times \log_{10} \times 1.65 = \mathbf{4.35\ db}$$

H-7. It is required to calculate the attenuation and characteristic impedance at a frequency of 8,800 Mc/s, of an air-filled coaxial cable line having the following: diameter of inner conductor, 0.113 cm; inner diameter of outer conductor, 0.794 cm; resistivity of inner conductor, 1.78×10^{-6} ohm-cm; resistivity of outer conductor, 6.5×10^{-6} ohm-cm.

ANSWER. Calculate attenuation in a coaxial cable from

$$\alpha = 9.95 \times 10^{-6} \sqrt{f} \sqrt{\varepsilon/\varepsilon_o}\,(1/a\,\sqrt{\sigma_a} + 1/b\,\sqrt{\sigma_a})/\log(b/a)$$
$$+ 9.10 \times 10^{-8} \sqrt{\varepsilon/\varepsilon_o}\, f \tan\sigma \text{ decibels per meter}$$

In the above equation

f = frequency, cps
$\varepsilon/\varepsilon_o$ = ratio of permittivity of the dielectric to air = 1
a = radius of inner conductor, 5.65×10^{-4} meter
b = radius of outer conductor, 3.97×10^{-4} meter
σ_a = conductivity of inner conductor, 5.62×10^{7} mhos/cu m
σ_b = conductivity of outer conductor, 1.54×10^{7} mhos/cu m
$\tan\sigma$ = power factor of the dielectric = 0

If we simply substitute in the equation, we obtain

$$\alpha = \mathbf{0.332 \text{ decibel per meter}}$$

The characteristic impedance is given by

$$Z_o = 138 \sqrt{\frac{\varepsilon}{\varepsilon_o}} \log \frac{b}{a} = \mathbf{117 \text{ ohms}}$$

H-8. An extra high voltage transmission line of negligible resistance delivers power at 60 cps to a load center 300 miles away from the sending end of the line. The load is 1,000 mw at 500 kv with 100 per cent power factor. The geometrical and electrical characteristics of the line are as follows: flat horizontal configuration; 38′-0″ between phases; phase conductors: two 1,780 MCM ACSR per phase, bundled, with 18 in. between conductors; series impedance per unit length per phase is z series = $j0.6214$ ohm per mile; shunt impedance per unit length, phase to neutral is:

$$z \text{ shunt} = -j0.1548 \times 10^{6} \text{ ohms per mile}$$

(*a*) Determine the characteristic impedance or surge impedance of the line.

(*b*) Determine the surge impedance loading of the line (s.i.l.).

(*c*) If this line is to have 70 per cent series compensation, what would be the ohmic reactance per phase of the series capacitors required to achieve this compensation?

(*d*) Assume the installed cost of series compensation is $5,000 per mvar. How much would it cost to install 70 per cent series compensation for a rated load of 1,000 mw at unity power factor?

(*e*) Mention at least three advantages of using series compensation on extra-high voltage lines.

ANSWER.

(*a*) In power system work characteristic impedance (Z_c) is sometimes called surge impedance of the transmission line.

$$Z_c = \sqrt{\frac{z}{y}}$$

where $z = 0.6214 \underline{/90°}$ ohm per mile

$$y = \frac{1}{z \text{ shunt}} = \frac{1}{-j0.1548 \times 10^6} = j6.46 \times 10^{-6}$$

$$y = 6.46 \times 10^{-6} \underline{/90°} \text{ mho per mile}$$

Therefore, $$Z_c = \sqrt{\frac{0.6214}{6.46 \times 10^{-6}}} \underline{\Big/\frac{90° - 90°}{2}} = \sqrt{0.0962 \times 10^6}$$

$$Z_c = \mathbf{310 \text{ ohms per mile}}$$

(*b*) Surge impedance loading (s.i.l.) of a line is the power delivered by the line to a purely resistive load equal to its surge impedance.

$$\text{s.i.l.} = \frac{V_{\text{load}}^2}{Z_c} = \frac{500^2}{310} = \mathbf{806 \text{ kw}}$$

(*c*) A series capacitor improves load voltage effectively by compensating directly for the line inductive reactance which causes the voltage drop. The shunt impedance in this problem should be ignored due to its high value; leakage can be neglected.

$$X_c \text{ per mile} = (70 \text{ per cent } Z_{\text{series}}) = 0.7 \times j0.6214 = j0.435$$

$$X_c = 0.435 \times 300 \text{ miles} = \mathbf{130.5 \underline{/-90°} \text{ ohm}}$$

(*d*) $$I_{\text{load}} = \frac{1{,}000{,}000}{\sqrt{3} \times 500} = 1{,}156 \text{ amp}$$

$$\text{mvar} = 3 \times 1{,}156^2 \times 130.5 = 520$$

$$\text{Cost} = 520 \times \$5{,}000 = \mathbf{\$2{,}600{,}000}$$

(*e*)

1. Series compensation reduces the net line reactance and therefore increases the line's power-transfer ability.

2. Series compensation gives instantaneous and automatic voltage drop (regulation) reduction after load has been changed. The reductive correction is proportional to the load change.

3. High electrical stresses on capacitors are avoided in series correction.

4. A shunt capacitor improves load voltage (reduces voltage drop) by neutralizing part of the lagging current in a circuit. A series capacitor improves load voltage more effectively, compensating directly for the line reactance which causes the voltage drop. Consequently, the same voltage correction is obtained with a smaller rating of series capacitors than shunt, usually in the ratio of one-half to one-fourth. However, the shunt capacitor corrects power factor to a greater extent.

H-9. A typical open-wire telephone line has a resistance of 10 ohms per mile, an inductance of 0.0037 henry per mile, a capacitance of 0.0083×10^{-6} farad per mile, and conductance of 0.4×10^{-6} mho per mile at a frequency of 1,000 cps. Calculate its characteristic impedance and propagation constant.

ANSWER.

$$\text{Characteristic impedance} = Z_0 = \sqrt{(R + jWL)/(G + jWC)}$$

$$= \sqrt{\frac{10 + j23.2}{(0.4 + j52.1) \times 10^{-6}}} = 697 \underline{/-11.4^\circ} = \mathbf{683 - j138\ ohms}$$

$$\text{Propagation constant} = \gamma = \sqrt{(R + jWL)(G + jWC)}$$
$$= \alpha + j\beta = \mathbf{0.0074 + j0.0356\ per\ mile}$$

where α = attenuation constant, neper per mile = 0.0074

β = phase constant = 0.0356 radian per mile

H-10. An electrically parallel, double-circuit transmission line is to be built in a heavy-loading district. There are two configurations to choose from: hexagonal or vertical flat. The former has a geometric mean radius (r') of 0.0403 ft; neglect transposition of phases in the latter. Refer to Fig. H-5. It is required to (*a*) choose one

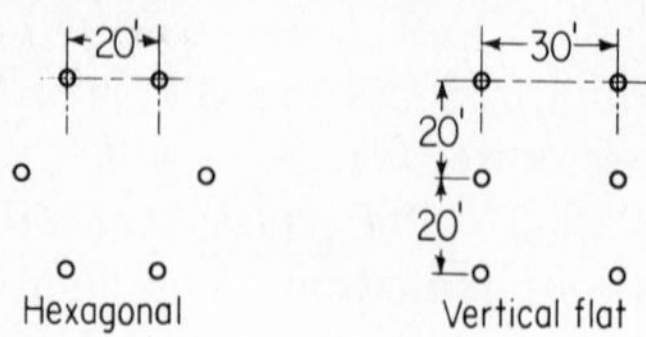

Fig. H-5

configuration and give your reasons, (*b*) calculate the inductive reactance per phase per mile. Identify the phases.

ANSWER.

(*a*) Select the hexagonal configuration. The line is to be built in a heavy-loading area, i.e., above 3,000 ft above sea level, which means heavy loads of ice and snow. When this load drops off, the line flips vertically, and if it were not a hexagonal configuration, the lines would come in contact and cause a line-to-line short. Another advantage of the hexagonal arrangement is the equilateral configuration. This means that transposition can be neglected.

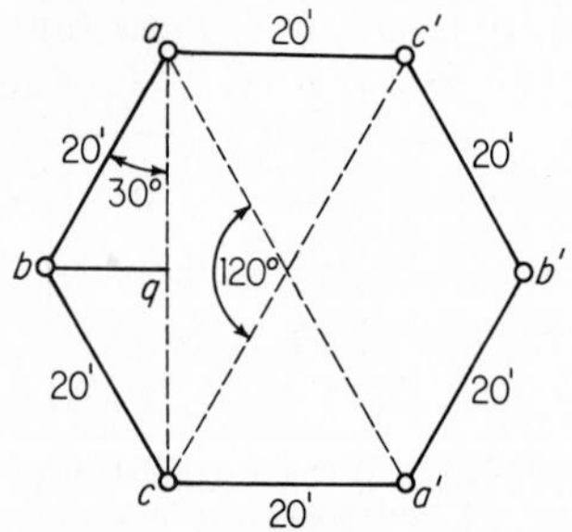

Fig. H-6

(*b*) It is recommended that the $abc - c'b'a'$ configuration be used. This way each phase, $aa' - bb' - cc'$, has the same conductor spacing of 40 ft. The inductance is lowered if the individual conductors of a phase are separated as widely as possible and if the distance between phases is kept small. This results in a low D_{eq} and a high D_s. Proceeding, refer to Fig. H-6. ac is a geometric chord;

$$ac = 2r \sin (120/2) = 2 \times 20 \times \sin 60° = 40 \times 0.866 = 34.64 \text{ ft}$$

Also $\quad bq = (34.64/2) \tan 30° = 17.32 \times 0.5774 = 10 \text{ ft}$

Therefore

$$bb' = 10 + 20 + 10 = 40 \qquad \text{And } aa' = cc' = \sqrt{34.64^2 + 20^2}$$

Using the geometric mean radius of 0.0403

$$D_s = D_{sa} = D_{sb} = D_{sc} = \sqrt[4]{(0.0403)^2(40)^2}$$
$$= \sqrt{0.0403 \times 40} = 1.27 \text{ ft}$$

To obtain D_{eq}: $D_{ba'} = D_{b'a} = D_{bc'} = D_{cb'} = 34.64 \text{ ft} = ac$

$$D_{bc} = D_{ab} = D_{ac} = \sqrt[4]{20^2 \times 34.64^2} = \sqrt{20 \times 34.64} = 26.32 \text{ ft}$$

$$D_{eq} = \sqrt[3]{D_{ab} \times D_{bc} \times D_{ac}} = \sqrt[3]{(26.32)^3} = 26.32 \text{ ft}$$

$$L = 0.7411 \log (26.32/1.27) = 0.7411 \log 20.72$$
$$= 0.976 \text{ millihenry per mile per phase}$$

$$X_L = 2\pi f L \times 10^{-3} = 377 \times 0.976 \times 10^{-3}$$
$$= \mathbf{0.368 \text{ ohm per mile per phase}}$$

Supplement **I**

NETWORK ANALYSIS

Questions and Answers

I-1. By use of nodal analysis, find the voltage across AB in the circuit shown in Fig. I-1.

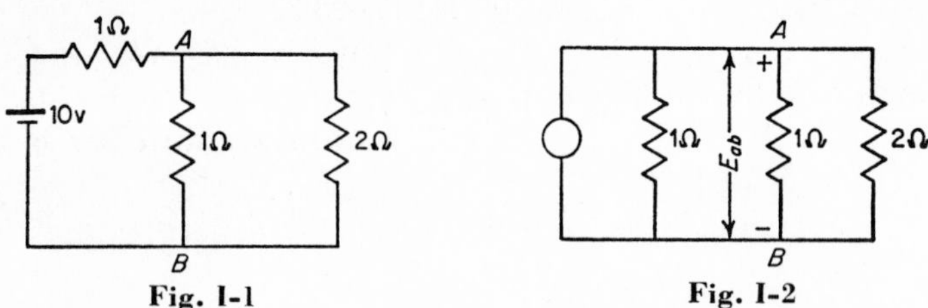

Fig. I-1 **Fig. I-2**

ANSWER. Draw the equivalent circuit shown in Fig. I-2. Then the nodal equation from application of Kirchhoff's first law at point A is: $10 = E_{ab}/1 + E_{ab}/1 + E_{ab}/2$, from which $E_{ab} = 4$ **volts.**

I-2. A circuit carries a direct current of 10 amp on which is superimposed a sinusoidal component of 5 amp maximum value. (*a*) What kind of ammeter will indicate the *rms* effective current in the circuit? (*b*) What is the value of the *rms* effective current?

ANSWER.

(*a*) Use a soft-iron vane-type ammeter.

$$(b)\ I_{\text{rms}} = \sqrt{\frac{1}{2\pi}\int_0^{2\pi} (5 \sin \omega t + 10)^2\, d\,\omega t}$$

$$I_{\text{rms}} = \sqrt{\frac{1}{2\pi}\int_0^{2\pi} (25 \sin^2 \omega t + 100 \sin \omega t + 100)\, d\,\omega t}$$

$$I_{\text{rms}} = \sqrt{\frac{1}{2\pi}\left\{\left[25\left(\frac{x}{2} - \frac{1}{4}\sin 2x\right)\right]_0^{2\pi} + 100(-\cos x)_0^{2\pi} + (100x)_0^{2\pi}\right\}}$$

$$I_{\text{rms}} = \sqrt{\frac{1}{2\pi}\left[\left(25 \times \frac{2\pi}{2}\right) + 200\pi\right]} = \sqrt{112.5} = \mathbf{10.6\ amp}$$

Note $x = \omega t$ in the calculations.

I-3. Calculate the current I_o in the output of the circuit in Fig. I-3, in which $R_1 = R_2 = 8$ ohms, $Z_3 = 5 + j5$ ohms, $Z_4 = 5 - j5$ ohms and $R_o = 1$ ohm.

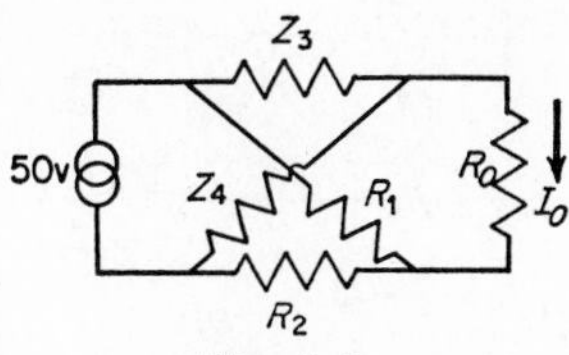

Fig. I-3

ANSWER. Convert to equivalent circuits Figs. I-4 through I-6.

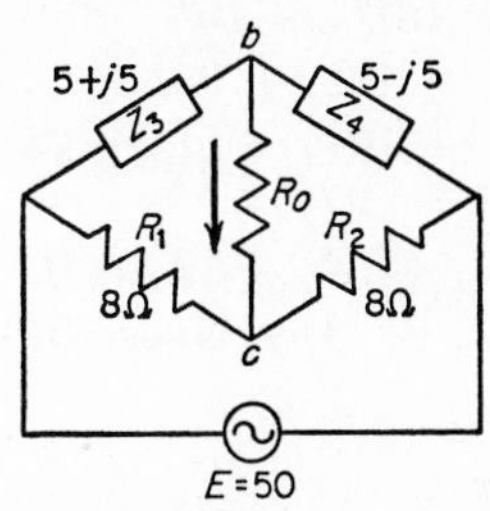

Fig. I-4

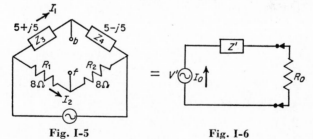

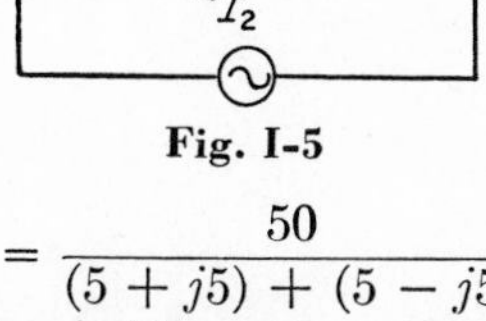

Fig. I-5

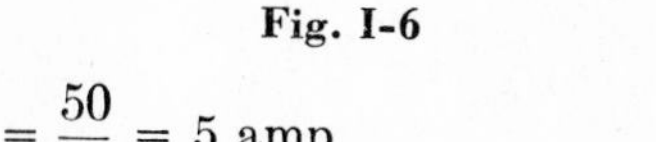

Fig. I-6

$$I_1 = \frac{50}{(5 + j5) + (5 - j5)} = \frac{50}{10} = 5 \text{ amp}$$

$$I_2 = 50/(8 + 8) = 3.12 \text{ amp}$$

$$Z' = \frac{(5 + j5)(5 - j5)}{(5 + j5)(5 - j5)} + \frac{8 \times 8}{8 + 8} = \frac{25 + 25}{10} + 4 = 9 \text{ ohms}$$

Assume $V_b > V_f$. Then

$$V' = I_2R_1 - I_1(Z_3) = 3.12 \times 8 - 5(5 + j5) = -j25$$

Therefore, $\dfrac{V'}{Z' + R_o} = \dfrac{-j25}{9 + 1} = -j25$ **amp** for I_o

I-4. Calculate the current I_{ab} in the circuit shown in Fig. I-7.

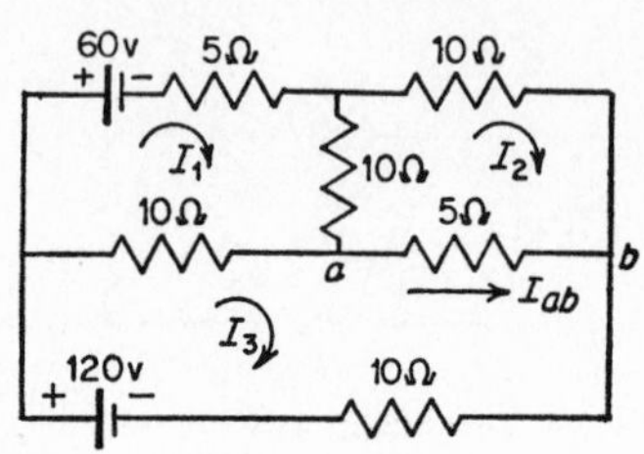

Fig. I-7

ANSWER.
$$(5 + 10 + 10)I_1 - 10I_2 - 10I_3 = -60 \qquad (1)$$
$$-10I_1 + 25I_2 - 5I_3 = 0 \qquad (2)$$
$$-10I_1 - 5I_2 + 25I_3 = 120 \qquad (3)$$

Dividing (1) by 5, $\quad 5I_1 - 2I_2 - 2I_3 = -12.$
Dividing (2) by 5, $-2I_1 + 5I_2 - I_3 = 0.$
Dividing (3) by 5, $-2I_2 - I_2 + 5I_3 = 24.$
Then

$$I_2 = \frac{\begin{vmatrix} 5 & -12 & -2 \\ -2 & 0 & -1 \\ -2 & 24 & 5 \end{vmatrix}}{\begin{vmatrix} 5 & -2 & -2 \\ -2 & 5 & -1 \\ -2 & -1 & 5 \end{vmatrix}} = \frac{96 - 24 + 120 - 120}{125 - 4 - 4 - 20 - 5 - 20}$$

$$= \frac{72}{72} = 1 \text{ amp}$$

$$I_3 = \frac{\begin{vmatrix} 5 & -2 & -12 \\ -2 & 5 & 0 \\ -2 & -1 & 24 \end{vmatrix}}{72} = \frac{12\begin{vmatrix} 5 & -2 & -1 \\ -2 & 5 & 0 \\ -2 & -1 & 2 \end{vmatrix}}{72}$$

$$= \frac{12}{12}(50 - 2 - 10 - 8) = \frac{1}{6}(30)$$

$$= 5 \text{ amp}$$

Therefore, $I_{ab} = 5 - 1 =$ **4 amp.**

I-5. A three-wire, 120/240-volt system is created by the use of a balancer set, the armatures of which are connected as shown in Fig. I-8. Assume efficiency of each balancer unit as 0.80. Calculate the currents in I_1, I_2, and I_3.

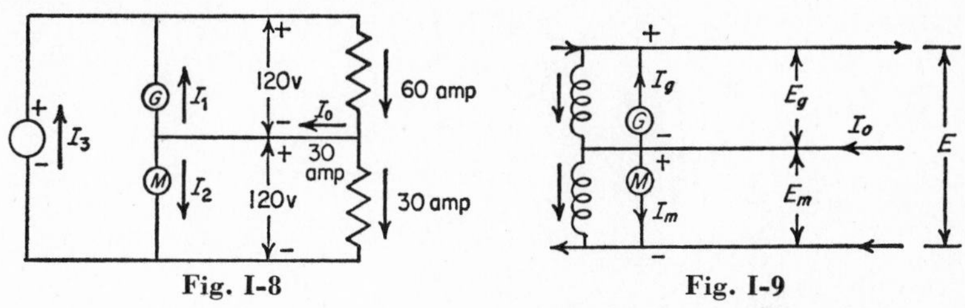

Fig. I-8 **Fig. I-9**

ANSWER. Balancer sets consist of two like units coupled together, each half-voltage, in series. On *balanced load,* they run "light" as motors. With an *unbalanced load,* the current flow is as shown in Fig. I-9. Motor acts as motor generator. $I_m > I_g$ by that amount required to supply the armature loss. Since motor output is equal to generator input, $120 \times I_2 \times 0.80 = (120 \times I_1)/0.80$, from which

$$I_1 = 0.64\, I_2 \tag{1}$$

Also from the diagram,

$$I_1 + I_2 = I_o = 30 \text{ amp} \tag{2}$$

From (1) and (2) it may be found that

$$I_1 = \mathbf{11.6\ amp} \qquad \text{and} \qquad I_2 = \mathbf{18.3\ amp}$$

Also $I_3 + I_1 = 60$, from which

$$I_3 = 60 - 11.6 = \mathbf{48.4\ amp}$$

I-6. The following circuit is used to find a contact fault on a pair of telephone lines which are 30 miles long. Resistors R_1 and R_2 are each 300 ohms, and the unit resistance per mile of line may be taken as 40 ohms. Find the distance y at which the fault occurs if at the balancing point R_3 is equal to 440 ohms.

ANSWER. Refer to Fig. I-10. Let R be resistance per mile and y and l be measured in miles. Then at the balancing point the

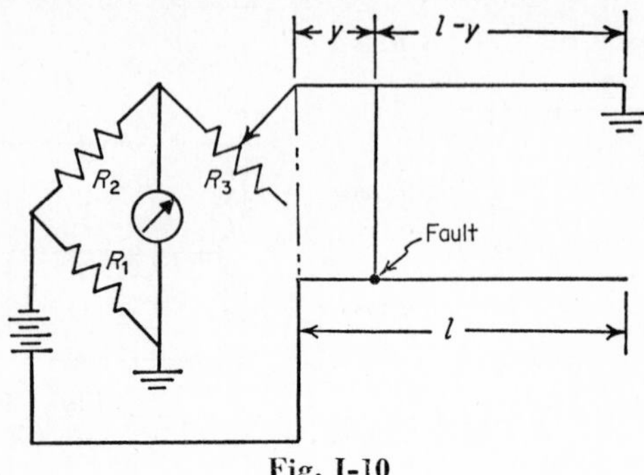

Fig. I-10

following relation exists

$$\frac{R_2}{R_1} = \frac{R_3 + yR}{R(l - y)}$$

Thus, if $R_2 = R_1$, $R(l - y) = R_3 + Ry$. And $y = Rl - R_3/2R$. However, in this problem $R = 40$ ohms, $l = 30$ miles, $R_3 = 440$ ohms. Finally,

$$y = (1{,}200 - 440)/80 = {}^{76}\!/_{8} = \mathbf{9.6\ miles}$$

I-7. Apply (*a*) the superposition-of-currents method, (*b*) Kirchhoff's laws, (*c*) Thévenin's theorem to find the current in branch AC of the network $ABCD$ shown.

ANSWER. Refer to Fig. I-11 as the basis for the problem.

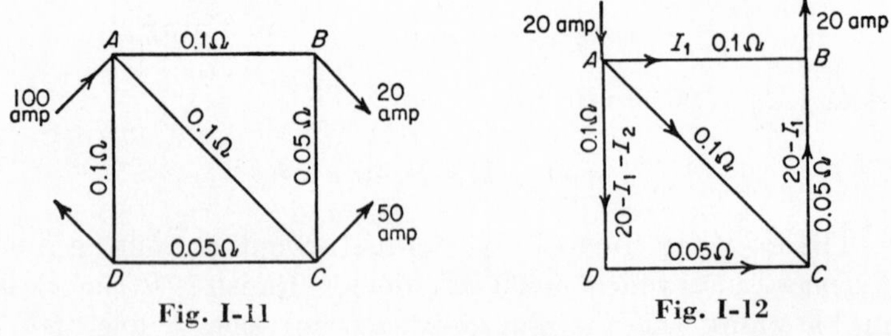

Fig. I-11 Fig. I-12

(*a*) Set up circuit as shown in Fig. I-12 and consider the 20-amp load acting alone. Assume the currents to flow as shown. Now find the current I_2 from A to C. *For loop ADC:*

$$(20 - I_1 - I_2)0.15 = 0.1I_2 \tag{1}$$

For loop ABC:

$$0.1I_1 - (20 - I_1)0.05 = 0.1I_2 \tag{2}$$

From (1) and (2),

$$I_2 = 40/7 \tag{3}$$

Now consider the 50-amp load acting alone. And permit the

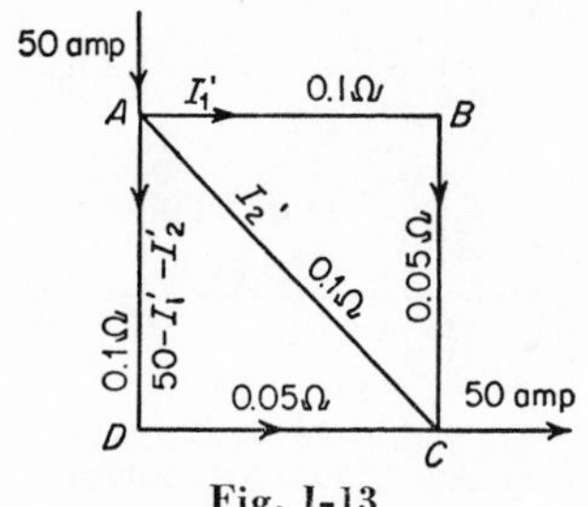

Fig. I-13

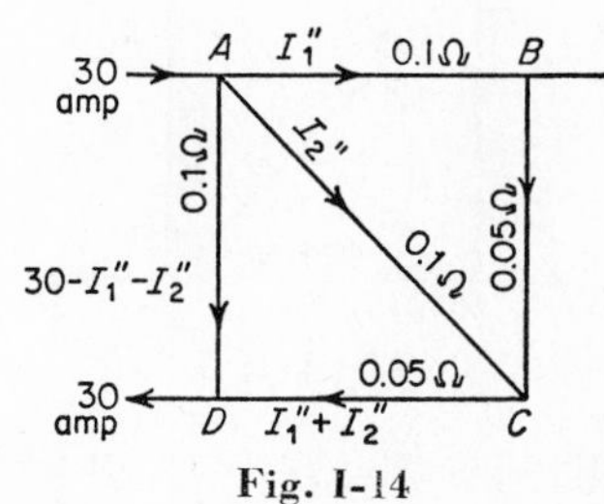

Fig. I-14

currents to flow as shown in Fig. I-13. Find I_2'. *For loop ABC:*

$$0.15I_1' = 0.1I_2' \tag{4}$$

For loop ADC:

$$0.15(50 - I_1' - I_2') = 0.1I_2' \tag{5}$$

Combining (4) and (5),

$$I_2' = 150/7 \tag{6}$$

Now consider the 30-amp load acting alone. And permit the currents to flow as shown in Fig. I-14. Purpose is to determine I_2''.

For loop ABC:

$$0.15I_1'' = 0.1I_2'' \tag{7}$$

For loop ADC:

$$0.1(30 - I_1'' - I_2'') = 0.1I_2'' + 0.05(I_1'' + I_2'') \tag{8}$$

Combining (7) and (8),

$$I_2'' = 60/7 \tag{9}$$

Combining (3), (6), and (9), total current in line AC due to all loads acting at same time is

$$I_2 + I_2' + I_2'' = (40/7) + (150/7) + (60/7) = \mathbf{35.8\ amp}$$

(*b*) *Application of Kirchhoff's laws.* Let the currents in AB and AC be I_1 and I_2 respectively. By means of Kirchhoff's first law,

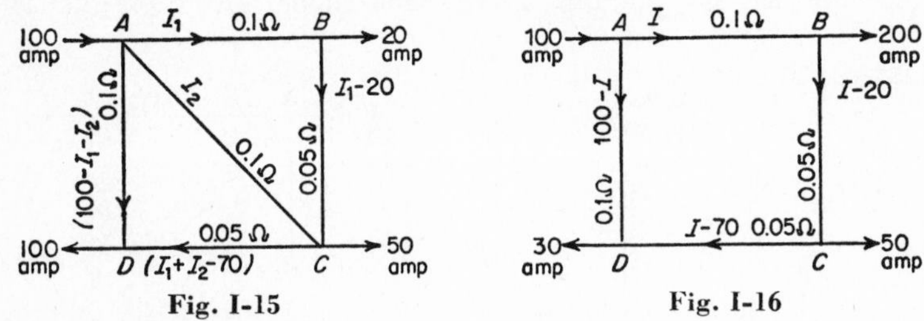

Fig. I-15

Fig. I-16

write down the currents in the remaining conductors in terms of I_1 and I_2 as shown in Fig. I-15. *For loop ABC:*

$$0.1I_1 + 0.05(I_1 - 20) - 0.1I_2 = 0 \tag{1}$$

from which

$$3I_1 = 20 + 2I_2 \tag{2}$$

For loop ADC:

$$0.1(100 - I_1 - I_2) - 0.05(I_1 + I_2 - 70) - 0.1I_2 = 0 \tag{3}$$

from which

$$270 - 3I_1 - 5I_2 = 0 \tag{4}$$

And from combining (2) and (4),

$$270 = 20 + 2I_2 + 5I_2 \qquad \text{and} \qquad I_2 = 250/7 = \mathbf{35.7\ amp}$$

(*c*) *Application of Thévenin's theorem.* Remove conductor AC as shown in Fig. I-16. Then first find voltage between A and C and the resistance between A and C with AC out. Current in AC is equal to $E_{ac}/(R_{ac} + 0.1)$. But $R_{ac} = (0.1 + 0.05)/2$. Let current in $AB = I$. Then for loop $ABCDA$

$$0.1I + 0.05(I - 20) + (I - 70)0.05 = (100 - I)0.1$$

from which $I = {}^{146}\!/_3$. Working out calculations for E_{ac},

$$E_{ac} = 0.1I + 0.05(I - 20)[0.1 \times ({}^{146}\!/_3)] + 0.05[({}^{146}\!/_3) - 20] = 6.25 \text{ volts}$$

Current in circuit AC

$$\frac{6.25}{(0.15/2) + 0.1} = {}^{250}\!/_7 = \mathbf{35.7\ amp}$$

I-8. Calculate the voltage across the 40-ohm resistor in the network

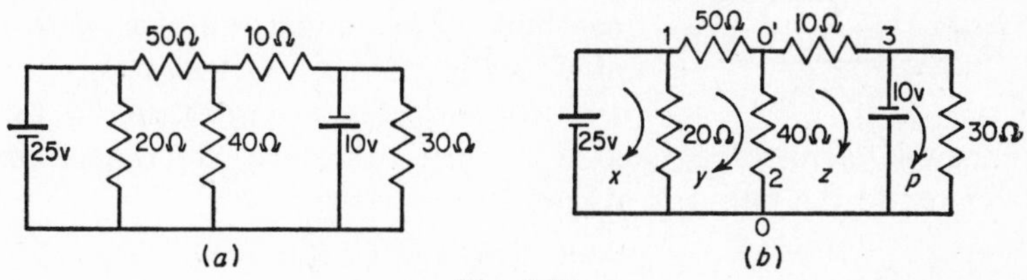

Fig. I-17

shown in Fig. I-17a by using (a) Millman's theorem, (b) generalized form of Norton's theorem.

ANSWER.

(a) *Application of Millman's theorem.* Refer to Fig. I-17b. Show points 0, 0′, 1, 2, and 3 and arrowed letters x, y, z, and p. Then

$$E_{01} = -25E \qquad E_{02} = 0 \qquad E_{03} = +10 \text{ volts}$$

$$Y_1 = {}^1\!/_{50} \text{ mho} \qquad Y_2 = {}^1\!/_{40} \text{ mho} \qquad Y_3 = {}^1\!/_{10} \text{ mho}$$

$$E_{00'} = \frac{E_{01}Y_1 + E_{02}Y_2 + E_{03}Y_3}{Y_1 + Y_2 + Y_3} = \frac{[-25({}^1\!/_{50}) + 10({}^1\!/_{10})]}{[({}^1\!/_{50}) + ({}^1\!/_{40}) + ({}^1\!/_{10})]} = \frac{100}{29} = \mathbf{3.45\ volts}$$

(b) *Application of Norton's theorem.* Now short out 0 and 0′. Under these conditions the short circuit current flowing with the 25-volt battery is ${}^{25}\!/_{50} = {}^1\!/_2$ amp. For the 10-volt battery system the short circuit current is $-(10/10) = -1$ amp. Likewise for the resistance

$$\frac{1}{R} = \frac{1}{50} + \frac{1}{40} + \frac{1}{10} \qquad \text{or} \qquad R = \frac{200}{29} \text{ ohm}$$

Thus, $E_{00'} = (1/2)(200/29) =$ **3.45 volts.**

I-9. The circuit shown in Fig. I-18 includes a generator with an internal resistance, $R_g = 1{,}000$ ohms; an emf of constant magnitude, $E = 150$ volts; and variable frequency. The coil in the tank circuit has an inductance, $L = 200$ microhenrys, and a Q factor, $Q = 10$, which may be considered constant. The tuning capacity is fixed at $C = 10^{-10}$ farad. Calculate (*a*) the frequency of parallel resonance, (*b*) the voltage E_o across the tank at parallel resonance, (*c*) the power delivered to the tank coil at parallel resonance.

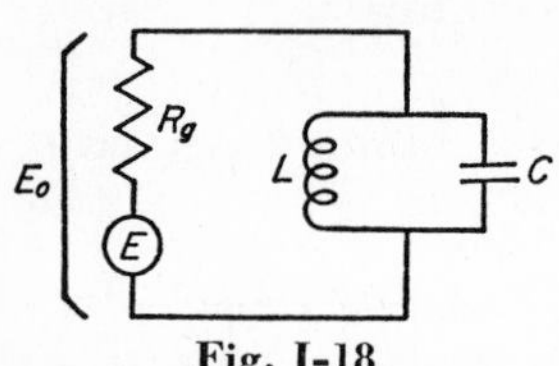

Fig. I-18

ANSWER.

(*a*)
$$Q = \frac{\omega L}{R_L} = 10 \qquad R_L = \frac{\omega L}{Q} = \frac{\omega L}{10}$$

For resonance

$$\frac{X_c}{R_c^2 + X_c^2} = \frac{X_L}{R_L^2 + X_L^2} \qquad \text{Also} \qquad R_c = 0$$

$$\omega C = \frac{\omega L}{R_L^2 + \omega^2 L^2} = \frac{\omega L}{(\omega L/10)^2 + \omega^2 L^2}$$

Therefore

$$\omega C = \frac{100}{\omega L(1 + 100)} \qquad \text{Also} \qquad \omega_o^2 = \frac{100}{101LC}$$

Accordingly

$$\omega_o = \sqrt{\frac{100}{101 \times 200 \times 10^{-6} \times 10^{-10}}} = 7.04 \times 10^6$$

$$f = \frac{\omega}{2\pi} = 7.04 \times \frac{10^6}{2\pi} = \mathbf{1.12 \times 10^6 \ cps}$$

$$(b)\ Y = \frac{1}{R + j\omega L} + j\omega C$$

$$= \frac{R}{R^2 + \omega^2 L^2} + j\omega C - \frac{\omega L}{R^2 + \omega^2 L^2}$$

$$Y = \frac{R}{R^2 + \omega^2 L^2}$$

$$Z = \frac{R^2 + \omega^2 L^2}{R} = R + R\left(\frac{\omega L}{R}\right)^2 = R + RQ^2 = R(1 + Q^2)$$

$$= R(1 + 10^2)$$

$$Z = 101\frac{\omega L}{Q} = 101 \times \frac{7.04 \times 10^6 \times 200 \times 10^{-6}}{10}$$

$$= 14{,}200 \text{ ohms}$$

Therefore

$$E_o = 150 \times (14{,}200/15{,}200) = \mathbf{140\ volts}$$

(c)

$$P_o = I_o^2 Z_R = \frac{E_o^2}{Z_R} = \frac{140^2}{14{,}200} = \mathbf{1.38\ watts}$$

I-10. A series circuit of R, C, and L is to resonate at $\omega_r = 10^6$ radians per sec, have a bandwidth between half-power points of $0.1\omega_r$, and draw 10 watts from a 100-volt source. (*a*) Calculate the necessary values of R, C, and L. (*b*) What is the bandwidth of this circuit between quarter-power points, as a fraction of ω_r?

ANSWER. $1/n$ power point is the frequency at which the power of the circuit is $1/n$th of the resonance.

(*a*) $\omega_r = 10^6$ radians per sec. Also $\omega_2 - \omega_1 = 0.1\omega_r = 10^5$ radians. Therefore, $Q = \omega_o/(\omega_2 - \omega_1) = 10^6/10^5 = 10$. Then at resonance $P = E^2/R$. Thus

$$R = \frac{E^2}{P} = \frac{100^2}{10} = \mathbf{1000\ ohms}$$

Finally

$$L = \frac{QR}{10^6} = \frac{10 \times 10^3}{10^6} = \mathbf{10^{-2}\ henry}$$

$$LC = \frac{1}{\omega_o^2} = \frac{1}{10^{12}}$$

Thus,

$$C = \frac{1}{10^{-2} \times 10^{12}} = \mathbf{10^{-10}\ farad}$$

(*b*) For quarter-point, $n = 4$. And the fractional bandwidth is

$$\frac{\omega_2 - \omega_1}{\omega_r} = \frac{R_T\sqrt{4-1}}{10^6 L_T} = \frac{1{,}000\sqrt{3}}{10^6 \times 10^{-2}} = \mathbf{0.1732}$$

I-11. In Fig. I-19, if $V = 100 + 50 \sin \omega t + 25 \sin 3\omega t$, find current and average power. Note $\omega = 500$ radians per sec.

ANSWER. At $\omega = 0$, $Z = 5$; at $\omega = 500$ radians per sec,

$$I_o = V/R = {}^{100}\!/_5 = 20 \text{ amp.}$$
$$Z_1 = 5 + j0.02(500) = 5 + j10 \qquad \tan^{-1} {}^{10}\!/_5 = 63.4°$$
$$i_1 = \frac{\bar{V}_1 \max}{|Z_1|}(\sin \omega t - \theta_1) = \frac{50}{\sqrt{5^2 + 10^2}} \sin(\omega t - 63.4°)$$
$$i_1 = 4.48 \sin(\omega t - 63.4°)$$

At $3\omega = 1{,}500$ radians per sec, $Z_3 = 5 + j\,0.02(1{,}500) = 5 + j30$

$$i_3 = \frac{V_3 \max}{|Z_3|} \sin(3\omega t - \theta_3) = \frac{25}{30.4} \sin(3\omega t - 80.5°)$$
$$i_3 = 0.823 \sin(3\omega t - 80.5°)$$

Therefore

$$i = 20 + 4.48 \sin(\omega t - 63.4°) + 0.823 \sin(3\omega t - 80.5°)$$
$$I_{\text{rms}} = \sqrt{20^2 + 4.48^2/2 + 0.823^2/2} = \sqrt{410.6} = \mathbf{20.25\ amp}$$

Average power $= I^2_{\text{rms}}R = 20.25^2 \times 5 =$ **2,053 watts.**

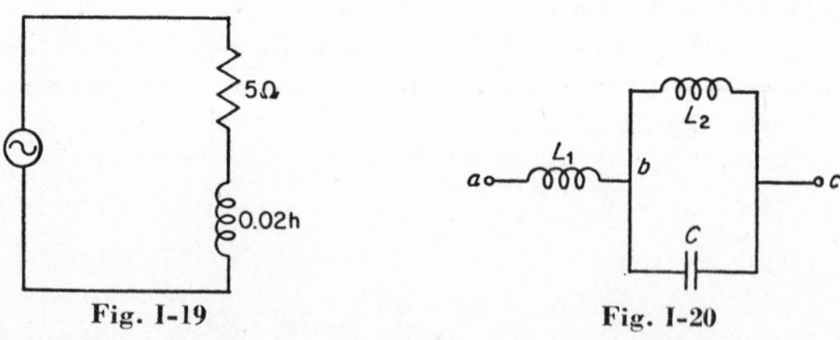

Fig. I-19 Fig. I-20

I-12. Calculate the size of the elements and the configuration of the two-terminal network of pure reactances which will have a positive reactance of 1,000 ohms at $\omega = 10^6$ radians per sec, infinite reactance at $\omega = 2 \times 10^6$ radians per sec and zero reactance at $\omega = 3 \times 10^6$ radians per sec.

ANSWER. Refer to Fig. I-20. Then

Z_{ac}	ω
$j1{,}000$	10^6
∞	2×10^6
0	3×10^6

$$Z_{ac} = j\omega L_1 + \frac{L_2/C}{j\omega L_2 + 1/j\omega C}$$

$$Z_{ac} = \frac{j\omega(L_1 + L_2 - \omega^2 L_1 L_2 C)}{1 - \omega^2 L_2 C}$$

Then

$$L_1 + L_2 - 9 \times 10^{12} L_1 L_2 C = 0 \qquad (1)$$

$$1 - 4 \times 10^{12} L_2 C = 0 \qquad (2)$$

From (2), $L_2 C = 1/(4 \times 10^{12})$.
Substituting in (1),

$$L_1 + L_2 = 9 \times 10^{12} L_1 \frac{1}{4 \times 10^{12}} = \frac{9}{4} L_1$$

That is,

$$L_2 = \frac{5}{4} L_1 \qquad (3)$$

Since $Z_{ac} = j1{,}000$ when $\omega = 10^6$,

$$j1{,}000 = \frac{j10^6[L_1 + (5/4)L_1 - 10^{12} L_1 \times 1/(4 \times 10^{12})]}{1 - 10^{12} \times 1/(4 \times 10^{12})} \qquad (4)$$

$$1 = \frac{1{,}000[(9/4)L_1 - (1/4)L_2]}{3/4} = (8{,}000/3)L_1$$

Finally,

$L_1 = 3/8{,}000 =$ **0.375×10^{-3} henry**
$L_2 = (5/4)L_1 =$ **0.469×10^{-3} henry**

$$C = \frac{1}{4 \times 10^{12}} \times \frac{1}{469 \times 10^{-6}} = 10^{-6}/1{,}876 = \mathbf{0.533 \times 10^{-9}\ farad}$$

I-13. The unbalanced network shown in Fig. I-21 is supplied with power from a 100-volt generator. The values of the bridge arms

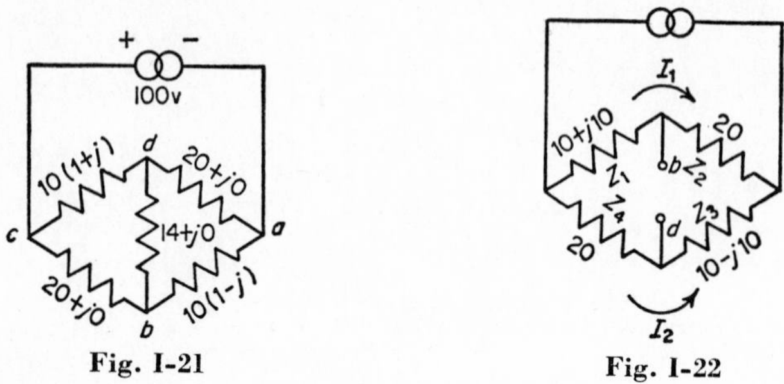

Fig. I-21

Fig. I-22

are expressed in ohms as shown. Calculate the current in the detector connected across *bd*.

ANSWER. Show the Z's and the current paths set up in Fig. I-22.

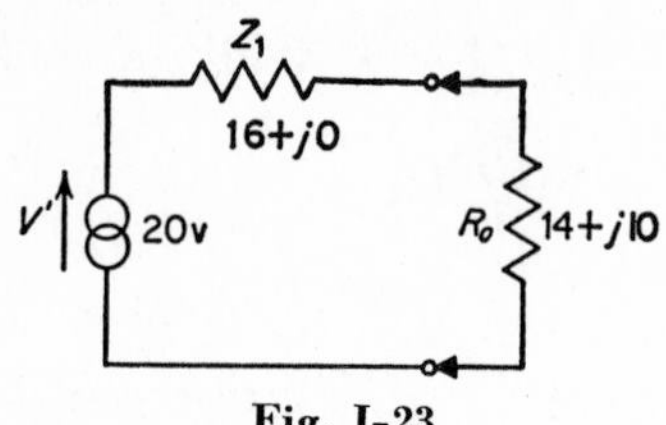

Fig. I-23

Then show the equivalent circuit as Fig. I-23.

$$Z' = \frac{Z_1Z_2}{Z_1 + Z_2} + \frac{Z_3Z_4}{Z_3 + Z_4} = \frac{20(10 + j10)}{(10 + j10) + 20} + \frac{20(10 - j10)}{20 + (10 - j10)}$$
$$= 16 + j0$$

$$I_1 = \frac{100}{20 + (10 + j10)} = \frac{100}{30 + j10} = \frac{10}{3 + j} = 3 - j$$

$$I_2 = \frac{100}{20 + (10 - j10)} = \frac{100}{30 + j10} = 3 + j$$

Therefore

$$V' = (I_2Z_4 - I_1Z_1) \qquad \text{assuming } V_b > V_d$$
$$V' = (3 + j)20 - (3 - j)(10 + 10j) = 60 + 20j - 10(3 - j)(1 + j)$$
$$= 60 - 20j - 10(3 + 2j + 1)$$
$$= 60 + 20j - 40 - 20j = 20V$$

Finally

$$I_{bd} = 20/(16 + 14) = 2/3 = \mathbf{0.667\ amp}$$

I-14. In Fig. I-24, find the voltage between N and P with switch K open.

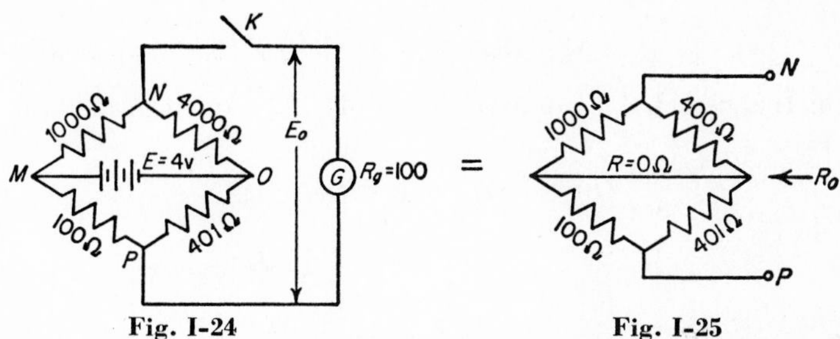

Fig. I-24 Fig. I-25

ANSWER. This shows the application of Thévenin's theorem to the Wheatstone Bridge. Set up the equivalent circuits shown in Fig. I-25 and I-26.

Then

$$V_{NP} = E_o = 4\left(\frac{1{,}000}{4{,}000 + 1{,}000} - \frac{100}{100 + 401}\right) = 1{,}597\mu\text{v}$$

$$R_o = \frac{1{,}000 \times 4{,}000}{1{,}000 + 4{,}000} + \frac{100 \times 401}{100 + 401} = 800 + 80 = 880 \text{ ohms}$$

$$I_g = \frac{1{,}579}{800 + 100} = \mathbf{1.63\ \mu a}$$

I-15. Calculate the current in the series circuit in Fig. I-27, at time t, from the subsidiary equation if there is no initial current and no

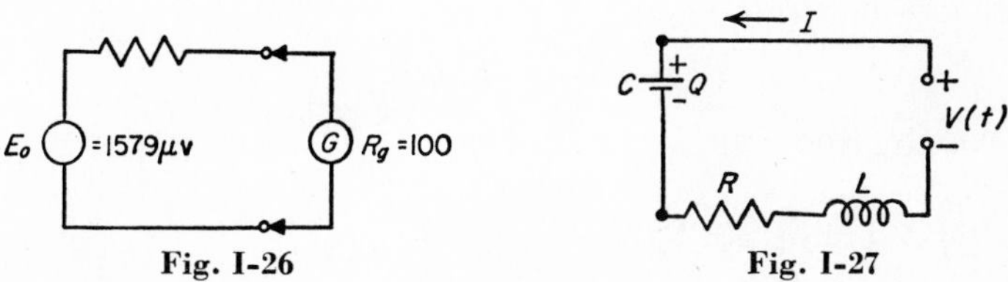

Fig. I-26 Fig. I-27

initial charge on the capacitor and the applied voltage E is constant. Assume $R = 2\sqrt{L/C}$.

ANSWER.

$$I = \frac{dQ}{dt} \tag{1}$$

Also

$$\frac{L\,dI}{dt} + RI + \frac{Q}{C} = V \tag{2}$$

Now solve (1) and (2) with initial values, $I = I_o$.

$$Q = Q_o \qquad \text{at} \qquad t = 0 \tag{3}$$

Form the subsidiary equations from (1), (2), and (3), resulting in

$$(Lp + R)\bar{I} + \frac{\bar{Q}}{C} = LI_o + \bar{V} \tag{4}$$

$$\bar{Q} = \frac{\bar{I}}{p} + \frac{Q_o}{p} \tag{5}$$

From combining (4) and (5), we obtain

$$\left(Lp + R + \frac{1}{Cp}\right)\bar{I} = \bar{V} + LI_o - \frac{Q_o}{Cp} \tag{6}$$

Now, since a constant voltage E is applied at $t = 0$, and since $I_o = Q_o = 0$, the subsidiary equation (6) becomes

$$\left(Lp + R + \frac{1}{Cp}\right)\bar{I} = \frac{E}{p} \tag{7}$$

Next

$$\bar{I} = \frac{E}{L}\,[(p + \alpha)^2 + \beta^2] \tag{8}$$

where $\alpha = R/2L$

$$\beta^2 = 1/LC - R^2/4L^2$$

And finally, from (8)

$$\mathbf{I} = \frac{\mathbf{Ete}^{-\alpha t}}{\mathbf{L}} \qquad \text{when } R = \sqrt{L/C}$$

I-16. A condenser (without losses) and a coil in series are connected to a 20-volt, 796-cycle source. The voltage across the condenser, 4 μf, is 20 volts, and the circuit as a whole has a power factor of 0.90, current leading. Compute the resistance and inductance of the coil.

ANSWER. See Fig. I-28. Then $\omega = 2\pi f = 6.28\ (796) = 5{,}000$. Also $\theta = -26°$. Then

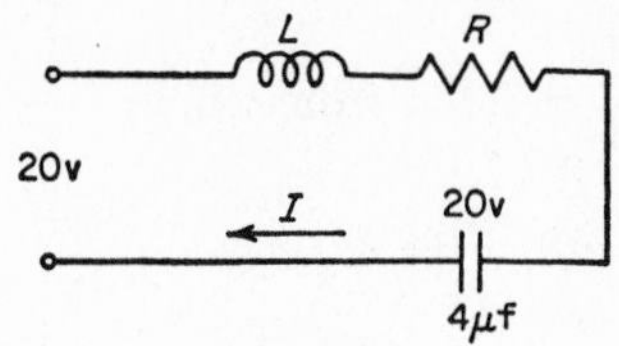

Fig. I-28

$$Z_c = 0 - j\frac{1}{\omega C} = \frac{1}{5{,}000 \times 4 \times 10^{-6}}$$

$$I = \frac{E}{Z_c} = 20 \times 5{,}000 \times 4 \times 10^{-6} = 0.4 \text{ amp}$$

Computing resistance required in problem,

$$P = EI\cos\theta = I^2R \qquad R = \frac{E\cos\theta}{I} = \frac{20 \times 0.9}{0.4} = \mathbf{45\ ohms}$$

Computing inductance required in problem,

$$\tan\theta = \frac{(1/\omega C) - \omega L}{R} \qquad \text{from which} \qquad L = \mathbf{5.6\ mh}$$

I-17. Find the equivalent resistance of the network shown in Fig. I-29 if the dotted resistor is omitted.

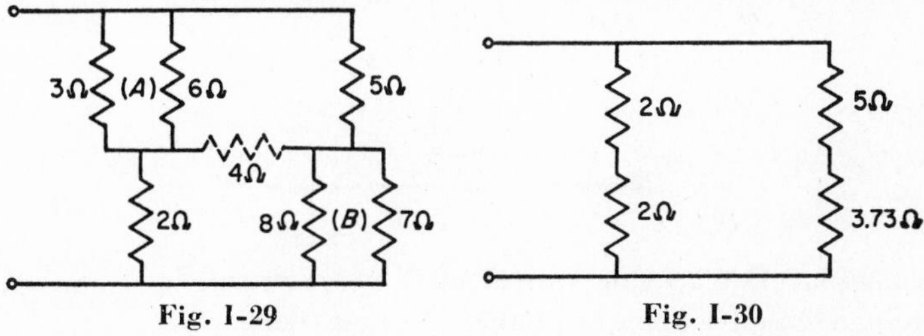

Fig. I-29

Fig. I-30

ANSWER. See Fig. I-30 for the equivalent circuit. Then Equivalent (*A*)

$$\tfrac{1}{3} + \tfrac{1}{6} = \text{conductance} = (3 + 6)/(3 \times 6) = \tfrac{1}{2}$$

$$G = \frac{1}{2} = \frac{1}{R} \qquad \text{also} \qquad R = 2$$

Equivalent (*B*)

$$\tfrac{1}{8} + \tfrac{1}{7} = \text{conductance} = (7 + 8)/(7 \times 8) = 15/56 = 1/3.73$$

$$G = \frac{1}{3.73} = \frac{1}{R} \qquad \text{from which} \qquad R = 3.73$$

Finally set up Fig. I-31. Then

$$\frac{1}{4} + \frac{1}{8.73} = G = \text{conductance} = \frac{12.73}{34.92} = \frac{1}{R}$$

from which $R = 34.92/12.73 =$ **2.745 ohms.**

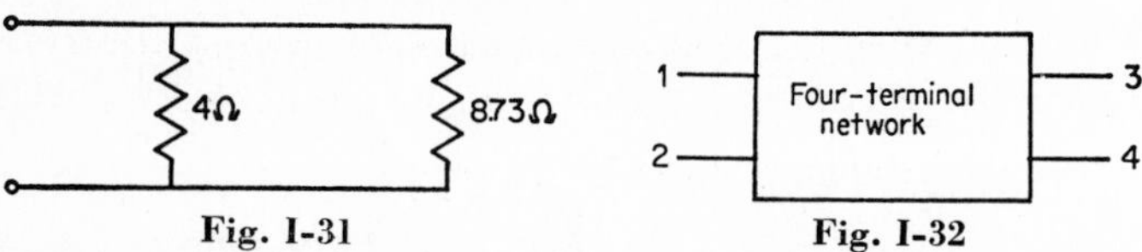

Fig. I-31 Fig. I-32

I-18. Open- and short-circuit tests are performed on the four-terminal network in Fig. I-32. An impedance of $(250 + j100)$ ohms is measured between terminals 1 and 2 with terminals 3 and 4 open-circuited. With terminals 3 and 4 shorted the impedance between terminals 1 and 2 is found to be $(400 + j300)$ ohms. An impedance of $(200 + j0)$ ohms is measured between terminals 3 and 4 with terminals 1 and 2 open. Determine and evaluate the impedances of the equivalent T network.

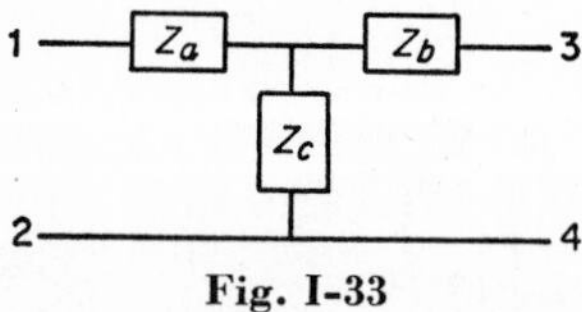

Fig. I-33

ANSWER. Set up the equivalent T network in Fig. I-33. With terminals 3 and 4 open, we have

$$Z_a + Z_c = (250 + j100) \text{ ohms}$$

With terminals 3 and 4 shorted, we have

$$Z_a + \frac{Z_b Z_c}{Z_b + Z_c} = (400 + j300) \text{ ohms}$$

When terminals 1 and 2 are open-circuited,

$$Z_b + Z_c = 200 \text{ ohms}$$

Combining the above equations and solving for Z_a, Z_b, and Z_c,

$$Z_a = (150 + j300) \text{ ohms} \qquad \textbf{Ans.}$$
$$Z_b = (100 + j200) \text{ ohms} \qquad \textbf{Ans.}$$
$$Z_c = (100 - j200) \text{ ohms} \qquad \textbf{Ans.}$$

I-19. (*a*) For the network shown, determine the driving point admittance, $Y(s)$. (*b*) If the network is driven by an ideal voltage generator, what natural modes are present in the input current? (*c*) If the network is driven by an ideal current generator, what natural modes are present in the input voltage? See Fig. I-34.

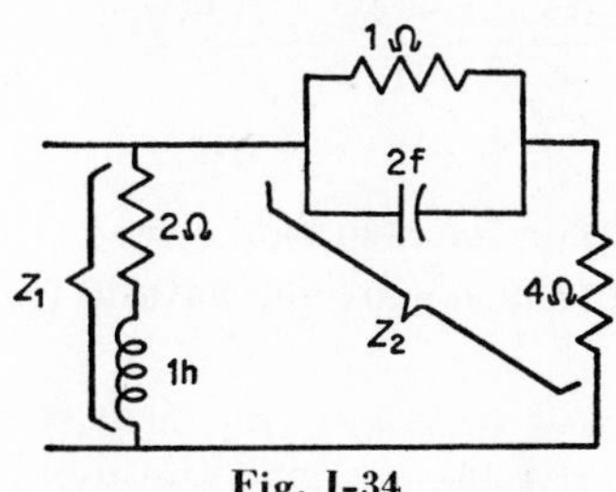

Fig. I-34

ANSWER. Using Laplace transformation,

(*a*) $$Z_1 = 2 + s$$

$$Z_2 = 4 + \frac{1/(2s)}{1 + 1/(2s)} = 4 + \frac{1}{2s + 1} = \frac{8s + 5}{2s + 1}$$

Therefore,

$$Z(s) = \frac{Z_1 Z_2}{Z_1 + Z_2} = \frac{(2 + s)(8s + 5)}{2s^2 + 13s + 7}$$

Then since $Y(s) = 1/Z(s)$,

$$Y(s) = \frac{2s^2 + 13s + 7}{(s + 2)(8s + 5)} \qquad \textbf{Ans.}$$

(*b*) For ideal voltage generator, $e(t)$ = constant. $F(s) = 1/s$ (1 unit).

$$I(s) = F(s)Y(s) = \frac{1}{s}\frac{2s^2 + 13s + 7}{(s+2)(8s+5)}$$

$$I(s) = \frac{1}{s}\frac{\frac{1}{4}s^2 + \frac{13}{8}s + \frac{7}{8}}{(s+2)(s+\frac{5}{8})} = \frac{A}{s} + \frac{B}{s+2} + \frac{C}{s+\frac{5}{8}}$$

Now calculate constants A, B, and C.

$$A = \left.\frac{\frac{1}{4}s^2 + \frac{13}{8}s + \frac{7}{8}}{(s+2)(s+\frac{5}{8})}\right|_{s=0} = \frac{\frac{7}{8}}{2 \times \frac{5}{8}} = 0.7$$

$$B = \left.\frac{\frac{1}{4}s^2 + \frac{13}{8}s + \frac{7}{8}}{s(s+\frac{5}{8})}\right|_{s=2} = -\frac{1}{2}$$

$$C = \left.\frac{\frac{1}{4}s^2 + \frac{13}{8}s + \frac{7}{8}}{s(s+2)}\right|_{s=\frac{5}{8}} = \frac{1}{20}$$

Thus $$I(s) = \frac{0.7}{s} + \frac{-\frac{1}{2}}{s+2} + \frac{\frac{1}{20}}{s+\frac{5}{8}}$$

And $$i(t) = 0.7 - \tfrac{1}{2}e^{-2t} + \tfrac{1}{20}e^{-\frac{5}{8}} \qquad \textbf{Ans.}$$

The transient solution for each loop current will consist of a series of terms $A_i e^{s_1 t}$, where the s_1's are the natural frequencies or natural modes.

(c) If the network is driven by an ideal current generator, use $Z(s)$ determined in part (a) and find the roots of its denominator

$$Z(s) = \frac{8}{2} \times \frac{(s+2)(s+0.625)}{(s+0.59)(s+5.91)}$$

and insert in equation

$$v(t) = K_1 e^{s_1 t} + K_2 e^{s_2 t}$$

Finally,

$$v(t) = K_1 e^{-0.59t} + K_2 e^{-5.91t} \qquad \textbf{Ans.}$$

It is not necessary to solve for K_1 and K_2.

Supplement J

TRANSISTORS

Questions and Answers

J-1. The h parameters of a transistor are given by $h_{ee} = 39$ ohms, $h_{ec} = 380 \times 10^{-6}$ ohm, $h_{ce} = -0.98$, $h_{cc} = 0.49$ μmho. Compute

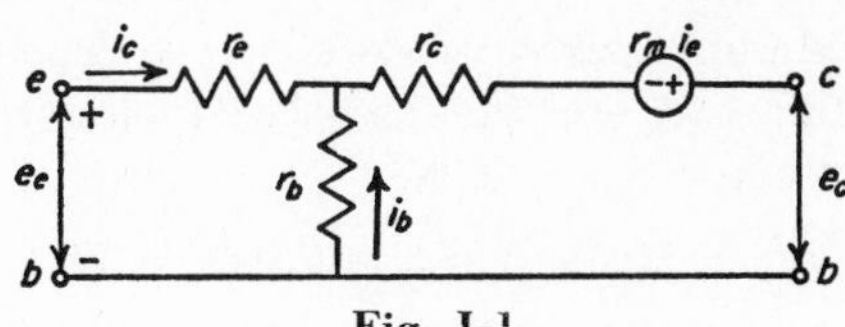

Fig. J-1

the constants of the equivalent circuits for Fig. J-1.

ANSWER.

$$r_e = h_{ee} - \frac{h_{ec}}{h_{cc}}(h_{ce} + 1) = 39 - \frac{380 \times 10^{-6}}{0.49 \times 10^{-6}}(-0.98 + 1) = \mathbf{23.5}$$

$$r_b = \frac{h_{ec}}{h_{cc}} = \frac{380 \times 10^{-6}}{0.49 \times 10^{-6}} = \mathbf{776}$$

$$r_c = \frac{1 - h_{ec}}{h_{cc}} = \frac{1 - 380 \times 10^{-6}}{0.49 \times 10^{-6}} = \mathbf{2.04 \times 10^6}$$

$$r_m = -\frac{h_{ec} + h_{ce}}{h_{cc}} = -\frac{380 \times 10^{-6} - 0.98}{0.49 \times 10^{-6}} = \mathbf{2 \times 10^6}$$

$$a = \frac{h_{ec} - h_{ce}}{h_{ec} - 1} = \frac{380 \times 10^{-6} - 0.98}{380 \times 10^{-6} - 1} = \mathbf{0.98}$$

J-2. Compute the stability factor of the transistor circuit in Fig. J-2.

ANSWER. In practical cases, the biasing circuits in Fig. J-2 are not adequate, since transistors possess a thermal instability which is of such a magnitude that additional circuits must be used to compensate for it. This thermal instability occurs mainly because the collector saturation current I_{co} varies markedly with temperature. Thermal runaway can take place and the transistor can destroy itself. A measure of these variations in I_{co} is called the *stability factor* S and is defined by $S = dI_c/dI_{co}$. Thus, a small

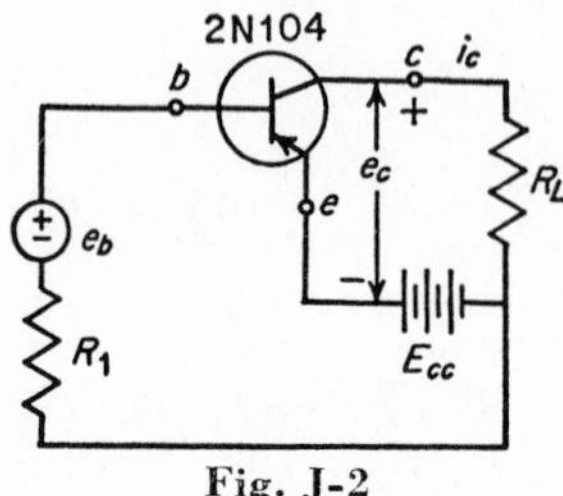

Fig. J-2

value of S implies a high degree of stability. S depends not only on the transistor but also on the associated circuitry. Also $S = 1/(1 - \alpha)$, and assuming $\alpha = 0.98$,

$$S = 1/(1 - 0.98) = 1/0.02 = \mathbf{50}$$

J-3. One unit of a 12AX7 dual triode is connected into the circuit shown in Fig. J-3, where $E_{bb} = 300$ volts, $E_{cc} = 1.5$ volts, $R_L = 100{,}000$ ohms, and $e_g = 1.5 \sin \omega t$ volts. Calculate the magnitude

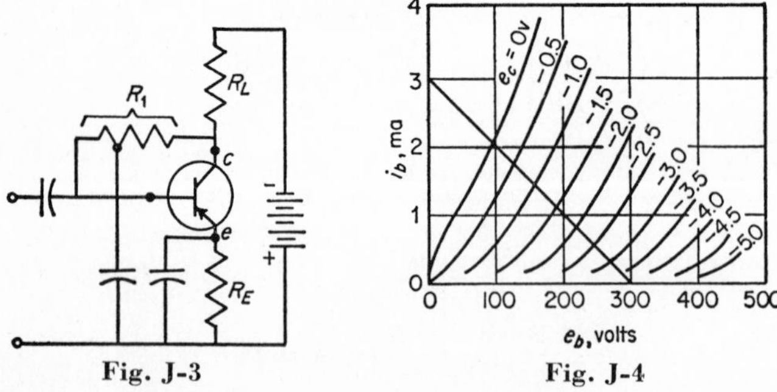

Fig. J-3 Fig. J-4

of the second harmonic distortion in the output. Fig. J-4 shows characteristics of the triode.

ANSWER. From load line we see that

$$I_o \text{ max} = 2.03 \text{ ma}$$
$$I_o \text{ min} = 0.25 \text{ ma}$$
$$I_{oT} = 1.05 \text{ ma}$$

Then

$$I_{o1} = \frac{2.03 - 0.25}{4\sqrt{2}} = 1.78/2.82 = 0.63$$

$$I_{o2} = \frac{2.03 + 0.25}{4\sqrt{2}} - \frac{1.05}{2\sqrt{2}} = 0.0319 \text{ ma}$$

Therefore per cent distortion is

$$(0.0319/0.63) \times 100 = \textbf{5.06 per cent}$$

J-4. Compute the voltage gain E_2/E_1 and the current gain $-I_2/I_1$ of the *n-p-n* junction transistor amplifier in Fig. J-5.

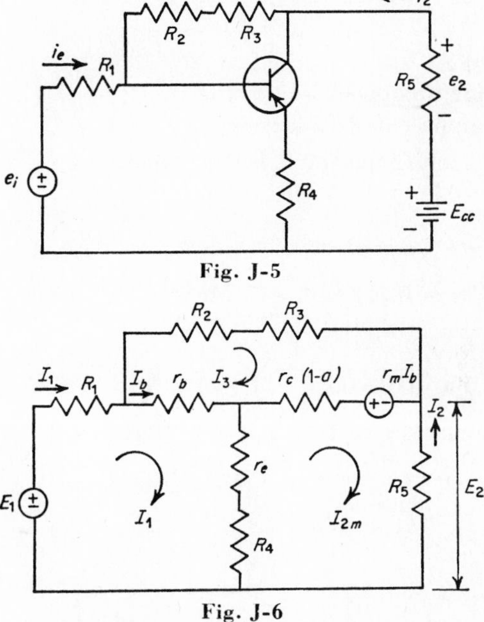

Fig. J-5

Fig. J-6

ANSWER. Set up equivalent circuit below, Fig. J-6. Then

$$E_1 = (R_1 + r_b + r_e + R_4)I_1 - (r_e + R_4)I_{2m} - r_b I_3$$

$$0 = (r_e + R_4 - r_m)I_1 + [r_e + R_4 + R_5 + r_c(1 - a)]I_2 - [r_c(1 - a) + r_m]I_3$$

$$0 = -(r_b + r_m)I_1 - [r_o(1 - a)]I_a + [r_a + r_m + r_c(1 - a) + R_2 + R_3]I_3$$

From these mesh equations we obtain

$$\frac{E_2}{E_1} = \frac{I_{2m}R_5}{E_1} = \frac{(-\Delta_{12}E_1)/\Delta \times R_5}{E_1} = -\frac{\Delta_{12}R_5}{\Delta} \qquad \textbf{Ans.}$$

$$-\frac{I_2}{I_1} = \frac{I_{2m}}{I_1} = \frac{-\Delta_{12}E_1\Delta}{\Delta\Delta_{11}E_1} = -\frac{\Delta_{12}}{\Delta_{11}} \qquad \textbf{Ans.}$$

J-5. A transistor has a current amplification factor of 0.96 at low frequencies and the cutoff frequency is 5 Mc/s. Calculate the current amplification factor at 10 Mc/s and calculate the frequency at which the current amplification factor falls to 0.7.

ANSWER. We have found that α varies with frequency as

$$\alpha = \alpha_o\left[\frac{1}{1 - j(f/f_\alpha)}\right]$$

where α_o = low frequency value of α
f_α = alpha cutoff frequency

Also, alpha cutoff frequency f_α is that where $\alpha = \alpha_o/\sqrt{2}$. Accordingly, if $\alpha_o = 0.96$, $f_\alpha = 5$ Mc/s, and $f = 10$ Mc/s, we find

$$\alpha = 0.96/\sqrt{1 + (10/5)^2} = 0.43$$

If $\alpha_o = 0.96$, $\alpha = 0.7$, and $f_o = 5$ Mc/s,

$$f = f_o\sqrt{(\alpha_o/\alpha)^2 - 1}$$
$$= 5\sqrt{(0.96/0.7)^2 - 1} = 5\sqrt{0.88} = \mathbf{4.65\ Mc/s}$$

J-6. The circuit shown in Fig. J-7 is triggered by a pulse of one

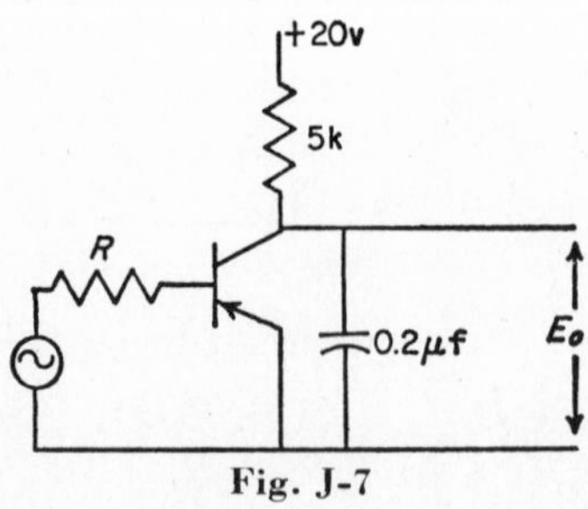

Fig. J-7

millisecond duration, i.e., the transistor is "off" for one millisecond. Consider the transistor to be ideal. (*a*) Sketch the output voltage wave form, (*b*) determine the maximum output of the voltage wave, (*c*) determine the deviation from linearity of the output voltage.

ANSWER.

(*a*) Transistor acts as an opening and closing switch. Fig. J-8*a* is the timing diagram and Fig. J-8*b* is the wave form sketch.

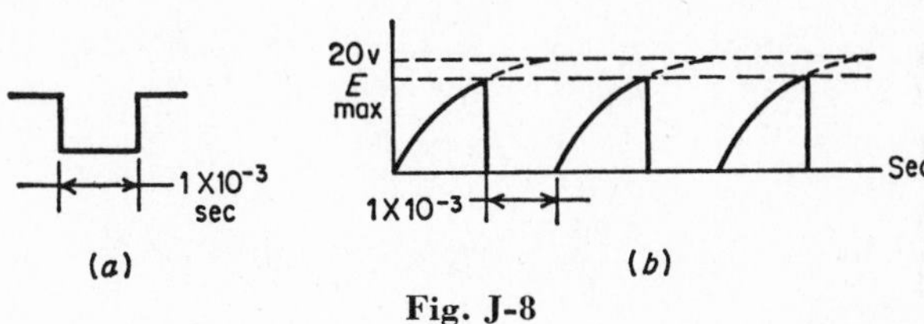

Fig. J-8

(*b*) The circuit diagram is shown in Fig. J-9. If the transistor conducts, it shorts out *C;* when the pulse comes along it removes the short and *C* can be charged through the 5K resistor. If the capaci-

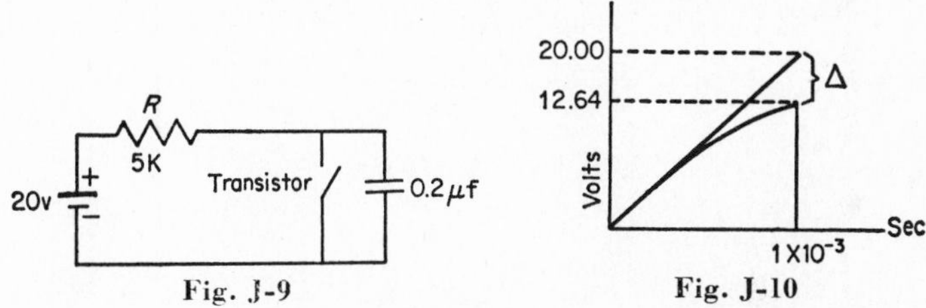

Fig. J-9

Fig. J-10

tor would be charged linearly, during the time constant or *RC* seconds the voltage would build up to 20 volts. Thus

$$RC = 5 \times 10^3 \times 0.2 \times 10^{-6} = 1 \times 10^3 \text{ sec}$$

This is equal to the given pulse length. Now, since the capacitor is not charged linearly, during the allotted *RC* time, the voltage only reaches to 63.2 per cent of the final value, which is the available $E_{\max}$. Thus

$$E_{\max} = 0.632 \times 20 = \mathbf{12.64\ volts} \qquad \text{(See Fig. J-10.)}$$

(*c*) Deviation from linearity

$$\Delta = 20.00 - 12.64 = 7.36\ \text{.olts}$$

It can be proved that during 1×10^{-3} sec the voltage reaches 20 volts if C is linearly charged.

$$e_o = K(1_1 - e^{-t/rc})$$

$$\text{At } t = 0,\ e_o = 0 = K_1(1 - 1).$$

$$\text{At } t = \infty,\ e_o = 20 = K_1(1 - 0). \quad \text{Thus, } K_1 = 20.$$

Then $\quad e_o = 20(1 - e^{-t/rc}) \quad$ and $\quad \dfrac{de_o}{dt} = \dfrac{20}{RC} e^{-t/rc}$

At $t = 0$,

$$\frac{de_o}{dt} = \frac{20}{RC} = \frac{20}{10^{-3}} = 20 \times 10^3 \text{ volts per sec}$$

But since we have only 1×10^{-3} sec,

$$e_o = 20 \times 10^3 \times 1 \times 10^{-3} = \mathbf{20\ volts}$$

Supplement K

FILTERS

Questions and Answers

K-1. Design a low-pass composite filter which is to work between resistive impedance of 500 ohms and which has a cutoff frequency of 4,000 cps. In addition to the peak of attenuation produced at $f_p = 1.25$ and $f_c = 5{,}000$ cps by the terminating half-section, a peak is desired at 4,500 cps.

ANSWER.

For the prototype, $R = 500$ ohms and $f_c = 4{,}000$ cps. Fig. K-1.

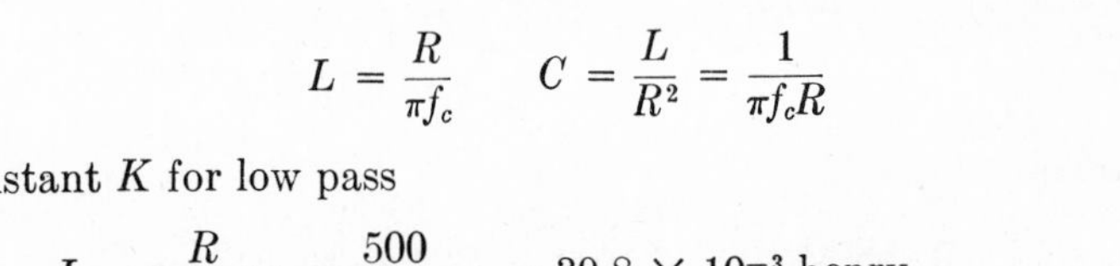

Fig. K-1

$$L = \frac{R}{\pi f_c} \qquad C = \frac{L}{R^2} = \frac{1}{\pi f_c R}$$

Constant K for low pass

$$L_K = \frac{R}{\pi f_c} = \frac{500}{\pi \times 4{,}000} = 39.8 \times 10^{-3} \text{ henry}$$

$$C_K = \frac{1}{\pi R f_c} = \frac{1}{\pi \times 500 \times 4{,}000} = 0.159 \times 10^{-6} \text{ farad}$$

For the terminating half-section. See Fig. K-2.

$$L_1 = mL_K = 0.6 \times 39.8 \times 10^{-3} = 23.9 \times 10^{-3} \text{ henry}$$

$$L_2 = \frac{1 - m^2}{4m} L_K = \frac{0.64}{4 \times 0.6} 39.8 \times 10^{-3} = 10.6 \times 10^{-3} \text{ henry}$$

$$C_2 = mC_K = 0.6 \times 0.159 \times 10^{-6} = 0.0954 \times 10^{-6} \text{ farad}$$

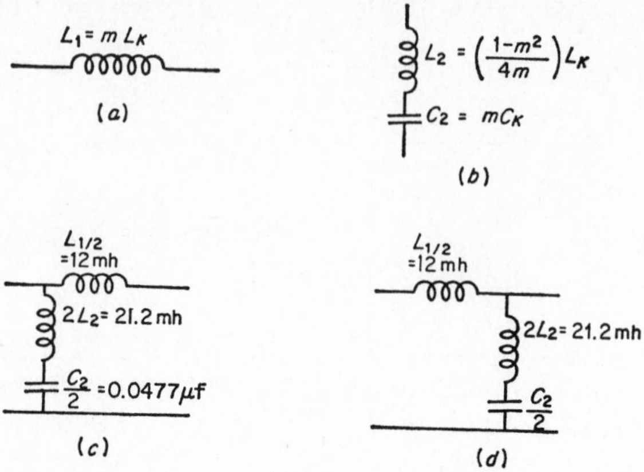

Fig. K-2

To design a section which provides a peak of attenuation at 4,500 cps. See Fig. K-3.

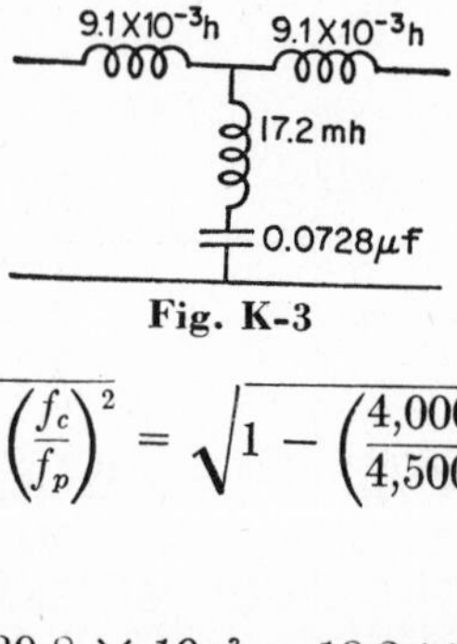

Fig. K-3

$$m = \sqrt{1 - \left(\frac{f_c}{f_p}\right)^2} = \sqrt{1 - \left(\frac{4{,}000}{4{,}500}\right)^2} = 0.458$$

Therefore

$$L_1' = mL_K = 0.458 \times 39.8 \times 10^{-3} = 18.2 \times 10^{-3} \text{ henry}$$

$$L_2' = \frac{1 - m^2}{4m} \times L_K = \frac{1 - 0.458^2}{4 \times 0.458} \times 39.8 \times 10^{-3} = 17.2 \times 10^{-3} \text{ henry}$$

$$C_2' = mC_K = 0.458 \times 0.159 \times 10^{-6} = 0.0728 \times 10^{-6} \text{ farad}$$

Individual section. See Fig. K-4. Values in microhenry and microfarad.

Completed section. See Fig. K-5. Values in microhenry and microfarad.

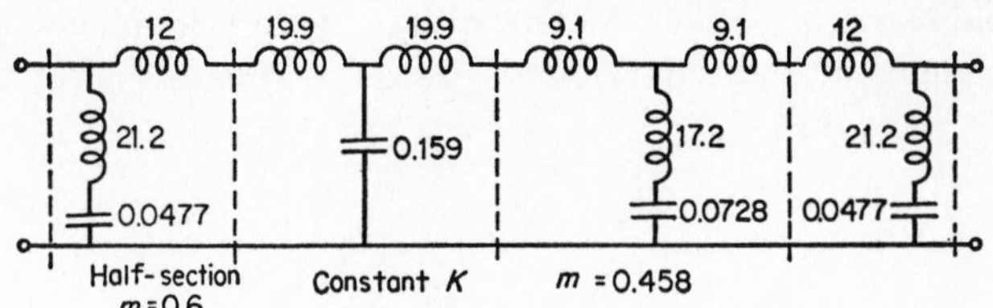

Fig. K-4

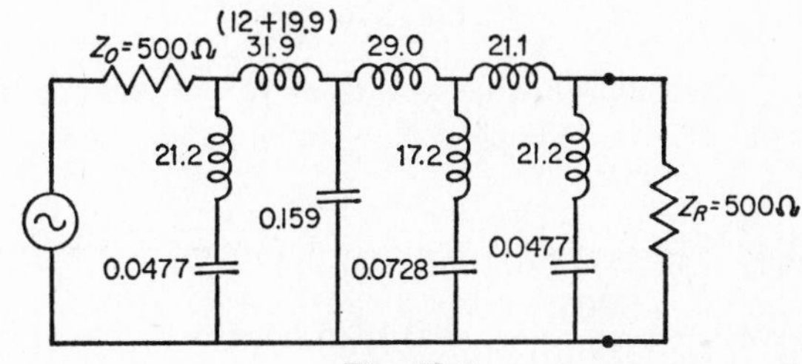

Fig. K-5

K-2. In a constant-K band-pass filter the ratio of the capacitances in the shunt and series arms is 100:1, and the resonant frequency of both arms is 1,000 cps. Calculate the bandwidth of the filter.

ANSWER. The upstream frequency f_1 of the pass band is

$$f_1 = f_c\left(\sqrt{\frac{C_1}{C_2} + 1} - \sqrt{\frac{C_1}{C_2}}\right)$$

The downstream frequency f_2 is given by

$$f_2 = f_c\left(\sqrt{\frac{C_1}{C_2} + 1} + \sqrt{\frac{C_1}{C_2}}\right)$$

where f_c = the resonant frequency of both arms at 1,000 cps

C_1 = the capacitance in the series arms

C_2 = the capacitance in the shunt arm

Thus

$$f_1 = 1{,}000(\sqrt{1.01} - \sqrt{0.01}) = 906 \text{ cps}$$

$$f_2 = 1{,}000(\sqrt{1.01} + \sqrt{0.01}) = 1{,}104 \text{ cps}$$

Finally, the bandwidth $f_2 - f_1 = 1{,}104 - 906 = 198$, say **200 cps.**

K-3. A high-pass filter of simple section is required to operate between resistance termination of 1,000 ohms, i.e., source resistance and load resistance are both 1,000 ohms. Cutoff frequency is 1,000 cps and maximum attenuation at 800 cps. Design the filter.

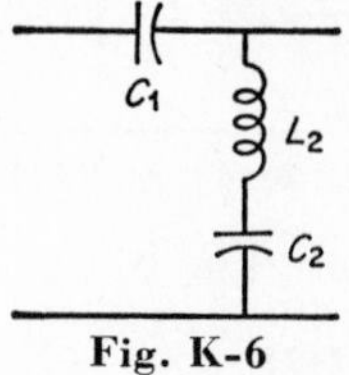

Fig. K-6

ANSWER. Use the following notation: $f_1 = 1{,}000$ cps, $R = 1{,}000$ ohms, $m = 0.6$, $f_{1p} = 800$ cps. Now refer to Fig. K-6.

$$C_1 = \frac{1}{4\pi f_1 m R} = \frac{1}{4\pi(1{,}000)(0.6)(1{,}000)} = 0.1325 \times 10^{-6} \text{ farad}$$

$$L_2 = \frac{R}{4\pi f_1 m} = \frac{1{,}000}{4\pi(1{,}000)(0.6)} = 0.1325 \text{ henry}$$

$$C_2 = \frac{m}{(1 - m^2)\pi f_1 R} = \frac{0.6}{(1 - 0.6^2)\pi(1{,}000)(1{,}000)} = 0.298 \times 10^{-6} \text{ farad}$$

Since the attenuation peak at 800 cps is required, an m-type filter

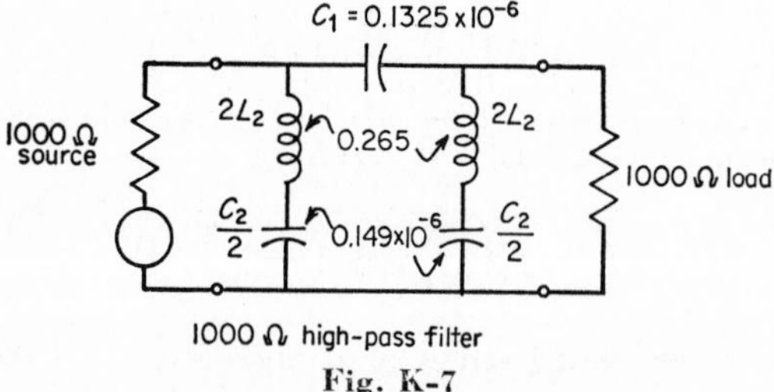

1000 Ω high-pass filter

Fig. K-7

is to be designed. Fig. K-7 depicts a 1,000-ohm high-pass filter with R equal to 1,000 ohms.

Supplement **L**

SHORT CIRCUIT CALCULATIONS AND FAULTS

Questions and Answers

L-1. Fig. L-1 shows a one-line diagram of a power system. What are the phase currents for phases A, B, and C at fault F for a line-to-

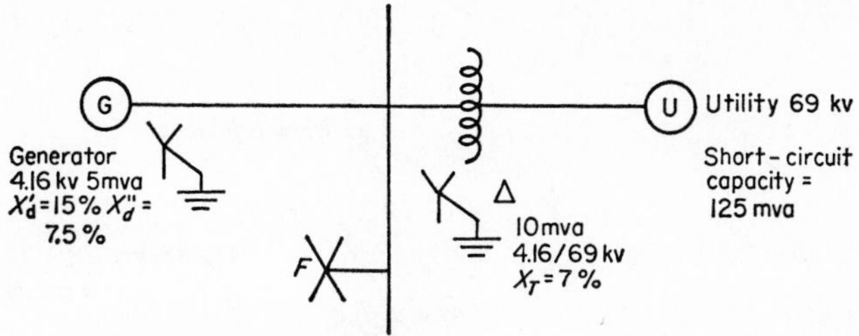

Fig. L-1

line fault? You are required to show sequence networks and interconnections and to neglect resistances.

ANSWER. Select bases of 5 mva and 4.16/69 kv. Therefore

Generator (subtransient reactance): $X_d'' = 0.075$ per unit (p.u.) and $E_a = 1.0$ p.u.

Transformer: $X_T = 0.07\ (5\text{ mva}/10\text{ mva}) = 0.035$ p.u.

Utility: $X_U = (69^2\text{ kv}/125\text{ mva})$ $\qquad Z_{\text{base}} = (69^2\text{ kv}/5\text{ mva})$

Then $$X_U\text{ p.u.} = \frac{X_U}{Z_{\text{base}}} = \frac{5}{125} = 0.04\text{ p.u.}$$

The sequence network is shown in Fig. L-2.

$$I_{a1} = \frac{1}{j0.075} = -j13.3$$
$$I_{a2} = -I_{a1} = +j13.3$$
$$I_{a0} = 0$$

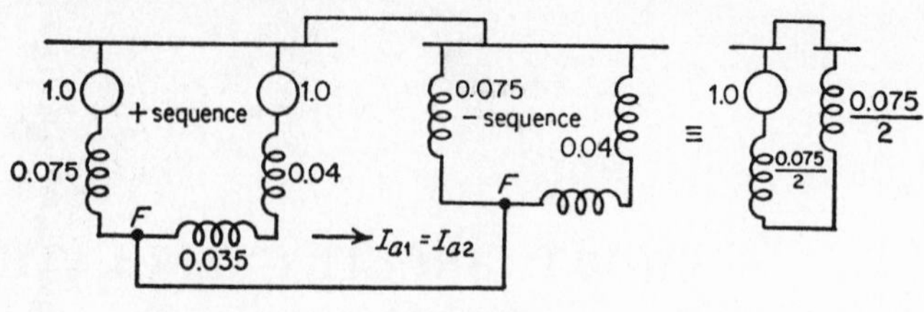

Fig. L-2

Therefore

$$
\begin{aligned}
I_a &= -j13.3 + j13.3 + 0 = 0 \\
I_b &= -j13.3(-0.5 - j0.866) + j13.3(-0.5 + j0.866) = -23.0 \text{ p.u.} \\
I_c &= -13.3(-0.5 + j0.866) + j13.3(-0.5 - j0.866) = +23.0 \text{ p.u.}
\end{aligned}
$$

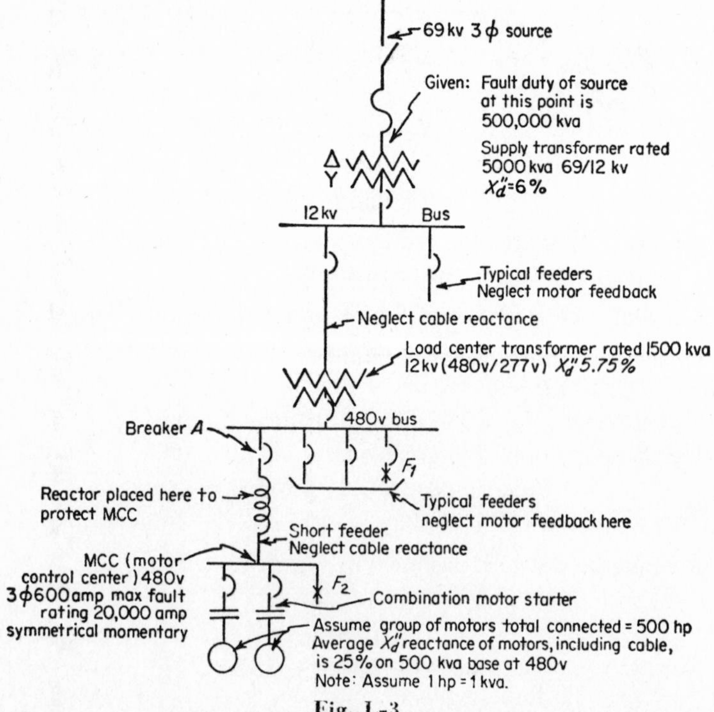

Fig. L-3

The actual phase currents are

$$I_a = \mathbf{0}$$

$$I_b = -23.0 \times I_{\text{base}} = -23.0 \times \frac{5{,}000 \text{ kva}}{3(4.16 \text{ kv})} = \mathbf{-15{,}900 \text{ amp}}$$

$$I_c = \mathbf{+15{,}900 \text{ amp}}$$

L-2. For Fig. L-3 shown, using 1,000 kva common base,

(*a*) Draw the reactance diagram.

(*b*) Solve for three-phase fault in symmetrical momentary amperes at F_1 when breaker A is open.

When breaker A is closed, the fault at F_1 must not exceed 20,000 amp symmetrical momentary, including feedback from connected motors.

(*c*) Draw reactance diagram.

(*d*) Find required size of a reactor, placed as shown, which limits total to 20,000 amp. Express rating in per-unit reactance on the required base, or in ohms per phase.

ANSWER.

(*a*) The positive sequence reactance diagram is shown in Fig. L-4, or drawn differently as in Fig. L-5.

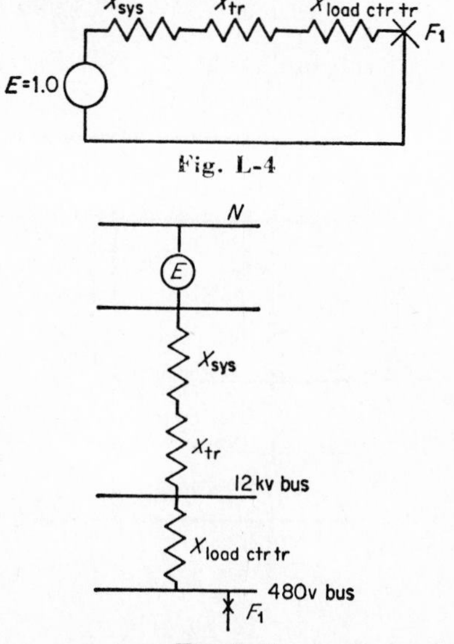

Fig. L-4

Fig. L-5

(*b*) Breaker *A* is open; fault is at F_1, using 1,000 kva common base.

$$X_{\text{system}} = \frac{1{,}000}{500{,}000} \times 1.0 = 0.002 \text{ p.u.}$$

$$X_{\text{tr}} = \frac{1{,}000}{5{,}000} \times 0.06 = 0.012 \text{ p.u.}$$

$$X_{\text{load center tr}} = \frac{1{,}000}{1{,}500} \times 0.0575 = 0.038 \text{ p.u.}$$

For a fault at F_1, the equivalent network diagram is shown in Fig. L-6.

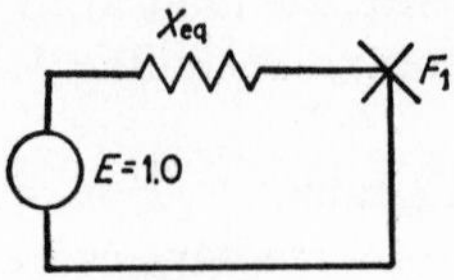

Fig. L-6

$$I_{\text{fault}} = \frac{1.0}{0.052} = 19.1 \text{ p.u.}$$

$$\begin{aligned} I_{\text{sym (moment)}} &= 19.1 \times I_{\text{base}} \text{ (480 volts)} \\ &= 19.1 = \frac{1{,}000}{\sqrt{3} \times 0.48} \\ &= 19.1 \times 1{,}203 = \mathbf{23{,}000\ amp} \end{aligned}$$

(*c*) Breaker *A* is closed; fault is at F_2. See Fig. L-7 for reactance diagram.

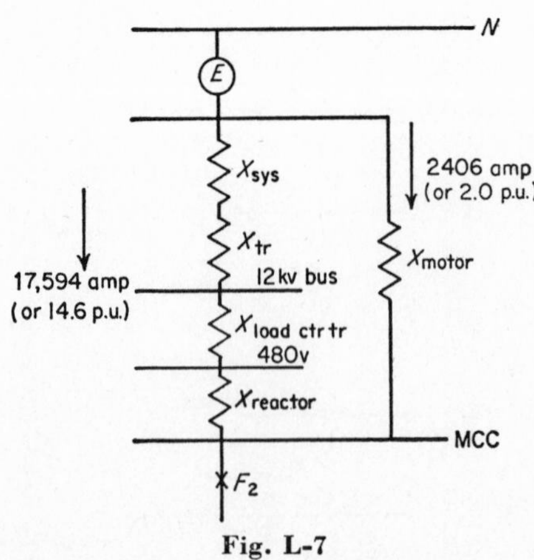

Fig. L-7

(*d*) Motor control center (MCC) bus symmetrical momentary rating is 20,000 amp.

$$X_{\text{motor}} = (1{,}000/500) \times 0.25 = 0.5$$
$$I_{\text{motor}} = 1.0/X_{\text{motor}} \times I_{\text{base}} \text{ (480 volts)}$$

Limit system contribution to 20,000 − 2,406 = 17,594 amp. This is equivalent to 17,594/1,203 = 14.6 p.u.

$$X_{\text{reactor}} \text{ p.u.} = 1/14.6 - 1/19.1 = 0.0162 \text{ p.u.}$$

or $0.162 \times (0.48^2/1.0) = \mathbf{0.0037\ ohm\ per\ phase}$

The alternate method for part (*d*) is similar to the above, but we obtain the result directly in ohm per phase.

$$X_{\text{total required with reactor}} = (480/\sqrt{3}) \times (1/17{,}594)$$
$$X_{\text{total required without reactor}} = (480/\sqrt{3}) \times (1/23{,}000)$$
$$X_{\text{reactor}} = (480/\sqrt{3})(1/17{,}594 - 1/23{,}000)$$
$$(480/\sqrt{3})(0.00005 - 0.000043)$$
$$(480/\sqrt{3})(0.000014) = \mathbf{0.0037\ ohm\ per\ phase}$$

Reactor size is the next higher standard size in ohms. The current rating of the reactor is 600 amp, the rating of the MCC bus.

L-3. Given the power system in Fig. L-8, determine fault currents at bus *F* by symmetrical components for the following conditions:

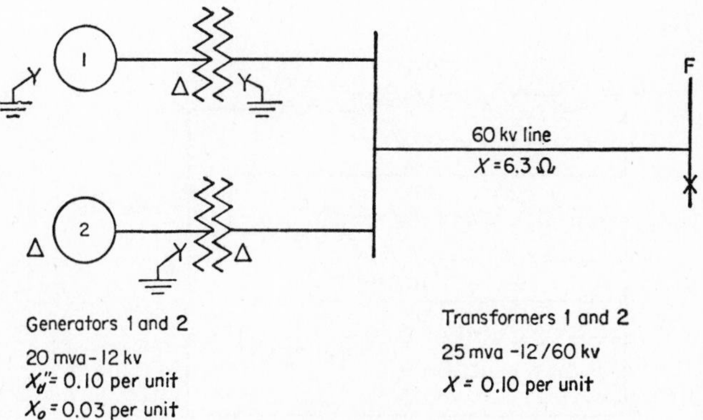

Fig. L-8

(*a*) three-phase fault, (*b*) single line-to-ground fault. You are required to show impedance diagrams, neglecting resistances.

ANSWER.

(*a*) Refer to Fig. L-9 for impedance diagram. Base kva = 25,000 = 25 mva; base kv = 60. Then

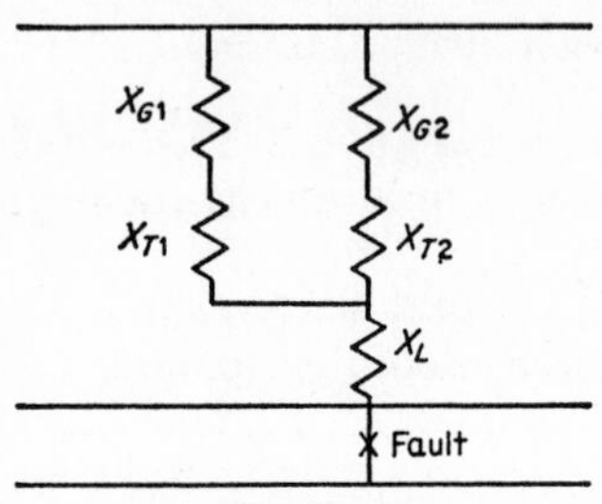

Fig. L-9

$$X_{G1} = X_{G2} = (25/20) \times 0.10 \times 100 = 12.5 \text{ per cent}$$
$$X_{T1} = X_{T2} = (25/25) \times 0.10 \times 100 = 10.0 \text{ per cent}$$
$$X_L = 25{,}000 \times 6.3 \times 100/(60^2 \times 1{,}000) = 4.38 \text{ per cent}$$
$$X_{G1} + X_{T1} = X_{G2} + X_{T2} = 12.5 + 10.0 = 22.5 \text{ per cent}$$
$$X_{\text{total}} = \frac{22.5 \times 22.5}{22.5 + 22.5} + 4.38 = \frac{22.5}{2} + 4.38 = 15.63 \text{ per cent}$$
$$\text{kva}_{\text{fault}} = (25{,}000 \times 100)/15.63 = 160{,}000$$
$$I_{\text{fault}} = 160{,}000/(\sqrt{3} \times 60) = \mathbf{1{,}541 \text{ amp}}$$

(*b*) The three sequence circuit diagrams can be connected as in Fig. L-10.

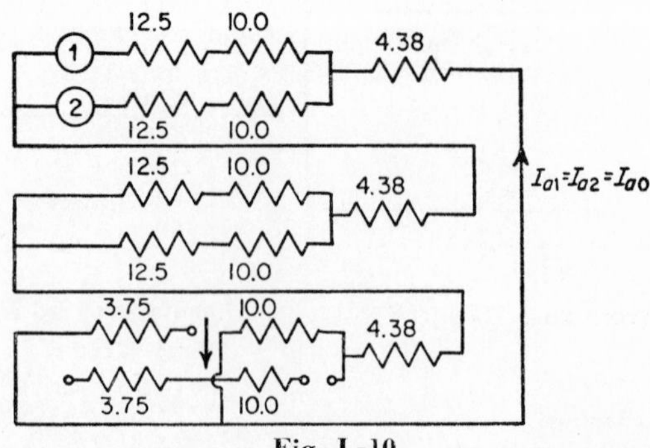

Fig. L-10

$X_1 = X_2 = 15.63$ per cent [from part (*a*) above]
$X_o = 10.0 + 4.38 = 14.38$ per cent
$X_o = (25/25) \times 0.03 \times 100 = 3.75$ per cent (not used because these elements are not connected and carry no current)

$$I_{a1} = I_{a2} = I_{ao} = \frac{100}{15.63 + 15.63 + 14.38} = \frac{100}{45.64} = 2.19 \text{ p.u.}$$

$$I_a = I_{a1} + I_{a2} + I_{ao} = 3 \times 2.19 = 6.57 \text{ p.u.}$$

$$I_a = 6.57[25{,}000/(\sqrt{3} \times 60)] = \mathbf{1{,}582\ amp}$$

L-4. Resistors are often placed in the neutral of grounded power systems to limit magnitude of ground-fault current, thus minimizing damage. When residual connected relays (51N-ground relay) are used, the selected resistor must permit enough current flow in a ground fault to operate protective relays. Resistors are selected so that the maximum ground-fault current bears a definite relationship to the rating of the largest current transformer.

(*a*) Select a suitable ohm rating for the neutral resistor.

(*b*) Assume a bolted single line-to-ground fault at the 4,160-volt bus. Assume zero ohms in all earth connections (disregard values on diagram). Calculate maximum ground-fault current at rated voltage.

(*c*) For a single line-to-ground fault at location F_1, and considering all resistance values given on diagram, calculate ground-fault current. (Note: Transformer and cable reactances are negligible.)

(*d*) For ground current in (*c*), what will be the secondary residual ground current in the current transformer's (*CT*) rates 800/5 and 200/5?

(*e*) Given a choice of relays, with current ratings 0.5 to 2.0 amp or 1.5 to 6.0 amp, which is your choice for relay 51N at both locations shown. Give a suitable tap setting for your selected relay.

ANSWER. Refer to Fig. L-11. CT = current transformer rating as marked. 51 = phase overcurrent standard inverse time relay. 51N = residual ground current standard inverse time relay. Assume no motor feedback.

(*a*) The largest CT is 800/5 amp; the resistor will be rated for 800 amp.

$$R_{\text{resistor}} = \frac{E_{ph}}{I} = \frac{4{,}160}{\sqrt{3} \times 800} = \mathbf{3.0\ ohms}$$

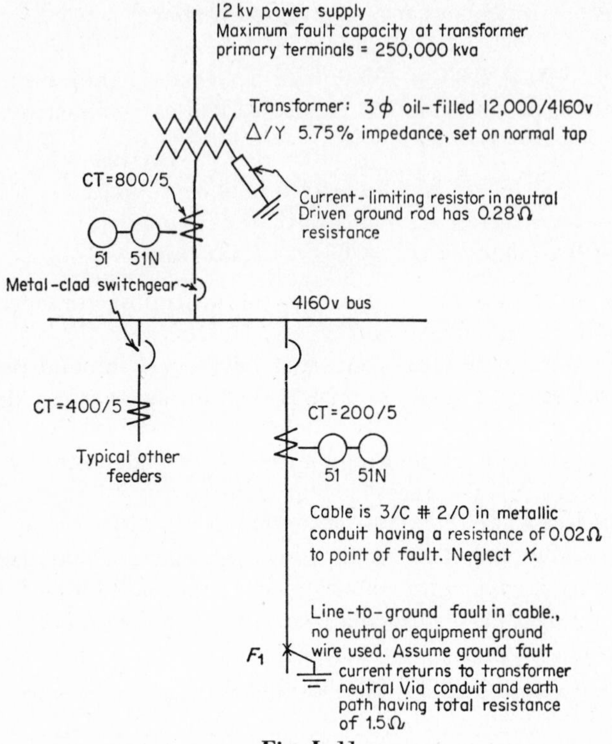
12 kv power supply
Maximum fault capacity at transformer
primary terminals = 250,000 kva
Transformer: 3 φ oil-filled 12,000/4160v
Δ/Y 5.75% impedance, set on normal tap
CT=800/5
Current-limiting resistor in neutral
Driven ground rod has 0.28Ω
resistance
51 51N
Metal-clad switchgear
4160v bus
CT=400/5
Typical other
feeders
CT=200/5
51 51N
Cable is 3/C # 2/0 in metallic
conduit having a resistance of 0.02Ω
to point of fault. Neglect X.
Line-to-ground fault in cable.,
no neutral or equipment ground
wire used. Assume ground fault
current returns to transformer
neutral Via conduit and earth
path having total resistance
of 1.5Ω
F1

Fig. L-11

The actual ohmic value of the resistor is 3.0 − 0.28 (resistance of the driven rod) = 2.72 ohms. However, in practice the resistance of the rod is neglected. Some textbooks claim that the ground return path resistance of 1.5 ohms includes the ground rod, but the ground rod is neglected in both cases.

(*b*) The maximum ground-fault current I_{gf} at rated voltage is

$$I_{gf} = 4{,}160/(\sqrt{3} \times 3.0) = 2{,}400/3.0 = \mathbf{800\ amp}$$

(*c*) Refer to Fig. L-12. Neglect cable resistance. By neglecting the ground and cable resistance, we obtain

$$I_{gf} = 2{,}400/(3 + 1.5) = \mathbf{533\ amp}$$

The same result could be obtained by the method of symmetrical components, which in this case would be very laborious. Also note the transformer rating is not given.

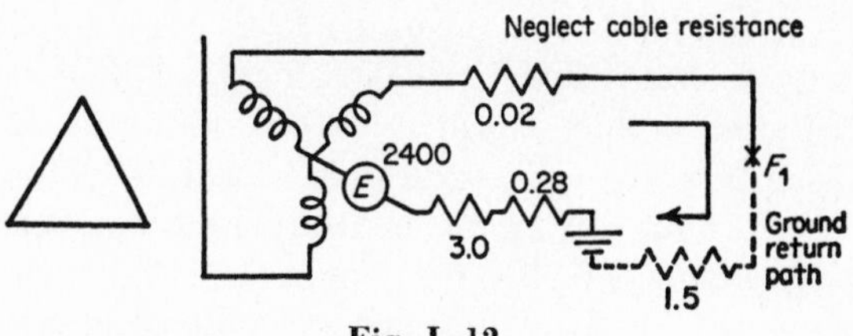

Fig. L-12

(*d*) The secondary current of CT 800/5 without saturation will be

$$I_{gf} \times 5/800 = 533 \times 5/800 = 3.3 \text{ amp}$$

The secondary current of CT 200/5 without saturation will be

$$533 \times 5/200 = \mathbf{13.3\ amp}$$

(*e*) Select 0.5 to 2.0 amp relay for feeder (0.5 amp tap); select 0.5 to 2.0 amp relay for main circuit breaker (0.5 amp tap). The selection is based on the CT secondary current and should be within the thermal rating of the relays. The 0.5 to 2.0 amp relay has a 1-sec thermal rating of 70 amp. The 0.5 tap gives $^{200}\!/_5 \times 0.5 = 20$ amp primary current for the feeder. The other 0.5 tap gives $^{800}\!/_5 \times 0.5 = 80$ amp primary current for the main. Reference: Industrial Power Systems Handbook, Beeman, *McGraw-Hill.*

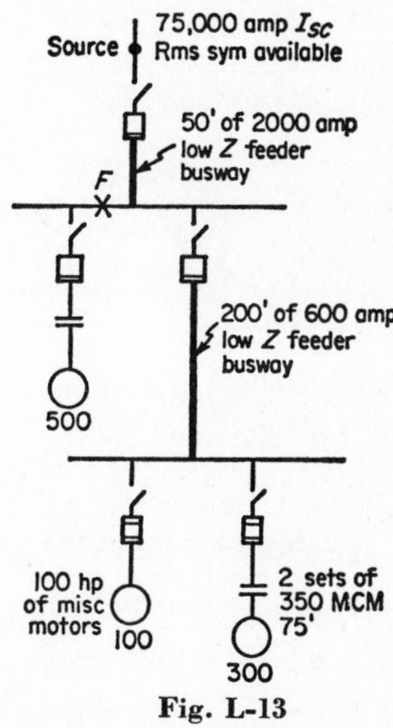

Fig. L-13

L-5. A simple radial system is served from a secondary $^{480}\!/_{277}$ Y network. Fig. L-13 shows a one-line diagram of the system. All the elements are not shown, just those necessary. The remaining load is lighting or small motors too far removed from the buses to enter into the calculation. Find the short-circuit current at point *F*.

ANSWER. Use 10,000 kva base. Voltage = 480. The following data will be used: X/M' 2,000-amp duct = 0.0031 ohm. Z/M' 2,000-amp duct = 0.0081 ohm. Since $\frac{2}{3}Z$ is greater than X, use

$\frac{2}{3}Z$. So $\frac{2}{3}Z = 0.0054$ ohm. Now similarly for the 600-amp duct: $X/M' = 0.0179$ ohm. $Z/M' = 0.0393$ ohm. $\frac{2}{3}Z = 0.0262$ ohm. For the 350 MCM duct: $X/M' = 0.0491$ ohm. $Z/M' = 0.0617$ ohm. $\frac{2}{3}Z$ is less than X, so use X. And $\frac{1}{2}X$ for paralleled cables $= 0.0245$ ohm. Now for motors: The per-unit reactance to their own kva rating is 0.25. Kva equals 1 per hp. Now diagram the system as in Fig. L-14. Assign per-unit reactance values as

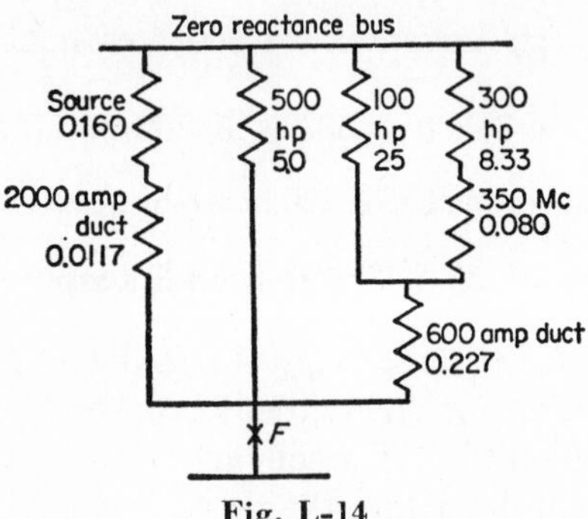

Fig. L-14

shown in Fig. L-14. The source per-unit reactance equals

$$\frac{\text{Base kva}}{I_{sc} \times \sqrt{3}\text{ kv}} = \frac{10{,}000}{75{,}000 \times 1.73 \times 0.48} = 0.160$$

The constant for converting ohms reactance for a 480-volt system to per-unit reactance on a 10,000 kva base is 43.4.

For 2,000-amp duct: $43.3 \times 0.0054 \times 0.05 = 0.0117$
For 600-amp duct: $43.4 \times 0.0262 \times 0.20 = 0.227$
For 350 MCM: $43.4 \times 0.0245 \times 0.075 = 0.080$

The per-unit reactance of 100 hp motor is given by

$$\frac{\text{Per-unit reactance on kva rating} \times \text{base kva}}{\text{kva rating}}$$

$$\frac{0.25 \times 10{,}000}{100} = 25$$

Per-unit reactance of 300-hp motor

$$\frac{0.25 \times 10{,}000}{300} = 8.33$$

Per-unit reactance of 500-hp motor

$$\frac{0.25 \times 10{,}000}{500} = 5$$

Now combine reactance values. Add series reactances. Fig. L-15.

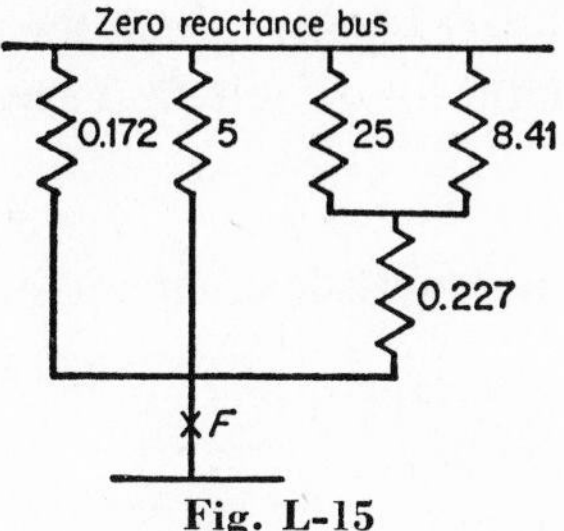

Fig. L-15

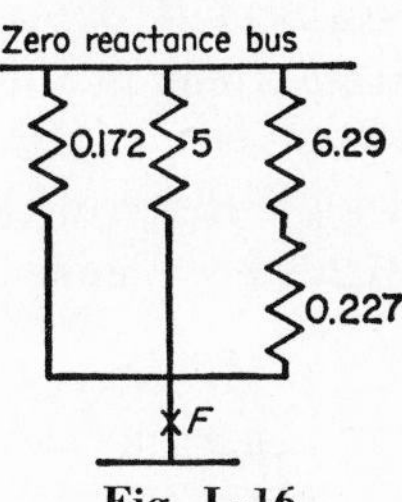

Fig. L-16

$$X_T = X_1 + X_2 = 0.160 + 0.0117 = 0.172$$
$$X_T = X_1 + X_2 = 8.33 + 0.080 = 8.41$$

Combine parallel reactances. Fig. L-16.

$$X_T = \frac{X_1X_2}{X_1 + X_2} = \frac{25 \times 8.41}{25 + 8.41} = 6.29$$

Add series reactances. Fig. L-17.

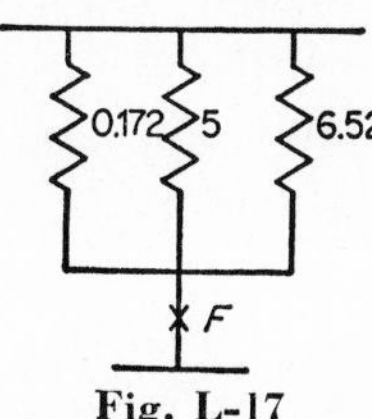

Fig. L-17

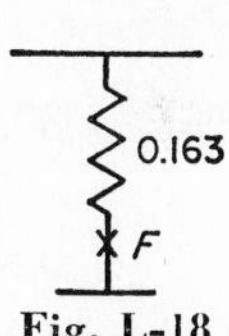

Fig. L-18

$$X_T = X_1 + X_2 = 6.29 + 0.227 = 6.52$$

Combine parallel reactances. Fig. L-18.

$$X_T = \frac{1}{1/X_1 + 1/X_2 + 1/X_3} = \frac{1}{1/0.172 + 1/5 + 1/6.62} = 0.162$$

Calculate short-circuit currents.

$$I_{sc}\text{rms sym} = \frac{\text{Base kva}}{PU \times \sqrt{3} \times \text{kv}} = \frac{10{,}000}{0.162 \times 1.73 \times 0.48}$$

$$= \mathbf{74{,}300\ amp}$$

This is the short-circuit current at point *F*.

L-6. An alternator and a synchronous motor are rated 30,000 kva, 13.2 kv, and both have subtransient reactances of 20 per cent. The line connecting them has a reactance of 10 per cent on the base of the machine ratings. The motor is drawing 20,000 kw at 0.8 power factor leading and a terminal voltage of 12.8 kv when a symmetrical three-phase fault occurs at the motor terminals. Find the subtransient current in the alternator, motor, and fault by using the internal voltages of the machines.

ANSWER. Refer to Fig. L-19*a* and *b*. Now choose as base 30,000 kva, 13.2 kv. Figure L-19*a* shows the equivalent circuit of the

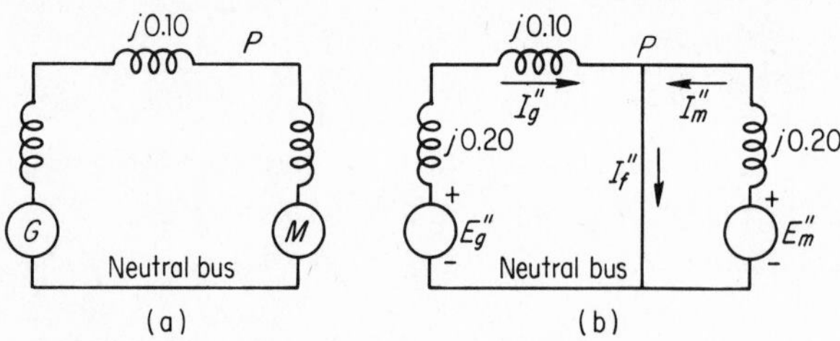

Fig. L-19. Equivalent circuits (*a*) Before the fault. (*b*) After the fault.

system described. Use the voltage at the fault V_f as the reference phasor.

$$V_f = 12.8/13.2 = 0.97 \underline{/0^\circ} \text{ per unit}$$

$$\text{Base current} = 30{,}000/(\sqrt{3} \times 13.2) = 1{,}310 \text{ amp}$$

$$I_L = \frac{20{,}000}{0.8 \times \sqrt{3} \times 12.8} = 1{,}128 \underline{/36.9^\circ} \text{ amp}$$

$$= 1{,}128/1{,}310 = 0.86 \underline{/36.9^\circ} \text{ per unit}$$

$$0.86(0.8 + j0.6) = 0.69 + j0.52 \text{ per unit}$$

For the generator,

$$V_t = 0.970 + j0.1(0.69 + j0.52)$$

$$0.970 + j0.069 - 0.052 = 0.918 + j0.069 \text{ per unit}$$

$$E_g'' = 0.918 + j0.069 + j0.2(0.69 + j52)$$

$$= 0.918 + j0.069 + j0.138 - 0.104 = 0.814 + j0.207 \text{ per unit}$$

$$I_g'' = (0.814 + j0.207)/j0.3 = 0.69 - j2.71 \text{ per unit}$$

$$= 1{,}310(0.69 - j2.71) = 904 - \mathbf{j3{,}550 \text{ amp}}$$

For the motor,

$$V_t = V_f = 0.97 \underline{/0^\circ} \text{ per unit}$$

$$E''_m = 0.97 + j0 - j0.2(0.69 + j0.52) = 0.97 - j0.138 + 0.104$$

$$1.074 - j0.138 \text{ per unit}$$

$$I''_m = (1.074 - j0.138)/j0.2 = -0.69 - j5.37 \text{ per unit}$$

$$1{,}310(-0.69 - j5.37) = -904 - \mathbf{j7{,}040\ amp}$$

In the fault,

$$I''_f = I''_g + I''_m = 0.69 - j2.71 - 0.69 - j5.37 = -j8.08 \text{ per unit}$$

$$-j8.08 \times 1{,}310 = \mathbf{-j10{,}590\ amp}$$

Figure L-19*b* shows the paths of I''_g, I''_m, and I''_f.

Supplement **M**

TRANSIENT ANALYSIS

Questions and Answers

M-1. The circuit shown in Fig. M-1 represents a modified ignition system. Assume the system to be at the steady state with switch

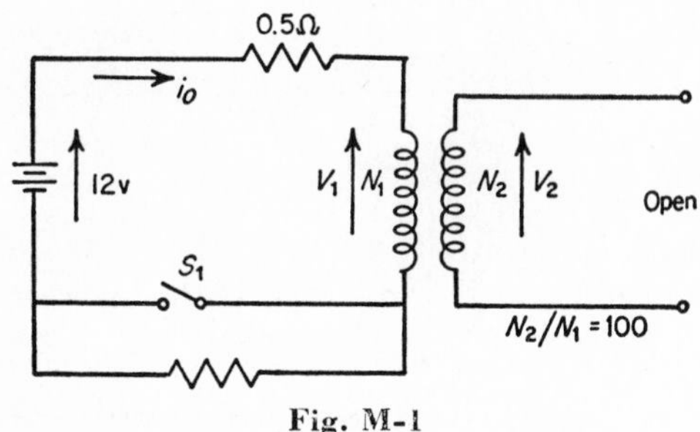

Fig. M-1

S_1 closed. What are the primary and secondary voltages V_1 and V_2 at $t = 0$, if the switch opens at $t = 0$?

ANSWER. At the steady state with switch closed.

$$V = i_oR \qquad 12 = i_o(0.5) \qquad i_o = 24 \text{ amp}$$

At $t = 0$ and switch open

$$L_1 \frac{di}{dt} + iR = E$$

$$L_1 \frac{di}{dt} = V_1 = E - iR = 12 - 24(0.5 + 10)$$

$$L_1 \frac{di}{dt} = V_1 = 12 - 252 = \mathbf{-240 \text{ volts}}$$

Mutual inductance is given by $L_1 N_2/N_1$

$$V_2 = L_1 \frac{di}{dt} \times \frac{N_2}{N_1} = -240 \times 100 = \mathbf{-24{,}000\ volts}$$

M-2. The field circuit of a synchronous motor has an inductance of 50 henrys (assumed constant) and a resistance of 100 ohms. The excitation voltage is 120 volts. Compute the time it takes the field current to build up to 99 per cent of its final value when excitation voltage is applied.

ANSWER. $L = 50$ henry $\quad R = 100$ ohms $\quad V = 120$ volts

$$i = i_{\max} (1 - e^{RT/L})$$

$$0.99i = i_{\max} (1 - e^{2t})$$

$$i_{\max} = \frac{E}{R} = \frac{120}{100} = 1.2 \text{ amp}$$

$$1.188/1.2 = 1 - e^{2t}$$

$$0.99 - 1 = -e^{2t} \qquad 0.01 = e^{2t} \qquad t = \mathbf{2.3\ sec}$$

M-3. The circuit shown in Fig. M-2 is initially in the steady state with the switch closed. At $t = 0$ the switch is opened (assume no

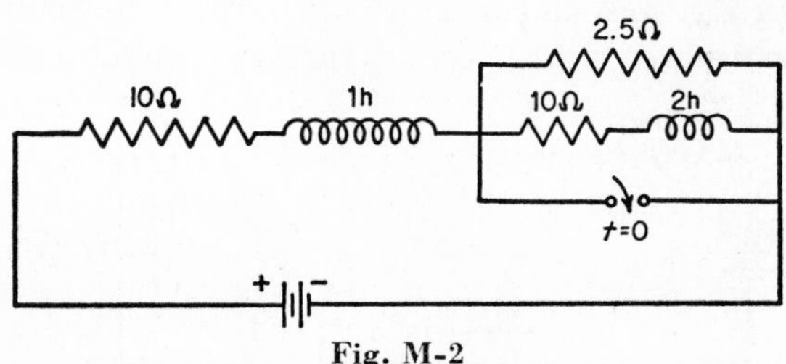

Fig. M-2

sparking). Just after the switch is opened:

(*a*) What is the current in the 1-henry inductance; in the 2.5-ohm resistor; in the 2-henry inductance?

(*b*) What is the voltage (magnitude and direction) across the 1-henry inductance?

(*c*) What is the rate of change of current in the 1-henry inductance; in the 2-henry inductance?

ANSWER.

(*a*) Current in the 1-henry inductance: for the steady state,

$$V = iR \qquad i = \frac{V}{R} = \frac{120}{10} = 12 \text{ amp}$$

Current through the 1-henry inductance: **12 amp**
Current through the 2-henry inductance: **0 amp**[1]
Current through 2.5-ohm resistor: **12 amp**
(*b*) Voltage across 1-henry inductance:

$$120 = 10(12) + V_1 + 2.5(12)$$
$$V_1 = -30 \text{ volts} \qquad V_1 = \mathbf{30\ volts}$$

(*c*) Rate of change of current through 1-henry inductance:

$$V_1 = i_1 \frac{dI}{dt} \qquad \frac{V}{L} = \frac{dI}{dt} = \mathbf{-30\ amp\ per\ sec}$$

Through the 2-henry inductance:

$$\frac{V}{L} = \frac{dI}{dt} = \frac{30}{2} = \mathbf{+15\ amp\ per\ sec}$$

M-4. (*a*) In Fig. M-3 all condensers are initially uncharged. The reversing switch is suddenly closed to the left. Find time expression

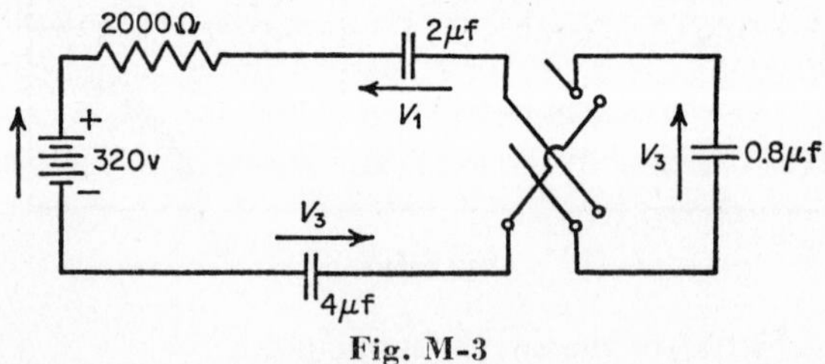

Fig. M-3

for the current. After steady-state conditions have been reached, what total charge will have been displaced through the circuit and what will be the final voltage across each condenser? (*b*) The switch is now thrown to the right.

[1] Because there cannot be a change in current for the conductor.

ANSWER. Time expression for current. Switch is to left.

$$(a) \qquad i = i_{\max}e^{-t/RC} \qquad i_{\max} = \frac{E}{R} = \frac{320}{2{,}000} = 0.16 \text{ amp}$$

$$\frac{1}{C} = \frac{1}{2} + \frac{1}{4} + \frac{1}{0.8} = 2\mu\text{f}$$

$$i = 0.16e^{-t/(2{,}000 \times \frac{1}{2} \times 10^{-6})}$$

$$i = 0.16e^{-1{,}000t} \qquad \textbf{Ans.}$$

(*b*) After steady state find value of total charge *q*, microcoulombs

$$q = it \qquad i = \frac{dq}{dt} = 0.16e^{-1{,}000t}$$

$$q = 0.16(-1/1{,}000)e^{-1{,}000t} = 0.16(-1 \times 10^{-3}) = \textbf{160 mc}$$

$$V_1 = \frac{q}{C} = \frac{160}{2} = \textbf{80 volts}$$

$$V_2 = (160 \times 10^{-6})/(4 \times 10^{-6}) = \textbf{40 volts}$$

$$V_3 = 160/0.8 = \textbf{200 volts}$$

M-5. An uncharged condenser in series with a 150-volt voltmeter of 20,000 ohms resistance is suddenly connected to a 120-volt battery. One second later the voltmeter reads 80 volts. Compute the following values at this instant.

(*a*) The rate of flow of electric charge through the circuit.
(*b*) The rate at which energy is dissipated in the voltmeter.
(*c*) The rate of change of stored energy in the condenser.
(*d*) The capacitance of the condenser.
(*e*) The rate at which the voltage across the condenser is changing.

ANSWER. Refer to Fig. M-4.

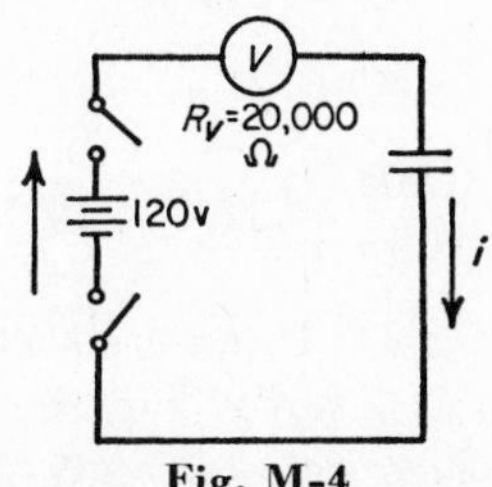

Fig. M-4

(*a*) At $t = 0$ switches are closed, and 1 sec later V reads 80 volts. Find dq/dt.

$$i = \frac{V}{R} = \frac{80}{2{,}000} = \mathbf{4 \times 10^{-3}\ amp}$$

(*b*)

$$w = \int_0^1 P\,dt = P = i^2R = (16 \times 10^{-6})(2 \times 10^4) = \mathbf{0.32\ watt}$$

(*c*) $w = \frac{1}{2}CV^2 \qquad q = CV \qquad C = \frac{q}{V} \qquad w = \frac{1}{2}qV$

$$\frac{dw}{dt} = \frac{1}{2}\frac{dq}{dt}V = i\frac{V}{2} = \mathbf{0.16\ watt}$$

(*d*)

$$Ri + \frac{q}{C} = 80$$

$$R\frac{di}{dt} + \frac{i}{C} = 0$$

$$\ln i = -\frac{E}{RC} + \ln k$$

$$i = ke^{-t/RC}$$

At $t = 0$, $i = E/R = 120/20{,}000 = 0.006$ amp. Then $i = 0.006^{-t/RC}$.

At $t = 1$, $i = 0.004$ amp. Then it follows

$$0.004 = 0.006e^{-(1/20{,}000C)}$$

$$\ln\frac{0.004}{0.006} = -\frac{1}{20{,}000}C$$

$$-0.405 = -\frac{1}{20{,}000}C$$

$$8{,}300 = \frac{1}{C}$$

Finally

$$C = \mathbf{123\ \mu f}$$

(*e*) $\frac{dV}{dt} \times C = i$

$$\frac{dV}{dt} = \frac{i}{C} = \frac{0.004}{123.5} = \mathbf{32.4 \times 10^{-6}\ volts\ per\ sec}$$

Transient and Steady State. In an a-c system it is assumed that each cycle of operation is exactly like the preceding one. This condition is called the *steady state.* In any system there must be a starting-up period during which there is a nonrepeating disturbance called a *transient.* Whenever an a-c circuit is brought from rest to a steady-state condition, or if it is brought from one steady state to another, a transient is involved. Under very special conditions, the transient can be zero. It is sufficient to realize that a transient must appear at one time in the history of the circuit and that it dies out, leaving the steady-state condition to prevail.

M-6. Switch S in the circuit shown in Fig. M-5 is closed at $t = 0$ and $E = 100 + 500t$ volts. Calculate the value of E_{ab} when $t = 0.1$ sec. Assume initial charge on condenser C is zero.

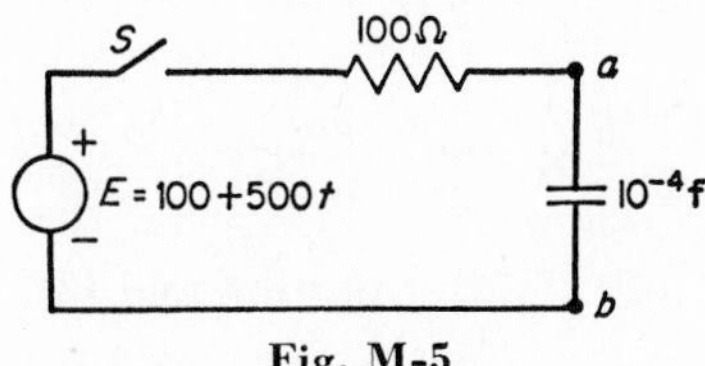

Fig. M-5

ANSWER. Solution by Laplace transformation. The Laplace transform method of circuit analysis provides for the solution of transient problems by simple algebraic means. In the classical method the solution of circuit equations is carried out in the time domain $f(t)$. The Laplace transform method converts the problem details from $f(t)$ into the complex frequency domain $F(s)$, whereby the solution can be obtained by purely algebraic means. It has also been established that another feature of the Laplace transform method is that the initial conditions are introduced at an early stage in the solution. This eliminates the long drawn out evaluation of arbitrary constants that is needed in the classical method. In such problems reference should be made to tables of Laplace transforms and operations, listed in "Mathematical Handbook for Scientists and Engineers" by Korn and Korn, McGraw-Hill and "Handbook of Laplace Transformation (Tables and Examples)" by Nixon, Prentice-Hall.

For the solution at hand the classical method for voltage is given by $f(t) = 100 + 500t$. The Laplace transform operation is

$$E(s) = \frac{100}{s} + \frac{500}{s^2} \tag{1}$$

Voltage across ab is E_{ab}, and in the complex frequency domain

$$E_{ab}(s) = \frac{E(s)}{R + 1/Cs} = \frac{100/s + 500/s^2}{RCs + 1} = \frac{100}{RC} \times \frac{s + 5}{s^2(s + 1/RC)} \quad (2)$$

The latter portion of the above equation in the complex frequency domain is solved for as a function of time. The inverse Laplace transformation converts the solution in the complex frequency domain back to the time domain $f(t)$.

$F(s)$:

$$\frac{s + d}{s^2(s + \alpha)} \quad (3)$$

$f(t)$:

$$\frac{d - \alpha}{\alpha^2} \times e^{-\alpha t} + \frac{d}{\alpha} \times t + \frac{\alpha - d}{\alpha^2} \quad (4)$$

From an inspection of (2) latter portion and (3),

$$\frac{1}{RC} = \frac{1}{10^2 \times 10^{-4}} = 100 = \alpha$$

The time domain for the voltage across ab:

$$E_{ab}(t) = \frac{100}{RC} \times \left(\frac{5 - 100}{10^4} \times e^{-100t} + \frac{5 \times t}{100} + \frac{100 - 5}{10^4} \right)$$

$$E_{ab} = \frac{100}{10^2 \times 10^{-4}} \left(-0.0095e^{-10} + \frac{0.5}{100} + 0.0095 \right)$$

Finally sine t is given as 0.1, by substitution

$$E_{ab} = 10^4(-0.0095 \times 0.000045 + 0.0145) = \mathbf{145+\ volts}$$

M-7. Write the differential equations that are necessary to solve for i_1, i_2 in the circuit Fig. M-6.

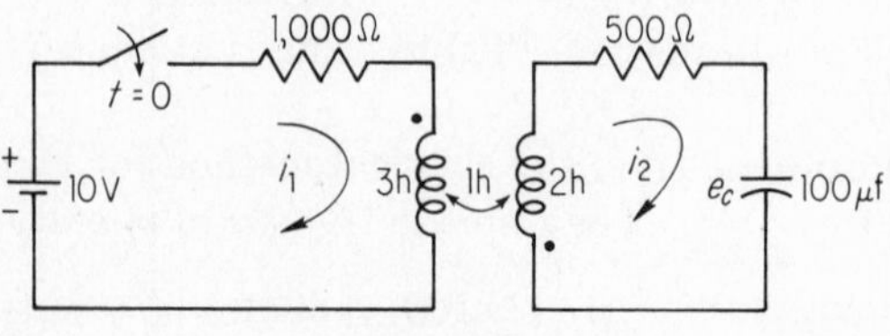

Fig. M-6

ANSWER. Loop 1: $10 = 1{,}000 i_1 + 3 \frac{di_1}{dt} + 1 \frac{di_2}{dt}$

$$\text{Loop 2: } 0 = 500 i_2 + \frac{10^6}{100} \int_0^t i_2\, dt + 2\, di_2/dt + dt_1/dt$$

$$i_1(0) = 0 \qquad i_2(0) = 0 \qquad \text{Also } e_c(0) = 0 \qquad \textbf{Answers}$$

Use tables of Laplace transforms.

M-8. Determine the current i as a function of time for the circuit shown in Fig. M-7 with $e = 10 \times$ (first-order unit impulse) and $e_0(0) = 5$ volts.

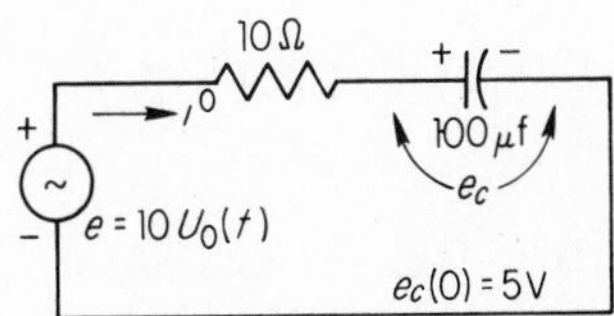

Fig. M-7

ANSWER. The voltage equation is

$$10 U_0(t) = 10 i + \frac{10^6}{100} \int_0^t i\, dt + 5$$

The Laplace transform is

$$10 = 10 I(s) + 10^4[I(s)/s] + 5/s = I(s) \left[10 + \frac{10^4}{s} \right] + 5/s$$

$$I(s) = \frac{10 - 5/s}{10 + 10^4/s} = \frac{10s - 5}{10s + 10^4} = \frac{s}{s + 10^3} - \frac{0.5}{s + 10^3}$$

$$= 1 - \frac{10^3}{s + 10^3} - \frac{0.5}{s + 10^5} = 1 - \frac{1{,}000.5}{s + 10^3}$$

Therefore,

$$i(t) = U_0(t) - 1{,}000.5 \epsilon^{-10^3 t} U_1(t) \qquad \textbf{Answer}$$

M-9. Solve for the current i, as a function of time, in the circuit Fig. M-8 without using the Laplace-transform method. $i(0) = 2$ amp.

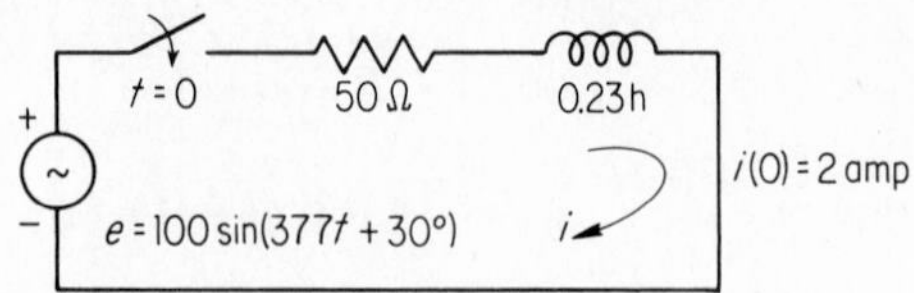

Fig. M-8

ANSWER. $Z(jw) = 50 + j377 \times 0.23 = 50 + j86.7 = 100 \underline{/60°}$ ohms

$$i_{ss} = 100/100 \sin (377t + 30° - 60°) = 1 \sin (377t - 30°)$$

The voltage equation is

$$100 \sin (377t + 30°) = 50i + 0.23\, di/dt$$

Let $i = i_{ss} + i_{tr}$; then

$$100 \sin (377t + 30°) = 50i_{ss} + 0.23 \frac{di_{ss}}{dt} + 50i_{tr} + 0.23 \frac{di_{tr}}{dt}$$

Since $50i_{ss} + 0.23 \dfrac{di_{ss}}{dt}$ satisfies 100 sin (377t + sin 30°), we get

$$50i_{tr} + 0.23\, di_{tr}/dt = 0$$

From which $i_{tr} = A_G{}^{-(50/0.23)t} = A\epsilon^{-217.2t}$
Hence $i = i_{ss} + i_{tr} = \sin (377t - 30°) + At^{-217.2t}$
At $t = 0$, $i(0) = 2$ amp.

$$2 = \sin (-30°) + A = -\tfrac{1}{2} + A. \qquad \text{From which } A = 2.5$$

Therefore, $i = \sin (377A - 30°) + 2.5t^{-217.2t}$. **Answer**

M-10. Show the time function which when combined will give the waveform shown in Fig. M-9 and derive the Laplace transform for the waveform given.

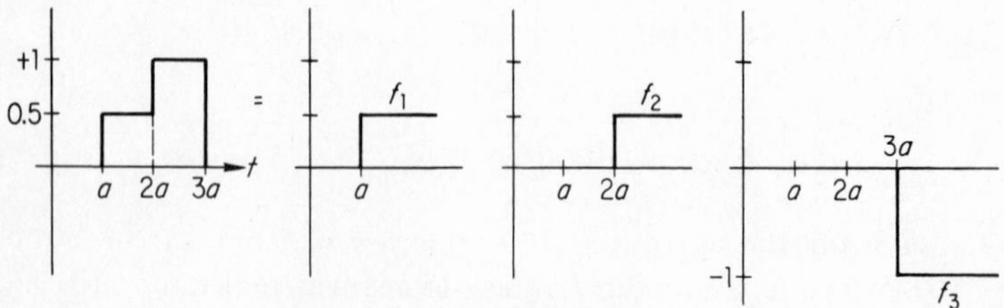

Fig. M-9

ANSWER. $f(t) = f_1 + f_2 + f_3$

$$\begin{aligned} f(t) &= 0.5u_{-1}(t - a) + 0.5u_{-1}(t - 2a) - 1u_{-1}(t - 3a) \\ F(s) &= F_1(s) + F_2(s) + F_3(s) \\ &= \frac{0.5}{s}\epsilon^{-as} + \frac{0.5}{s}\epsilon^{-2as} - \frac{1}{s}\epsilon^{-3as} \end{aligned}$$

or $$= \frac{0.5}{s}\epsilon^{-as}(1 + \epsilon^{-as} - 2\epsilon^{-2as}) \qquad \textbf{Answer}$$

M-11. Derive the Laplace transform of the first-order unit impulse by starting with the waveform given in Fig. M-10. Show all your work.

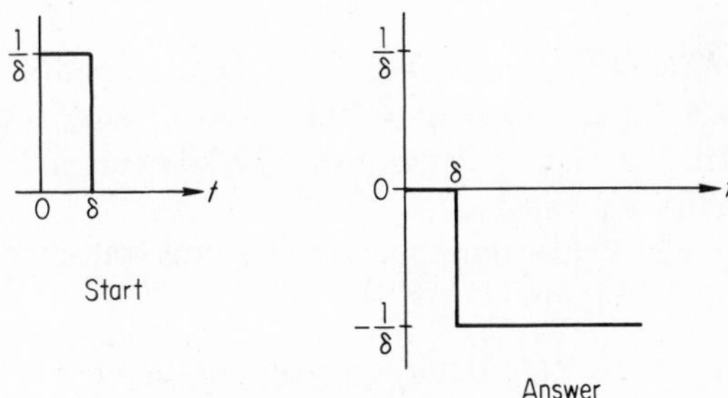

Fig. M-10

ANSWER. From Fig. M-10 the Laplace transform of the square wave is

$$(Fs) = \frac{1}{\delta} \times \frac{1}{s} - \frac{1}{\delta} \times \frac{1}{s}\epsilon^{-\delta s} = 1 - \left[\frac{\epsilon^{-ds}}{\delta s}\right]$$

The transform of the first-order unit impulse is then

$$\mathcal{L}[U_0(t)] = \lim_{\delta\to 0} \frac{1 - \epsilon^{-ds}}{\delta s} = 0/0 \qquad \text{Indeterminate}$$

By l'Hospital's rule finally

$$\mathcal{L}[U_0(t)] = \lim_{\delta\to 0} \frac{d/d\delta(1 - \epsilon^{-ds})}{d/d\delta\ (\delta s)} = \lim_{\alpha\to 0} \frac{s\epsilon^{-ds}}{s} = s/s = \mathbf{1}$$

Supplement N

AERIALS

Questions and Answers

N-1. If 100 kw of energy is radiated from an aerial of 100 meters effective height at a frequency of 60 kc per sec, what is the strength of the electric field at a distance of 100 kilometers? Assume no absorption effects present.

ANSWER. The field strength due to a distant transmitting station, assuming no absorption effects, is

$$E = 377(hI/\lambda d) \qquad \text{volts per meter}$$

where h = effective height of transmitting aerial, meters
d = distance, meters
λ = wavelength, meters
I = aerial current, amp

Then, power radiated from aerial is $W = 1.58h^2I^2/\lambda^2$, kw. Therefore,

$$E = 300\sqrt{W/d} \qquad \text{volts per meter}$$

Finally, $W = 100$, $d = 100 \times 10^3$, meters. So that $E =$ **0.03 volt per meter.**

N-2. A certain aerial has an effective height of 100 meters and the current at the base is 450 amp rms at a frequency of 40 kc per sec. Calculate the power radiated. Total resistance of aerial circuit being 1.12 ohms, determine efficiency of the aerial.

ANSWER. $W = 1.58 \times 100^2 \times 450^2/7.5^2 = 56.9$ kw.

$$E = 300\sqrt{100/100 \times 10^3} = \textbf{0.03 volt per meter}$$

N-3. Choose a suitable antenna for a communications system operating at 100 MHz. What dimensions should the antenna have? Determine the system wavelength.

ANSWER. Table N-1 summarizes the names, types, uses, bandwidths, and wavelengths of a variety of antennas in common use.

TABLE N-1*

Antenna name	Antenna type	Typical uses	Bandwidth	Typical wavelengths
Dielectric rod	Surface-wave	Radar feeds and arrays	10 %	1–6 GHz
Yagi	Surface-wave	TV/FM reception	10 %	1–5 GHz
Half-wave dipole	Resonant	Communications, navigations, radar, etc.	5–40 %	10 MHz–5 GHz
Half-wave slot	Resonant	Aircraft and missiles	....	100 MHz–35 GHz
Rhombic	Traveling-wave	Short-wave transmitting and receiving for long ranges	2–1	2–30 MHz
Axial mode helix	Traveling-wave	Tracking, telemetry, aerospace, ground stations	1.7–1	100 MHz–3 GHz
Log periodic	Frequency-independent	Ecm and direction finding	10–1	10 MHz–12 GHz
Equiangular spiral	Frequency-independent	Ecm, telemetry, aircraft, missiles, arrays	10–1	100 MHz–35 GHz
Paraboloidal reflector	Aperture	Radar, communications, radio, astronomy; other high-gain uses	Determined by feed	300 MHz–70 GHz
Conical horn	Aperture	Radar, communications	1.6–1	300 MHz–70 GHz
Pyramidal horn	Aperture	Radar, communications	1.6–1	300 MHz–70 GHz
Dielectric lens	Aperture	Radar, communications, radio, astronomy; other high-gain uses	Determined by feed	300 MHz–70 GHz

* For additional data see R. S. Gordon and K. W. Duncan, Ready-reference Data Simplifies Antenna Design, *Electronics*, Dec. 21, 1962.

Study of this table shows that the half-wave dipole is a suitable antenna for communications in the 10-MHz to 5-GHz wavelength range. This is the type we will work with. Now let us compute the antenna dimensions. Use the relation dipole length, ft = 467.4/frequency, MHz. Or, dipole length = 467.4/100 = **4.67 ft** for half-wave operation.

The system wavelength may be determined by using the relation $\lambda = 300 \times 10^6/f$, where λ = system wavelength, meters; f = system frequency, Hz. Thus, $\lambda = 300 \times 10^6/100 \times 10^6 =$ **3 meters.**

The length of a quarter-wave antenna is $\lambda/4.2$ ft. A grounded quarter-wave antenna is termed a basic Marconi antenna.

To compute the power input to an antenna, use the relation $P = I^2R$, where P is power input to the antenna, watts; I is antenna current, amp; R is antenna resistance, ohms. Note the properly tuned antenna is considered to be pure resistance. Hence, the power, voltage, current, and resistance of a tuned antenna can be computed using Ohm's law and power equations.

RECOMMENDED REFERENCE LIST

Basic Electrical Principles

Elements of Electricity, Timbie, *Wiley*

Engineers' Manual, Hudson, *Wiley*

Standard Handbook for Electrical Engineers, Fink and Carroll, *McGraw-Hill*

Electrical Engineers' Handbook, Pender and Del Mar, *Wiley*

Basic Electrical Principles, Suffern, *McGraw-Hill*

A Course in Electrical Engineering, vols. 1 and 2, Dawes, *McGraw-Hill*

Differential Equations for Electrical Engineers, Franklin, *Wiley*

Electrical Measurements, Laws, *McGraw-Hill*

Basic Electrical Measurements, Stout, *Prentice-Hall*

Electric Machines

Catechism of Electrical Machinery, *Fairbanks, Morse & Co.*

Electrical Circuits and Machinery, Hehre and Horners, *Wiley*

Electric Power Equipment, Tarboux, *McGraw-Hill*

Principles of Direct-current Machines, Langsdorf, *McGraw-Hill*

Elements of Electrical Machine Design, Still and Siskind, *McGraw-Hill*

Guide to Selecting Polyphase Motors, Product Engineering Magazine, vol. 29, no. 33, p. 58, Aug. 18, 1958, *McGraw-Hill*

Electrical Engineering Laboratory Experiments, Ricker and Tucker, *McGraw-Hill*

Electric Motor Repair, Rosenberg, *McGraw-Hill*

Connecting Induction Motors Dudley, *McGraw-Hill*

Theory of Alternating-current Machinery, Langsdorf, *McGraw-Hill*

Alternating-current Machines, Puchstein and Lloyd, *Wiley*

Design of Electrical Apparatus, Kuhlmann, *Wiley*

E. M. Synchronizer Series, Electric Machinery Manufacturing Company

D-C and A-C Machines, Liwschitz, Garik, and Weil, *Van Nostrand*

Electric Machinery, Fitzgerald and Kingsley, Jr., *McGraw-Hill*
Electric Machinery and Control, Kosow, *Prentice-Hall*

Electric Circuits

Electric Circuits, MIT, EE Staff, *Wiley*
Alternating-current Circuits, Kerchner and Corcoran, *Wiley*
Principles of Electric Power Transmission, Woodruff, *Wiley*
Capacitors for Industry, Bloomquist and Partington, *Wiley*
Power Factor Economics, Rogers, *Wiley*
Electrical Distribution Fundamentals, Sanford, *McGraw-Hill*
Electric System Operation, Strotzki, *McGraw-Hill*
Transmission Line Theory, King, *McGraw-Hill*
Alternating Current Circuits, Tang, *International*
Bus Conductors—Technical Bulletin, *Kaiser Aluminum Company*
Westinghouse Distribution Digest, cat. 80-000, *Westinghouse Electric Co.*
Copper for Bus Bars, *Copper Development Assoc.*, London
Analysis of Alternating Current Circuits, Le Page, *McGraw-Hill*
Electric Circuit Theory, Benson and Harrison, *Arnold*
Analysis of Electric Circuits, Middendorf, *Wiley*
Linear Circuits, Scott, *Addison-Wesley*

Illumination

Illumination Engineering, Schilling, *International*
Illumination Engineering, Boast, *McGraw-Hill*
25 Industrial Lighting Solutions, *General Electric Co.*
Essential Data for General Lighting Design, Bull. LS-173, *General Electric Co.*
Foot Candles in Modern Lighting Practice, Bull. LS-119, *General Electric Co.*
Lamp Bulletin, *General Electric Co.*
Westinghouse Data Book, *Westinghouse Staff*
Standard Handbook of Engineering Calculations, Hicks, *McGraw-Hill*

Networks

General Network Analysis, Le Page and Seely, *McGraw-Hill*
Alternating-current Circuit Theory, Reed, *Harper*
Introduction to Electric Power Systems, Tarboux, *International*
Active Networks, Rideout, *Constable*
Elements of Power System Analysis, Stevenson, *McGraw-Hill*
Transmission Lines and Networks, Johnson, *McGraw-Hill*

Symmetrical Components Applied to Electric Power Networks, Calabrese, *Ronald Press*

Communications

The Electrical Fundamentals of Communication, Albert, *McGraw-Hill*
Electrical Communications, Albert, *Wiley*
Electronic and Radio Engineering, Terman, *McGraw-Hill*
Radio Engineers Handbook, Terman, *McGraw-Hill*
Essentials of Radio, Shursburg and Asterfield, *McGraw-Hill*
RCA Transmitting Tubes, Bull. TT4, *Radio Corp. of America*
Electrical Engineers' Handbook—Communications and Electronics, Pender and McIlwan, *Wiley*
Antennas, Kraus, *McGraw-Hill*
Acoustics, Beranek, *McGraw-Hill*
Communication Engineering, Everitt and Anner, *McGraw-Hill*
Radio Engineering Handbook, Henney, *McGraw-Hill*

Electronics

Electronic Circuits and Tubes, Cruft Laboratory War Training Staff, *McGraw-Hill*
Applied Electronics, Gray, *Wiley*
RCA Receiving Tube Manual, RC-18, *Radio Corp. of America*
Theory and Application of Electron Tubes, Reich, *McGraw-Hill*
Radiotron Designers Handbook, Smith, *Wireless Press*
Electronics in Industry, Bendz, *Wiley*
RCA Transistors and Semi-conductor Diodes, Bull. SCD-18, *Radio Corp. of America*
Electronic Engineering Principles, Ryder, *Prentice-Hall*
Electronics, Millman and Seely, *McGraw-Hill*
Electronic Fundamentals and Applications, Ryder, *Pitman*
Junction Transistor Electronics, Hurley, *Wiley*
Electronics, Starr, *Pitman*
Handbook of Electronics Circuits, Feinberg, *Chapman and Hall*
Electronic and Radio Engineering, Terman, *McGraw-Hill*
Vacuum and Solid State Electronics, Harris and Robson, *Pergamon*
Vacuum Tubes, Spangenberg, *McGraw-Hill*
Electronic Engineering, Seely, *McGraw-Hill*
Fundamentals of Engineering Electronics, Dow, *Wiley*
Electronics, Parker, *Arnold*
Electronics, Chirlian and Zemanian, *McGraw-Hill*

Engineering Electronics, Ryder, *McGraw-Hill*
Electron Tube Circuits, Seely, *McGraw-Hill*
Theory and Application of Industrial Electronics, Cage, *McGraw-Hill*
Electronics in Engineering, Hill, *McGraw-Hill*
Electronics: Discrete and Integrated, Schilling and Belove, *McGraw-Hill*
Semiconductor Devices and Applications, Greiner, *McGraw-Hill*
Fundamentals of Electric and Electronic Circuits, Mandl, *Prentice-Hall*
Basic Automatic Control Theory, Murphy, *Van Nostrand*

Transistors

Vacuum Tube Circuits and Transistors, Arguimbau and Adler, *Wiley*
Transistor Circuit Analysis, Joyce and Clarke, *Addison-Wesley*
Electrons and Holes in Semiconductors, Shockley, *Van Nostrand*
Transistor Electronics, Lo et al., *Prentice-Hall*

Wave Analysis

Wave Guides, Lamont, *Methuen*
Wave Filters, Jackson, *Methuen*
An Introduction to Fourier Analysis, Stuart, *Science Paperbacks*

Laplace Transforms

An Introduction to the Laplace Transformation, Jaeger, *Span*
Handbook of Laplace Transformation, Nixon, *Prentice-Hall*
Mathematical Handbook for Scientists and Engineers, Korn and Korn, *McGraw-Hill*
Transients in Electrical Circuits, Lago and Waidelich, *Ronald*

Electrical Engineering Practice

Today's Electrical Practice, Power Magazine, *McGraw-Hill*
National Electrical Code, *National Fire Protection Association*
National Electrical Code Handbook, Abbott and Steka, *McGraw-Hill*
American Electricians Handbook, Croft, *McGraw-Hill*
Electrical Wiring Specifications, Whitehorne, *McGraw-Hill*
American Institute of Electrical and Electronic Engineers, *Conference Papers*
Electrical Drafting and Design, Bishop, *McGraw-Hill*
Electrical Drafting, Van Gieson, *McGraw-Hill*
Industrial Electricity, Nadon and Gelmine, *Van Nostrand*
Electrical Code Diagrams, vols. 1 and 2, Segall, *McGraw-Hill*

Engineers Relay Handbook, Pender, *Wiley*
Practical Electrical Wiring, Richter, *McGraw-Hill*
NFPA Handbook of the National Electrical Code, Steka and Brandon, *McGraw-Hill*
Operational Electricity, Hubert, *Wiley*
Industrial Power Systems Handbook, Beeman, *McGraw-Hill*
Transformer Principles and Practice, Gibbs, *McGraw-Hill*
Motor Selection and Application, Libby, *McGraw-Hill*
Electrical Systems for Commercial Buildings, *IEEE*
Electric Power Distribution for Industrial Plants, *IEEE*
Industrial Electrical Systems, Power Magazine Editors, *McGraw-Hill*
Standards for Motors and Generators, *NEMA*
Standards for Industrial Control, *NEMA*
Industrial Machines, *American National Standards Institute* (ANSI)
Synchronous Motors, *ANSI*
Electrical Systems Design, McPartland and Novak, *McGraw-Hill*
Electrical Design Details, McPartland and Novak, *McGraw-Hill*
Electrical Equipment Manual, McPartland and Novak, *McGraw-Hill*
Electrical Systems for Power and Light, McPartland and Novak, *McGraw-Hill*
Electric Current Abroad, *Government Printing Office*
Industrial Electricity, Timbie and Willson, *Wiley*

In addition to the above references, much examination help may be obtained from the various technical magazines such as: *Power, Product Engineering, Machine Design, Power Engineering, Plant Engineering, Electrical Engineering, Electrical World, Nucleonics, ISA Journal, Electrical Construction and Maintenance, Electronics, Edison Electric Institute Monographs, Specifying Engineer.*

PROBLEM INDEX

PROBLEM INDEX

Part 1

Direct Current

Battery Circuits

Condensers

D-C Machinery

D-C Motors

D-C Transmission

Electrical Units and Ohm's Law

Magnetism and the Magnetic Circuit

Measurement of D-C Electricity

Power, Energy, and Thermal Units

Temperature Coefficient of Resistance

Units of Area and Resistance

Wire Sizing

Part 2

Alternating Current

Aerials

Alternating Current

A-C Transmission Lines

Electron Tubes and Circuits

Equivalent Circuits

Filters

Induction Motors

Lighting

Network Analysis

Power Circuits

Short-Circuit Calculations

Synchronous Motors

Transformers

Transient Analysis

Transistors

Waveform Analysis

SUBJECT INDEX

SUBJECT INDEX